Computer-Integrated Manufacturing

Third Edition

James A. Rehg
Penn State Altoona

Henry W. Kraebber
Purdue University

PEARSON

Prentice
Hall

Upper Saddle River, New Jersey
Columbus, Ohio

Library of Congress Cataloging-in-Publication Data
Rehg, James A.
   Computer-integrated manufacturing / James A. Rehg, Henry W. Kraebber.—3rd ed.
      p. cm.
   Includes bibliographical references and index.
   ISBN 0-13-113413-2
   1. Computer integrated manufacturing systems.   I. Kraebber, Henry W.   II. Title.

TS155.63.R45 2005
670'.285—dc22

                                          2004001895

Editor in Chief: Stephen Helba
Executive Editor: Debbie Yarnell
Development Editor: Kate Linsner
Editorial Assistant: Jonathan Tenthoff
Production Editor: Louise N. Sette
Production Supervision: *The GTS Companies*/York, PA Campus
Design Coordinator: Diane Ernsberger
Cover Designer: Jeff Vanik
Production Manager: Matt Ottenweller
Marketing Manager: Jimmy Stephens

This book was set in Zaph Calligraphic 801 by *The GTS Companies*/York, PA Campus. It was printed and bound by R.R. Donnelley & Sons Company. The cover was printed by Phoenix Color Corp.

Pearson Education Ltd.
Pearson Education Singapore Pte. Ltd.
Pearson Education Canada, Ltd.
Pearson Education—Japan

Pearson Education Australia Pty. Limited
Pearson Education North Asia Ltd.
Pearson Educación de Mexico, S.A. de C.V.
Pearson Education Malaysia Pte. Ltd.

10 9 8 7 6 5 4 3 2 1
ISBN   0-13-113413-2

About the Authors

James A. Rehg, CMfgE, is an Associate Professor of Engineering at Penn State Altoona, where he teaches automation controls courses in the BS program in Electromechanical Engineering Technology. He earned both a Bachelor of Science degree and a Master of Science degree in Electrical Engineering from St. Louis University and has completed additional graduate study at Wentworth Institute, University of Missouri, South Dakota School of Mines and Technology, and Clemson University. Before coming to Penn State Altoona, he was the CIM coordinator and department head of CAD/CAM/Machine Tool Technology at Tri-County Technical College. Prior to that, he was the Dean of Engineering Technology and Director of Academic Computing at Trident Technical College in Charleston, South Carolina. He held the position of Director of the Robotics Resource Center at Piedmont Technical College in Greenwood, South Carolina, and was department head of Electronic Engineering Technology of Forest Park Community College in St. Louis, Missouri. In addition, he was a Senior Instrumentation Engineer for Boeing in St. Louis. Professor Rehg has authored five texts on robotics and automation and has presented numerous papers on subjects related directly to training in automation and robotics. He has also been a consultant to nationally recognized corporations and many educational institutions. He has led numerous seminars and workshops in the areas of robotics and microprocessors and has developed extensive seminar training material. In addition, he has received numerous state awards for excellence in teaching and was named the outstanding instructor in the nation by the Association of Community College Trustees.

Henry W. Kraebber, PE, CPIM, is a Professor of Mechanical Engineering Technology at Purdue University in West Lafayette, Indiana. He has fifteen years of experience and leadership in manufacturing operations, engineering, quality, and management. He has worked at the Collins Avionics and Missiles group of Rockwell International, the Plough Products Division of Schering-Plough Corporation, and Flavorite Laboratories, Inc. His work has supported the production of industrial, consumer, and military products in the food, consumer products, and electronics areas. He currently teaches courses in manufacturing operations, manufacturing quality control, and integrated systems in the Computer-Integrated Manufacturing Technology degree program. Mr. Kraebber earned a Bachelor of Science degree in Industrial Engineering from Purdue University and a Master of Engineering degree in Industrial Engineering from Iowa State University. He is President of the CIM in Higher Education (CIM/HE) Alliance, a nonprofit corporation that supports CIM and manufacturing education. He is a senior member of the Society of Manufacturing Engineers (SME) and the Institute of Industrial Engineers (IIE). He is an active member of the American Production and Inventory Control Society (APICS) and has served as President and Vice President of Education for the Wabash Valley Chapter.

This text is dedicated to three very special young men—Jim and Richard Rehg and Karl Kraebber—and to our families, students, and friends, who have helped make this possible.

Contents

Preface **xiii**

PART 1 **INTRODUCTION TO CIM AND THE MANUFACTURING ENTERPRISE** **1**

1 **The Manufacturing Enterprise** **3**

1–1 Introduction 3

1–2 External Challenges 7

1–3 Internal Challenges 10

1–4 World-Class Order-Winning Criteria 15

1–5 The Problem and a Solution 22

1–6 Learning CIM Concepts 26

1–7 Going for the Globe 28

1–8 Summary 35

Bibliography 35

Questions 36

Problems 37

Projects 38

Appendix 1–1: The Benefits of a CIM Implementation 40

Appendix 1–2: Technology and the Fundamentals of an Operation—Authors' Commentary 41

2 **Manufacturing Systems** **43**

2–1 Manufacturing Classifications 45

2–2 Product Development Cycle 52

2–3   Enterprise Organization   54

2–4   Manual Production Operations   59

2–5   Summary   62

Bibliography   63

Questions   63

Projects   64

Case Study: Evolution and Progress—One World-Class Company's
Measurement System   65

Appendix 2–1: CIM as a Competitive Weapon   68

PART 2   THE DESIGN ELEMENTS AND PRODUCTION ENGINEERING   **69**

3   Product Design and Production Engineering   **71**

3–1   Product Design and Production Engineering   72

3–2   Organizational Model   74

3–3   The Design Process: A Model   75

3–4   Concurrent Engineering   86

3–5   Production Engineering   93

3–6   Summary   102

Bibliography   102

Questions   103

Projects   104

Case Study: Repetitive Design   105

4   Design Automation: CAD and PDM   **107**

4–1   Introduction to CAD   107

4–2   The Cost of Paper-Based Design Data   110

4–3   CAD Software   111

4–4   CAD: Yesterday, Today, and Tomorrow   121

4–5   Application of CAD to Manufacturing Systems   129

4–6   Selecting CAD Software for an Enterprise   129

4–7   Product Data Management   134

4–8   Summary   136

Bibliography   137

Questions   137

Projects   138

Appendix 4–1: Web Sites for CAD Vendors   139

Appendix 4–2: B-Splines to NURBS   140

Appendix 4–3: Web Sites for Computer Companies   142

5   Design Automation: CAE   **143**

5–1   Design for Manufacturing and Assembly   144

5–2   CAE Analysis   152

5–3   CAE Evaluation   163

5–4   Group Technology   177

5–5   Production Engineering Strategies   184

5–6   Design and Production Engineering Network   204

5–7   Summary   210

Bibliography   211

Questions   212

Problems   213

Projects   213

Appendix 5–1: Ten Guidelines for DFA   215

Appendix 5–2: Web Sites for CAE Vendors   216

Appendix 5–3: Web Sites for Rapid Prototyping Vendors   217

PART 3   CONTROLLING THE ENTERPRISE RESOURCES   **219**

6   Introduction to Production and Operations Planning   **221**

6–1   Operations Management   222

6–2   Planning for Manufacturing   223

6–3   MPC Model—Manufacturing Resource Planning (MRP II)   228

6–4 Production Planning 234

6–5 Master Production Schedule 240

6–6 Inventory Management 245

6–7 Planning for Material and Capacity Resources 248

6–8 Introduction to Production Activity Control 252

6–9 Shop Loading 254

6–10 Input-Output Control 257

6–11 Automating the Planning and Control Functions 258

6–12 Summary 259

Bibliography 259

Questions 260

Problems 261

Projects 264

Appendix 6–1: Priority Rule Systems 265

7 Detailed Planning and Production-Scheduling Systems 271

7–1 From Reorder-Point Systems to Manufacturing Resource Planning (MRP II) 271

7–2 Material Requirements Planning 273

7–3 Capacity Requirements Planning 285

7–4 Manufacturing Resource Planning 288

7–5 Features of Modern Manufacturing Planning and Control Systems 291

7–6 Summary 300

Bibliography 302

Questions 303

Problems 304

Projects 305

Appendix 7–1: Wight's Bicycle Example 306

Appendix 7–2: ABCD Checklist 308

Appendix 7–3: An ERP Example Using WinMan 318

8 **Enterprise Resources Planning, and Beyond** **323**

8–1 MRP II: A Driver of Effective ERP Systems   326

8–2 Information Technology   327

8–3 The Decision to Implement an ERP System   330

8–4 Identifying ERP System Suppliers   332

8–5 Developing Technologies: Converging and Enabling   335

8–6 Integrating Systems to Manage Design Data   341

8–7 Summary   348

Bibliography   350

Questions   351

Projects   351

9 **The Revolution in Manufacturing** **352**

9–1 Just-in-Time Manufacturing   353

9–2 Synchronized Production   364

9–3 The Emergence of Lean Production   366

9–4 Modern Manufacturing Systems in a Lean Environment   369

9–5 Summary   374

Bibliography   375

Questions   375

Projects   376

Case Study: Production System at New United Motor Manufacturing, Part 1   376

Case Study: Production System at New United Motor Manufacturing, Part 2   378

PART 4 **ENABLING PROCESSES AND SYSTEMS FOR MODERN MANUFACTURING** **383**

10 **Production Process Machines and Systems** **385**

10–1 Material and Machine Processes   387

10–2 Flexible Manufacturing   405

10–3 Fixed High-Volume Automation   413

10–4   Summary   417

Bibliography   419

Questions   420

Projects   421

Appendix 10–1: History of Computer-Controlled Machines   422

11   Production Support Machines and Systems   **425**

11–1   Industrial Robots   426

11–2   Program Statements for Servo Robots   448

11–3   Programming a Servo Robot   454

11–4   Automated Material Handling   461

11–5   Automatic Guided Vehicles   465

11–6   Automated Storage and Retrieval   476

11–7   Summary   480

Bibliography   481

Questions   481

Projects   483

Case Study: AGV Applications at General Motors   484

12   Machine and System Control   **486**

12–1   System Overview   487

12–2   Cell Control   493

12–3   Proprietary Versus Open System Interconnect Software   497

12–4   Device Control   499

12–5   Programmable Logic Controllers   500

12–6   Relay Ladder Logic   502

12–7   PLC System and Components   506

12–8   PLC Types   511

12–9   Relay Logic Versus Ladder Logic   513

12–10   Computer Numerical Control   521

12–11 Automatic Tracking 525

12–12 Network Communications 529

12–13 Summary 531

Bibliography 532

Questions 533

Projects 534

Appendix 12–1: Turning G Codes 535

13 **Quality and Human Resource Issues in Manufacturing** **537**

13–1 Quality Foundations 538

13–2 Total Quality Management 543

13–3 Quality Tools and Processes 547

13–4 Defect-Free Design Philosophy 557

13–5 The Changing Workforce 561

13–6 Self-Directed Work Teams 562

13–7 Summary 567

Bibliography 569

Questions 569

Projects 570

Index **571**

Preface

The global economy and technological innovations bring many new issues and twists to the subject of computer-integrated manufacturing (CIM). It remains as broad as the complex manufacturing enterprises it attempts to model. Some persons would suggest that CIM is too broad for a single course or textbook. However, the essence of CIM is in the integration of the enterprise elements: physical integration through the linking of hardware and software systems, logical integration through shared common enterprise information and data, and philosophical integration based on a new sense of purpose and direction in every entity in the enterprise. Therefore, the integration so critical to a CIM implementation is best introduced in a single course so that links between the enterprise elements can be explored. This book was written to support such an introductory course.

Understanding the operation of a comprehensive CIM solution requires some study of traditional manufacturing practice, a look at the current state of CIM, and consideration of how technology and operating procedures may change in the future. The integration of product design techniques and fundamental manufacturing principles, along with a look at changing operations and information systems that support CIM throughout the enterprise, makes this book unique. In the book, we do the following:

- Describe the different types of manufacturing systems or production strategies used by industries worldwide. This description is important because no two CIM solutions are the same.

- Go beyond the description of automated machines and software solutions because a successful CIM implementation demands more than technology. In practice, ordering hardware and software is the last step in a CIM implementation; the preliminary work is what guarantees a successful CIM project.

- Discuss the impact of CIM on all the major elements in an enterprise: product design, shop-floor technology, and manufacturing production and operational control systems.

- Provide a convincing argument for implementing CIM so that the enterprise will be competitive in the global market. In practice, the technologies available to manufacturers around the globe open every market to world-wide competition.

■ Look at the computer-based systems of the CIM enterprise that support the growing just-in-time and lean production initiatives.

In addition, the third edition has the following significant changes: Work-cell-design case studies have been added at the end of chapters 1 through 4, 5, 10, and 11, with the work keyed to the concepts presented in the chapters. The chapter on CAD, chapter 4, was changed extensively to an overview of the CAD function in an enterprise and an introduction to product data management (PDM). The enterprise networking concepts were updated and expanded. The finite-element analysis and rapid prototyping sections in chapter 5 were updated and expanded. Numerous new figures have replaced older images, and many new images have been added.

Also new to this edition is a CD-ROM containing the demo version of the WinMan software. This software provides its users with an opportunity to work with a fully functional computer-based enterprise resources planning (ERP) system. This tool allows users to see how a modern data-driven system can help companies better manage their operations and the related data. The CD includes a fully functional single-user system that can be installed on a PC. The demonstration takes users through the basics of the system—from the building of item and structure databases to the functions needed to manage customer orders, material management, manufacturing, and accounting information.

To provide a complete overview of the computer-integrated enterprise, we divided the book into four parts. In the first part, chapters 1 and 2, we provide an overview of global competition, describe an internal manufacturing strategy, discuss in detail the problem facing manufacturing and the development of an effective solution, and characterize the operation of different types of enterprises. In the characterization, we furnish a classification and description of the manufacturing systems and production strategies used by manufacturing, provide an explanation of the product development and engineering change cycle, and give an overview of the enterprise organization. At the end of part 1, the need for change in manufacturing is made clear and a basic strategy for change in the organization is established. In addition, the description of the enterprise organization in part 1 provides a framework for the CIM concepts introduced in the rest of the text. Part 1 provides the critical introduction to manufacturing and the enterprise that is necessary for a course designed to teach CIM.

In part 2, which includes three chapters, we examine the three major design and engineering process segments that take a product from concept to production. Chapter 3 introduces design and production engineering concepts and issues. The use of CIM technology to design and produce world-class products with enhanced enterprise productivity is emphasized. The old design model is compared to a recommended new process that incorporates a concurrent engineering focus to product design. This part of the text concludes with an in-depth description of production engineering functions and the opportunities for productivity gains through integration of technology and data in the enterprise. Computer-aided design (CAD) is the focus of chapter 4. Since design is the starting point for development of the product database, a full chapter is devoted to the integration of CAD into the enterprise operation. To emphasize this integration, we changed the chapter title and the content to include PDM (product data management). The function of CAD technology in the product design process is discussed, and the systems used to develop the product models are described. The importance of PDM and its link to the CAD technology and design department are covered. In chapter 5, we explore the relationships between the concurrent engineering product design model and the computer-aided engineering (CAE) technology available to support every step of the design process and production engineering. We include a complete definition of CAE, design for manufacturing and assembly, finite-element and mass-properties analysis, rapid prototyping, group technology, computer-aided process planning, computer-aided manufacturing, production and process modeling and simulation, maintenance, automation, and product cost analysis. In the final section of chapter 5, we describe the computer network used to tie the design and production engineering functions to the common enterprise database and other business functions.

Part 3 of the text shifts the CIM focus to controlling the enterprise resources. CIM is alive and growing in applications that support the management and control of the enterprise. The first chapter in the sequence, chapter 6, describes the concept of manufacturing planning and control (MPC) with a model of a typical MPC system. The function of manufacturing planning and a high-level look at the systems and technologies available for CIM implementations are presented. Attention is given to the high-level system elements of production planning and the master production schedule (MPS). Chapter 6 provides an overview of the critical concepts that are explored in more detail in the following two chapters.

In chapter 7, three key elements in the MPC model are discussed in more detail. These include material requirements planning (MRP), capacity requirements planning (CRP), and the production activities that execute the material and capacity plans. Automation software used to implement CIM in this critical part of the enterprise is introduced and explained. Software solutions for the manual MPC functions are included at the end of each section. At the conclusion of chapters 6 through 8, the reader will understand the operation of an MPC system and will be able to follow the logic and calculations of each function in MPC and describe key features of application software capable of automating the MPC functions.

Chapter 8, "Enterprise Resources Planning, and Beyond," develops the links between the concepts from MRP and MRP II systems that are essential parts of the new ERP systems. The pace of change in technology and new systems at the end of the 1990s has been extraordinary. There is no way to predict the future, but it is clear that new systems and system features will continue to be developed. Technologies for design, processing and control, information systems, and communication are rapidly converging. The emerging technologies offer substantial new opportunities and benefits, but also present new challenges for the manufacturing enterprise.

The quiet revolution in manufacturing coming from just-in-time (JIT) manufacturing and lean production methods is the focus of chapter 9. These methods are based on the elimination of waste throughout the enterprise. The operations and management techniques that are used in JIT and lean systems have deep roots in classic industrial engineering. The methods provide surprising results when they are woven into an integrated system for the operation of the enterprise. Small lot sizes approaching a lot size of 1 unit, visual signals, and expanded work by employee teams are at the heart of this revolution. Computer-based systems still have an important role to play in the support of the JIT and lean production initiatives.

Part 4 concentrates on the processes and systems that lay the foundation for modern manufacturing and enterprise-wide concepts critical to a successful CIM implementation. Chapter 10 covers the commonly used production process machines used in manufacturing. In addition, manufacturing systems including one or more machines, called *flexible manufacturing cells* and *flexible manufacturing systems*, are addressed in the chapter. Chapter 11 covers machines and systems that support production, including coverage of industrial robots, material-handling systems, automatic guided vehicles, and automatic storage and retrieval systems. The techniques used for the control of production systems are the focus of chapter 12. The control systems discussed include cell control hardware and software, device control hardware and software, programmable logic controllers, and computer numerical controllers. The operation and the management of enterprise networks and common databases are also discussed. A successful implementation of any high technology requires a change in the management viewpoint on manufacturing management and human resource development. As a result, a discussion of a broad range of quality issues and the effective use of human resources are included in chapter 13.

In summary, part 1 begins with a global view of manufacturing. In the second and third parts, we focus on the activities required to convert raw material into finished goods and introduce technology to aid in the conversion and the management of the enterprise. The last part of the text shifts back to systems that enable the enterprise to manufacture products competitively, with the discussion centered on the services and support functions required for successful CIM implementation. Common products (hardware, software, and systems) are included throughout the book to demonstrate the technology and to stress the integration issues.

We tried to include important trends and real industrial practices in this text. The inputs from colleagues in industry have contributed directly to the improved content of this edition. Special thanks go out to Patrick Delaney, President of SIBC Corporation; Rick Anderson, President of TTW, Inc.; Kurt Freimuth, President of Factory Floor Solutions; Charlie Colosky, President of Operations Development Associates; and Joel Lemke, President of ENOVIA Corporation, for important inputs and background information.

The logical order of topics and chapter content was tested in a series of workshops at Trident Technical College offered to college faculty and industrial employees. In addition, the text has been used in numerous university courses at Purdue and other institutions. The insight gained through discussions in these settings was critical to the development of this book. We would especially like to thank John Sjolander, Jerry Bell, and Alan Kalameja for their help with the design automation and control elements. Special thanks to Marci Rehg for her help in developing the CIM workshop material, where many of the presentation ideas were tested. Donald Lucas and Hugo Ramos, former graduate students in the School of Technology at Purdue University, worked on research projects on product lifecycle management that have contributed to this text. Thanks also to all the students who have helped us develop and test instructional materials related to CIM over the years.

Finally, thanks to the IBM Corporation, founders of the initial CIM in Higher Education Alliance program, for support in developing the CIM workshops and the CIM capability at two- and four-year colleges. The CIM in Higher Education Alliance is now an independent, nonprofit corporation that continues to encourage and support CIM and education for manufacturing. Thanks also to the reviewers, Don Arney (Ivy Tech State College, IN) and Dr. Michael Costello (Southern Illinois University at Carbondale).

James A. Rehg (jar14@psu.edu)

Henry W. Kraebber (kraeb@purdue.edu)

Introduction to CIM and the Manufacturing Enterprise

PART GOALS

The goals for this part are to introduce the reader to the challenges faced by manufacturing organizations operating in today's global economy and to clearly define the types of manufacturing systems, production strategies, and enterprise organizations used to produce consumer goods.

Therefore, after you complete part 1, the following will be clear to you:

- Manufacturing is vital to the economic health of the United States and our current high standard of living.
- Manufacturing companies must meet serious internal and external challenges to succeed in the face of increasing global competition.
- The operational philosophy of computer-integrated manufacturing (CIM) is necessary at some level for the survival of most manufacturing companies.
- All manufacturing systems and production strategies will benefit from the application of the CIM philosophy.

CAREER INSIGHTS

This text describes the operation of an enterprise that uses a combination of people and technology to bring products to market that are equal or superior in quality to those produced elsewhere, timely in delivery, competitive in price, and able to meet the customer's need for a pleasing and attractive appearance. To meet these criteria, the organization has employees with a range of educational experience from high school graduates to employees with university graduate degrees. However, the application of technology in the design, production, and management of the enterprise products skews the need for employees' education beyond the high school level. An organization able to meet these criteria must have employees with a broad cross section of skills learned in a number of postsecondary education disciplines. Some of the disciplines in highest demand are those from engineering and business. The engineering curricula include electrical, mechanical, industrial, environmental,

CAREER INSIGHTS (contd.)

and manufacturing, whereas the business degrees are in finance, marketing, accounting, and management. The text describes a broad mix of technologies and manufacturing systems used to produce consumer goods for sale across the globe. Each of these technologies and systems must have well-educated and knowledgeable employees who can implement and apply the technology in the most effective manner to achieve the optimum results. As a result, organizations like those described in the text offer exciting employment opportunities with significant financial reward. If you find topics that are interesting and challenging in the different production disciplines covered in this text, then you have identified areas in manufacturing that could offer you a rewarding career.

The Manufacturing Enterprise

OBJECTIVES

After completing this chapter, you should be able to do the following:

- Define *manufacturing*
- Describe the fall and rise of U.S. manufacturing
- List and describe the external challenges facing manufacturing companies
- Describe a manufacturing strategy that can be used to overcome the external challenges
- Describe and give examples of order-winning criteria and order-qualifying criteria
- Describe the Terry Hill model
- Describe a technique for changing the product lifecycle
- List and describe the seven world-class manufacturing standards
- Given system parameters, calculate the setup time, number of bad parts, manufacturing-space ratio, and inventory turns
- Describe the results of doing nothing to improve manufacturing
- Write the definition of *computer-integrated manufacturing,* or *CIM*
- Describe all of the elements of the enterprise wheel
- Describe the function of the process segments
- Describe the three-step process for implementing CIM

1–1 INTRODUCTION

This book is about the manufacturing enterprise and the operational principles that are necessary for manufacturers to be competitive in the present global market–driven economy. Defining *manufacturing* is a logical starting point.

Manufacturing is a collection of interrelated activities that includes product design and documentation, material selection, planning, production, quality assurance, management, and marketing of goods.

The fundamental goal of manufacturing is to use these activities to convert raw materials into finished goods on a profitable basis. The ability to produce this conversion efficiently determines the success of the enterprise. History is rich with examples of manufacturing enterprises, large and small, unable to deliver products that were competitive in cost, quality, and timely delivery to the marketplace. Manufacturers never intentionally set a course that leads to product or company failure, so why do some succeed and prosper while others fade and die? No doubt, many reasons account for these failures, but studies of U.S. industries indicate that some common denominators influenced the failure. U.S.-based manufacturers were most successful in the captive markets associated with postwar years in the 1950s and 1960s. The successes in these glory years brewed complacency in U.S. companies that resulted in an inability to compete in the open markets of the 1970s and 1980s.

The lessons learned in the 1970s and 1980s resulted in changes across U.S. industries. The changes started in the multinational companies and trickled down to smaller industry groups. As a result of these improved manufacturing practices, U.S. industries reclaimed a leadership role by the mid-1990s and will continue that leadership role in the current century.

Foreign competition in Japan and Germany rose from the postwar ashes to build industrial giants that competed on an equal basis with their North American counterparts. When Japan began to penetrate world markets in the 1960s, U.S. manufacturers refused to make the necessary changes in manufacturing methodologies to retain their leadership position. A review of U.S. manufacturing during the last fifty years provides a useful base from which to view the changes necessary for manufacturing to remain strong and competitive in the global market.

Three Stages of Manufacturing Retreat

The first stage in the retreat of U.S. production dominance started during the Vietnam War with the emergence of small electronic consumer goods such as radios and tape decks. The retreat continued in the years that followed. Products with brand names such as Panasonic and Sony began to appear more frequently in U.S. markets and replaced products from General Electric and RCA. U.S. manufacturers mistakenly assumed that cheap labor and central government support were the reasons that Japan could produce these low-cost consumer goods. Although Japan did have a labor-cost advantage, its manufacturers were also developing efficient manufacturing methods and quality systems.

The second stage in the retreat was marked by the Japanese practice of copying successful U.S. products. During this period, the Japanese copied U.S. products and offered similar, higher-quality products to U.S. customers at lower cost. In addition to the examples listed in Figure 1–1, the shift was especially evident in the automotive industry. Other countries followed this practice as a method of gaining entrance

Figure 1–1 Technology Lost: Invented Here, Made Elsewhere. In the Early 1970s, the United States Was the Leading Inventor and Manufacturer in the World. *(Source: Courtesy of Council on Competitiveness, U.S. Department of Commerce.)*

U.S.-invented technology	*U.S. producers' market share (%)*			
	1970	*1975*	*1980*	*1987*
Phonographs	90	40	30	1
Color televisions	90	80	60	10
Machine tools	99	97	79	35
Telephones	99	95	88	25
Semiconductors	89	71	65	64
Computers	N.A.	97	96	74

into the U.S. market. The oil shortage of 1973 opened up the automotive market to Japanese and German automotive manufacturers when U.S. customers turned to cheaper, more-fuel-efficient imports. What consumers discovered, however, was a product that in many respects was far superior to the alternative built in the United States.

The third stage in industrial competition from offshore companies started in the late 1980s and featured rapid product development and manufacturing of upscale products with a much greater profit margin. In producing these products, offshore companies continued to deliver world-class quality, design, and performance while extracting a premium price. The products, for example, included cars from Toyota and Mercedes-Benz, sound and video equipment from Sony and Phillips, medical and metalworking equipment—the list goes on. These competitive world-class companies proved that they could now play at any level in the production game.

Return to Power

The United States regained the status of a manufacturing leader in the 1990s for several reasons, some business, some economic, and others political. A list of the most significant factors in the rebirth of U.S. manufacturing dominance follows.

Economic Factors

- Deregulation of energy and communications markets
- Low inflation as a result of downward pressures on wages, the prices of raw materials, and the deregulated energy markets
- Falling interest rates during the last decade
- The collapse of the Asian economy owing to the excesses of financial institutions in managing real-estate portfolios and corporate loans

Business Factors

- Consolidation of competitive companies and companies with complementary products in most markets

- Restructuring of corporate America
- New and expanding technological leadership
- Partnerships between the United States and offshore companies
- Adoption of CIM concepts in many industry groups
- Increased productivity as a result of consolidations, restructuring, technology, CIM, and better labor-management relations

Political Factors

- The consolidation of the European Union
- Pressures to open closed markets

Taken together, these factors permitted the U.S. industry base to enter the next century well prepared to compete in the global market.

Product Versus Process Goals

The success of U.S. manufacturers following World War II was largely due to the technology and industrial base spawned by the war and the captive markets associated with the postwar economy. In this period, two-thirds of U.S. research dollars were allocated to product development, while the balance was spent on developing technologies to improve the way in which products were manufactured. As a result, innovative products in transportation and consumer electronics flowed from U.S. industries. In contrast, most European and Japanese industries were neutralized by the war effort and could not compete with the innovative resources in the United States. As a result, the offshore countries, rebuilding their industrial base, focused the majority of their development efforts on improved manufacturing processes. Conceding product innovation to the United States, industries in Japan and Germany cultivated manufacturing technology and production systems that could produce products at *higher quality* and *lower cost*. At the same time, U.S. industry continued to emphasize *product development research* at the expense of better production processes and improved manufacturing technologies. As a result, few U.S. manufacturers were able to take product innovations to market dominance because they were not prepared to be the lower-cost, higher-quality producers when the product reached market maturity. The items listed in Figure 1–1 are good examples of products invented in the United States but "lost" to Japanese and other offshore industries because of inferior manufacturing technology and production systems in U.S. companies. The standards for world-class performance are based on excellence in production and manufacturing systems because innovation alone is insufficient to guarantee market dominance.

To be competitive, a manufacturer must recognize and meet two types of challenges: *external* and *internal*. If manufacturers fail to meet these challenges, the nation becomes a net user of consumable and durable manufactured goods instead of a net supplier. At that point, the standard of living *falls* in comparison to that of other superior industrialized nations.

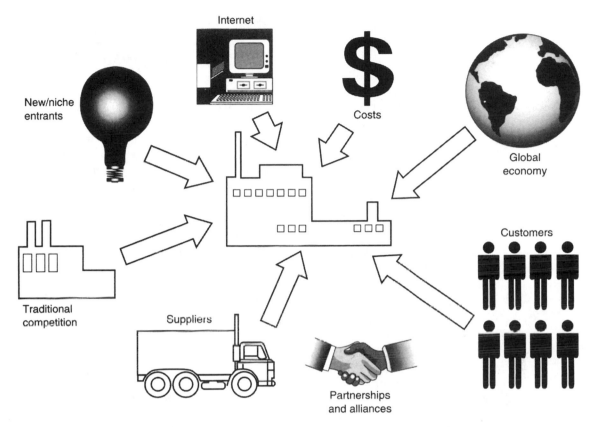

Figure 1–2   External Challenges.

1–2 EXTERNAL CHALLENGES

External challenges result from forces and conditions *outside* the enterprise. The major external challenges, illustrated graphically in Figure 1–2, include *niche market entrants, traditional competition, suppliers, partnerships and alliances, customers, global economy, cost of money,* and *the Internet.* Let's examine each item in more detail.

Companies have always had product lines called *cash cows.* These are high-revenue-generating products that carry the remainder of the product line. The success of the cash cow is a result of market dominance, technological superiority, patent protection, or the absence of competition. Through the late 1950s and 1960s, IBM's mainframe computer line was an example of this phenomenon. In the late 1960s and early 1970s, however, an employee of IBM recognized the profit potential of this market niche and set up a company, Amdhal, to compete with IBM in this market area. The Windows operating system from Microsoft is another example of a cash cow. In the late 1990s a small company, Red Hat Software, started shipping a competing operating system called *Linux.* While it is not clear that Linux will take a significant market share from Microsoft, it is an example of niche marketing. The production and design technology available today and the global reach of the Internet make it possible for small companies, such as Amdhal and Red Hat Software, to develop a product in a market niche and compete with an established producer.

Traditional competition continues to provide a challenge to manufacturing. Manufacturers often counted on the traditional competitor's ignorance as a buffer; however, those days are gone. For example, a study conducted by a midwestern product design and development company indicated that products it got to market six months after the competition reduced its product's profits after taxes by 33 percent. The technology to increase market share is affordable and available to everyone, so the challenge to stay ahead of the traditional competition is stronger today than ever before.

Another external challenge is presented by *suppliers.* Traditionally, many companies have purchased parts and subassemblies on the basis of price variance, where vendors are selected on a low-bid basis. In this price-variance model, the pool of suppliers can change frequently and is often large. Price continues to be a factor; for example, the same midwestern study indicated that a product purchase price that was 9 percent higher than that of the competition resulted in a 22 percent decrease in profits during the product's life. However, in addition to setting a competitive product price, today's suppliers must *meet minimum defect levels,* have *predictable delivery,* and *shorten product design time.* The challenge for companies is to establish fewer but more collaborative purchaser-supplier relationships to meet the needs of the producer and the end customer. The supplier challenge is often addressed through a process called *supply chain management.*

Some of the challenges facing manufacturing have been present for several years; the *partnerships and alliances* element is relatively new. Born from the intense global competition of the 1990s and 2000s, this external challenge results from companies with common interests who join forces to increase their competitive advantage and thus gain more market share. The mechanisms for combining assets and resources include *mergers, friendly* and *hostile buyouts, partial ownership* through stock purchases, and many types of *business agreements.* Examples of this challenge appear in every market sector. An example is the merger between Daimler-Benz, the German automotive manufacturer, and Chrysler, the number three U.S. automotive producer, to create a powerful international company: DaimlerChrysler. With a partnership and alliance, two companies with lower market share can move ahead of the nearest competitor through their combination of customer accounts. The combined resources give the new organization a competitive advantage. In addition, a company can add technology faster through acquisition than by the traditional route of developing it in-house. Cisco Systems, the leading producer of network products, is an example. Cisco's numerous purchases of smaller communication companies add new products faster than Cisco could by using existing development resources. When companies within market sectors consolidate, the companies left without a strategic alliance are often not as competitive and run the risk of losing market share and profits.

The most difficult external challenge for manufacturing is provided by the *customers*. In the captive automotive market of the 1920s, Henry Ford could advertise that the public could have a Model T Ford in any color desired, as long as it was black. Today's sophisticated shopper buys on the basis of *quality, service, cost, performance*, and *individual preference*. If a manufacturer cannot meet all the needs of the customer, another manufacturer is ready to step in for the sale. The proliferation of businesses on the Internet provides access to products from across the world from the customer's home.

Another recent challenge is the business change created by the *Internet*. The development of the World Wide Web (WWW) is changing the way companies do business. The WWW is providing companies with direct access to customer households. Through either the proliferation of the home computer or through the development of the *set-top box* for the cable TV market, the WWW has entered many homes around the globe. The challenge for companies is to determine the best strategy in this quickly changing market area. A good example is the electronic commerce (e-commerce) area. Before Amazon.com initiated the sale of books over the Web, the future for small, traditional bookstores appeared to be challenged only by the large retail book companies like Barnes & Noble. Now, all of the major booksellers are scrambling for a Web presence because the percentage of books purchased over the Internet continues to increase. In addition to facilitating e-commerce, the Internet will affect manufacturers and vendors by

- Providing vendors easy access to manufacturing data
- Creating a window into the manufacturing environment
- Making many of the transactions paperless

Companies that embrace Internet technology intelligently will gain an advantage over their competitors who ignore the power of the WWW or misjudge their correct relationship with this new technology.

Most manufacturers recognize that the present *global economy* provides both an expanded base for marketing products and increased competition from every corner of the world. The challenge for companies is to provide products that meet world-class standards and to market the goods and services to the global economy just as effectively as the competitors both here and offshore.

Another external challenge presented to manufacturers is the *cost of money*. Even when money is borrowed at a low interest rate, the company that uses capital resources the most effectively will be the long-term winner. The cost to introduce and sustain a product in the marketplace has reached staggering proportions. Manufacturing methods that reduce the cost of doing business, such as inventory reduction, become tools for survival. In addition, greater emphasis is placed on how money is used during the design phase to reduce production and delivery costs during the product's lifetime. For example, the results of the study by the midwestern design company indicate that a 100 percent overrun in development cost reduces profit for the product by only 4 percent. These data indicate that spending the money to ensure the best design is wise because money invested in product design up front can maximize profits during the product's lifetime.

Companies must respond to the external challenges (niche market entrants, traditional competition, suppliers, partnerships and alliances, customers, global economy, the cost of money, and the Internet) by

■ Recognizing that these challenges exist and admitting that the problems they create must be solved because they will never go away

■ Developing an internal manufacturing strategy to minimize the negative impact of the external challenges on the success of the business

1–3 INTERNAL CHALLENGES

The internal challenge is to develop a *manufacturing strategy* that will conform to the following description:

> *A manufacturing strategy is a plan or process that forces congruence between the corporate objectives and the marketing goals, and the production capability of a company.*

Satisfying this definition is a major problem for many companies. In many organizations a *great manufacturing divide* separates the marketing and production sides of the enterprise (Figure 1–3). On one side of the divide, marketing views orders from a *total dollar* viewpoint, while on the production side, orders are judged from a *volume* and *product-mix* standpoint. The marketing area seldom considers what the order product mix will do to production efficiency because it is not measured against that standard but by *sales dollars*. In contrast, the production area is often judged on *capacity utilization* and *shop productivity measures* and not on total order value. As a result, manufacturing is frequently not aligned with the product needs of the marketplace, and the orders accepted by marketing provide a bad fit with

Figure 1–3 The Great Manufacturing Divide.

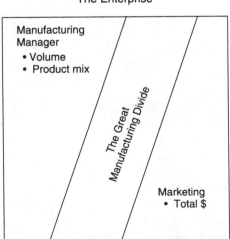

the production processes available. A look at a typical corporate *planning process* (listed next) is the starting point for an understanding of this problem.

1. Define corporate goals.
2. Develop the necessary marketing strategies to satisfy the corporate goals.
3. Analyze the marketplace and determine how the products will fit with market conditions and competition.
4. Determine the process to be used for manufacturing the products.
5. Provide the manufacturing infrastructure necessary for the production.

Corporate planners use an *iterative* process in developing the first three planning steps because each step affects the other three. Too often, the last two are viewed as simply a linear extension of the first three. Experience indicates that steps 4 and 5 have a significant impact on the first three steps because the manufacturing process is often *constrained* and can change only over a limited range. In addition, the degree of *flexibility* in the production infrastructure has limits. However, including manufacturing (steps 4 and 5) in the planning iterations is frustrated by the complexity of the manufacturing processes and differences in the enterprise measurement systems. Read over the preceding five planning steps until you see the great manufacturing divide located between steps 3 and 4.

A procedure for recognizing both manufacturing and marketing in developing a corporate strategy while providing an ordered and analytic approach is needed. Such a procedure is provided by the *Terry Hill model*.

Terry Hill Model

The Terry Hill model furnishes a framework (Figure 1–4) that *links* manufacturing and marketing decisions so that a common executable strategic plan is possible. The five steps in the planning process just defined are listed across the top of the matrix in Figure 1–4, but the third step is renamed *manufacturing order-winning criteria*. Note, for example, that marketing is interested in *product markets and segments, range of products, standard versus custom products,* and *innovation*. In contrast, manufacturing strategies focus on *process choice* and *production infrastructure*. The model framework helps to stimulate corporate debate about the business so that manufacturing can assess the degree to which it can produce the products demanded by the marketplace. However, a *common language* understood by both marketing and manufacturing is necessary to debate current and future market needs. The common language is based on the *order-winning criteria* for the product.

Order-Winning Criteria

The *order-winning criteria*, linking the marketing focus and manufacturing strategies, provide a vocabulary to describe product market requirements that can be translated into process choices and infrastructure requisites by manufacturing.

Corporate goals	Marketing strategy	Manufacturing order-winning criteria	Manufacturing strategies	
			Structure	Infrastructure
What the company is going to do: • Growth • Profit margin • Other financial measures	How the company will reach desired goals: • Product markets and segments • Mix • Volumes • Standardization versus customization • Level of innovation • Leader versus follower	• Price • Quality • Lead time • Delivery/reliability • Flexibility • Innovation ability • Size • Design leadership	• Capacity • Facilities • Technology (processes used) • Vertical integration (degree to which all parts are produced internally)	• Workforce (pay, skill level) • Quality achievement process • Manufacturing planning and control system • Organization (control, measurement, motivation)
External focus		*Common language*	*Internal focus*	

Figure 1–4 Order-Winning Criteria Model.

Figure 1–5   Product Lifecycle
Curve.

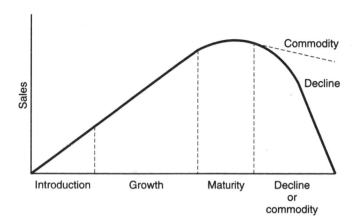

Typical order-winning criteria include *price, quality, delivery speed,* and *innovation ability.* Therefore, the term *order-winning criteria* is defined as follows:

> *Order-winning criteria are the minimal operational capabilities required to get an order.*

Note that the order-winning criteria provide direction to every unit or group in the enterprise. For example, if quality is an order-winning criterion, the manufacturing unit must provide that feature; however, when design innovation is the criterion, the design group is responsible. Order-winning criteria such as *after-sales service, delivery reliability,* and *flexible financial policies* touch areas outside manufacturing. The order-winning criteria provide every area with clear directions.

Although one criterion may be dominant, a mix of order-winning criteria is often present for a product. In addition, the order-winning criteria change during the lifecycle of the product (Figure 1–5). The chart in Figure 1–5 indicates that products move through four stages: *introduction, growth, maturity,* and *decline.* However, some products exhibit the characteristics of a commodity, and the decline phase levels to a relatively constant market share. Breakfast cereals, for instance, exhibit that type of market performance. Note that product sales are at their *maximum* at maturity. In the early part of the product lifecycle, innovation is frequently the dominant order-winning criterion. For example, in the early days of the transistor radio, RCA was one of the dominant companies because it had the research and design talent to satisfy the dominant order-winning criterion: innovation. When a product reaches maturity, price frequently becomes the key order-winning criterion. In the previous example, Japanese companies took over the transistor-radio market during maturity because they could satisfy the dominant order-winning criterion: price.

Changing the Product Lifecycle

Many U.S. manufacturers try to lengthen the maturity stage of the product lifecycle in Figure 1–5 because that part of the curve provides maximum profits. Costs associated with product development are considered an *investment,* and a longer maturity stage represents a better return on that investment.

In contrast, some Japanese competition views the development cost of the initial product as *sunk cost*—just the cost of doing business. With this view, the business pressure for a long model life is not present. However, a shorter model life is not synonymous with design obsolescence; the product continues to have high quality standards for design and operational life. The model life is intentionally shortened by the design and introduction of a new and better model by using one of three methods: *kiazen, leaping,* or *innovation.* The first method, *kiazen,* focuses on improvement of the current model. For example, the improvement team may be challenged to create a new model with enhanced features at a 10 percent reduction in cost within twelve months of the original product introduction. The second method, *leaping,* attempts to develop a different product with similar form, fit, and function characteristics as those of the initial product. The Sony Walkman is an example of this technique, with the initial product's being a standard tape recorder. The last method, *innovation,* seeks to use genuine new product invention to identify follow-up merchandise. As a result of the success of this technique, many U.S. companies now follow a similar path.

This three-pronged approach to shorten model life is frequently accomplished with three cross-functional teams, each team working on one of the three methods. When this technique is executed successfully, three distinctly different models are introduced as a follow-up to the current product. As a result, profits grow through *increased market share* in the product area as consumers stay with the manufacturer and trade up from the original product or abandon an older model from a competitor.

Regardless of the length of the product life, the match between what marketing is selling during the life of a product and the ability to design and manufacture the desired product is strengthened by dialogue in a language both manufacturing and marketing understand. The *order-winning criteria* for the product provide this common language.

Order-Winning Versus Order-Qualifying Criteria

It is important to make the distinction between criteria that *win product orders* and those that *qualify a product* to compete in the marketplace. An example best describes this difference. In the 1970s, the order-winning criterion for the color TV market in the United Kingdom was *price.* The Japanese entered the market with products far superior in quality and reliability to the existing UK products. The buyers of TVs in the United Kingdom established a new quality and reliability standard for TV purchases based on the imported TVs. The price could not drop low enough for the UK companies to sell their TVs that were below the new quality and reliability standard. Therefore, *quality and reliability* became the new order-qualifying criteria for selling TVs in the UK market. By the 1980s, UK companies matched the quality and reliability standards set by the Japanese, and the order-*winning* criterion reverted to price. At this point, the order-*qualifying* criterion became quality.

The enterprise must, therefore, provide the order-qualifying criteria to get into the market or maintain its market share. After the qualifying criteria are satisfied, the enterprise attends to the winning criteria on which orders are won. Market share is *increased* when the order-winning criteria are understood and executed better than the competition understands and executes them. Market leaders must also have the ability to anticipate when the order-winning criteria are changing during the product's lifecycle.

The Terry Hill model provides a process and language that permit marketing and manufacturing to discuss the needs of the marketplace and the degree to which manufacturing process choices and infrastructure can satisfy these market needs. The intention in this text is not to describe the Terry Hill model in the detail required for implementation, but to emphasize that the internal challenge for every company is the development of a process that permits congruence between the corporate marketing goals and the production capability.

Meeting the Internal Challenge

In simplistic terms, the four steps for achieving internal consistency are as follows:

1. Analyze every product and agree on the order-qualifying and order-winning criteria for the product at the current stage in the product's life.
2. For every product, project the order-winning criteria for the future stages in the product's life.
3. Determine the *fit* between the process capability required to meet the criteria in steps 1 and 2 and the current capability in manufacturing.
4. On the basis of the results of steps 1, 2, and 3, either change or modify the marketing goals, or upgrade the manufacturing processes and infrastructure to force internal consistency.

A final word of caution is needed about setting the order-winning criteria necessary to meet marketing's goals. A tendency exists to select too many different order-winning criteria, which results in failure to identify the criterion most critical to winning orders. Despite the imperfections in the process, it is critical for marketing to identify the order-winning criteria as a prerequisite to developing a corporate marketing strategy.

1–4 WORLD-CLASS ORDER-WINNING CRITERIA

A review of the product order-winning criteria most often present (i.e., price, quality, delivery) indicates that the responsibility for the criteria most often falls on the manufacturing unit in the enterprise. Most of the criteria listed in the third column in Figure 1–4 must be addressed by the manufacturing unit. Manufacturing also shares the responsibility for overcoming the external challenges. For manufacturing to meet this responsibility, improvement in several shop-floor standards is

Figure 1–6 Internal
Challenges.

Attribute	World-class manufacturing standards
Setup time	
System	<30 min.
Cell	<1 min.
Quality	
Captured	1,500 ppm
Warranty	300 ppm
Cost of quality	3–5%
Manufacturing/total space	>50%
Inventory	
Product velocity	>100 turns
Material residence time	3 days
Flexibility	270 parts
Distance	300 ft.
Uptime	95%

necessary. A table of world-class manufacturing standards compiled by Coopers and Lybrand is given in Figure 1–6. How well a company compares with the best in the world is an indication of how ready it is to provide the criteria necessary to win orders. It is important for every company to determine how well it compares with the process capability of other world-class manufacturing units. In the following sections, we *describe* each attribute and *indicate* the order-winning criterion affected.

Setup Time

Setup time is the time required to get a machine ready for production. A long setup time causes the size of the minimum number of parts produced, or *lot size,* to be large. The larger lot size results because the cost associated with the long setup time must be spread across more products so that product cost is not significantly raised. As a result, the long setup time leads to higher manufacturing costs because of larger inventories. With more time allocated to setting up the machine, less time is spent in production on the machine, so capital expenditures increase because more equipment and facilities are required to make up for lost production during the long setup time.

Calculating setup-time cost requires a value for the burden rate. The *burden rate* is a cost roll-up or total that includes recovery of the *original cost* of the process machine and associated support hardware, such as robots and material-handling equipment; *maintenance costs* (fault and preventive); *labor costs* (direct and indirect); *programming costs* for computer numerical control (CNC) machines and robots; and any *other costs* associated with putting the machine into the production system. A typical range of burden cost for an automated production system might be $200–$500 per hour, depending on the level of automation present. These costs accumulate every hour of the day. One part of the cost of manufactured items produced on the machine is set by the burden rate of the production machine. The following exercise describes the impact that setup time and burden rate have on production and part cost.

Exercise 1–1

A machine burden rate (the cost per hour to have the machine in production) is $150 per hour. Production requires three setups of 45, 90, and 50 minutes during three shifts. What is the annual cost of the setup time for a 6-day production operation?

Solution

$$
\begin{aligned}
\text{Total lost time per day} &= 45 + 90 + 50 \text{ min.} \\
&= 185 \text{ min.} = 3.08 \text{ hr.} \\
\text{Lost time per year} &= 3.08 \text{ hr.} \times 6 \text{ days} \times 52 \text{ weeks} \\
&= 961 \text{ hr.} \\
\text{Annual cost} &= 961 \text{ hr.} \times \$150 \text{ per hr.} \\
&= \$144{,}150
\end{aligned}
$$

Thus, the manufacturer starts $144,150 in the hole on parts produced on this machine owing to setup-time cost. The order-winning criteria affected by setup time include price, delivery speed, and flexibility.

Exercise 1–2

A production system has a burden rate of $300 per hour. The setup time for a machined casting is 3.6 hours, and the system can produce a finished part every 6 minutes.

a. For a lot size of 500 parts, how much of the part cost is associated with setup cost?

b. Reducing inventory levels requires the lot size to be reduced to 100 parts. What new setup time is required to maintain the same setup cost for the parts?

c. The finished-part cost is the sum of raw material cost, part production cost, setup cost, and a 30 percent margin. Assume the raw material cost is $5 per part. What percentage of the finished-part cost results from the setup time for the 500-part lot-size production?

Solution

a.
$$
\begin{aligned}
\text{Setup-time cost} &= \text{setup time} \times \text{burden rate} \\
&= 3.6 \text{ hr.} \times \$300/\text{hr.} \\
&= \$1{,}080
\end{aligned}
$$

$$
\begin{aligned}
\text{Setup cost per lot-size part} &= \frac{\text{setup-time cost}}{\text{lot size}} \\
&= \frac{\$1{,}080}{500} \\
&= \$2.16
\end{aligned}
$$

b. Determine the total setup-time cost for a lot size of 100 at a per-part setup cost of $2.16.

$$\text{Total setup-time cost} = \text{lot size} \times \text{setup cost per part}$$
$$= 100 \times \$2.16$$
$$= \$216$$

$$\text{Setup time (new)} = \frac{\text{setup-time cost}}{\text{burden rate}}$$
$$= \frac{\$216}{\$300/\text{hr.}}$$
$$= 0.72 \text{ hr.}$$
$$= 43.2 \text{ min.}$$

c. Determine the percentage of final-part cost resulting from setup cost for a 500-part lot size.

$$\text{Part production per hr.} = \frac{60 \text{ min.}}{\text{part production time (min.)}}$$
$$= \frac{60 \text{ min./hr.}}{6 \text{ min./part}}$$
$$= 10 \text{ parts/hr.}$$

$$\text{Part production cost} = \frac{\text{burden rate}}{\text{part production per hr.}}$$
$$= \frac{\$300/\text{hr.}}{10 \text{ parts/hr.}}$$
$$= \$30/\text{part}$$

$$\text{Finished- part cost} = [\text{raw material cost} + \text{part production cost} + \text{setup cost (from part a)}] \times 1.3$$
$$= (\$5 + \$30 + \$2.16)1.3$$
$$= \$48.31$$

$$\text{Setup percentage} = \frac{\text{setup cost}}{\text{finished-part cost}} \times 100$$
$$= \frac{\$2.16}{\$48.31} \times 100$$
$$= 4.5\%$$

Quality

Quality is expressed as a *percentage of defective parts produced* or as a *percentage of total sales* in Figure 1–6. *Captured quality*, defects found before the product is shipped to the customer, has units of parts per million (ppm). *Warranty quality* represents defective parts discovered after the product is shipped to the customer. *Total quality* represents the total cost in percentage of sales that quality costs the enterprise. A useful rule of thumb is that the total cost of poor quality in percentage of sales is usually *three times* the captured cost *plus* the warranty-quality cost.

Exercise 1–3

Quality is often represented as a percentage and with units of ppm, as illustrated in Figure 1–6. Determine the number of bad parts for the following lot sizes, using the world-class standard values for captured quality listed in Figure 1–6.

a. 50,000 parts

b. 5,000 parts

c. 500 parts

d. 50 parts

Solution

a. Determine the number of bad parts for quality represented as a percentage of parts and in ppm notation.

$$\text{Bad parts} = \text{total parts} \times \frac{\text{ppm bad parts}}{10^6}$$

$$= 50{,}000 \text{ parts} \times \frac{1{,}500}{1{,}000{,}000}$$

$$= 75(\text{world-class standard})$$

b.

$$\text{Bad parts} = 5{,}000 \times 0.0015$$

$$= 7.5 \text{ parts (world- class standard)}$$

c.

$$\text{Bad parts} = 0.75 \text{ part (world-class standard)}$$

d.

$$\text{Bad parts} = 0.075 \text{ part (world-class standard)}$$

The order-winning criteria affected by this standard are quality and price.

Manufacturing-Space Ratio

The *manufacturing-space ratio* standard is a measure of how efficiently manufacturing space is utilized. The total *footprint* of the machines, plus the area of workstations where *value is added* to the product, is divided by the total manufacturing space. Value is added at any workstation that processes the part (i.e., machining) or assembles the part with another part. The *larger* this ratio, the more efficient the production operation. Manufacturing shop-floor space allocated to material handling, material transport, storage of raw materials and finished goods, in-process work queues, inspection, expediting, and supervision decreases this ratio. Therefore, the less floor space devoted to these non-value-added elements, the higher the ratio. The order-winning criterion affected by this standard is price.

Exercise 1–4

An automated work cell used by a manufacturer covers 300 square feet of factory floor space. The CNC process machine in the cell has a footprint that measures 75 inches by 80 inches. The remaining space includes pallets for raw materials and finished parts, an operator's desk, walking space around the machine, and an inspector's bench.

a. Calculate the manufacturing-space ratio for the work cell.

b. Calculate the total work-cell area if the cell conforms to the world-class standard listed in Figure 1–6.

Solution

a. Convert the machine footprint to square feet. Then find the ratio.

$$\text{Machine footprint (ft.}^2) = \frac{75 \text{ in.}}{12 \text{ in.}} \times \frac{80 \text{ in.}}{12 \text{ in.}}$$
$$= 6.25 \text{ ft.} \times 6.67 \text{ ft.}$$
$$= 41.69 \text{ ft.}^2$$

$$\text{Manufacturing-space ratio} = \frac{\text{process machine footprint}}{\text{total work-cell area}} \times 100$$
$$= \frac{41.69 \text{ ft.}^2}{300 \text{ ft.}^2} \times 100$$
$$= 13.9\%$$

b.
$$\text{Total work-cell area} = \frac{\text{process machine footprint}}{\text{manufacturing-space ratio (world-class)}}$$
$$= \frac{41.69 \text{ ft.}^2}{50\%}$$
$$= 83.4 \text{ ft.}^2$$

Inventory: Velocity and Residence Time

Raw materials, partially completed parts, subassemblies, and finished goods that are waiting to be shipped to the customer represent inventory. Manufacturing *inventory* can be grouped into two broad categories: *movement* and *organization*. *Movement*, or *transport, inventories*, sometimes called *in-transit* or *pipeline*, are necessary because more than one machine is required to manufacture products, so they must be transported from one location to another. Movement inventories in manufacturing are called *work-in-process (WIP)* inventories. The design and type of manufacturing system used dictates the level of movement inventory present in the system.

Organization inventories uncouple successive stages in the production and distribution systems to neutralize the disturbances in the system. Organization inventories are classified into the following three groups: *cycle stock, safety stock,* and *anticipation stock*. *Cycle stock* inventories exist when the quantity of produced or purchased parts is larger than the current requirement. The inventory costs associated with this excess number of manufactured or purchased parts are less than the costs of manufacturing the parts in smaller quantities or of purchasing them in small lot sizes. Using the calculation from exercise 1–1, for example, you could argue that a larger-than-ideal lot-size production is justified because the inventory costs are less than the costs associated with more frequent machine setup time.

Safety stock provides protection against irregularities and uncertainties in the demand or supply stream. The most common example is the stocking of extra items of raw materials because of the uncertainty attached to production quality. In many manufacturing systems, it is difficult to predict accurately how many production parts will be scrapped as a result of poor quality.

Anticipation stock is associated with products whose markets exhibit seasonal patterns of demand and whose supply stream is relatively constant. It is often not economically feasible to produce a sufficient number of products on a weekly or monthly basis to supply a seasonal demand. For example, it may not be economical to have a production system that can produce the number of lawn mowers sold in the spring months. The inventory costs suggest that it is more economical to produce the products all winter and stock them for the spring sales. Other examples are Christmas toys and calendars for the following year.

The inventory needed to reduce disturbances in manufacturing and in filling customer orders is a *necessary cost*. However, the *goal* is to reduce the inventory level continuously while maintaining continuous and smooth production plus on-time and complete order shipments to the customer. Inventory levels are measured in terms of *inventory turns*, or *velocity*, and in *residence time*. The number of *inventory turns* for a product is equal to the *annual cost of goods sold* (COGS) divided by the *average inventory value*. The *residence time* is the average number of days that a part spends in production. If the goal is reduced inventory, then the number of inventory turns should *increase* while residence time must *decrease*. If this standard is performed well, parts and raw materials spend less time in the plant. As a result, the order-winning criteria that benefit are usually *price, quality, delivery speed,* and *delivery reliability*.

Exercise 1–5

A manufacturer has an annual sales volume of $500 million with the COGS $0.90 per sales dollar. The average inventory value is $25 million. What are the inventory turns for this manufacturer?

Solution

$$\text{COGS} = \$500 \text{ million} \times \$0.90$$
$$= \$450 \text{ million}$$
$$\text{Inventory turns} = \frac{\$450 \text{ million}}{\$25 \text{ million}}$$
$$= 18$$

Flexibility

Flexibility is a measure of the number of different parts that can be produced on the same machine. Improved efficiency in this standard results from good part design, innovative fixture design (*fixtures* are devices used to hold parts and raw material while they are in production), and well-planned production machine procurement. Excellence in this standard aids primarily the price order-winning criterion. Flexibility also measures the ability to produce new product designs in a short time. To force a shorter product lifecycle, the enterprise must have flexibility as an order-winning criterion.

Distance

The *distance* standard measures the total linear feet of a part's travel through the plant from raw material in receiving to finished products in shipping. A high value for this standard indicates more non-value-added time for the part and decreased quality because of handling: therefore, a higher part cost results. Keeping the value of this standard low helps the price and quality order-winning criteria.

Uptime

Uptime is the percentage of time a machine is producing to specifications compared to the total time that production can be scheduled. A reduction in the uptime standard means that more equipment is required for the same level of production. For example, industry data indicate that every 1 percent improvement in uptime reduces capital equipment cost by 10 percent. As a result, improvement in this metric helps the *price* order-winning criterion.

1–5 THE PROBLEM AND A SOLUTION

An enterprise must recognize the external challenges (niche market entrants, traditional competition, suppliers, partnerships and alliances, customers, global economy, the cost of money, and the Internet) that are present and develop a manufacturing strategy to win orders on the basis of the criteria present in the marketplace. The problem is that the order-winning criteria drive the market. The enterprise cannot change these criteria, and the environment that creates the external challenges will not go away. Therefore, the enterprise must change.

The Cost of Doing Nothing

Figure 1–7 illustrates the consequences to a product for a manufacturer who chooses to ignore the forces present in the marketplace. The next product is less competitive than its predecessor, so the margins continue to drop. The southeastern United States had numerous examples of textile manufacturers that followed

Figure 1–7   Cost of No Action.

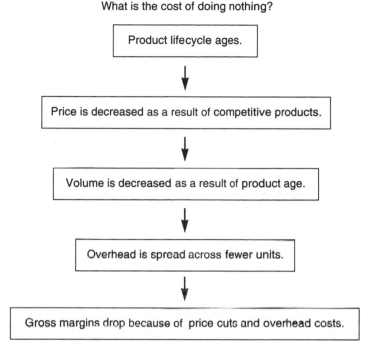

What is the cost of doing nothing?

> Product lifecycle ages.

> Price is decreased as a result of competitive products.

> Volume is decreased as a result of product age.

> Overhead is spread across fewer units.

> Gross margins drop because of price cuts and overhead costs.

this downward spiral in the 1960s and 1970s. One case is that of a mill that produced first-quality cloth in the 1950s and could sell in markets of its own choosing. With time, the production machines and processes aged and did not stay even with world competition; as a result, the manufacturer was relegated to producing only low-market-value muslin before closing its doors in the 1970s. A similar story can be told in countless other market sectors, steel production being another example. The turbulence in manufacturing from the mid-1960s to the early part of the 1980s caused a dramatic change in U.S. manufacturing. In general, the pressures created a stronger manufacturing base in most market sectors. However, the challenge from on- and offshore competition requires that a company be alert. With the life of the enterprise at risk, what should be done?

A Solution

The following seven world-class measures must be addressed by manufacturers:

1. Inventory turns by product
2. Setup times on production equipment
3. Total quality and rework
4. Design and manufacturing lead time by product
5. Production efficiency by product

6. Employee output or productivity by product
7. The number of product improvement suggestions per day per employee

In general, these seven basic standards are often not adequately addressed by companies. The first three were addressed in detail in the previous sections, but the last four are new. Manufacturers looking to improve market share or reduce product cost frequently improve one area with an *island of automation*. Improvements in design lead time using computer-aided design (CAD) software are a good example of an island of automation that is frequently built. However, fixing one problem while ignoring others often reduces overall productivity. Manufacturing presents an integrated set of problems that require an integrated solution. That solution is *computer-integrated manufacturing*. The late Joseph Harrington, Jr., recognized the value of integration and in 1973 introduced the term *computer-integrated manufacturing* in his book of that title. Although he intentionally avoided developing the acronym CIM (pronounced "sim"), the abbreviation was in common use by the early 1980s.

CIM Defined

The Computer and Automation Systems Association (CASA) of the Society of Manufacturing Engineers (SME) defined *CIM* as follows:

> *CIM is the integration of the total manufacturing enterprise through the use of integrated systems and data communications coupled with new managerial philosophies that improve organizational and personnel efficiency.*

Appendix 1–1, at the end of this chapter, the *Industry Week* survey titled "The Benefits of a CIM Implementation," supports the use of CIM to deliver a host of order-winning criteria.

The manufacturing enterprise wheel (Figure 1–8) illustrates the integration called for in the definition and shows the interrelationship among all parts of an enterprise. Take a few minutes to study all the segments present in the wheel, and then read the CIM definition again.

CIM describes a new approach to manufacturing, management, and corporate operation. Although CIM systems can include many advanced manufacturing technologies such as *robotics, computer numerical control* (CNC), *computer-aided design* (CAD), *computer-aided manufacturing* (CAM), *computer-aided engineering* (CAE), and *just-in-time* (JIT) production, it goes beyond these technologies. CIM is a new way to do business that includes a commitment to *total enterprise quality, continuous improvement, customer satisfaction,* use of a *single computer database* for all product information that is the basis for manufacturing and production decisions in every department, removal of *communication barriers* among all departments, and the *integration* of enterprise resources. The enterprise wheel shown in Figure 1–8 is an update from the original SME/CIM Wheel and has the following six defined areas:

1. The hub of the wheel, titled *Customer,* is the primary target for all marketing, design, manufacturing, and support efforts in the enterprise. Only with a clear understanding of the marketplace and the customer can the enterprise be successful.

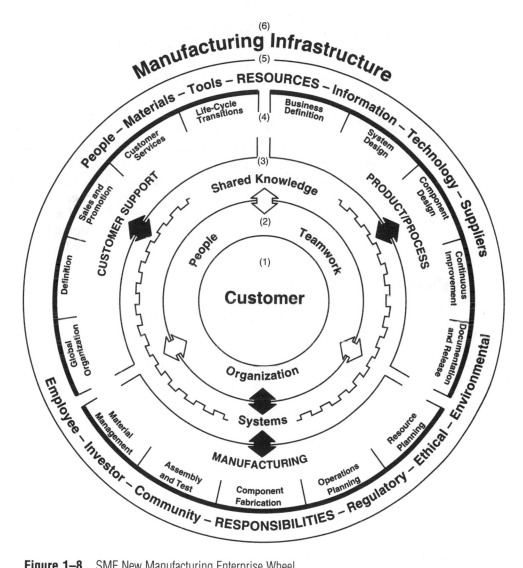

Figure 1–8 SME New Manufacturing Enterprise Wheel.
(Source: Courtesy of the Society of Manufacturing Engineers, Dearborn, Michigan, copyright 1993, Third Edition.)

2. The next layer on the wheel focuses on the means of organizing, hiring, training, motivating, measuring, and communicating to ensure teamwork and cooperation in the enterprise. The techniques used to achieve this goal include self-directed teams, teams of teams, organizational learning, leadership, standards, rewards, quality circles, and a corporate culture.

3. This section focuses on the shared corporate knowledge, systems, and common data used to support people and processes. The resources used include manual and computer tools to aid research, analysis, innovation, documentation, decision making, and control of every process in the enterprise.

4. The three main categories of processes, *product/process definition, manufacturing,* and *customer support,* make up this section of the wheel. Included in this group are fifteen key processes that form the product lifecycle.

5. The enterprise has resources that include capital, people, materials, management, information, technology, and suppliers. It also has responsibilities to employees, investors, and the community, as well as regulatory, ethical, and environmental obligations.

6. The final part of the wheel is the manufacturing infrastructure. This infrastructure includes customers and their needs, suppliers, competitors, prospective workers, distributors, natural resources, financial markets, communities, governments, and educational and research institutions.

The CIM principles covered in this text address most of the areas in the CASA/SME new manufacturing enterprise wheel.

1–6 LEARNING CIM CONCEPTS

Just as the enterprise wheel divides the operational aspects of CIM into major segments, the study of CIM follows a similar path. Section 1–5 introduced and defined CIM. Next, the three process segments are described. Later in this book, the automation options are explored and implementation is addressed, together with management of the information systems and human resources. Let's start with the three process segments.

Figure 1–9   The Enterprise Areas.

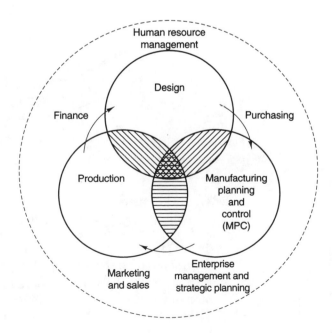

Process Segments

Figure 1–9 illustrates another view of the enterprise, including the three process segments with overlapping areas that indicate shared data and resources. The product design data, for example, are needed in manufacturing planning and control (MPC) to plan process routings. The degree of the overlap varies because some segments share more information than do others. The development of a new product usually starts in the design circle and moves in the direction of the arrows in Figure 1–9. The CIM concepts associated with each segment are covered in that same order in this book. As a result, part 2 starts with the design and documentation work required when a new product is produced. In the enterprise wheel, this is called "product/process" and includes *business definition, system design, component design, continuous improvement,* and *documentation and release.* Discussions about this area in the literature usually reference CAD, the automation software frequently used to improve design and drawing productivity. However, implementing CIM in the product/process segment requires more than just using CAD software. In chapters 3 through 5, we describe the CIM requirements and benefits in this segment of the enterprise wheel and enterprise.

The second process segment required for product development, labeled *manufacturing planning and control (MPC)* in Figure 1–9, includes all the *process planning, production scheduling, inventory management,* and *capacity planning* required for efficient manufacturing. In the enterprise wheel (Figure 1–8), the segment called *manufacturing* includes elements from MPC and from the production circle in Figure 1–9. The enterprise wheel elements include *material management, assembly and test, component fabrication, operations planning,* and *resource planning.* Manufacturing resource planning (MRP II) concepts and software are frequently applied to manage this area of the production process. Implementation of a software package for MPC does not guarantee that this segment complies with the CIM definition, however. The requirements for a CIM implementation in this area of manufacturing are detailed in chapters 6 through 9.

The last process segment, labeled *production* in Figure 1–9, includes all the activity associated with the production, or shop floor. This includes some of the elements found in the manufacturing segment of the enterprise wheel. The application of CIM principles for the shop-floor, or production areas is covered in chapters 10 and 11. Frequently, manufacturers confuse the installation of flexible manufacturing cells (FMCs), CNC machines, smart conveyor systems, automatic tracking, or industrial robots with the implementation of CIM. However, a careful examination of the CIM definition and the enterprise wheel indicates that CIM is *not* just

- Automated hardware and software
- A manufacturing system bought from a vendor, installed in the enterprise
- Manufacturing strategies such as just-in-time (JIT) and simultaneous engineering

An enterprise that embraces CIM concepts has a managerial philosophy that

- Uses customer satisfaction as the basis for decisions
- Espouses total quality (TQ) principles
- Values the ideas of every employee
- Does not accept the status quo but works toward continuous improvement

In addition, a successful CIM implementation

- Shares data across all the departments
- Uses automation hardware and software to integrate enterprise operations effectively so that product data are created only once and used many times

However, the process segments produce a world-class product only when the enterprise resources are well managed.

1–7 GOING FOR THE GLOBE

The ultimate goal of the enterprise is to develop an internal strategy that leads to world-class levels of manufacturing performance. Order-winning criteria are used to identify and rank the world-class performance measures most critical for manufacturing success in a given market. The bar graph of CIM benefits in appendix 1–1 (Figure 1–14) implies that CIM is synonymous with world-class measures such as *lower manufacturing costs, higher product quality, better production control, better customer responsiveness, reduced inventories, greater flexibility,* and *smaller lot-size production.* Therefore, a CIM implementation starts an enterprise on a journey toward acquiring these world-class standards. CIM is implemented with the following three steps.

The CIM Process: Step 1

Implementation of a successful CIM system follows a three-step process.

STEP 1: Assessment of the enterprise in three areas:

- *Technology*
- *Human resources*
- *Systems*

Building an enterprise-wide CIM system requires extensive planning, many months of hard work, and a substantial investment in people, hardware, and software. Assessment must be the first step because it prepares the enterprise for step 2, simplification, and step 3, implementation. The assessment of the enterprise technology, human resources, and systems includes a study to determine the following:

- The current *level* of technology and process sophistication present in manufacturing

Figure 1–10 Obstacles to a CIM Implementation.

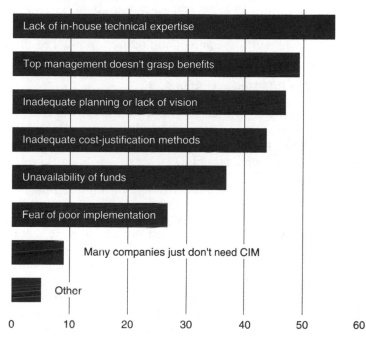

(Percentage of survey respondents)

- The current *state* of employee readiness for the adoption of CIM automation across the enterprise
- The reason *why* the production systems function as they do

In each of the three areas—technology, human resources, and enterprise systems—the capability, strengths, and weaknesses are identified, measured, and documented. The assessment process consists of an internal self-study with a large educational component. The critical nature of education is illustrated in Figure 1–10 by a survey of 139 companies planning a CIM implementation. Note that 55 percent of the respondents listed *lack of in-house technical expertise* as a major obstacle to CIM implementation. In-house expertise in the design and construction of automation systems is often not present in smaller companies. Overcoming this deterrent is accomplished only when the education component for every employee focuses on the following:

- The necessity for enterprise change to remain competitive in a national and world marketplace
- The need to support a new model in enterprise operations that includes teamwork, TQ, improved productivity, reduced waste, continuous improvement, common databases, and respect and consideration for all ideas and suggestions regardless of the level from which they are initiated
- The hardware and software necessary to implement a CIM system, and the management strategy required to run the system successfully

CIM is not hardware and software; CIM is a way to manage the new technologies for improved market share and profitability. From the start of the implementation, everyone in the organization must understand how CIM relates to his or her job. As a result, assessment and education must be first.

The CIM Process: Step 2

In step 1, the enterprise was *studied* department by department, process by process, and activity by activity. Personnel were *educated* about the company and the technology was planned for the CIM implementation. With this base established, step 2 is begun.

> *STEP 2: Simplification—elimination of waste*

In most departments, CIM moves the manual or functionally automated operations into an integrated enterprise-wide solution. For example, the manual record-keeping process for tracking the location and quantity of parts and products moves to a computerized system, with all inventory information stored in a common electronic database or data warehouse. The software is often called *product data management,* or *PDM.* Installing automation, for example, without first eliminating the unneeded operations in the inventory process would just automate many of the poor practices present in the manual system. *Simplification* is defined as follows:

> *Simplification is a process that removes waste from every operation or activity to improve the productivity and effectiveness of the department and organization.*

What is waste, and where is it found? *Waste* is every possible operation, move, or process that does not *add value* to the final product. If an activity log was made for raw material as it passed through manufacturing, the time in manufacturing would be divided into five general categories: *moving, waiting in queue, waiting for process setup, being processed,* and *being inspected.* Only one of the five, *being processed* by the production machine, adds value to the part, and then only if the process produces a good part. The remaining four are classified as *cost-added* operations because the part is *not* increasing in value as a result of the activity, and the enterprise continues to pay overhead costs during that time. In short, any operation or activity that does not add value to a product adds cost and is waste. Removing all waste from operations is often not possible; however, all possible cost-added processes must be eliminated.

The search for waste often focuses on direct labor cost. Yet in most manufacturing operations, direct labor usually accounts for only 2–10 percent of the COGS. Therefore, eliminating waste by direct labor cost alone would not provide significant savings. However, case studies show that cost-added operations account for as much as 70 percent of all activities in an enterprise. Eliminating as many of the cost-added operations as possible requires significant change in the daily business and manufacturing process. The rule in Figure 1–11 offers

Figure 1–11 Rule for Elimination of Cost-Added Operations.

Three-Step Rule for Eliminating Waste

Reduction	Total eliminated (%)
Reduce by 50%	50
Reduce by 50% again	75
Make it 10% of what it originally was	90

A world-class company works the three-step rule to arrive at a 90 percent reduction in waste.

a process designed to stimulate enterprise change and reduce the level of cost-added operations significantly. A comprehensive list of cost-added operations for an area allows the system or manufacturing operation to be changed to eliminate half of the waste, a 50 percent reduction. Reducing waste in these significant amounts requires that the enterprise *attack waste fundamentally*. For example, a plant uses a forklift to move material between production machines in a batch manufacturing operation. To reduce the transit time between machines, an identified cost-added activity, management links the production machines with a material-handling conveyor. *Can you recognize the problem with this waste reduction solution?* The process is quicker and some direct labor has been eliminated, but material is still moved, WIP inventory is not less, and the batch sizes have not changed significantly. The conveyor is a *superficial improvement*. A *fundamental improvement*, elimination of transportation, is achieved by moving the production machines close together. Identifying fundamental waste reduction requires that every employee be trained in the fundamentals of the business during the first step in the CIM implementation process.

The reduction of waste by 50 percent in Figure 1–11 is the first step in the process of eliminating 90 percent of the identified cost-added processes. As the figure indicates, the 50 percent of the waste still remaining is reduced by 50 percent, so the total waste present is only 25 percent of the original amount. The process continues until no more than 10 percent of the original waste is present. At this point, the manufacturing process is sufficiently stripped of unnecessary cost-added operations so that a successful CIM implementation is possible. However, the war on waste is never over because employees must use the *continuous-improvement technique* to continue searching for cost-added operations to eliminate.

An aggressive simplification exercise on the shop floor to reduce cost-added activity ensures that what gets automated by the shop-floor hardware and software is not bad production processes. When the simplification process is applied in every office and department in the enterprise, performance improves and installed automation is effective. World-class manufacturing requires the elimination of as much of the cost-added activity from the entire enterprise as possible during the simplification step in the CIM implementation.

The CIM Process: Step 3

The first two steps prepared the enterprise for CIM implementation; step 3 builds the system.

STEP 3: Implementation with performance measures

Implementation with *performance measures* has two requirements: (1) measuring the success of CIM implementation at regular intervals, and (2) recording the changes in key manufacturing and business parameters. The objective for the enterprise is to become world class, so the enterprise must track the level of achievement in moving to the new world-class performance. One tracking technique uses four key standards from Figure 1–6, along with checks on productivity and continuous-improvement efforts. The seven key measurement parameters are as follows:

1. Product cycle time
2. Inventory turns by product
3. Production setup times
4. Manufacturing efficiency
5. Quality and rework
6. Employee output or productivity
7. Employee continuous-improvement suggestions

Initially, operational data from all seven measurements are recorded to establish a *baseline* before implementation of any automation or process improvements begins. In enterprises with successful ongoing CIM programs, the performance measurements are reviewed monthly so that changes in the performance can be recorded. However, for the initial CIM implementation, companies often start by measuring just two or three of the first five measurement parameters listed previously. They focus first on the parameter(s) that would provide the greatest initial payback. *Inventory turns* is often selected first because most manufacturing operations have some excess inventory, and dollar savings are easy to identify. As control of the processes is established, additional performance measures are added to the improvement project. The results listed in Figure 1–12 show the changes achieved during a six-month period by a valve manufacturer with $25 million in sales. The remarkable improvements in *floor space* (manufacturing efficiency) are a result of a shift from job shop and repetitive production environments to a *structured-flow system* that groups machines by product (production environments are covered in chapter 2). The case study at the end of chapter 2 illustrates a similar improvement process for another company. Compare the final results for the valve manufacturer (Figure 1–12) and those for the company described in the case study at the end of chapter 2 (Figure 2–14) with the world-class standards shown in Figure 1–6.

Another performance measurement technique, called the *ABCD checklist*, was developed by Oliver Wight Companies. The checklist measures enterprise

Valve Manufacturer's Performance Report Card Case History

Measure	Baseline	Six months	Eighteen months
Cycle time	18 weeks	6 weeks	1 week
Inventory turns	4	8	48
Quality (finished part)	85%	95%	99.8%
Floor space	800 ft.^{2}	400 ft.^{2}	80 ft.^{2}

Figure 1–12   Results of Improved Business Operations.
(Source: Stickler, "Going for the Globe Part III," P and IM Review,
December 1989, p. 42. Reprinted by permission of APICS.)

progress toward world-class performance in the following five basic business functions: *strategic-planning processes, people or team processes, total quality and continuous-improvement processes, new-product development processes,* and *planning and control processes.* A company uses the rating scale in Figure 1–13 to answer a series of questions. On the basis of the average of the numeric score, the company is classified as class A, class B, class C, or class D. The overview questions are

Level	Planning and control processes	Continuous improvement processes
Class A	Effectively used companywide; generating significant improvements in customer service, productivity, inventory, and costs	Continuous improvement has become a way of life for employees, suppliers, and customers; improved quality, reduced costs, and increased velocity are contributing to a competitive advantage.
Class B	Supported by top management; used by middle management to achieve measurable company improvements	Most departments are participating and are actively involved with some suppliers and customers; substantial contributions are being made in many areas.
Class C	Operated primarily as better methods for ordering materials; contributing to better inventory management	Processes are utilized in limited areas; some departmental improvements have occurred.
Class D	Information inaccurate and poorly understood by users; providing little help in running the business	Processes are not established.

Scoring process from overview questions:
Class A: average greater than 3.5
Class B: average between 2.5 and 3.49
Class C: average between 1.5 and 2.49
Class D: average less than 1.5

Figure 1–13   ABCD Scoring Process.
(Source: Courtesy of Oliver Wight Publications, Inc.)

included in appendix 7–2. *Take several minutes and review the questions.* Determining the response for some overview questions is difficult without additional clarification. This clarification is provided by a group of detailed questions for some of the overview questions.

The audit process to determine the ABCD classification level is usually performed by a team of company managers who know current production performance, along with a representative of Oliver Wight Companies who facilitates completion of the ABCD checklist instrument. The general characteristics of companies operating at the class A, B, C, and D levels are also provided in the appendix. *Review the characteristics of a class A company.* As you can see, achieving a class A designation implies world-class performance. Establishing the class of operation is not the critical factor; however, identifying the areas in the enterprise that caused a score of less than class A is valuable. On the basis of the overview question results, processes can be improved, required training for personnel can be identified, and continuous improvement can be tracked. Reread the case study at the end of chapter 2 and note the journey to class A status and world-class performance.

The *final step* in the CIM implementation process consists of the acquisition and installation of hardware and software to the specifications developed in steps 1 and 2. When assessment and simplification are complete and a performance measurement system is in place, successful implementation of hardware and software is ensured. Unfortunately, many companies implement CIM automation starting with step 3. As a result, they automate all the disorder of the poor production processes and produce waste at a record rate.

Managing the Resources

Meeting internal and external challenges with a strategy built on CIM requires change throughout the enterprise. The installation of new automation hardware and software, the development of communications networks, and the establishment of a central enterprise database are some of the significant changes under CIM. For the CIM implementation to be successful, however, the most significant change must be in the human resources area. A new awakening to the power of the employee must occur. Traditional corporate managers must make the difficult but necessary transition to team concepts, horizontal management structures, and the idea of a workforce empowered to make decisions and solve problems. Chapter 13 provides a view of the new role for managers and the complex systems they manage.

Manufacturing has changed significantly since 1973 when Dr. Harrington coined the phrase *computer-integrated manufacturing.* Computer technology, especially, has made exponential advances in speed, function, and computational power. Despite the changes in world markets and technology, the principles outlined by Harrington continue to ring true. His arguments for a system-oriented approach to the enterprise and against highly fragmented manufacturing operations that produce only localized optimization are still valid. Manufacturing continues to evolve and technology continues to progress, but the need for an integrated solution to the many problems across the enterprise remains.

1–8 SUMMARY

Much has been written and said about the value of a manufacturing base in the United States. Our current standard of living depends on our manufactured goods' remaining competitive in world markets. To be competitive today requires that manufacturers meet the external and internal challenges and produce products to world-class standards. Meeting these standards requires a manufacturing strategy that forces internal consistency between marketing and manufacturing. One model, offered by Terry Hill, uses product order-winning criteria as a common denominator for all marketing and manufacturing decisions. Application of the model frequently exposes an integrated set of problems present in the organization. Computer-integrated manufacturing (CIM) offers a solution for this integrated problem set. The dependence on CIM concepts for survival of manufacturers in world markets becomes clearer every day.

In basic terms, CIM is the integration of all enterprise operations and activity around a common corporate database. Although the CIM concept is simple, application and implementation are difficult and complex. Application and implementation always start with education, and in the chapters that follow, the basic concepts associated with CIM are explained in detail.

BIBLIOGRAPHY

BARRIS, R. R. *Justifying Automation—The New Realities.* Burlington, MA: Coopers and Lybrand, 1990.

CLARK, P. A. *Technology Application Guide: MRP II Manufacturing Resource Planning.* Ann Arbor, MI: Industrial Technology Institute, 1989.

GROOVER, M. P. *Automation, Production Systems, and Computer-Integrated Manufacturing.* 2nd ed. Upper Saddle River, NJ: Prentice Hall, 2001.

HARRINGTON, J., Jr. *Computer-Integrated Manufacturing.* New York: Industrial Press, 1973.

HILL, T. *Manufacturing Strategy.* Homewood, IL: Richard D. Irwin, 1989.

OLIVER WIGHT INTERNATIONAL, INC. *The Oliver Wight ABCD Checklist for Operational Excellence.* 5th ed. New York: John Wiley & Sons, Inc., 2000.

OWEN, J. V. "Flexible Justification for Flexible Cells." *Manufacturing Engineering,* September 1990.

ROWEN, R. B. *A Manufacturing Engineer's Introduction to Supply Chain Management.* CASA/SME Blue Book Series, 1999.

SHERIDAN, J. H. "Toward the CIM Solution." *Industry Week,* October 16, 1989.

SHRENSKER, W. L. *CIM: Computer-Integrated Manufacturing: A Working Definition.* Dearborn, MI: CASA of SME, 1990.

SOBCZAK, T. V. *A Glossary of Terms for Computer-Integrated Manufacturing.* Dearborn, MI: CASA of SME, 1984.

STICKLER, M. J. "Going for the Globe Part II." *P and IM Review,* November 1989, 32–34.

STICKLER, M. J. "Going for the Globe Part III." *P and IM Review,* December 1989, 41–43.

VOLLMANN, T. E., W. L. BERRY, and D. C. WHYBARK. *Manufacturing Planning and Control Systems.* 4th ed. Homewood, IL: Richard D. Irwin, 1997.

QUESTIONS

1. Define *manufacturing.*
2. Describe the types of economic conditions that made U.S. manufacturers most successful in the 1950s and 1960s.
3. Describe the retreat of U.S. production from the 1950s to the present day.
4. What are the external challenges faced by manufacturers worldwide?
5. Identify the external challenge that is the most difficult to overcome, and describe why.
6. How should companies respond to external challenges?
7. What is the internal challenge faced by manufacturers?
8. What causes the great manufacturing divide?
9. What is the difference between order-winning criteria and order-qualifying criteria?
10. How does the Terry Hill model help to overcome the internal challenge faced by manufacturers?
11. Describe how the lifecycle of a product affects the order-winning criteria.
12. What steps are necessary to achieve the internal consistency required for a congruence between marketing and management, and manufacturing operations?
13. What are the world-class order-winning criteria compiled by Coopers and Lybrand?
14. What are inventory turns, and how are they defined?
15. Why does a 1 percent increase in machine uptime translate into a 10 percent savings in capital equipment?
16. What are seven requirements for a solution to the primary problem facing manufacturing?
17. What contribution did Joseph Harrington, Jr., make toward solving the manufacturing problem?
18. How is *computer-integrated manufacturing* defined?
19. Identify and describe the segments that form the enterprise wheel.

20. Compare the enterprise wheel in Figure 1–8 with the enterprise representation in Figure 1–9. What are the similarities and differences?
21. What is the CIM managerial philosophy?
22. What constitutes a successful CIM implementation?
23. What resource in an enterprise must change most when a CIM system is implemented? Why is that necessary?
24. Describe the three-step process for implementing CIM.
25. What three areas are studied in the assessment stage, and what information is gathered?
26. Describe the focus of CIM education during the implementation.
27. Define *manufacturing waste* and *simplification*.
28. Compare fundamental process improvements with superficial improvements.
29. What are performance measures, and how are they used?
30. Give one example of a performance measure that could be used to determine the effectiveness of a CIM implementation.
31. What is the ABCD checklist, and how is it used?

PROBLEMS

1. True Bore Machine Company has annual sales of $120 million and an average annual inventory valued at $5 million. With a COGS of $0.85 per sales dollar, what is the inventory turn value for the company? What average inventory turn value would produce a turn ratio of 300?
2. The plastic-injection-molding machines at Great Plastics Molding, Inc., are down as a result of setups for an average of 16.7 percent of the time in three shifts. If machine downtime cost is $135 per hour, how much does setup time cost for each machine on an annual basis?
3. Metal Enclosures, Inc., has machine footprints that total 5,780 square feet and 8,240 square feet of value-adding assembly area. If the plant has 18,420 total square feet of product and inventory area, what is the manufacturing-space ratio?
4. A production system has a burden rate of $200 per hour and a setup time of 4.4 hours. What is the setup cost per part for a lot size of 200 parts?
5. The system described in problem 1 must move to a 75-part lot size. How much must the setup time be reduced to keep the setup cost per part constant?
6. A casting costs $15, the burden cost on the machine that finishes the casting is $300 per hour, time for the part on the machining center is 15 minutes, and the setup time is 3 hours. If the castings are machined in lot sizes of 50, what percentage of the part cost is due to setup time?
7. A study to replace the operator in problem 6 with automation indicates that the burden rate for the cell would increase to $345 per hour, the part machining time would drop to 10 minutes, and the setup time would drop by 0.5 hour.

Would this additional automation decrease the part cost? What is the part cost for the ideal lot size of 25 on this new cell?

8. Repeat the calculations in exercise 1–3, using the warranty-quality standard values from Figure 1–6.

9. Compare the results of the defective-part calculations from example 1–3 and problem 8. What is the significance in manufacturing when the projected defect rate in a lot-size production is less than 1 part?

10. A work cell includes two 48-inch-square pallets for raw casting and finished castings, an operator's work area used to insert bushings into the machined casting that is 30 inches by 48 inches, a tool rack that covers 12 square feet, wide walk-ways and machine clearances that cover 45 square feet, and a production machine that measures 58 inches by 97 inches. Determine the manufacturing-space ratio for the work cell.

11. A company has 10 work cells like the one described in problem 10. Because of increased demand, another cell must be added, but there is no factory floor space. What manufacturing-space ratio would be needed in each cell to put 11 work cells in the space that now supports 10? What two areas in the work cell could most likely be reduced to allow for an 11th work cell?

12. What is the significance to product cost or manufacturing profit of the two inventory turn values in problem 1 when the cost of money is considered?

PROJECTS

1. Use primary and secondary research techniques to make a list of the names, addresses, phone numbers, and products of manufacturers in your city, region, or state, using the following guidelines: five manufacturers with more than 500 employees in the plant location, five manufacturers with more than 100 but fewer than 500 employees at the location, and five manufacturers with 100 or fewer employees at the location.

2. Determine the current order-winning criteria and stage in the lifecycle for one product from each group of manufacturers identified in project 1. Identify and describe an order-qualifying criterion for the products examined.

3. Prepare a report on the concept of order-winning criteria described in the book *Manufacturing Strategy* by Terry Hill.

4. Prepare a report on the definition of *CIM* provided by Harrington in his book *Computer-Integrated Manufacturing*. Compare this with the SME definition given in this chapter.

5. Identify up to five local or national companies or products that experienced loss of market share due to the process illustrated in Figure 1–7. What was the significant mistake made in each case?

6. A company that manufactures kitchen products wants to add a set of plastic storage bowls or containers to its product line. The set would have four plastic bowls with approximate diameters of 6, 8, 10, and 12 inches, with matching

snap-on lids. The company wants to distribute the product through Wal-Mart, Target, or a chain of stores in your region. Do the following: visit stores to document competitive products by documenting the features, determining the selling price, and obtaining some indication of the number of units sold per week.

7. A company that manufactures kitchen products wants to add a high-end manual can opener to its product line. The company wants to distribute the product through Wal-Mart, Target, or a chain of stores in your region. Do the following: visit stores to document competitive products by documenting the features, determining the selling price, and obtaining some indication of the number of units sold per week.

8. A company that manufactures kitchen products wants to add a manual handheld eggbeater to its product line. The company wants to distribute the product through Wal-Mart, Target, or a chain of stores in your region. Do the following: visit stores to document competitive products by documenting the features, determining the selling price, and obtaining some indication of the number of units sold per week.

9. A company that manufactures kitchen products wants to add a manual handheld garlic press to its product line. The company wants to distribute the product through Wal-Mart, Target, or a chain of stores in your region. Do the following: visit stores to document competitive products by documenting the features, determining the selling price, and obtaining some indication of the number of units sold per week.

APPENDIX 1–1: THE BENEFITS OF A CIM IMPLEMENTATION

Industry Week conducted a survey of managers and executives in manufacturing industries to determine the status of CIM implementations. When asked what the benefits of a CIM implementation meant for their companies, the 139 respondents, including forty-nine CEOs and presidents and fifty-three vice presidents, indicated that the biggest payoff is in manufacturing cost and quality. A bar graph that shows the survey responses for this question appears in Figure 1–14. An important point is that the manufacturing industry's list of benefits from CIM includes a high percentage of the key order-winning criteria just identified. Stated more directly, if a firm needs to improve on the standard order-winning criteria, CIM will deliver results.

Figure 1–14   CIM Benefits.

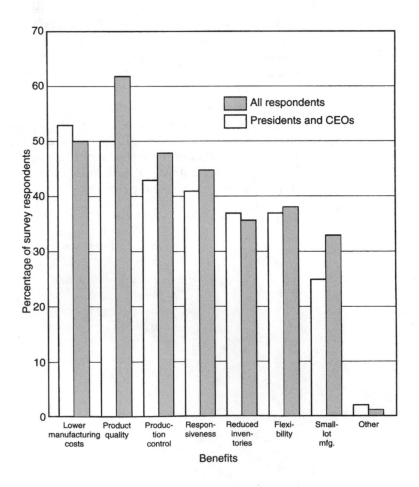

APPENDIX 1–2: TECHNOLOGY AND THE FUNDAMENTALS OF AN OPERATION—AUTHORS' COMMENTARY

There are a lot of activities and tasks taking place every day in manufacturing companies. The best companies, the ones that are truly world-class operations, understand the activities and tasks needed to be performed to satisfy their customers. World-class companies know what they need to do and have plans and systems in place that give them the desired results. There is nothing magic about it. The best-run companies understand the fundamentals of their operations and develop systems that help perform these fundamentals.

You can be sure that the operations of manufacturing companies are filled with distractions. Problems in any part of the operation can cause a company to lose its focus and lose sight of the fundamentals. Changing products, markets, and processes and increased competition can cause problems for even the best-run companies.

Consider the way computers and technologies are changing the workplace. In the early 1980s, some people thought the new technology would solve all the problems of manufacturing. Advanced "lights out" factories were expected to use computers and software programs to drive and control all parts of the operation. CIM at this time was often considered a product that a company bought and had installed. Human intervention in manufacturing operations was expected to become minimal. Technology has had a significant positive effect on manufacturing; however, technology by itself has not become the solution to all the problems facing manufacturing companies. Technology can be effective only if it helps a company perform the fundamentals of its operations better.

Companies that have spent large amounts of resources to try to automate all of their operations have generally not been very successful. During the late 1980s, the General Motors (GM) Corporation spent huge amounts of money on automation for its factories. The hope was that these investments would transform the plants from some of the most inefficient into the world leaders. Some of the technology was well conceived but only for an isolated cell or limited functions, and this resulted in the creation of isolated pockets or islands of automation. The improvement in productivity in isolated cells or even across some departments did not transfer to the total organization or systems of the plant. The isolated changes and optimizations often did not produce improvements in the overall operation of the business. Therefore, the results, as seen at GM, did not turn out the way people wanted or intended. The productivity of the GM plants did not improve significantly. The continuing problems that came with the automation led to additional problems and, in many cases, the eventual removal of excess automation and computer controls.

When IBM moved its printer manufacturing operation from Boca Raton, Florida, to the Technology Triangle in North Carolina, it left much of the automation behind. The Florida facility had a heavily automated assembly line that used many robots to assemble the initial ProPrinter line of dot matrix printers. Before the move, the printer was totally redesigned to reduce the number of individual parts, eliminate difficult assembly operations, and permit all assembly easy access from the top. As a result, a large number of robots were replaced with human operators. The human operators could assemble the redesigned printer faster and with increased productivity. The new assembly process did use new technology to aid in the process, but only when the use of technology was shown to better the productivity of a human operator. Technology should never be used as the first step to overcome a poor design.

The Toyota Motor Manufacturing Company is recognized as a leader in manufacturing and one of the most effective producers of automobiles in the world. Toyota keeps its focus on the critical work elements that add value. Waste of all kinds is eliminated. Work is kept as simple as possible. Toyota does use automation and computers, but only when it sees that doing so makes sense and the work of the people in the systems is complemented.

The effective use of technology requires companies to understand what they do and what is needed to satisfy their customers. Automation and the application of computers to a poorly designed and planned system can become a disaster. Spending limited resources on computers and automation that do not help achieve the fundamentals can drive a company out of business. Spending on technology without a careful understanding of the business and the needs of the customers and the marketplace can lead to waste and the creation of expensive processes that do not get the job done. To avoid such outcomes, managers of businesses should do the following:

- Use technology to solve problems and improve operations only when doing so is needed and makes sense.
- Keep a clear focus on the fundamentals of the business and use technology to make the operations more effective.
- Think of the complete system; do not add automation and computers in isolation.

"Taking care of business" was the motto used by a Memphis music icon to get things done efficiently. Every industry should follow this motto.

Manufacturing Systems

OBJECTIVES

After completing this chapter, you should be able to do the following:

- Describe a production system model and the two criteria for classifying manufacturing
- Describe the manufacturing system model and the two categories used to classify manufacturing
- Name and define two manufacturing systems used to classify all manufacturing
- Name the four production strategies and describe how they affect manufacturing lead times
- Describe the product development cycle for new products and for the existing products
- Given the model of a manufacturing organization, describe the function of each unit within the model
- Describe a typical production sequence for a manual manufacturing system

Manufacturing was defined in chapter 1 as activities that eventually lead to the marketing of goods. In the process, raw materials are converted into finished products. The complete input-output model is described in Figure 2–1. The five inputs required are *raw materials, equipment, tooling and fixtures, energy,* and *labor.* The traditional outputs are finished goods and scrap.

A visit to manufacturing sites confirms that no two companies or manufacturing operations are the same. Even plants built by the same corporation to produce the same products have some differences. The reason for the major differences between manufacturing sites is both *product* and *technology* based. In the first case, the *product* being manufactured dictates the manufacturing process required. For example, assembling cars is very different from refining gasoline. As a result, manufacturing automobiles requires a production facility different from the refinery used for producing gasoline.

Figure 2–1 Production Model.

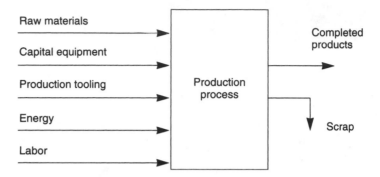

Changes in *technology* cause significant differences in manufacturing plants as well. The differences are evident in production facilities producing the same product. For example, an automotive assembly plant built five years ago would be different from one built today. The new plant would take advantage of any technological advancement in the last five years that improves production efficiency.

If every production facility is different, it naturally follows that the computer-integrated manufacturing (CIM) systems installed in these production facilities will each be unique. Therefore, the variation in CIM implementations is also *product* and *technology* based. For example, production-scheduling software for the automotive assembly plant would not be effective for the refinery. Thus, the ability to recognize the different types of production facilities is critical for the correct recommendation of a CIM solution. Learning the description of the various types of manufacturing facilities currently in use is the first step and the subject covered in this chapter.

Technology *change* also influences the CIM solution. In the same example, production control software installed five years ago would not have the features available in similar software installed today. The continuous improvements in technology are tolerable only if they can be easily incorporated into the current system. Passing product data between manufacturing software packages from two different vendors is a good example. If both software solutions transfer the data to a common database using an industry standard format for the data, future enhancements to the software are acceptable from a system integration standpoint. However, increased software capability may require upgrading of the hardware platform where the software resides. Learning the detailed operation of current hardware and software applications is just a temporary solution. It is more important to understand *why* different CIM technologies are used and *how* they affect the total integrated system. After the manufacturing systems are studied in this chapter, in the remainder of the book we answer the *why* and *how* questions regarding the CIM technology required to integrate these systems.

2–1 MANUFACTURING CLASSIFICATIONS

Manufacturing is classified according to two criteria:

1. How merchandise is produced in the *manufacturing system*
2. How customer demand is satisfied by a *production strategy*

Classification of manufacturing operations is necessary for CIM implementation. Much of the hardware and software used in a solution is tailored to a specific type of manufacturing system.

Manufacturing System Classification

Classification by manufacturing system divides all production operations into the following five groups: *project, job shop, repetitive, line,* and *continuous.* Overlap cannot be avoided between some of the categories, and most manufacturers use two or more of the manufacturing systems in the production of an entire product line. Classification of companies into these groups requires a detailed analysis and evaluation of the production operations. As a result of the classification process, the activity in the three major process segments (design, manufacturing planning and control, and production) in Figure 1–9 is identified and understood. Therefore, success in matching enterprise needs with automation hardware and software is easier to achieve. Distinguishing characteristics of each classification category are described next.

Project.   The most distinguishing characteristic of the project category is that products are *complex,* with many parts, and are most often *one of a kind.* For example, project-type companies build oil refineries, large office buildings, cruise ships, and large aircraft. In each case, the products may be similar but usually are not identical. The plant layout, another discriminating factor for this category, is called *fixed position* (Figure 2–2a). Because of their size and weight, products such as ships and large aircraft remain in one location, with the equipment and parts being moved to them. In addition, the design drawings are complex, the lead time is long, the customer is identified before production starts, and production scheduling usually uses project management techniques.

Job Shop.   Job shops are also distinguished by *low volume* and *production quantities,* called *lot sizes,* that are small. However, compared to the products in the project category, the size and weight of the parts in the job shop group are very small. As a result, the parts are moved or routed between fixed production work cells for manufacturing processing. The classic machine shop with lathes, mills, grinders, and drill presses is the example most often cited for this category. The plant layout for the job shop (Figure 2–2b) is frequently called a *job shop,* or *process, layout.* Other distinguishing features include less than 20 percent repeat production

Figure 2–2 Types of Plant Layouts: (a) Fixed-Position Layout; (b) Process Layout; (c) Product-Flow Layout.

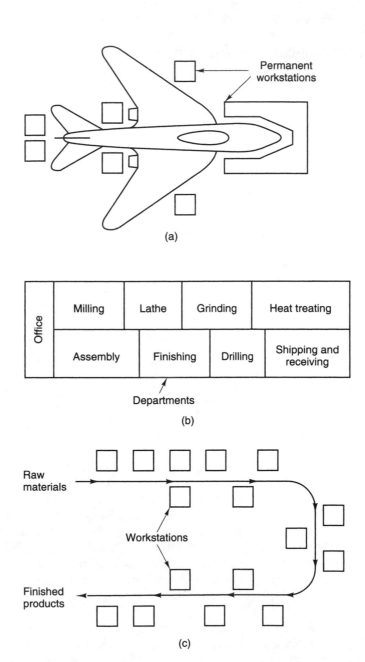

on the same part, noncomplex products, and intensive scheduling and routing on the shop floor. In addition, the raw material is usually purchased as needed for each project.

Repetitive. The repetitive manufacturing system has the following four unique characteristics: (1) orders for repeat business approach 100 percent, (2) blanket contracts with customers for multiple years occur frequently, (3) orders have a moderately high volume with lot sizes varying over a wide range, and (4) orders have fixed routings for the production machines. The plant layout could be either the process layout shown in Figure 2–2b or more like the product-flow layout shown in Figure 2–2c. The production machines are frequently special purpose machines, called *transfer machines*, built to produce a specific product or family of products. Automotive subcontractors are representative of this type of manufacturing. A supplier to Ford, for example, could have an order to supply 3,000 water pumps per week made up from three different model types. The weekly quantity for each model may vary over some range, and the contract might span three years.

Line.   The line manufacturing system has three distinguishing characteristics: (1) the delivery time (often called *lead time)* required by the customer is often shorter than the total time it takes to build the product, (2) the product has many different options or models, and (3) an inventory of subassemblies is normally present. Car and truck manufacturing is an example of this category. If a customer or dealer order for a car triggered the start of production for all of the car's parts, it would take months to build the car. For example, if the bolts were not made, the seat material not woven, the raw material for the tires not produced, the engine not built, and the CD player not assembled when the order arrives, the car could not be built as fast as the customer or dealer requires. Manufacturing facilities in this group usually use the product-flow plant layout (Figure 2–2c).

Continuous.   Continuous manufacturing systems have the following five characteristics: (1) manufacturing lead time is greater than the lead time expected by the customer, (2) product demand is predictable, (3) an inventory of finished products is maintained, (4) volume is high, and (5) products have few options. This type of production system always uses the product-flow plant layout (Figure 2–2c), with the production line limited to one or just a few different product models. Examples of this industry type include the production of nylon carpet yarn, the production of breakfast cereal, and the production of petroleum and chemical products. For products like these, the production process is often continuous, with raw material, such as chemical compounds, entering on one end of the system and a finished product, such as nylon filament thread, flowing out the other end.

Figure 2–3 offers comparative data on the characteristics of the manufacturing systems just described. Two of the items, *labor content* and *design component,* may be misunderstood without additional clarification. *Labor content* refers to the size of the human labor component or the difficulty that automation faces in replacing the human element. The *design component* element indicates the relative number of hours devoted to product design. Study the chart. You should begin to see how the type of manufacturing operation dictates where the productivity gains will occur as automation is introduced with CIM. For example, computer-aided design (CAD) technology uses computers to automate the design and documentation of a product. Therefore, it would apply when the design component

	Project	Job shop	Repetitive	Line	Continuous
Process speed	Varies	Slow	Moderate	Fast	Very fast
Labor content	High	High	Medium	Low	Very low
Labor skill level	High	High	Moderate	Low	Varies
Order quantity	Very small	Low	Varies	High	Very high
Unit quantity cost	Very large	Large	Moderate	Low	Very low
Routing variations	Very high	High	None	Low	Very low
Product options	Low	Low	None	Very high	Very low
Design component	Very large	Large	Very small	Moderate	Small

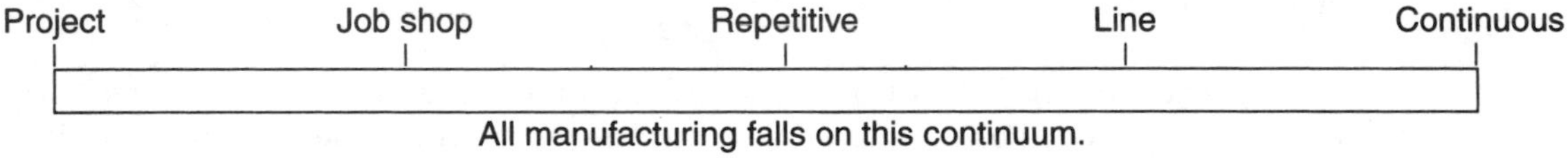

Project	Job shop	Repetitive	Line	Continuous

All manufacturing falls on this continuum.

Figure 2–3 Manufacturing Characteristics.

is large. The last element on the chart, design component, indicates that in a project type, the manufacturer has an excellent opportunity for an impact from CAD, but CAD's effect on a continuous manufacturing operation would probably be minimal because the design component (the need for CAD) is small.

Exercise 2–1
Use your general knowledge about the following products to classify (P, project; J, job shop; R, repetitive; L, line; C, continuous) each by the manufacturing system(s) used.

a. Nylon carpet yarn
b. Gillette Plus razors
c. Space shuttle
d. Personal computers
e. Pontiac Sunbirds
f. Cheerios
g. Electric fan motors

h. Replacement pump part
i. Automotive alternators
j. Oil refinery
k. Hard-disk drives
l. Stainless steel dinnerware
m. Televisions
n. Special metal bracket

Solution
The first letter represents the most common type of system used: (a) C; (b) C; (c) P; (d) L, R; (e) L; (f) C; (g) R; (h) J; (i) R; (j) P; (k) R, L; (l) R, L; (m) L; (n) J.

The number of different groups in the manufacturing classification system is not critical. Some literature lists only three categories: job shop, batch production, and mass production. In that case, the five categories just described would be mapped into these three.

The number of categories is not critical; understanding what makes automotive production different from nylon production is, however. The continuum at the bottom of Figure 2–3 represents the different types of manufacturing systems just described. "Project" and "Continuous" are opposite in function and represent the two possible extremes on the left and right, respectively. The other three fall somewhere between. Every manufacturing system fits on the continuum, with some companies using several production technologies.

It should be clear by now that a single manual process or software package cannot be used to handle the production-scheduling requirements of all five manufacturing system categories. Selecting a process or software application requires knowledge about the production operation. Implementing a CIM solution requires a complete knowledge of the manufacturing system to be improved or automated, and classification is the first step.

Production Strategy Classification

The *production strategy* used by manufacturers is based on several factors; the two most critical are *customer lead time* and *manufacturing lead time*. Knowing the definition of each is important.

> *Customer lead time identifies the maximum length of time that a typical customer is willing to wait for the delivery of a product after an order is placed.*

For example, the consumer expects preferred brands of commodities, such as toothpaste, to be available on the store shelf whenever a purchase is desired. Rarely will the consumer wait for a delivery if the brand selected is not available. When the brand is not available, a competitor's product is chosen or the product desired is purchased at a different store. In this example, immediate delivery or satisfying a customer lead time of zero is the order-qualifying criterion.

> *Manufacturing lead time identifies the maximum length of time between the receipt of an order and the delivery of a finished product.*

Manufacturing lead time and customer lead time must be matched. When a new car with specific options is ordered from a dealer, for example, the customer is willing to wait only a few weeks for delivery of the vehicle. As a result, automotive manufacturers must adopt a production strategy that permits the manufacturing lead time to match the customer's needs.

The production strategies used to match the customer and manufacturer lead times are grouped into four categories: *engineer to order (ETO)*, *make to order (MTO)*, *assemble to order (ATO)*, and *make to stock (MTS)*. A description of each follows.

Engineer to Order. A manufacturer producing in the ETO category has either a product that is in the first stage of the lifecycle curve or a complex product with a unique design produced in single-digit quantities. Examples of ETO include construction industry products (bridges, chemical plants, automotive production lines) and large products with special options that are stationary during production (commercial passenger aircraft, ships, high-voltage switchgear, steam turbines). Because of the nature of the product, the customer is willing to accept a long manufacturing lead time because the engineering design is part of the process.

Make to Order.   The MTO technique assumes that all the engineering and design are complete and the production process is proven. Manufacturers use this strategy when the demand is unpredictable and when the customer lead time permits the production process to start on receipt of an order. New residential homes are examples of this production strategy.

Assemble to Order.   The primary reason that manufacturers adopt the ATO strategy is that customer lead time is less than manufacturing lead time. An example from the automotive industry was used in the preceding section to describe this situation for line manufacturing systems. This strategy is used when the option mix for the products can be forecast statistically: for example, the percentage of four-door versus two-door automobiles assembled per week. In addition, the subassemblies and parts for the final product are carried in a finished-components inventory, so the final assembly schedule is determined by the customer order. John Deere and General Motors are examples of companies using this production strategy. Also, some computer companies make a personal computer to customer specifications, so they follow ATO specifications.

Make to Stock.   The last strategy, MTS, is used for two reasons: (1) the customer lead time is less than the manufacturing lead time, or (2) the product has a set configuration and few options so that the demand can be forecast accurately. If positive inventory levels (the store shelf is never empty) for a product is an order-qualifying criterion, this strategy is used. When this order-qualifying criterion is severe, the products are often stocked in distribution warehouses located in major population centers. This option usually applies to mature products and commodity-type products such as toothpaste and food products.

Figure 2–4 compares the production activities, such as design, to the four production strategies. The manufacturing lead times are set by the three major activities in product development: design or engineering, manufacturing, and assembly. ATO has two lead times; the use of subassemblies produces the shortest ATO time. MTS has a zero manufacturing lead time because the customer is not willing to wait and delivery must be immediate. ATO assumes manufacturing is complete, and MTO assumes that the design is finished. Compare the figure with the definitions of the four production strategies.

The relationship between the product lifecycle curve described in the last chapter and the production strategies just described is illustrated in Figure 2–5. When a product enters the market, the demand is usually low, so an MTO strategy does not stress the resources of the company. As demand builds, an ATO strategy keeps the product delivery time even with competitive products without carrying

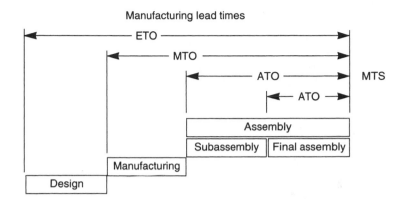

Figure 2–4 Production Strategies and Manufacturing Lead Times.

inventories of finished goods. At peak demand, a company often chooses an MTS strategy to meet customer orders and maintain market share. The cost of the finished-goods inventory is reduced by the efficiencies of scale because of higher volume and is more advantageous than increasing production capacity.

The relationship between the manufacturing systems described earlier and the production strategies just covered is illustrated in Figure 2–6. Study the chart before continuing. Customer demand and lead time still determine the production strategy shown in Figure 2–6, but in two cases (MTO and MTS) the manufacturer has a choice of the type of manufacturing system to be used.

The key point from the discussion of classification of manufacturing systems and production strategies is that industries are all different. The classification process simply helps you understand some of the similarities and differences that are present. Armed with this knowledge, a company can much more easily do the following:

■ Meet the external challenges
■ Identify the order-winning criteria

Figure 2–5 Production Strategies.

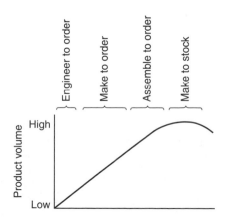

Manufacturing system	Engineer to order	Make to order	Assemble to order	Make to stock
Project	X			
Job shop		X		
Repetitive		X		X
Line			X	X
Continuous		X		X

Figure 2–6   Comparison of Manufacturing Systems and Strategies.

- Implement the CIM principles
- Adjust the management philosophy
- Integrate the hardware and software systems
- Compete in the world marketplace

Exercise 2–2
Use your general knowledge about the following products to classify (E, engineer to order; MO, make to order; A, assemble to order; MS, make to stock) each by the production strategy(ies) used.

a. Nylon carpet yarn
b. Gillette Plus razors
c. Space shuttle
d. Personal computers
e. Pontiac Sunbirds
f. Cheerios
g. Electric fan motors

h. Replacement pump part
i. Automotive alternators
j. Oil refinery
k. Hard-disk drives
l. Stainless steel dinnerware
m. Televisions
n. Special metal bracket

Solution
The first letter represents the most common production strategy used: (a) MS, MO; (b) MS; (c) E; (d) MS, A; (e) A, MS; (f) MS; (g) A; (h) MO; (i) MS, A; (j) E; (k) MS; (l) MS; (m) MS; (n) E.

2–2 PRODUCT DEVELOPMENT CYCLE

Despite the differences that exist across manufacturing, the product development cycle is generally the same. The cycle, illustrated in Figure 2–7, shows the linear process used to bring a product to market. Remember, however, that the customer continues to set the order-winning and order-qualifying criteria that influence every step in the process. Also, the type and level of activity in each block is directly

Figure 2–7 Product
Development Cycle.

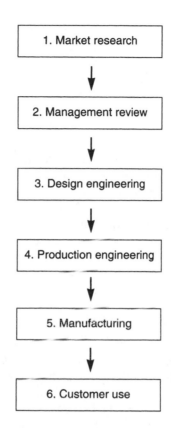

proportional to the type of manufacturing system and production strategy present. For example, the level of activity in the design engineering block would be high in a project system using an ETO strategy.

Do the names in blocks 3, 4, and 5 look familiar? They are the three major process segments found in Figure 1–9. Stop and compare the two figures. As a result, the product development cycle is consistent with the CIM model described earlier. Two types of product development are supported by the cycle: new-product development and existing-product changes.

New-Product Development

For a new product, the market research in block 1 of Figure 2–7 is provided by the *customer support* area in the enterprise wheel shown in Figure 1–8. Customer support determines the form, fit, and function that the product must satisfy, along with the order-winning and -qualifying criteria necessary to succeed in this new market sector. The form, fit, and function concept is described in detail in the next chapter, but for now it is defined as the technique used to specify a product. The *management review* process (Figure 2–7) looks at all the market data, product specifications, and current order-winning and -qualifying criteria and makes a *go* or *no go* decision on the new-product initiative. With the project approved, *design engineering* starts adding detail to the new product with an eye focused on the order-winning and -qualifying criteria present in the current market. In the next step, *production engineering* initiates the manufacturing planning and control functions. For example, processes, machines, and routings are selected that will ensure that a product is consistent with the order-winning and -qualifying criteria. The last enterprise activity in the cycle is *manufacturing* the product.

The process appears to be linear, with manufacturing at the end getting the demands from decisions made far upstream. The order-winning and -qualifying criteria should force congruence between the development of new products and a production system with the order-winning and -qualifying attributes identified by marketing. However, use of order-winning and -qualifying criteria is no guarantee that the design, for example, is optimized to the machines that must make the product in manufacturing. In chapter 3, a process called *concurrent engineering* is superimposed on the product development cycle to overcome the problems created by the apparent sequence of Figure 2–7.

Existing-Product Changes

The process of handling changes to existing products effectively is almost identical to the procedure just outlined for a new product. The only exception is the point where the activity starts in the development cycle (Figure 2–7). With a product change, the process can start at any block and then must ripple through the system in an orderly fashion.

The product development cycle identifies the need for an orderly flow of product data that the CIM system must support for the departments and the design and manufacturing process. New-product development emphasizes the need for a top-to-bottom data interface to support information flow. However, changes to existing products require an interface that works from any starting point in the cycle. Since design and product changes are inevitable, the CIM system must accommodate product changes, initiated at any point in the development cycle, in an orderly and controlled fashion.

2–3 ENTERPRISE ORGANIZATION

The enterprise model presented in Figure 2–8 shows the functional blocks found in most manufacturing organizations. The lines connecting the areas indicate formal communications that occur regularly between enterprise functions. There are as many different representations of manufacturing organizations as there are books discussing them. This model has sufficient detail to demonstrate the information and data flow normally present and provides the framework needed to discuss the functions of the various areas involved in getting a product to the customer. Study the model until you are familiar with the names in each box and

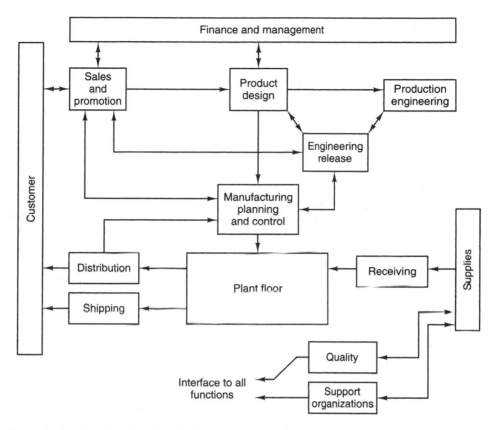

Figure 2–8 Manufacturing Organization.

compare the block names with the element list in the enterprise wheel shown in Figure 1–8.

A CIM implementation affects every part of an enterprise; as a result, every block in the organizational model is affected. Therefore, a successful CIM implementation across the enterprise requires an understanding of the functions performed by each block in the model. The basic operation of the enterprise is described in the following sections.

Sales and Promotion

The fundamental mission of *sales and promotion (SP)* is to create customers. To achieve this goal, many companies have nine internal functions: *sales, customer service, advertising, product research and development, pricing, packaging, public relations, product distribution,* and *forecasting.* An analysis of the enterprise model shown in Figure 2–8 indicates that sales and promotion interfaces with several other areas in the business. The *customer services* interface supports three major customer functions: *order entry, order changes,* and *order shipping and billing.* The order-change interface usually involves changes in product specifications, change in product quantity (ordered or available for shipment), and shipment dates and requirements.

Sales and marketing provide strategic- and production-planning information to the finance and management group, product specification and customer feedback information to product design, and information for master production scheduling to the manufacturing planning and control group. The interface with engineering release ensures that any proposed changes to the product are reviewed and approved by SP before implementation.

Finance and Management

Finance and management has the responsibility of setting corporate goals, performing financial functions, and performing medium- and long-range planning. In the financial area, four major internal functions are performed: *cash management, financial planning, financial analysis,* and *strategic planning.* The financial planning is long-term planning for the operations of the business. The financial analysis unit generates a direction for the future that is based on past and current financial conditions. This unit usually draws data from three sources: *balance sheet, income statement,* and *statement of change* in the financial position. The strategic-planning unit works with two time frames: *medium term* and *long range.* The medium term includes yearly budget projections for production objectives, corporate structure, and manufacturing infrastructure. The long-range planning includes long-term objectives for the corporation and the strategies necessary to achieve these objectives.

The *accounting* unit is responsible for three areas: *general accounting, cost accounting,* and *related functions.* The areas supported by the accounting unit include accounts payable, accounts receivable, general ledger, and the three cost accounting areas: manufacturing, product, and overhead.

If this is a book about CIM, why study finance and management or sales and marketing? The answer is that every part of the enterprise uses the shared database and must be considered in the integration process. As a result, the CIM integrator must understand how these units function in the organization. This reference to the function or operation of these two units is the last because the operation of these functions is well documented in business books. However, when production control software is discussed in a subsequent chapter, the links to these units are explored.

Product and Process Definition

The units that share a formal interface with the *product* and *process definition* unit are illustrated in Figure 2–9. Keep in mind that this unit is one of the three process segments in the enterprise wheel (Figure 1–8) and is the design component in Figure 1–9. The unit includes *product design, production engineering,* and *engineering release.* The product design portion provides three primary functions: (1) product design and conceptualization, (2) material selection, and (3) design documentation. The production engineering area establishes three sets of standards: *work, process,*

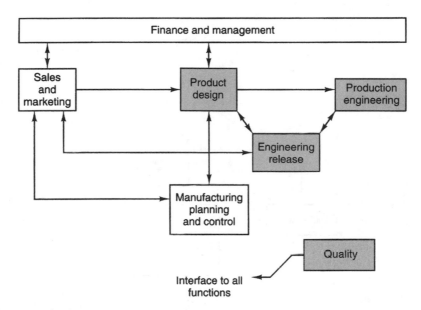

Figure 2–9   Product and Process Definition Engineering.

and *quality.* The engineering-release area manages engineering change on every production part in the enterprise. Engineering release has the responsibility of securing approvals from departments across the enterprise for changes made in the product or production process. The operation and automation of this area of the enterprise model are covered in considerable detail in later chapters.

Manufacturing Planning and Control

The *manufacturing planning and control (MPC)* unit (Figure 2–10) has a formal data and information interface with several other units and departments in the enterprise. This unit is the second process segment identified on the enterprise wheel (Figure 1–8) and in Figure 1–9. The MPC unit has responsibility for the following:

- Setting the direction for the enterprise by translating the management plan into manufacturing terms.
- Providing detailed planning for material flow and production capacity to support the overall plan.
- Executing these plans through detailed shop scheduling and purchasing action.

This unit is the most complex in the enterprise because of the quantity of data and the large number of interfaces into and out of the area. As a result, the data and

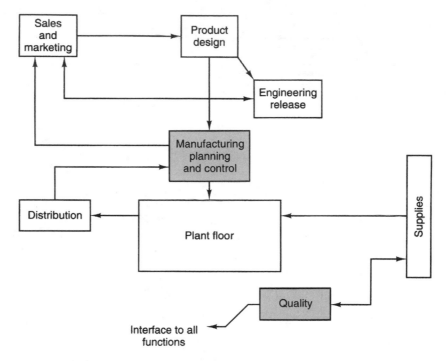

Figure 2–10   Manufacturing Planning and Control.

information flow is expressed through an MPC model such as that shown in Figure 2–11. Study the illustration and learn the names of the various functions listed. Each function is explored in detail in later chapters.

Shop Floor

Interfaces with the *shop-floor* unit are illustrated in Figure 2–12. This unit is the third process segment and is identified in Figure 1–9 as production. Compare the two figures. Although the use of different terms to identify this area may seem confusing, several terms are used in the literature to refer to the production unit. Shop-floor activity often includes *job planning and reporting, material movement, manufacturing process, plant-floor control,* and *quality control.* The operation and automation of this area of the enterprise model are covered in considerable detail in later chapters.

Support Organizations

The *support* organizations, indicated by the box in Figure 2–8, vary significantly from firm to firm. The functions most often included are *security, personnel, maintenance, human resource development,* and *computer services.* Basically, the support organization is responsible for all of the functions not provided by the other model elements.

Figure 2–11 MPC Model for
Information Flow.

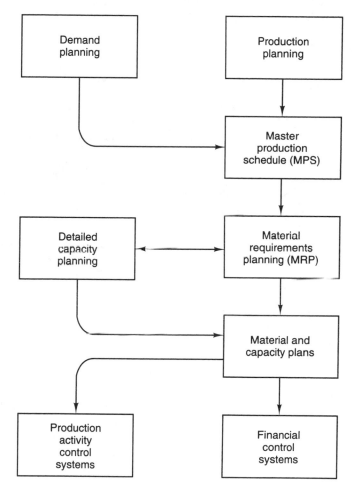

In the last six sections, we described the general organization of an enterprise engaged in manufacturing. In addition, we provided an overview of the functions performed in each of the six major units, and we compared this model to the enterprise wheel shown in Figure 1–8 and the process model shown in Figure 1–9. In the remainder of the book, we investigate the CIM management and automation issues present in these units. Before starting this more detailed study, however, we need to look at how the production flow occurs in the enterprise model shown in Figure 2–8.

2–4 MANUAL PRODUCTION OPERATIONS

The previous discussion, centered on the enterprise organization, identified the links between each unit and the functions provided. However, we have not yet provided a sense of how production flows in a weakly integrated manufacturing

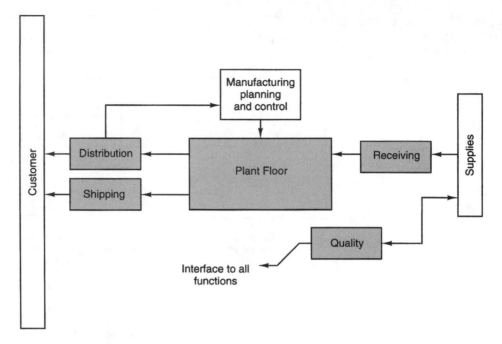

Figure 2–12 Shop Floor.

organization. Figure 2–13 shows one possibility for the flow required to bring a product to a customer. Study the figure and note the symbol for a paper operation and the symbols for computer-driven processes and a database. Refer frequently to the illustration as the product flow through the system is described next.

Activity enters the system as either a *design* or a *request for engineering action* (*REA*). An REA could come from marketing and result from a product change requested by a customer, or the REA could be a change requested by an internal department to correct a problem. Note that currently both are usually paper-based operations.

The *product design* group, using CAD, makes the drawings to describe the design or REA. Although this unit is usually automated with CAD, the drawing file is saved on the computer in the product design area. The paper drawing of the design or REA is thrown over the office wall, figuratively speaking, for the next activity.

In the *product definition* group, all the different parts in the design or REA are listed in a bill of materials so that it is easy to understand what it takes to manufacture the product. Again, even though this stage may be a computer-based application in the department, the work rarely goes to the *engineering-release* management group for approval as an electronic file or as computer-based data.

The *manufacturing definition* group starts to work after the design or REA is approved through the engineering-release process and is delivered to production.

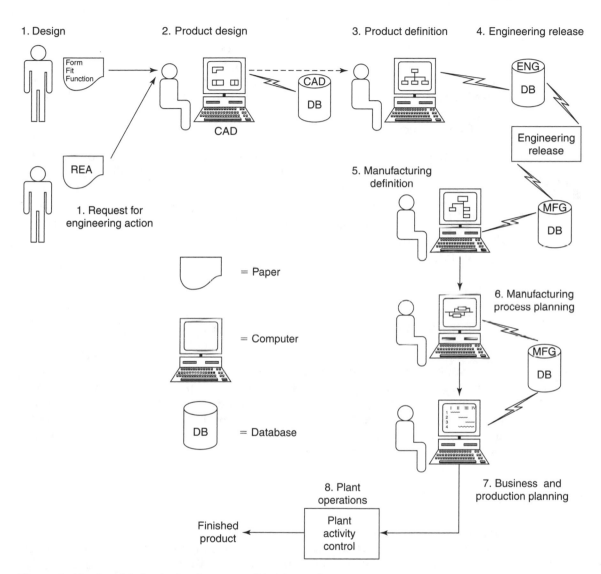

Figure 2–13 Possible Production Sequence—Weak Integration.

Here, a manufacturing bill of materials is produced that separates the parts in the design into different categories, such as those that are purchased from external vendors and those that are manufactured in-house. Internally, this process is frequently manual, so the solution is passed to the next area in a paper format.

Manufacturing process planning determines the types of machines required to process the parts and the production sequence, called the *routing*, to be used. Establishing the routing is most often a manual process; as a result, information is passed to the production area in a paper format.

The next step, *business and production planning,* is frequently automated with a computer and scheduling software. The production process and machine sequence information from the handwritten manufacturing process plan is keyed into the software, and production schedules are produced. Shop orders are released across the manufacturing area, and production activity control is established to control the shop floor.

The production flow just described has the following characteristics:

- There is unidirectional flow from a design or an REA to the shop floor. The system does not provide feedback to the department on the status of the product after it leaves a unit. Most critically, the status of the part in production is available only by going to the shop and checking.
- A mixture of manual operations and islands of automation prevents communication with other units.
- Productivity is lost by reentry of the same product information in two or more units.
- The system is inflexible.
- There is a sense that the operation is a collection of independent units, each operating on its own with little regard for how that operation affects the remainder of the enterprise.

Unfortunately, this description fits many manufacturing organizations to some degree. Intervention is required to bring these organizations in line with the production strategies that win orders and overcome the external challenges described earlier. The intervention to bring about the desired change comes from the implementation of CIM principles. A description of that process forms the topics for the remainder of this book.

2–5 SUMMARY

Each company is unique as a result of the product manufactured or the technology level used in the production process. The differences among the companies require that the CIM implementation for each be different. To specify the type of CIM solution that is appropriate, you must know where a company fits in the manufacturing systems and production strategy classification scheme. Classification under manufacturing systems divides all production operations into the following five groups: *project, job shop, repetitive, line,* and *continuous.* The production strategies in use are grouped into four categories: *engineer to order, make to order, assemble to order,* and *make to stock.* The production strategy changes at different points in the product lifecycle. Knowing where a company fits into the relationship between the manufacturing system and production strategies helps in identifying successful CIM solutions.

The product development cycle (Figure 2–7) describes the activities required for the development of a new product and for making changes to existing products. Analysis of the development cycle reveals the interfaces that the CIM system must support in the enterprise. It becomes apparent that the CIM system must control product changes in an orderly and controlled fashion regardless of the point in the development cycle where they are initiated.

The enterprise organization needs a model to support an understanding of the functions performed by every department and group. The model in Figure 2–8 is used to describe these functions. The production flow of a product through the enterprise is shown in Figure 2–13. A design or a request for engineering action (REA) triggers the activity, which then moves through the groups in sequential fashion.

BIBLIOGRAPHY

CLARK, P. A. *Technology Application Guide: MRP II Manufacturing Resource Planning.* Ann Arbor, MI: Industrial Technology Institute, 1989.

GROOVER, M. P. *Automation, Production Systems, and Computer-Integrated Manufacturing.* 2nd ed. Upper Saddle River, NJ: Prentice Hall, 2001.

SHRENSKER, W. L. *CIM: Computer-Integrated Manufacturing: A Working Definition.* Dearborn, MI: CASA of SME, 1990.

SOBCZAK, T. V. *A Glossary of Terms for Computer-Integrated Manufacturing.* Dearborn, MI: CASA of SME, 1984.

STICKLER, M. J. "Going for the Globe Part II." *P and IM Review,* November 1989, 32–34.

STICKLER, M. J. "Going for the Globe Part III." *P and IM Review,* December 1989, 41–43.

VOLLMANN, T. E., W. L. BERRY, and D. C. WHYBARK. *Manufacturing Planning and Control Systems.* 4th ed. Homewood, IL: Richard D. Irwin, 1997.

QUESTIONS

1. Why is every CIM implementation different?
2. Describe the two classification techniques used to differentiate between manufacturers.
3. Define the following manufacturing systems: project, job shop, repetitive, line, and continuous.
4. Describe why the comparative data shown in Figure 2–3 support the fact that every CIM implementation is different.
5. Use your general knowledge about the following products to classify (P, project; J, job shop; R, repetitive; L, line; C, continuous) each by the manufacturing system(s) used.

   a. Antifreeze   d. Electric typewriters
   b. Bic lighters   e. Jeep Cherokee
   c. Cruise ships   f. Nachos

g. Automotive water pumps

h. Special coupling

i. Automatic transmission fluid

j. Concrete

k. Computer power supplies

l. Plastic picnic forks

m. CD players

n. Special motor mount

6. Define the following production strategy terms: *engineer to order, make to order, assemble to order,* and *make to stock.*

7. Describe how the production strategies change during the lifecycle of a product.

8. Use your general knowledge about the products listed in question 5 to classify (E, engineer to order; MO, make to order; A, assemble to order; MS, make to stock) each by the production strategy(ies) used.

9. What is the product development cycle, and how is it supported by the enterprise wheel (Figure 1–8)?

10. What is the primary distinction between new-product development and product changes in the development cycle?

11. Study the product development cycle in Figure 2–7 and list some of the data and information interfaces that CIM must provide.

12. What is the function of the enterprise organization model shown in Figure 2–8?

13. What is the function of each block in the organizational model?

14. Briefly describe the production flow through the manual production system.

15. What is the difference between a design and an REA?

16. What problems in the manual production system described in question 14 must CIM solve?

PROJECTS

1. Using the list of companies developed in project 1 in chapter 1, determine the manufacturing system and production strategies used for each product produced.

2. Compare the enterprise model (Figure 2–8) with the structure used by one company from the small, medium-sized, and large groups and describe how their operations differ from that of the model.

3. Compare the production operation sequence described in the text with that used by the companies selected in project 2. Describe the differences that are present.

4. View the SME video *CIM: Focus on Small and Medium-Size Companies* and prepare a report that describes the manufacturing system and production strategy used by the Ex-Cell-O Corporation and the Modern Prototype Company.

The following projects are a continuation of projects 6 through 9 at the end of chapter 1.

5. A company that manufactures kitchen products wants to add a set of plastic storage bowls or containers to its product line. The company wants to distribute

the product through Wal-Mart, Target, or a chain of stores in your region. Do the following:

a. Determine the order-winning and -qualifying criteria for the product.

b. Select the store chain(s) that would sell the products.

c. Determine the total number of units sold per week by the store chain(s).

d. Calculate the number of products that must be produced per hour to meet store shipment requirements. Assume two shifts working 5 days a week.

6. A company that manufactures kitchen products wants to add a high-end manual can opener to its product line. The company wants to distribute the product through Wal-Mart, Target, or a chain of stores in your region. Do the following:

a. Determine the order-winning and -qualifying criteria for the product.

b. Select the store chain(s) that would sell the products.

c. Determine the total number of units sold per week by the store chain(s).

d. Calculate the number of products that must be produced per hour to meet store shipment requirements. Assume two shifts working 5 days a week.

7. A company that manufactures kitchen products wants to add a manual handheld eggbeater to its product line. The company wants to distribute the product through Wal-Mart, Target, or a chain of stores in your region. Do the following:

a. Determine the order-winning and -qualifying criteria for the product.

b. Select the store chain(s) that would sell the products.

c. Determine the total number of units sold per week by the store chain(s).

d. Calculate the number of products that must be produced per hour to meet store shipment requirements. Assume two shifts working 5 days a week.

8. A company that manufactures kitchen products wants to add a manual handheld garlic press to its product line. The company wants to distribute the product through Wal-Mart, Target, or a chain of stores in your region. Do the following:

a. Determine the order-winning and -qualifying criteria for the product.

b. Select the store chain(s) that would sell the products.

c. Determine the total number of units sold per week by the store chain(s).

d. Calculate the number of products that must be produced per hour to meet store shipment requirements. Assume two shifts working 5 days a week.

CASE STUDY: EVOLUTION AND PROGRESS—ONE WORLD-CLASS COMPANY'S MEASUREMENT SYSTEM

A manufacturing company with over $450 million in annual sales was losing market share because of the following problems:

- The delay in shipping customer orders was getting longer.
- The production time for new orders was increasing.
- The development time for new products was increasing.
- Inventory levels for most products were high.
- Product rework was high as a result of poor production quality.

The company had a manufacturing resource planning (MRP II) software system, but the inaccuracy of the inventory and product parts list along with poor vendor delivery performance made the system dysfunctional. While the company used a cycle-count system to keep track of inventory, it was not enforced across all areas in the company. In addition, the company carried higher inventory levels as a buffer against a vendor on-time delivery rate of only 30 percent. To regain the competitive edge, management developed a CIM plan designed to achieve the following aggressive business goals:

- Class A MRP II greater than 95 percent with a continuous increase in standards and levels of performance
- Average order turnaround of less than 6 days from the receipt of order to shipment, with a process in place to ensure that this level of performance could be sustained
- Delivery performance rate greater than 95 percent to promised delivery and greater than 80 percent to requested delivery with a shift in philosophy from "delivery when we say we can" to "delivery when the customer requests it"
- A quality performance of fewer than 7,500 defective parts per million, with outgoing product quality measured in terms of the customers' expectations and standards
- A 7 percent reduction in overall real cost for the current fiscal year with continuous improvement in material, labor, and overhead costs
- A run-rate inventory turnover of five or greater per year
- A time to market of less than 15 days measured from the receipt of the job by operations to the release date

Then the company developed a performance measurement system to track and report progress toward these ambitious goals. Baseline readings were taken in the current year, and improvement was rapid, as indicated by the consolidated report card, which is shown in Figure 2–14.

	Current year	Three months later	Six months later	Nine months later	One year later
1. Class A MRP II	D– (51%)	B (84%)	A (93%)	A (95%)	A+ (97%)
2. Order turnaround time	18 days	10 days	10 days	10 days	7.6 days
3. Delivery performance (weekly)	26%	90%	92%	95%	96%
4. Order backlog (orders)	11,000	446	486	191	269
5. Production operations management performance (slips)	325	0	0	1	1
6. Direct ship performance	72%	94%	89%	95%	95%
7. New products released on schedule (weekly)	40%	95%	100%	100%	100%
8. Shortage area (orders)	4,800	271	570	220	174
9. Vendor delivery performance	31%	81%	98%	94%	96%
10. Inventory record accuracy	87%	91%	98%	96%	98%
11. Customer service posture	Fire fighter	Maytag repairperson	Minimum class A	Class A	Class A
12. Morale	Rotten	Positive	Good	Better	Even better

Figure 2–14   Documentation of Enterprise Improvements.

(*Source: Stickler, "Going for the Globe Part III," P and IM Review, December 1989, p. 43. Reprinted by permission of APICS.*)

APPENDIX 2–1: CIM AS A COMPETITIVE WEAPON

How important is CIM as a competitive weapon for U.S. industries? This question was answered by 139 respondents—including company presidents, CEOs, and vice presidents—to an *Industry Week* survey. Their responses to this survey are illustrated by the pie charts shown in Figure 2–15. Two key indications can be drawn from these results: (1) a large majority, 85 percent, consider CIM *essential* or *very important*; and (2) the small number of *no answers* indicates that management understands the question in sufficient depth to have an opinion. The second indication is significant because it says that the decision makers are committed and willing to listen to recommendations from the organization.

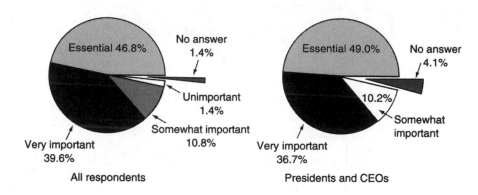

Figure 2–15 Importance of CIM.

The Design Elements and Production Engineering

PART GOALS

The goals for this part are to introduce the reader to the techniques and technologies used to design competitive products for distribution anywhere in the world. In addition, the methods used to manage the design data are addressed, along with a typical organizational structure that would support efficient product design and development.

After you complete part 2, the following will be clear to you:

- A design model with concurrent engineering must be used for successful development of competitive products for a global market.
- The design process should use state-of-the-art design and analysis tools and technologies for development of a competitive product with a minimum of development time.
- Graphics-based systems are the primary design tools used to translate the designer's product concepts into the production database.
- Production engineers should use state-of-the-art process planning and manufacturing engineering tools to create an efficient and effective system for producing the product.

CAREER INSIGHTS

The technologies described in part 2 are most often taught in engineering and engineering technology curricula. The technologies are learned in courses such as computer-aided drafting or design (CAD), but the application of the technology is often learned in courses focused on design, such as mechanical system design, machine design, and a capstone design course. To work in this segment of the manufacturing industry, you need to have a good background in these graphics-related technologies. Industry has many career slots in this area, with the educational requirements ranging from a high school diploma to graduate study in engineering. For employees with a high school diploma or an Associate of Science (AS) degree, the work responsibilities would most often be characterized as computer-aided *drafting*. The

CAREER INSIGHTS (contd.)

employees are primarily responsible for the creation of the production drawings used to describe the product to the production people responsible for manufacturing the parts. This characterization may be a little overly simplified, but the work is mostly drafting and little design. Graduates with AS degrees and in some cases just high school diplomas with training in the computer-aided drafting software are frequently recruited for these positions. However, it is possible for a person with only an AS degree to be promoted to a position that has design responsibility after he or she acquires some good industrial experience.

The more traditional path to a design career is through the Bachelor of Science (BS) degree in engineering or engineering technology. The disciplines studied—for example, mechanical, electrical, or aerospace—determine the type of design work the individual performs for the employer. The work required in the design area is more varied since it includes both creation of the product and analysis of the finished design. The criterion for the design has changed significantly in the last ten years. For example, the term *cradle to grave* is used to describe environmentally good product designs that will be earth friendly after their useful life is over. Also, a strong focus is placed on designs that are efficient to manufacture (this process is called *design for manufacturing*, or *DFM*) and on designs that are efficient to assemble (this process is called *design for assembly*, or *DFA*). There are also software products and computer tools used to analyze the product designs for a broad range of operational parameters, such as *stress* and *temperature* limits. Product designers today must be equally versed in the graphical software used to render the initial product design and in the wide variety of analysis software used to predict product performance.

Graduates trained in the design process and in the application software used to create and analyze product designs have an opportunity for an interesting and rewarding career in the product development segment of the CIM wheel. If you find the material presented in part 2 interesting, then this is a career area that you should explore.

Product Design and Production Engineering

OBJECTIVES

After completing this chapter, you should be able to do the following:

- Describe why product design is a good starting point for the study of computer-integrated manufacturing (CIM)
- Draw the product design and production engineering organizational model and describe the function of each block
- List the five steps in the design model and describe the function of each step
- Explain with examples the differences between concept-design and repetitive-design problems
- Explain the function of the synthesis filter in the design model
- Define *concurrent engineering* and describe how it fits into the design model
- Draw the improved product development process flow and describe the changes from the traditional approach
- List and describe the seven elements or responsibilities of production engineering

To produce a world-class product, product process personnel must meet four requirements:

1. They must follow a design model that takes the product through all the critical steps in the design process.
2. They must use computer software and other analytic tools to create, analyze, and test design options.
3. They must support a *concurrent engineering* process that brings everyone involved in the design and production process together during the design.
4. They must apply product and process design principles that lead to defect free manufacturing.

In the next three chapters, we provide an overview of the design process and the automation used to support all the stages of product development.

3-1 PRODUCT DESIGN AND PRODUCTION ENGINEERING

The product design and production engineering area is an appropriate starting point for a detailed study of CIM for the enterprise for the following two reasons:

1. Design and production engineering were the first areas in the enterprise to embrace technology as a means for improving efficiency and productivity by eliminating many manual tasks.
2. The initial data in the product database are a result of the design process and the development of the production system.

When designs are created graphically with computer software, the resulting file is the start of the product's electronic database. Moving away from manual drafting that is costly to change and labor intensive to create enhances productivity. Although the productivity gain in the design area through automation is good news, this success is tempered by the fact that the technology is not always implemented effectively. In many cases, the design automation solution falls far short of the standards set for CIM because the implementation has only a departmental focus.

Functional automation, the narrow view of automation isolated to a single function or departmental area, appears to provide significant productivity gains; however, under careful analysis the benefits for the total enterprise are often negative. For example, a case study of a manufacturer producing riding lawn mowers uncovered three locations where functional automation was used to produce product and part drawings: *product design* (part and product design drawings), *marketing* (service manual drawings and customer information), and *production* (drawings to aid in the programming of metal-cutting machines). The three areas, which *worked in isolation* with no enterprise-wide plan, *chose different computer* hardware and software technology solutions with *no compatibility* among the computer files created in each department. As a result, the benefit for the enterprise was *negative* for several reasons:

- Time was wasted by entering redundant data.
- Three separate computer images of the same product had to be maintained.
- The number of drawing errors for the product increased by a factor of three.
- Product quality suffered as a result of the drawing errors.
- Production costs increased because of more part-list errors and obsolete parts.

The second reason for starting the CIM study with product design and production engineering is because the initial creation of product data starts there. Generation of design drawings for a new product is one of the first activities in the design area. However, the format for creation and storage of the design must be consistent with a basic CIM axiom that demands the establishment of a *shared enterprise database with a single image of all product information.* The organization of this initial product data is lumped under a general category called *product data management (PDM)*. The software used to support this activity is described in the next chapter.

The implementation of a shared database for all product parts and specifications requires an enterprise CIM network where employees access and share product data electronically. A CIM local area network (LAN) to support product design is illustrated in Figure 3–1. *Local area networks,* are the electronic links that allow computers to exchange information. There are a number of standards, but the technology most commonly used to exchange data between computer systems is called *Ethernet,* which is covered in chapter 12. Study the network and find the storage location for the shared data. Note that the database is not restricted to the enterprise mainframe computer. Current network technology permits archiving of common part files on storage devices across the network. In this example, product design drawings are saved on a server on the LAN where most of the design activity occurs. In addition, the network technology permits all nodes or

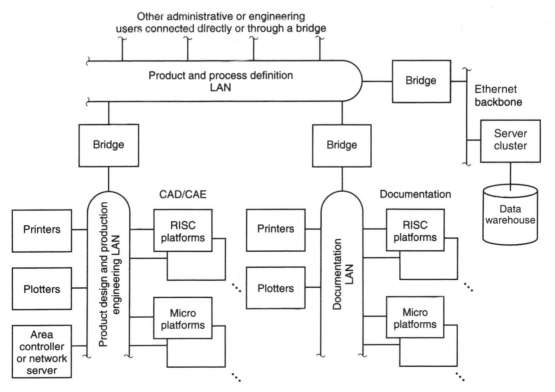

Figure 3–1   Product and Process Definition LAN.

users on the network to access the data distributed around the enterprise system through the bridges that link the networks. In Figure 3–1, the group responsible for design documentation and creation of working drawings for production uses a LAN in the group's work area. However, all of the computers could be on the same LAN.

When this principle of common enterprise data is applied to the riding-lawn-mower case study described earlier, the three different drawing systems are affected. Each system used to develop new mowers must be able to share a single drawing file in the enterprise database. Typically, the product drawing originates in the design area and is used or modified by other departments. Building CIM systems around a shared data precept changes *functional automation* into *enterprise data integration,* the topic of the remainder of the book.

3–2 ORGANIZATIONAL MODEL

It is important to start with an understanding of the fit between *product design* and *production engineering* in the organizational model presented in Figure 2–8. The product design and production engineering portion of the organizational model is illustrated in Figure 3–2. The formal interfaces, illustrated by the connecting lines and arrows, represent information and data flow that occur on a regular basis and are dictated by the information flow for normal operations. No formal interface is indicated (Figure 2–8) between the design and production engineering blocks and the shop floor. Certainly, informal communications links between

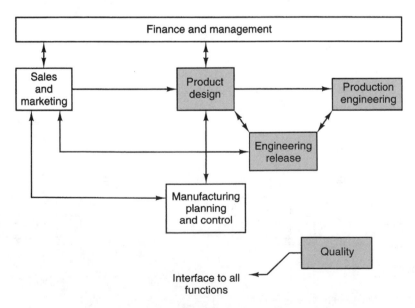

Figure 3–2   Product Design and Production Engineering.

these units exist, and in some organizations a formal link is established between the production engineering area and the shop floor.

Design Information Flow

The design activity starts with either a *new-design creation* or a *request for engineering action (REA)* in the *product design* block in Figure 3–2. Most frequently, the new-design information comes from *sales* and *marketing*. Internal REAs can originate from any department in the enterprise that identifies a product problem; external REAs originating from customer problems are usually initiated by the sales and marketing area. The product design area is responsible for *product design and analysis, material selection,* and *design and production documentation.*

The production engineering area adds *production standards* for *labor, process,* and *quality* to the shared product data from the design area. The term *production standard* means that details required to manufacture the part are identified and developed. *Labor* standards identify the *time* required to produce each part in the product. If the make-versus-buy analysis indicates that in-house production is most economical, then *process* standards are developed that specify the machines and processes to be used for production. Finally, *quality* standards and *verification* techniques for the product and production process are established.

Engineering release is responsible for *product change control.* This control mechanism ensures that all proposed changes are consistent with customer needs and production capability. Therefore, all changes to existing products and all new designs are normally reviewed by the four enterprise areas highlighted (plus sales and marketing) in Figure 3–2. This change-control process ensures congruence between the customer needs and the production capability demanded by order-winning and -qualifying criteria.

The separate block for quality does not imply that quality is external to the design and production engineering departments in the enterprise. In fact, world-class companies practice *quality at the source* to move toward a goal of near-defect-free operation for every action performed in the organization. *Quality at the source* implies that all areas implement defect-free management in a way that is unique to each area. For example, product design must have no design or documentation errors, and the product must be designed in compliance with the order-winning and -qualifying criteria and manufacturing capability.

In the remainder of this chapter, we develop the design process in greater detail and describe why the application of automation to the design segment is critical for world-class manufacturing.

3–3 THE DESIGN PROCESS: A MODEL

The general process for design is characterized by five basic steps:

1. Conceptualization
2. Synthesis

3. Analysis

4. Evaluation

5. Documentation

These five steps are modified from a six-step process described by Groover (2001) and Shigley and Mitchell (1983). In this new five-step model, conceptualization replaces *recognition of need* and *definition of the problem* in the six-step process. The *conceptualization* process, developed subsequently, defines the problem on the basis of the stated need and then divides it into two categories: *typical* and *atypical*. In the *synthesis* step, detail is added to the aggregate solution produced in conceptualization. When the product or part leaves the synthesis step, the design has sufficient detail for us to determine how it will perform. In the next step, *analysis*, the design is tested and the performance data are collected on as many phases of operation as possible. *Evaluation* of the performance data occurs in the fourth step. If evaluation of a product indicates that any part of the product does not meet performance and design specifications, then alternatives to the design are considered. The last step is *documentation;* here the final part details are added that permit manufacturing to produce a product that matches design specifications. The design process, just described, is used for all five manufacturing systems (*project, job shop, repetitive, line,* and *continuous*) and the four production strategies (*engineer to order, make to order, assemble to order,* and *make to stock*) described in chapter 2.

The design process is illustrated by a sequence of blocks in Figure 3–3. The illustration implies a *linear* process with *reiteration* between the blocks. The top three blocks are highly interactive, especially in the early stages of a product design. The new-product concept is divided into subsystems or parts. Potential solutions for the product and each part whirl through the conceptualization, synthesis, analysis, and evaluation steps with iterations generated whenever a better alternative is uncovered.

Figure 3–3   Design Model.

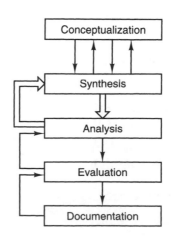

Eventually, the best alternatives for the product emerge and documentation is prepared to support the manufacturing requirements. Integrated design, synthesis, analysis, evaluation, and documentation software packages are used by CIM-driven companies in the design model shown in Figure 3–3. The software has integrated applications to support the entire design model and works from a shared electronic database of the product.

The *origin* of product and part information is the *design process*. For a CIM implementation, product design is the start of the enterprise-wide shared database. Industry studies in automated assembly indicate that up to 70 percent of a product's cost are set by the time a product leaves the evaluation stage. Therefore, revolutionary decreases in product cost can come only from the initial design group. As a result, the design process is the *starting point* for improving the cost order-winning criterion. The development of an effective design process is addressed in the remainder of this section.

Conceptualization

The design process and *conceptualization* starts with the characteristics of the product defined in terms of *form, fit,* and *function*. The desired *shape, style,* and *character* of the product are defined by the term *form*. The *fit* characteristic refers to the *marketing fit* or the order-winning criteria necessary to be successful in the market. The fit also describes the relationship of the desired product to other products in the company's line and the degree to which the product matches the target population. The *function* characteristic defines the product in terms of *performance, reliability, maintainability,* and any *other specific order-winning criteria* present. The form, fit, and function standards for the desired product establish the target limits for the finished design. The data for these three characteristics for new products often come from the marketing area. However, changes to the form, fit, and function for existing products come from a host of internal and external sources, including customer marketing, sales, and design. With the three characteristics identified, the design problem is defined and the design process is started.

Conceptualization in the design process involves defining the design problem as either *typical* or *atypical*. A typical design problem is one that is *similar* to previous product designs. The atypical design statement defines a product need that is *totally new* or different from that of any previous product. The atypical problems require a *concept-design approach;* the typical problems are handled with a process called *repetitive design. Repetitive design* is defined as follows:

> *Repetitive design is the application of the design process to a new product by using pieces of previously designed items or small variations from previous designs.*

Repetitive design applies when the designs are primarily collections of some standard parts or have large sections that are similar from design to design.

Conceptualization Using Repetitive Design.   The repetitive-design process is described by the following three-step sequence:

1. Establish all the form, fit, and function information and data for the desired product.
2. Categorize current products with similar form, fit, and function characteristics and then apply the following guidelines:

   ■ Apply parametric analysis to families of parts or assemblies that are similar in form and function but vary in size and detail.
   ■ Develop a set of standard parts for use across similar products.

3. Model the product design graphically and analytically to communicate the design configuration effectively.

Parametric analysis, referenced in the second step in the repetitive-design sequence, is a powerful repetitive-design tool offered with some computer-aided design (CAD) programs. A simple example of a roller manufacturer illustrates the parametric-design concept. The company produces machines that use rollers with the general shape shown in Figure 3–4. Dimensions A, B, C, and D vary from machine to machine, with about 1,500 sizes produced. Manufacturing the rollers without a CAD system and parametric design requires a drawing for all 1,500 standard sizes to be produced. To add a new-size roller, a draftsperson must draw the complete new part. When a parametric-based CAD solution is used, the part, shown in Figure 3–5a, is drawn once in the CAD system, with the dimensions that change represented by variables. The variable values for the standard rollers are listed in a table (Figure 3–5b) in the parametric-design software. When a standard size is selected, the CAD software creates the roller to the correct dimensions, plots a drawing of the part, and creates a program file to produce the part on a computer-controlled production machine. To create a new non-standard-size

Figure 3–4   Roller Shape.

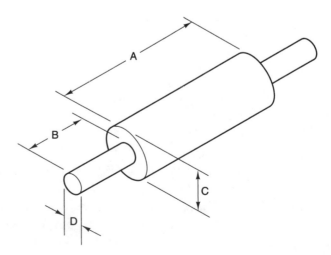

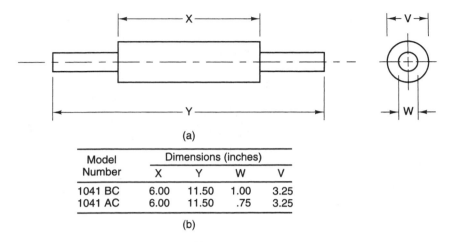

Model Number table and figure:

Model	Dimensions (inches)			
Number	X	Y	W	V
1041 BC	6.00	11.50	1.00	3.25
1041 AC	6.00	11.50	.75	3.25

(b)

Figure 3–5   Roller: (a) Part Dimensions; (b) Parameter Table.

roller, the draftsperson must enter the desired values for the variables into the table, and the CAD software produces the drawings and other necessary files for the new roller. The power of parametric-design software extends beyond this simple example. For example, the input data for the new roller could be a required minimum force on the roller surface; the parametric software would calculate the minimum dimension for W in Figure 3–5a on the basis of the force and the tensile strength of the roller metal.

The case study at the end of the chapter illustrates how repetitive design using the second step in the repetitive-design sequence increases the profitability of a power tool manufacturer. Study the case carefully.

Conceptualization Using Concept Design.   The *concept-design process* is defined as follows:

> *Concept design is the application of the design process for the creation of a new product that is unique, with no similarity to any product currently produced.*

The concept-design process draws heavily on the *creative nature*, or *creativity*, of the designer, who must see, in the mind's eye, a totally unique solution to the present problem. This technique, often called *thinking outside the box*, requires the *divergent-thinking* skills that are associated with an inventor. The creative process is a mental process where *past experiences, previous designs,* or *products* are changed, distorted, or combined to form a new combination that will satisfy an identified need. Often, the change, distortion, or combination is initially judged as not possible, not feasible, not workable, crazy, or ridiculous. This is because divergent thinkers think laterally and often without the logic seen in *convergent thinking,* which is usually taught and performed in engineering. However, making the leap in conceptualization often requires moving beyond the limits generally established.

History has many examples of products that resulted from true innovation on the part of the designer. One such product was the Sony Walkman cassette player. At the initial design meeting, the project manager placed on the table a wooden block that represented the size of the desired wearable tape player and listed the operational capability that needed to be included in the design. Knowing the size and capability of current portable recording and playback cassette devices (the "box" in "thinking outside the box"), the design team initially reacted by saying that the request was crazy and not possible. While the concept-design team borrowed many ideas from existing recording and playback products, they had to find new ways, "outside the box" ways, to make this product design possible.

Therefore, the conceptualization step in the design process includes a process defined by the following subset of steps:

1. Problem statement (C)
2. Removal of artificial limits and boundaries (D)
3. Problem definition (C)
4. Brainstorming (D)
5. Design selection (C)
6. Design acceptance (D)
7. Moving on to synthesis (C)

The (C) after steps 1, 3, 5, and 7 indicates that vertical convergent thinking is used predominantly in these steps. *Convergent thinking,* taught in mathematics and science classes, is often called *analytic thinking* or *deductive reasoning.* It is vertical because the solution process moves along a linear path until the first workable solution is reached. It is characterized as analytic, judgmental, critical, and selective.

The (D) after steps 2, 4, and 6 indicates that lateral divergent thinking is used predominantly in these steps. *Divergent thinking* is characterized as random, sporadic, nonjudgmental, and generative. Lateral thinking moves in many different directions until several solution possibilities are identified. The movement is sometimes linear, but it usually also includes jumping from one solution to another.

Step 1 is to establish all the form, fit, and function information associated with the desired product. Convergent thinking is used here to logically translate the customer needs into a specific series of design and product requirements. In the case of the Sony Walkman, it had to be a cassette player, with high-quality audio output, not exceeding a weight limit, fitted to a belt, easy to operate, with accessible controls, below a set price.

Step 2, a divergent-thinking activity, involves removing artificial limits and boundaries imposed by a narrow view of the product and the operating environment and results in an expanded problem statement. Often customers or design departments are so focused on finding a solution to a problem that the problem statement is demanding a solution more restrictive than necessary. When this occurs, only one solution will work. When this step is applied to the Sony Walkman example, the requirement "fitted to a belt" may be too restrictive. "Worn in a variety of ways" may be better. Also, perhaps a range of models and prices should be considered, and the product could be larger if it weighed less.

In step 3, the initial problem statement (step 1) and the expanded statement (step 2) are combined into a work set of design specifications by using the analytic-thinking process. The move from cognitive to divergent back to cognitive is intentional because it imposes a different type of thinking on the conceptualization process. Different individuals or the same individuals intentionally switching thinking modes may exercise the cognitive- and divergent-thinking processes.

In step 4, brainstorming is used to generate multiple ideas and solutions. Some solutions are silly or absurd while others are logical or practical, but all ideas are accepted as equal at this point in the conceptualization process. This is definitely a divergent-thinking process because there are many solutions, not just one. In the Sony example, some proposed Walkmans were too heavy, others had too many control features, but several were workable solutions.

In step 5, the cognitive process (analytic, judgmental, critical thinking) is used to logically analyze the proposed solutions (step 4) and identify the best solution from the set proposed. In the case of the Walkman, several designs were sent forward.

In step 6, the proposed design or designs are presented to the client and the fit within the client's population is explored and tested. Divergent thinking is used because this step is not simply accepting or rejecting the design but exploring a common fit between the solution and the user.

In step 7, synthesis in the design model starts with the results obtained in steps 1 through 6. As with all of the design process, iteration occurs throughout all of the steps so that the best design can be achieved.

At the start of every concept-design process, the potential for using standard parts, parametric parts, and results from previous designs must be considered along with the requirement for concept-design work dictated by a unique form, fit, or functional requirement. Data indicate that pure concept design is required less than 10 percent of the time, with the remaining 90 percent divided between repetitive design and some combination of the two design processes.

Synthesis

The second step in the design process is *synthesis.* The synthesis activity shown in Figure 3–3 includes the specification of *material,* addition of *geometric features,* and inclusion of greater *dimensional detail* to the aggregate design emerging from conceptualization. However, at this point the original product design model (Figure 3–3) needs some modification. The new model (Figure 3–6) identifies the second step as a synthesis filter. The synthesis filter is analogous to a paper filter that removes impurities from a liquid; however, in this case the filter adds *design enrichment* with the filtering process. The synthesis design filter removes *geometric features* and *material specifications* that add cost to the product but not market value. As the model illustrates, the filter uses two processes to achieve design enrichment: *design for assembly* and *design for manufacturing.* Conceptualization and synthesis are closely tied and

Figure 3–6   New Design Model.

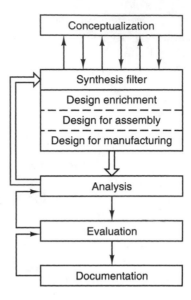

highly iterative (many up and down arrows show the information flow), with activities in these two processes often inseparable. As preliminary product ideas are enriched with features and details early in the design process, the design engineer uses both conceptualization and synthesis skills. However, as the product design becomes firm, more time is spent in synthesis—adding and verifying product features and details. With 70 percent of the manufacturing cost fixed in these early activities, the time spent getting the product *right* in the synthesis step is well justified. The procedures used to get the product *right*—design for assembly and design for manufacturing—are covered in chapter 5.

Analysis

Analysis is a method of determining or describing the nature of something by separating it into its parts. In the model of the design process (Figure 3–6), the analysis stage involves studying a single design solution or several alternative design choices by using mathematical and other scientific procedures. In the process, the elements, or nature of the design, are analyzed to determine the fit between the proposed design and the original design goals. The highly interactive nature of the synthesis and analysis stages is best illustrated with a simple example. The aluminum bracket in Figure 3–7a was part of a larger design. In the synthesis stage, the thickness of the metal was specified. However, when each part of the bracket (horizontal plate, junction, and vertical plate) was analyzed, the strength of the junction was not adequate for the typical bracket load. After several iterations between synthesis and analysis, the correct thickness at the junction was determined (Figure 3–7b), and the optimum design for the bracket was produced.

Figure 3–7 Aluminum
Bracket: (a) Original Design;
(b) Optimum Design.

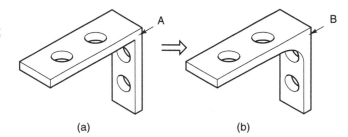

(a)   (b)

The types of analysis frequently used fall into two categories: *mass properties* and *finite element.* Although the analysis calculations could be performed manually, the use of a computer increases the analysis capability and significantly reduces the time required for such analysis. The *mass-properties analysis* includes calculations of the following two factors:

1. Solid-object features such as volume, surface area, weight, center of mass, and center of gravity
2. Mechanical tolerance requirements and assembly interference and fit (to ensure that the individual parts can be assembled properly even if all the tolerances go to the worst-case condition)

In *finite-element analysis (FEA),* four typical calculations are as follows:

1. The limits of stress and strain for a part
2. The heat-transfer capability of the part for a specific material
3. Airflow and fluid-flow analysis
4. The theoretical limits of operation of an electrical or electronic circuit

FEA and mass-properties analysis often require that the product and/or parts in the product be represented electronically by a solid model. This solid model has all the characteristics of the actual product or part, so the FEA or mass-properties analysis software uses the model to perform the analysis.

Study the design model shown in Figure 3–6 and note how the relationship between the top three blocks is illustrated. The many arrows in both directions indicate that conceptualization and synthesis are highly interactive and often inseparable. However, the line between synthesis and analysis is quite clear. A design is developed and then analysis is performed. Analysis is a critical link to ensure that the product meets the order-winning criteria required by the customer.

Evaluation

The *evaluation* step (Figure 3–6) involves checking the design against the original specifications. The optimum design delivered from the synthesis and analysis process is compared to the form, fit, and function requirements for the new product. Additional evaluation is performed to determine the match between the manufacturing and assembly requirements of the design and the capability of the manufacturing facility. If the design satisfies the evaluation criteria in every case, the product is passed to documentation. If any areas fall short, however, the design is returned to either the conceptualization, synthesis, or analysis process for more work. When a design is passed from evaluation, it meets all of the form, fit, and function requirements of marketing and sales plus it can be produced and assembled with the technology and skills of the people in manufacturing. Therefore, Terry Hill's model has been satisfied.

Evaluation often requires the construction of a prototype to test for conformance with critical order-winning criteria such as operational performance, reliability, compatibility, user-friendly operation, and other criteria. *Rapid prototyping*, a technique to produce a sample product quickly, is frequently used in this design step. Evaluation and rapid prototyping techniques are described in detail in chapter 5.

Documentation

The last step in the design process is the presentation of the design through a *documentation* process. The following items are frequently part of the documentation process:

- Creation of all necessary product and part views in the form of *working drawings* and *detail and assembly drawings*
- Addition of all design details, such as standard components, special manufacturing notes, and all dimensions and tolerances
- Creation of required engineering documents, such as part number assignments, detailed part specifications, design bill of materials (BOM), and a part and part number *where-used list*
- Creation of product electronic data files shared by the following departments: manufacturing planning and control, production engineering, marketing, and quality control

Documentation examples are provided in Figures 3–8 through 3–12.

This concludes the description of the design model process (Figure 3–6) required to bring a product design from initial concept to completed design. One problem remains with this design model, however. The model process was conducted in isolation because representation from manufacturing, external parts and equipment vendors, and other areas in the enterprise affected by the design were not integrated into the design process and flow. If the design is to be optimal for both the initial form, fit, and function requirements and the manufacturing capability of the enterprise, the design process or model must be enlarged. The required new model has several names: *simultaneous engineering, early manufacturing involvement,* and *concurrent engineering.* In this book, the term *concurrent engineering* is used.

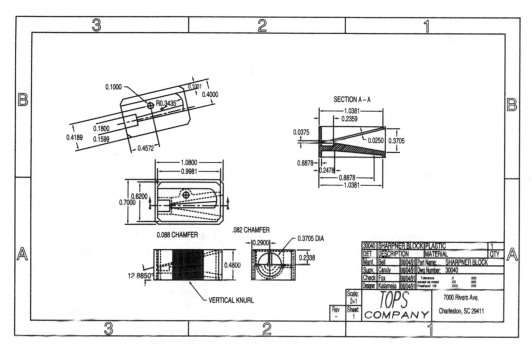

Figure 3–8   Working Drawing for Pencil Sharpener Block.

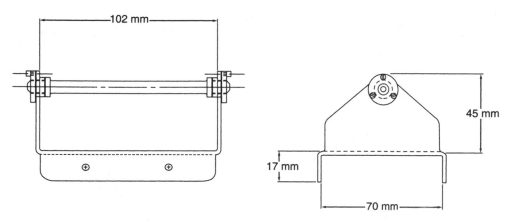

Figure 3–9   Assembly Drawing.

(Source: Boothroyd/Dewhurst, "Product Design for Manufacturing and Assembly," © April 1988, Manufacturing Engineering Magazine, *p. 42. Reprint by permission of the Society of Manufacturing Engineers, Dearborn, MI.)*

Figure 3–10 Bill of Materials for Spindle/Housing Assembly.

Bill of Materials			
Quantity	Part number	Description	Material
1	23301	Sheet metal base	Stainless steel
1	23302	Spindle	Stainless steel
2	23303	Bushings	Nylon
6	23304	Screws	Purchased

3–4 CONCURRENT ENGINEERING

Concurrent engineering is defined as follows:

> *Concurrent engineering implies that the design of a product and the systems to manufacture, service, and ultimately dispose of the product are considered from the initial design concept.*

Concurrent engineering (CE) is not a new concept; it was used in Japan in the early 1960s. As the definition indicates, CE is a design philosophy dealing with the process that a company uses to bring a new product to market. A comparison between CE and the traditional approach produces some interesting conclusions. First, let's look at the traditional process.

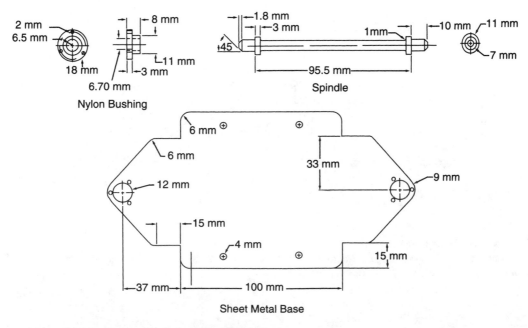

Figure 3–11 Working Drawing for Spindle/Housing Assembly.

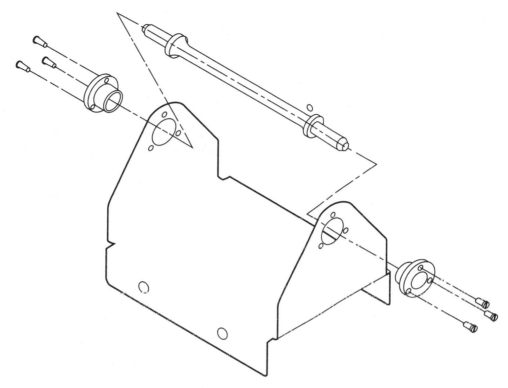

Figure 3–12   Exploded Assembly View of Spindle/Housing Assembly.

The Traditional System

A traditional design system still used by some companies is illustrated in Figure 3–13. Study the figure until all of the blocks are familiar.

The information flow in the sequence is linear, with little interaction between the many groups responsible for new-product development. In this traditional system, the product design process, using the model in Figure 3–6, is often completed in the engineering design block without input from any other group. As a result, production engineers get product information after the product design is completely finished and documented on engineering drawings. The production system is frequently designed by production engineers in the same type of isolation. Manufacturing and the personnel on the shop floor frequently receive completed product drawings and production system specifications from the production engineers shortly before production must begin. The group responsible for customer documentation and product service is often not involved with the product until shortly before the product is shipped from the factory. In most cases, the disposal of the product after its useful life is never considered.

This type of product development creates several problems, including delays in getting the product to market, poor quality, poor customer service, and high

Figure 3–13 Traditional
Product Development Process.

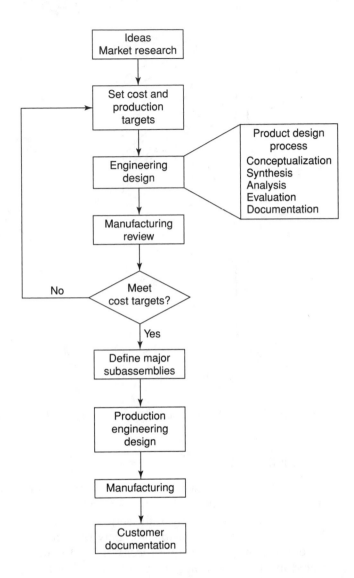

product cost. The problems occur because the design produced in isolation by engineering often cannot be manufactured to the required specifications in the shop. Even with the feedback to check cost targets, the manufacturing divide described in chapter 1 has been created. As a result, the merchandise produced usually does not have the order-winning or -qualifying criteria necessary for a successful product.

What is needed is a process that brings design engineering, marketing, production engineering, external vendors, production control, and the shop floor together throughout the design process. The design that results from this collective endeavor is usually optimal for the enterprise and matched to the order-winning and -qualifying criteria present in the marketplace. The design process described in Figure 3–6 must be modified to include *concurrent engineering* (CE).

Figure 3–14 New Model for
Product Design.

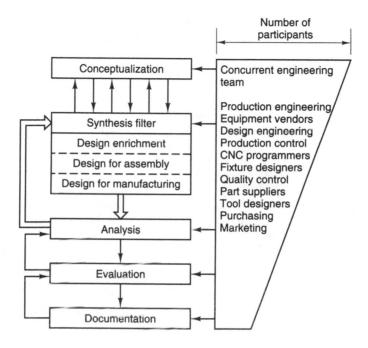

The New Model for Product Design

The design process model shown in Figure 3–14 includes CE. With this change, participation in the design process is not limited to product design engineers but includes all the specialties listed on the right side of the figure. Study the new model until you become familiar with all the participants.

The width of the concurrent block indicates the level of participation from the initial concept design through documentation. Notice that the level of participation from across the enterprise drops as the product design becomes firm. As the crucial design features are locked in at the conceptualization, synthesis, and analysis stages, there are fewer critical decisions that require broad input from across the enterprise.

The traditional product development chart shown in Figure 3–13 is changed to include CE, and the result is illustrated in Figure 3–15. Note that the CE design process shown in Figure 3–14 is embedded into the process, and the CE process includes product design, manufacturing system design, and customer documentation. With the CE product design process embedded into the product development flow in Figure 3–15, the design of the manufacturing system is concurrent with the design of the product. The following results:

- The likelihood that manufacturing can produce a product that satisfies the form, fit, and function set by marketing and that meets the order-winning and -qualifying criteria is enhanced with this process.
- The manufacturing system is ready to produce products much sooner than in the traditional product development process.

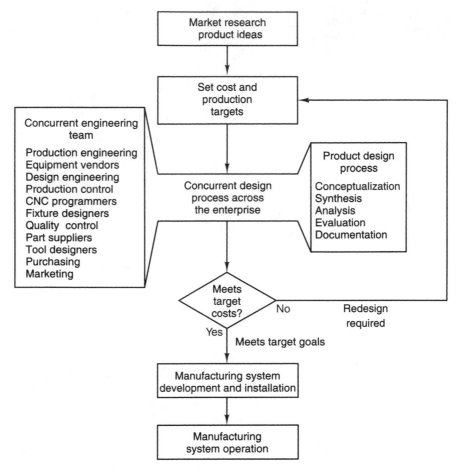

Figure 3–15   Improved Product Development Process.

■ The design portion of the process is longer as a result of the expanded CE team, but the total time for product development is shorter with the new model.

Concurrent Engineering: An Operational Model

There are many opinions about how CE should be managed and executed, but one of the clearest processes was proposed by James Nevins, Daniel Whitney, and four collaborators from the Draper Laboratory. The process is general enough to apply to any type of manufactured product. The process steps are consistent with the design process model shown in Figure 3–14 and the product development process shown in Figure 3–15.

The product design begins with a market plan in which rough concepts are developed and price and production targets are set. CE follows the approval of the market plan and proceeds in five phases:

1. *Concept phase.* In this phase most of the major decisions about the form, fit, and function of the product are tentatively established. Example areas include style, materials, performance, and energy consumption.

   - Decisions in this phase account for 70 percent of the overall product cost and impact marketability.
   - The best designers and the most sophisticated design tools are required.
   - Analysis of alternative designs requires the use of performance simulation software and cost-estimating models.

2. *Major subassembly design phase.* Following approval of the concept design, the product is divided into major subassemblies for more detailed design specification.

   - Model variations in the product are identified and isolated to the fewest possible subassemblies. Modular design practices that reduce the number of different subassemblies reduce cost and permit a quick response to changing market conditions.
   - The number of different departments involved in the design increases significantly in this phase.

3. *Single-part design phase.* The design of all the parts of the main product and all subassemblies is completed in this phase.

   - The number of people participating in the design is at a maximum.
   - Major design changes at this time cause major delays in design completion and are expensive.

4. *Design of part-pairs phase.* Following the design of detailed parts, the emphasis shifts to manufacturing processes instead of product performance. For mechanical designs, individual parts are analyzed for fit by employing part-mating theory (geometric tolerances), tolerance analysis, and tolerance simulation. For electronic products, the development of test programs and acceptance criteria for individual integrated circuits is an example of the activity.

5. *Grouping of parts and subassemblies.* The final phase focuses on additional refinements for efficient assembly. The assembly sequence is determined and the assembly equipment, jigs, and fixtures are designed.

   - Frequently, previous designs for individual parts and subassemblies must be modified. For example, tolerances may be refined; lifting eyes and gripping surfaces may be added.
   - Test strategies and programs are developed.

At several steps throughout the concurrent design process, the development of ISO 9000 process descriptions would be developed to ensure that the production process satisfies this important standard. The process just described may require the commitment of more time and money to the design process than in the traditional method. However, CE will reduce the total time to market because the production tooling and manufacturing systems are ready sooner.

Automating the CE Process

The CE process is well supported by product design software running primarily on engineering workstations and PCs. The products permit two or more designers located in different offices or even different cities to work jointly on the same part or product drawing. In some cases, video feedback between the sites is provided with audio links. Using this technique, every person involved in the selection of certain design features can witness the changes and can provide input to achieve a better, lower-cost product. Another feature of the design software, called *full associativity*, links all the related drawing files that have been created for a product. For example, the working drawings; solid-model, three-dimensional (3-D) drawings; and section views are linked by this feature. As a result, any change made on any of the drawings is changed automatically on every drawing of the part.

CE Success Criteria

Companies experienced in the application of the CE model have identified some important general guidelines:

- Construction of tooling and manufacturing systems does not begin until all five phases of the Nevins and Whitney model of CE are completed.
- The CE philosophy suggests that prototype tooling and production tooling be as similar as is practical so that production tooling produces no unexpected surprises.
- Each step in the CE process involves experienced designers of manufacturing systems, service systems, and disposal systems, not just product designers.

The involvement of participants requires more than simply a design review. Production engineers may need to begin preliminary design of manufacturing systems to ensure that the product can be made for the amount budgeted. Such a design may include the following:

- Outside suppliers of tooling and production machinery
- Designs of repair and maintenance systems to evaluate serviceability
- Preliminary design and cost estimates of tools for repair, assembly, and disassembly
- Design of ultimate disposal techniques, called the *cradle-to-grave process*, consistent with environmental regulations for the product area

This definition and operational model for CE is broad because of the *cradle-to-grave* emphasis placed on the design process. CE is also demanding because it requires exceptionally good, thorough engineering throughout the design process. However, the demands of the marketplace for world-class products require a CE process.

3–5 PRODUCTION ENGINEERING

Production engineering (PE) has the responsibility for developing a plan for the manufacture of the new or modified product that was generated through the CE process. The PE plan has seven elements:

1. Process planning
2. Production machine programming
3. Tool and fixture engineering
4. Work and production standards
5. Plant engineering
6. Analysis for manufacturability and assembly
7. Manufacturing cost estimating

In some organizations these PE activities are grouped under the *industrial engineering* department; in others, they would be assigned to the *manufacturing engineering* area. The name of the department where these functions are located is not critical. However, the use of CE to bring these seven areas together in the product design team is crucial for a successful product. When the classic method is used, the design is passed through one or more groups responsible for these seven activities. If this occurs, the design becomes less and less optimal as it passes through the groups. For example, if tooling to produce the product is not considered until last, then the part design may make it difficult to hold the material in a fixture during some manufacturing processes. The graphic in Figure 3–16 illustrates how the PE function could be structured. Note that process planners, production planners, industrial engineers, manufacturing engineers, and part programmers are all present. In any given implementation, the functions divided among these individuals could be shifted, or not all of these job titles may be present. What is important is that as the product design parts are described in working drawings, the PE component of the CE team begins the process of engineering the production system. All of the seven elements are present, and all would be active as needed to ensure a design that could be successfully manufactured while satisfying the market demands. Regardless of the location of these activities, it is important to understand the function they perform on the product. Therefore, the remainder of this chapter is devoted to a description of these activities.

Process Planning—Element 1

Process planning is often called *manufacturing planning*, *material processing*, and *machine routing* by different industry groups. The people who carry out this

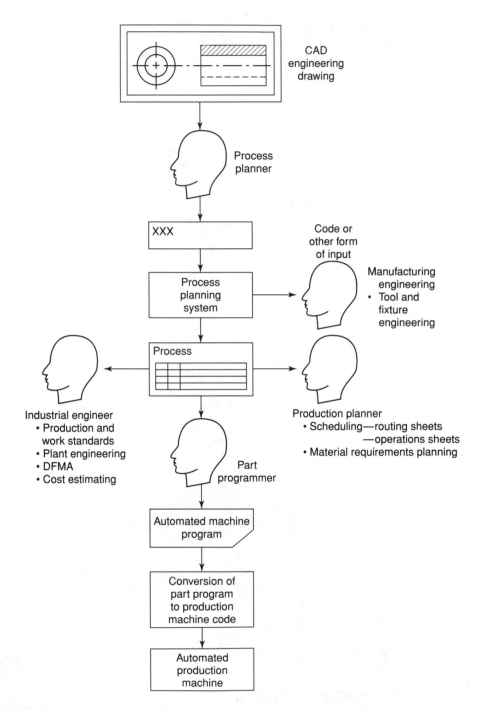

Figure 3–16 Typical Process-Planning System.

process are called *process planners, material processors,* or just *planners. Process planning* is defined as follows:

Process planning is the procedure used to develop a detailed list of manufacturing operations required for the production of a part or product.

Figure 3–16 indicates the order of events in process planning in the traditional company. The process starts with the documentation for a part or a family of parts for a completed product design arriving in planning. In Figure 3–17, the top left box contains a list of the typical detailed product documentation generated to support a design. The first activity performed on each part is a make-versus-buy decision made by PE. Next, a *routing sheet* is prepared for every part to be made and assembled in-house. *Routing sheets,* sometimes called *process plans* or *operations sheets,* describe the sequence of operations or manufacturing processes required to produce the finished product. A typical set of operations and machines for the machining of surfaces is listed in Figure 3–18. Take a few minutes to study the chart.

An example routing sheet for 100 of the stainless steel spindles shown in Figure 3–11 is illustrated in Figure 3–19. Note that in Figure 3–19 the work center (a specific production facility that includes one or more people and/or machines) is specified, along with the ID number and a brief description of the operation. The time data provided for each operation often include *setup* time, *unit run* time,

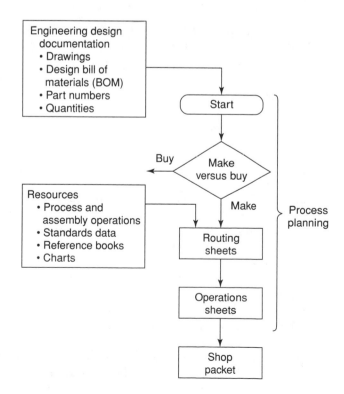

Figure 3–17 Process-Planning Sequence.

Machining operation	Description	Machine tool used
Milling	One of the two most frequently used material-removal operations. The tool is attached to either a vertical or horizontal drive. The rotary cutter is fixed in position and the material is moved past the cutter to remove material. The cutting teeth are placed either around the sides of a barrel-shaped tool or on the circular end face.	Vertical and horizontal manual mills and machining centers
Turning	Turning is the removal of material from the sides of a cylindrical-shaped part. The part is rotated in a lathe while the chisel-like cutting tool moves across the rotating cylindrical side, removing material.	Lathe
Facing	Facing is the removal of material from the end face of a cylindrical-shaped part. The part is rotated in a lathe while the chisel-like cutting tool moves across the end face.	Lathe
Drilling	Generally the drilling operation produces circular holes in parts. If the hole must be very precise, then a boring machine is used. For less precision, a mill or lathe is used, and for the least precision, a drill press or chemical mill is used.	Drill press, boring mill, vertical and horizontal mill, lathe, electro-chemical mill
Sawing	Sawing is used to remove large volumes of material to get material in the general shape of the part.	Band and cutoff saws
Grinding	Grinding is used for cylindrical- and rectangular-shaped parts to provide a precision finish to the surface or to achieve precision dimensions.	Centerless and cylindrical grinders

Figure 3–18   Manufacturing Processes.

queue time, and *other standard* times. Identify the operations listed on the routing sheet that appear on the process chart shown in Figure 3–18. The resources used to complete the routing and operations sheets are listed in the box adjacent to the routing sheet box shown in Figure 3–17.

The *operations sheet* (Figure 3–17), the next document in the process-planning sequence, describes in detail each of the operations on the routing sheet for the part. These operations often include tooling, jigs and fixtures needed to hold the part, sketches of setups, semifinish dimensions, machine settings, assembly instructions, handling requirements, inspection and testing requirements, and operator skill levels. Everything not covered in the working drawings that is necessary to produce the part is included on the operations sheet. Note in Figure 3–17 that the planner has process, assembly, and standards data; reference books; and charts to use as a guide

Machining operation	Description	Machine tool used
Shaping	A metal-removal operation in which the part is fixed and a chisel-shaped tool moves parallel with the part, peeling away material. Capable of removing a large amount of the material in each cut.	Horizontal and vertical shaper
Planing	Similar to the shaper operation but with a finer cutting operation.	Planer
Boring	The boring operation usually follows a drilling operation so that the boring tool does not have to remove as much material. Precise circular holes with precise locations are produced by boring operations.	Horizontal boring machine, boring mill, lathe
Reaming	Reaming operations must always be preceded by a drilling or boring operation to prepare the circular hole. The ream removes little material but adjusts the dimensions to a very tight tolerance.	Lathe, drill press, boring mill, horizontal boring machine
Broaching	Broaching is used to remove material from the inside of a square hole in a part. It has the same precision as reaming does for circular holes.	Broaching machine
Electrochemical milling	Electrochemical milling (ECM) is used to cut irregular-shaped holes in parts.	ECM machine
Wire electric discharge	Wire electric discharge uses a fine wire to cut a metal plate to an irregular shape.	Wire electric discharge machine (EDM)

Figure 3–18   (*cont.*)

in setting up the routing and operation sheets. In some organizations, the routing sheet and operation sheet are combined into one document.

All the information and documents prepared in process planning go into a *shop packet.* The shop packet, a package of documents used to plan and control the movement of an order on the shop floor, normally includes the following: a *manufacturing order* called a *traveler, operation and routing sheets, engineering documentation, pick list, move tickets, inspection tickets,* and *time tickets.* This process used to develop routing and operations sheets for products is most often associated with the job shop–type manufacturing system, but similar processes are used in the other four manufacturing systems (*project, repetitive, line,* and *continuous*) and in the four production strategies (*engineer to order, make to order, assemble to order,* and *make to stock*) described in chapter 2.

Figure 3–19 Routing Sheet for Stainless Steel Spindle Illustrated in Figure 3–9.

Part number: 252601   Lot size: 100
Part description: Spindle—stainless steel

| Work center | Operation | | Setup time (min.) | Unit run time (min.) |
	No.	Description		
10	01	Saw 3/4-in. stainless steel bar to length	10	0.47
30	01	Deburr	5	0.25
20	01	Turn shaft ends	30	1.17
20	02	Turn center	10	0.74
30	01	Deburr	5	0.15
40	01	Grinding	25	0.37
50	01	Degrease	5	0.20
60	01	Inspection	10	0.10

In the CE model, the process planners are working with the CE design team to produce routing information during the part design process as soon as the design is finalized. The process-planning procedure just described does not change, but the activity is moved up into the design process. As a result, the designs produced are optimized for the production machines on which they will be manufactured. In CIM enterprises, computer software can be used to generate most of the data in the packet. In addition, the packet need not move physically from work center to work center in paper form; rather, operators view the data files containing packet information on computer terminals in the work center.

In the fully manual planning system, the process plan is produced by the planner, who studies the part drawings and then establishes the appropriate procedures to manufacture the part. It requires a person knowledgeable in the processes listed in Figure 3–18 who has many years of experience in planning production. It is as much an art as it is a science.

Production Machine Programming—Element 2

The second item on the PE responsibility list is production machine programming. In the past, when there were only manually operated machines in the production area, this activity was not required. Automation of production machines started about the same time in the discrete- and continuous-production industries. In the discrete manufacturing area, automation started with metal-turning and metal-cutting machines such as the lathe and mill described in Figure 3–18. At the same time, the process industries, producing products such as gasoline and nylon, started introducing pneumatic process controllers to regulate the continuous-production processes.

As automation on the plant floor evolved to computer-driven machines and processes, the need for machine programming developed. Today, the use of computer-controlled production machines, often called *numerical control* (*NC*) or *computer numerical control* (*CNC*), in discrete- and continuous-process manufacturing is the rule rather than the exception. In many organizations, the responsibility for the programming of the production machines falls in the PE area in the enterprise. As a result, the automation of the factory floor created a new term, *computer-aided manufacturing* (*CAM*). CAM is discussed in chapter 12.

Tool and Fixture Engineering—Element 3

The third responsibility of PE is the *specification* and *design* of the tooling required in the production of the part. In metal and nonmetal cutting, the tooling includes *jigs* and *fixtures* to hold and position raw material and parts in machines while tools shape and finish the parts. Square-shaped vertical fixtures are visible in Figure 3–20 in front of the machining center. Parts to be machined are attached to

Figure 3–20   Production Work Cell Uses Multiple Pallets to Hold a Variety of Tooling for Many Different Parts.

(Source: Courtesy of Cincinnati Milacron, Inc.)

Figure 3–21   Stamping Die.

all four sides of these fixtures, often called *tomestones*. In the forming area, the tooling often includes the design of dies (Figure 3–21) and molds to shape parts. The requests for new production machinery and other support equipment such as stamping and forming dies for the new product are made by PE.

In the traditional manufacturing system, the tooling process begins after the part or product is completely designed. As a result, it is often impossible to design fixtures and tooling that allow the manufacture of a part under optimum conditions. However, in CE the tooling is specified as a part of the product design process; as a result, the parts and production tooling are developed in parallel. The result is tooling and parts that fit like "a hand in a glove."

Work and Production Standards—Element 4

The fourth part of the PE plan is the establishment and application of *work* and *production* standards. To determine the manufacturing cost, the standards group must establish the time required for the production of every part. The standards group uses two methods to satisfy this requirement. In the first and oldest method, *direct time studies,* an operator is timed with a stopwatch as he or she goes through a production operation. After the job is timed, the percentage of normal speed, or the rate that the worker used in the production process, is estimated between 80 and 120 percent. With the time-study data, an average production rate can be established for every step in the manufacturing process. Time studies are often used in purely manual operations.

The second and preferred method, *motion time measurement* (MTM), uses standard time data developed for basic work elements for manual tasks. All the individual time elements required to perform a new job are summed, and the result is a theoretical average time to perform the job when a worker is working at an average rate. This process can be automated easily with a computer and MTM software.

Plant Engineering—Element 5

The fifth element in the PE plan is plant engineering. In some cases, the production requirements for a new product require a new manufacturing facility. *Plant engineering* addresses the design of a production facility. Many companies go through this process when the initial production facility is designed and built to meet the needs of a new product line. However, some manufacturers that have frequent model changes, such as those in the automotive industry, would have a large plant engineering group as part of PE to make major facility changes annually.

Analysis for Manufacturability and Assembly—Element 6

The concept of *design for manufacturing and assembly* (DFMA), another responsibility usually embedded in PE as item 6, was introduced earlier in the synthesis filter stage of the design model. DFMA ensures that the finished design is optimal for both the manufacturing processes required and the assembly techniques needed. DFMA is effective only if it is performed by PE as part of the design process using CE techniques. DFMA concepts are addressed in greater detail in chapter 5.

Manufacturing Cost Estimating—Element 7

A crucial role for PE described in item 7 is the estimation of the product cost based on design-drawing data and work and production standards information. Consider an example from a U.S.-based robot manufacturer. When a new model was designed, the complete set of design drawings and part documentation was delivered to PE. The 300 or more parts were distributed to the planners. Before the manufacturing cost could be determined, the production process had to be established. With the routing and operations for every part established, the labor and overhead for each planned operation could be calculated. The cost for each part was then determined by summing the cost of individual operations. The total robot manufacturing and assembly cost was calculated by adding the cost of all the individual parts on the engineering or design BOM. The cost-estimating process is much quicker now because of software packages that do a *cost roll-up* for an entire BOM with hundreds of purchased and manufactured parts. Establishing the product cost during the design process is critical for marketing.

Therefore, integration of this process into the design activity through CE concepts is essential to obtaining an order-winning or -qualifying criterion in the cost area.

Using a CE model effectively to address these seven activities in PE improves the price, quality, and customer-response order-winning and -qualifying criteria. Therefore, PE is extremely important to the success of the enterprise and the development of a prosperous CIM implementation. Automation techniques frequently used in CIM to implement design engineering and PE are described in the next two chapters.

3–6 SUMMARY

The development of new products and modifications to existing products begins in design and production engineering. This engineering group is the bridge between the external focus of sales, marketing, finance, and management and the internal focus present in production planning and manufacturing. Using form, fit, and function specifications from marketing, design engineering creates a design that satisfies the product requirements. The design process is modeled as an interactive process that includes five steps: conceptualization, synthesis, analysis, evaluation, and documentation. Conceptualization is divided into concept design and repetitive design, and synthesis is expanded to include design for manufacturing and design for assembly. A major problem exists with the conventional design model; the process is performed in isolation so that the design frequently cannot be produced economically in the shop. The conventional process reinforces the "great manufacturing divide" and does not focus on satisfying the order-winning criteria present in the market.

Concurrent engineering, a team process that brings all critical departments together from across the enterprise to the design activity, is required to remove the design isolation problem. Concurrent engineering, often called *simultaneous engineering* or *early manufacturing involvement,* is described as an iterative five-step process that is embedded into the design model.

Production engineering, frequently termed *industrial engineering* or *manufacturing engineering* in organizations, has seven fundamental areas of responsibility: process planning, production machine programming, tool and fixture engineering, work and production standards, plant engineering, analysis for manufacturability and assembly, and manufacturing cost estimating. With a good understanding of the product design and documentation process, the next step is a study of the automation that supports the product design.

BIBLIOGRAPHY

ARNSDORF, D. R. *Technology Application Guide: CAD—Computer-Aided Design.* Ann Arbor, MI: Industrial Technology Institute, 1989.

GRAHAM, G. A. *Automation Encyclopedia*. Dearborn, MI: Society of Manufacturing Engineers, 1988.

GROOVER, M. P. *Automation, Production Systems, and Computer-Integrated Manufacturing*. 2nd ed. Upper Saddle River, NJ: Prentice Hall, 2001.

MITCHELL, F. H., Jr. *CIM Systems: An Introduction to Computer-Integrated Manufacturing*. Upper Saddle River, NJ: Prentice Hall, 1991.

SHIGLEY, J. E., and L. D. MITCHELL. *Mechanical Engineering Design*. 4th ed. New York: McGraw-Hill Book Company, 1983.

SOBCZAK, T. V. *A Glossary of Terms for Computer-Integrated Manufacturing*. Dearborn, MI: CASA of SME, 1984.

QUESTIONS

1. Describe how product design fits into the organizational model for an enterprise.
2. What is the difference between a new design and an REA?
3. What are the responsibilities of design engineering?
4. What are the responsibilities of production engineering?
5. What are the responsibilities of the engineering-release area?
6. Describe the design process model and describe briefly the function of each step in the process.
7. What is significant about the data and information that flow from the design process?
8. Why is the product design process critical for a price order-winning criterion?
9. Describe form, fit, and function.
10. Describe the difference between concept design and repetitive design.
11. What is parametric analysis?
12. Compare the interaction between conceptualization and synthesis in the design model.
13. Describe the two major types of analysis frequently used.
14. Define *concurrent engineering* in your own words.
15. What is the cradle-to-grave concept of product design?
16. What major flaw is present in the traditional design process?
17. Describe how concurrent engineering helps to eliminate the "great manufacturing divide" and to promote a congruence in the manufacturing strategy in the enterprise.
18. Describe the responsibilities present in production engineering.
19. Describe the function of routing sheets and operations sheets.
20. What is a shop packet?
21. What is the difference between direct time studies and motion time measurement?

PROJECTS

1. Using the list of companies developed in project 1 in chapter 1, identify the companies that use concurrent engineering in the design process.
2. Compare the product design model developed in this chapter with the design processes used by one company from the small, medium-sized, and large groups and describe how their processes differ from the design model in the text.
3. Compare the production engineering responsibilities listed in this chapter with those used in the companies selected in project 2. Describe the differences that are present.
4. View the SME video *Simultaneous Engineering* and prepare a report that describes the concurrent engineering model and product design processes used by Motorola and IBM.

The following projects are a continuation projects 5 through 8 at the end of chapter 2.

5. A company that manufactures kitchen products wants to add a set of plastic storage bowls or containers to its product line. The company wants to distribute the product through Wal-Mart, Target, or a chain of stores in your region. Do the following:
   a. Determine the manufacturing processes required to make each part of the product.
   b. Adjust the number of products needed per week by adding in replacement products for products discarded because of quality issues. Use the captured-quality value from Figure 1–6 for the calculations.
   c. Find the number of parts produced per hour when quality is considered.
6. A company that manufactures kitchen products wants to add a high-end manual can opener to its product line. The company wants to distribute the product through Wal-Mart, Target, or a chain of stores in your region. Do the following:
   a. Use the repetitive-design process on the competitor's product to develop a design for your product.
   b. Determine the manufacturing processes required to make each part of the product.
   c. Adjust the number of products needed per week by adding in replacement products for products discarded because of quality issues. Use the captured-quality value from Figure 1–6 for the calculations.
   d. Find the number of parts produced per hour when quality is considered.
7. A company that manufactures kitchen products wants to add a manual handheld eggbeater to its product line. The company wants to distribute the product through Wal-Mart, Target, or a chain of stores in your region. Do the following:
   a. Use the repetitive-design process on the competitor's product to develop a design for your product.

   b. Determine the manufacturing processes required to make each part of the product.

   c. Adjust the number of products needed per week by adding in replacement products for products discarded because of quality issues. Use the captured-quality value from Figure 1–6 for the calculations.

   d. Find the number of parts produced per hour when quality is considered.

8. A company that manufactures kitchen products wants to add a manual hand-held garlic press to its product line. The company wants to distribute the product through Wal-Mart, Target, or a chain of stores in your region. Do the following:

   a. Use the repetitive-design process on the competitor's product to develop a design for your product.

   b. Determine the manufacturing processes required to make each part of the product.

   c. Adjust the number of products needed per week by adding in replacement products for products discarded because of quality issues. Use the captured-quality value from Figure 1–6 for the calculations.

   d. Find the number of parts produced per hour when quality is considered.

CASE STUDY: REPETITIVE DESIGN

A manufacturer of handheld power tools for the construction industry has a broad product line of tool sizes and types. In the initial development of each product line and in the yearly updates of the products, the process used for each power tool design was totally independent. Each new product was treated as a new-design product and little reference was made to products designed earlier. Isolating each design was a result of the design department structure and a belief that a fresh start would produce new-design ideas.

The broad product line had most of the characteristics of a repetitive-design process, but the design specifications for current products with similar form, fit, and function characteristics were never reviewed. As a result, the company had to order and stock a total of 208 different washers to produce all the power tools. In many cases, the washers differed only in minor features such as surface finish.

Recognizing that repetitive-design guidelines were required, the company analyzed every product design and grouped the washers into two categories:

1. Washer designs that were common to more than one power tool product
2. Washer designs that performed the same function as the common washers but were different in shape, size, material, or finish

After the analysis, the company used the repetitive-design process to modify power tool designs around a small set of washers. It was able to build all the power tools with only 8 different washers. The difference in the cost of the washers may not be a significant part of the product cost. However, the dollars saved in managing the inventory of 8 washers versus 208 is significant.

In another study of repetitive-design savings, NCR found the hidden costs enormously out of proportion to the part price. In one case, NCR estimated that the material, labor, and overhead costs associated with one small screw used in a point-of-sale terminal actually equaled $12,500 during the life of the product. These two examples illustrate that the use of repetitive design to minimize the number of component parts can pay big dividends.

Design Automation: CAD and PDM

OBJECTIVES

After completing this chapter, you should be able to do the following:

- Define *CAD* and *PDM*
- Describe how the CAD technology is integrated into the product design model and why it is critical to the design process
- List the hidden cost of a paper-based design process
- Describe the evolution of CAD technology
- Describe the current state of CAD technology
- List and describe the four ways a part can be represented in CAD software
- List and describe the advantages offered by solid-model software
- List the major vendors supplying CAD software
- Describe the function of PDM and list the major software vendors

4–1 INTRODUCTION TO CAD

The initial question that needs attention is Why spend time on a very specific technology, like computer-aided design (CAD), when your background and career interest are in areas other than mechanical engineering, mechanical engineering technology, or product design? The short answer is that CAD is a critical technology for an integrated product development process in a computer-integrated manufacturing (CIM) organization. The longer response is that a study of the integration of operations across a manufacturing enterprise starts with the concept for a new product. It then continues through all of the corporate departments responsible in some way for the delivery of the finished product to the retailer or customer. If you work at any point on this continuum, then an understanding of how the other parts work makes you a better employee and usually leads to advancement and promotion within the organization or industry. CAD is one of the fundamental building blocks of a CIM system.

A Definition of CAD

The design department was one of the first areas in the enterprise to receive automation hardware and software. The technology most frequently implemented in the design area was called *CAD* and is defined as follows:

> *CAD is the application of computers and graphics software to aid or enhance the product design from conceptualization to documentation.*

As the definition indicates, CAD technology supports all levels in the product design process. For example, *computer-aided drafting* (*CAD*) automates the drawing or product documentation process, while *computer-aided design* (also *CAD*) is used to increase the productivity of the product designers. The parts, machines, and products illustrated in Figures 3–8 through 3–12 are examples of designs and documentation created with CAD software. The capabilities of the software for the CAD system to design a new product are quite different from the minimum resources required to document the product design. However, in every case, the user refers to the technology as *CAD,* a practice that leads to considerable confusion in defining the technology. Throughout the remainder of the book, the terms *computer-aided design* and *CAD* are used interchangeably when we are referring to the technology used to support all the stages in the design process. In this chapter, we fully describe CAD technology and indicate how CAD systems fit into the CIM enterprise system.

CAD in the Design Model

To understand the importance of CAD in the design model, study Figure 4–1, which indicates how CAD relates to the design process and the processes used in production and manufacturing engineering. Note that human input is a critical part of the design process. Six teams are identified: *product design, analysis, evaluation, documentation development, production engineering,* and *manufacturing engineering.* However, these teams may not be unique because some individuals may be on more than one team.

The process starts with conceptualization of the design by the product design team, which uses CAD to document and display the initial design concepts. The CAD system is also used to expand the product details in the synthesis process. The drawing files and design data are stored in a *product data management* (*PDM*) system, which is the interface between the CAD design and all of the other elements that need to use design data. PDM is introduced in this chapter, but a complete description of the management process is covered in chapter 8. In chapter 8, PDM is tied to other enterprise product and management systems.

The analysis team uses analysis software tools, two-dimensional (2-D) drawing files, and solid-model data of the product to determine the viability of the design. The evaluation team uses drawing data to create operational models of the new product for testing and to develop product cost. The CAD system data are also used in production engineering and manufacturing engineering to create processes

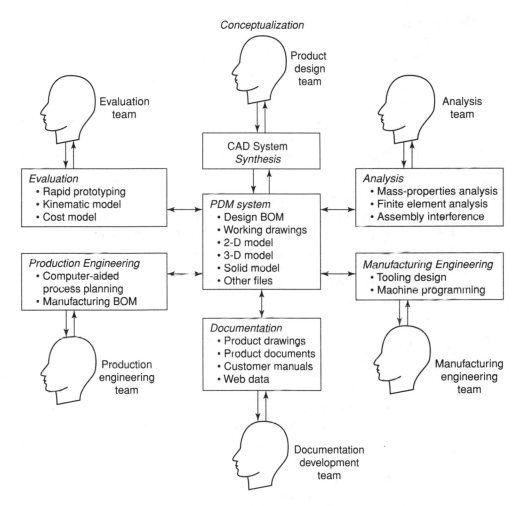

Figure 4–1   CAD Integration of the Design and Manufacturing Elements in the Enterprise.

required for production of the product. Data created in these areas, such as the manufacturing bill of materials (BOM), are saved in the PDM system for use by other departments in the enterprise. For example, purchasing personnel use the design BOM to order parts not manufactured in-house. In the documentation block, product documentation is created with CAD data for in-house needs in marketing and customer service and for external users such as customers. The software needed to support the integration illustrated in this model is described in this chapter and in chapter 5.

It is easy to see from this illustration that CAD is the glue that holds together many of the other elements in the design and production of the product. CAD technology enhances productivity because data are *created once but used many times.*

4–2 THE COST OF PAPER-BASED DESIGN DATA

While CAD has been in use for over thirty years, it is estimated that there are still more than 8 billion paper drawings worldwide, with fewer than 15 percent in a CAD format. These paper-based drawings are a result of companies who are currently not using CAD technology and of drawing documents produced BC (before CAD). Regardless of the reason for their existence, it is clear that there are hidden costs in paper-based design. Handling, storing, and maintaining paper design drawings is difficult, time consuming, and costly for the following reasons:

- All non-CAD media are susceptible to aging.
- Revisions are costly.
- Reproduction and distribution are slower for paper-based documents than for electronically generated CAD documents.
- Many transactions between companies and subcontractors are inefficient because one of the partners does not use CAD, so the process is reduced to a paper-based operation.
- Searching for specific information in paper-based designs is slow and cumbersome because electronic searching of files is not an option.
- Paper-based designs are limited to graphics and text and cannot include hyperlinks, audio, and video.
- Paper-based designs can be out of date even before the distribution is complete because of lengthy release cycles.
- Frequently, departments are working from different revisions of the same paper-based design.
- Facility costs for the storage and maintenance of paper archives can be substantial.
- Product data management software cannot be effectively implemented with a paper-based system.
- Industry estimates indicate that 5–7 percent of the designs are lost or misfiled as a result of the manual procedures for handling paper drawings.

So that these hidden costs can be avoided, all new designs should use the CAD electronic format, and existing paper-based designs that are still active should be converted to an electronic format. The options for converting paper drawings to an electronic format include these:

- Manually redrawing the document in a CAD system is most accurate but also slow and the most costly option.
- Using a digitizing tablet attached to a CAD system to trace over the original design with a puck is prone to errors and is still slow and costly.
- Using a service bureau to make the conversion by using one of the methods listed here reduces internal resource requirements but may pose security problems.

■ Scanning paper drawings into an electronic format compatible with the current CAD environment offers the most control in the transition from paper and has been successfully implemented in both large and small companies to solve integration problems between paper and CAD.

An organization will realize tremendous savings by moving away from paper because 7–10 percent of a company's operating expenditure is spent on manual document management processes and maintaining the paper trail.

4–3 CAD SOFTWARE

The historical time line shown in Figure 4–2 illustrates the significant developments for the CAD technology from its inception in the Sage Project at Massachusetts Institute of Technology (MIT) in 1963. The Sage Project, which focused on the development of CRT (cathode ray tube) displays and computer operating systems, created some of the first interactive computer graphics with a system called *Sketchpad*. The early CAD systems were basic graphics editors with few design symbols; by the end of the 1960s, however, CAD software could create two-dimensional (2-D) and three-dimensional (3-D) part drawings with the bounding edges represented by lines and curves.

CAD software falls into two broad categories—2-D and 3-D—according to the number of dimensions visible in the finished geometry. CAD packages that represent objects in two dimensions are called *2-D software*, while *3-D software* permits the parts to be viewed with the three-dimensional planes—height, width, and depth—visible. In addition, the software at the 3-D level classifies the part geometry as a *wire-frame, surface,* or *solid* model. Each of these classifications is described in the following sections. Use the Web sites for CAD vendors listed in appendix 4–1 to view many examples of the major CAD applications.

CAD: 2-D Wire-Frame Systems

A *wire-frame model* is a CAD drawing in which the intersection of planes in the object is represented by lines and arcs. In a 2-D wire-frame model of a part, only two dimensions of the object are visible in a view. Figures 3–8 and 3–11 are examples of 2-D wire frames. Some 2-D CAD software permits a 2-D surface to be assigned an *extruded depth* value. The result is often called 2½-D because it gives the impression of a third dimension when the object is displayed in an isometric view. The bracket shown in Figure 4–3 is an example of a part illustrated in 2½-D. CAD software limited to 2-D or 2½-D has two distinctive characteristics:

1. The software is just an electronic emulation and extension of basic board drafting and is understood by most people familiar with manual drafting.
2. Stored geometric entities are limited to points, lines, and arcs, so the drawing file size is small.

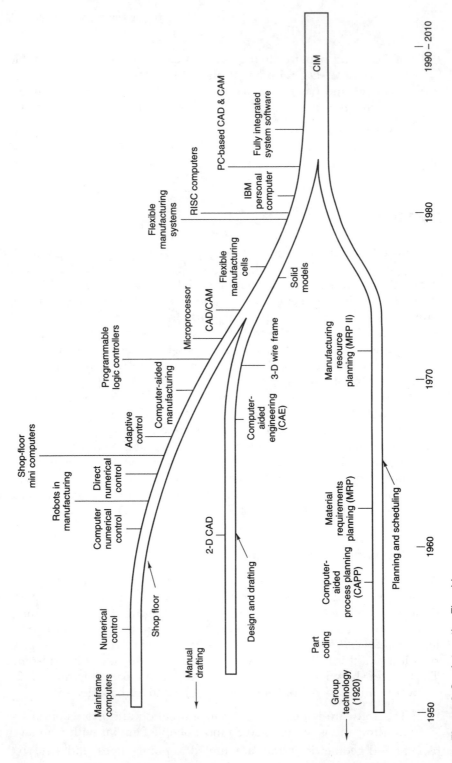

Figure 4–2 Automation Time Line.

Figure 4–3   2½-D CAD Part.

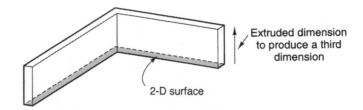

Extruded dimension
to produce a third
dimension

2-D surface

Applications for 2-D software occur in two broad areas: *basic drafting* operations and *part manufacturing.* As a result, estimates from industry sources indicate that over 2 million 2-D CAD systems are in use, with the number increasing each year. Manual drafting has disappeared from most industry sectors because CAD is more productive. CAD wins the productive battle because parts change frequently and drawings are always being modified. The power of CAD to enhance productivity becomes apparent when modifications of the drawings are required.

Two-dimensional part geometry is used to create working drawings for use in manufacturing applications. As mentioned previously, examples of 2-D CAD applications are illustrated in Figures 3–8 and 3–11. In addition, many 2-D CAD files are still used to generate computer programs for automated cutting tools called *computer numerical control* (*CNC*) machines.

CAD: 3-D Wire-Frame Systems

The *3-D wire-frame* system is an extension of the 2-D concept. The primary difference between 2-D and 3-D wire-frame CAD software is the association of a third dimension—usually a z data value—with each point, line, and arc. The 3-D model can be viewed from any angle by rotating the object's geometry around the x-, y-, or z-axis. Additional characteristics of the 3-D model include the following four:

1. A database for 3-D part geometry that is not significantly larger than that for the 2-D model
2. An easy-to-use command set with primitives (points, lines, and arcs) and construction techniques similar to those of the 2-D application
3. A data set that can be used to build a surface model
4. A data set that can be used to generate process programs for automatic cutting machines

The creation of a 3-D wire frame of a part can start with two or three views of the object in 2-D. The operator selects edges of surfaces in the 2-D views, and the 3-D software creates a 3-D wire frame. With the use of AutoCAD software, the plastic pencil sharpener of Figure 3–8 is illustrated in Figure 4–4 as a 3-D wire frame. A second approach to creating a 3-D wire frame uses the drawing commands to create a 3-D wire frame directly—without the aid of 2-D views. Building part geometry requires the application of 3-D lines, arcs, and curves. Notice how

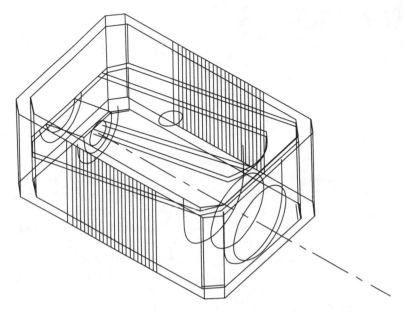

Figure 4–4   3-D Wire Frame of a Pencil Sharpener.

difficult it is to visualize the shape of the object (Figure 4–4) when all the lines in the object are present. The same drawing, with *hidden lines* removed, is illustrated in Figure 4–5; the shape of the object is now easy to see.

The application areas best suited for 3-D wire-frame CAD software include the following:

Figure 4–5   3-D Drawing with Hidden Lines Removed.

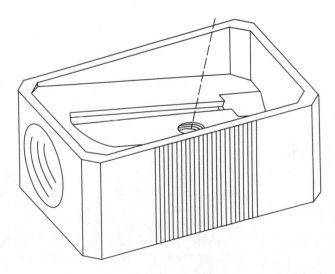

Figure 4–6   Ruled Surface.

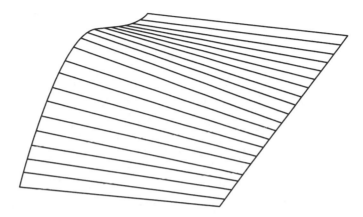

- The drawing of objects with only *planar* surfaces. If the object has a *sculptured* surface, location of the edges is difficult unless the part is viewed from specific angles.
- The drawing of parts with geometric and surface features that make it easy to distinguish open spaces from solid material in the part.
- The drawing of early conceptual designs when rapid creation of the overall part model without many details is desired.
- The drawing of large models when the use of solid models would create a database too large for system resources.

CAD: Surface-Model Software Systems

Objects created with surface-model systems combine the points, lines, and arcs from the 3-D wire frames with surfaces. The surfaces are defined mathematically in three ways:

1. *Ruled surface.* A *ruled surface* (Figure 4–6) is a surface formed by connecting two lines or arcs with straight lines.
2. *Surface of revolution.* A *surface of revolution* (Figure 4–7) is a surface formed by rotating a 2-D drawing of the cross section of the object around the *x-*, *y-*, or *z*-axis.
3. *Sculptured surface, or free-form surface.* A *sculptured surface*, like the fillet connecting the aircraft wing to the fuselage in Figure 4–8, is formed by approximating the complex surface shape with a *surface patch*. The technique used most frequently to form surface patches is based on *Bezier* and *B-spline* curve-generation mathematics. The nonuniform rational B-spline (NURBS) technique is used most frequently for complex surface-patch generation. A full description of B-spline operations is provided in appendix 4–2. Other surface types used by some CAD systems include *C poles*, *S poles*, and *Coons patches*.

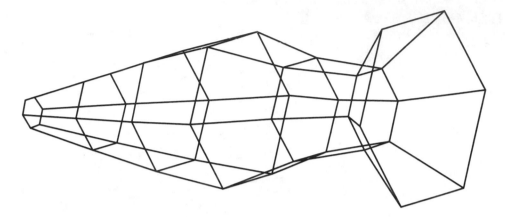

Figure 4–7   Surface of Revolution.

The surfaces of the model are frequently shaded, and hidden lines are removed to highlight features of the complex object to make it easier to visualize. Surface-modeling products may exist as an adjunct to wire-frame systems. Graphic systems used in automotive, aerospace, and in most design applications can provide 2D and wire frame views as needed from the solid model.

     The major application for surface-modeling systems is the design and drawing of objects formed from a collection of irregular surfaces. Examples include automotive body design, television cabinets, liquid soap bottles and containers, and appliances.

CAD: Solid-Model Software Systems

The *solid model* is a *mathematically complete* and *unambiguous* representation of part geometry. This means that the solid model of the object has all the properties of the actual part, including *physical* (size, mass, and material), *mechanical* (strength and elasticity), *electrical* (resistance, capacitance, and inductance), and *thermal*

Figure 4–8   Sculptured Surface.

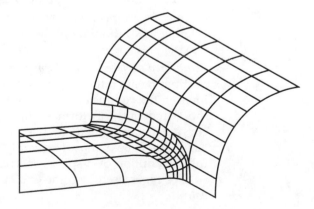

Figure 4–9 CAD Solid Model
Design of the SuperSoaker Water
Gun Designed and Built from ABS
Plastic.
(Source: Courtesy of Stratasys, Inc.)

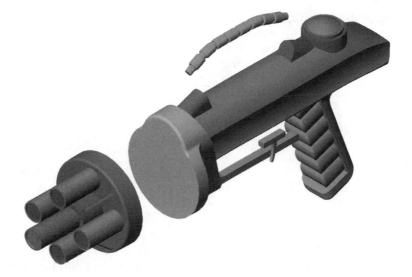

(conductivity and expansion coefficient) properties. Therefore, the part model
has all the characteristics of the real-world object. As a result, a test performed on
the solid-part model provides information on the performance of the actual part.
As a result of the benefits of solid models, industry data indicate that 60 percent
of the over 2 million 2-D CAD stations currently used will be moving to 3-D tools,
mostly solid-modeling software. The computer-aided engineering (CAE) applications described in chapter 5 illustrate how the properties of solid models are
tested in simulated systems. An example of a CAD solid model is provided in
Figure 4–9.

Two techniques—*constructive solid geometry* (*CSG*) and *boundary representation* (*B-rep*)—are used by a majority of the software vendors to create solid-model
CAD images. The product designer can use the two techniques individually on a
part or in a hybrid, or mixed, mode where the benefits of each technique are available. A third solid-modeling technique, *sweeping,* is used for certain types of geometry. Less frequently used techniques include *pure primitives, spatial occupancy,* and
cell decomposition. The operation of each type of the three most frequently used
solid modelers is described next.

Constructive Solid Geometry. The *CSG software,* sometimes called *explicit ma-
nipulations,* builds a solid geometry from a primitive-shape library. The library includes many basic object shapes, including *blocks, cubes, cylinders, cones, spheres,
wedges,* and *tori.* The shapes are combined by using the *union, intersection,* and *difference* Boolean operators. The example shown in Figure 4–10 illustrates this
process. Note that the two full boxes are combined by using the union operator to
create the base of the part. The cylinder and the block use the same union operator to create the pin on the part. Finally, the difference operator subtracts the

Figure 4–10 CSG Solid
Modeler.

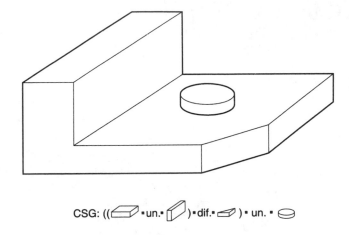

CSG: ((⬡ • un. • ⬡) • dif. • ⬡) • un. • ⬭

wedge from the base for the complete solid-part model. The CSG tree structure and compact code (Figure 4–11) produce an efficient database for the solid object. The CSG technique is the quickest and easiest method to use to produce solid-part models. However, many shapes, such as automotive body designs, cannot be created with a CSG modeler.

Boundary Representation.   The *B-rep solid modelers,* sometimes called *parametric modelers,* represent objects by describing the bounding faces. The edges, curves, and points in the faces are then defined. The same object modeled in Figure 4–10 with the CSG techniques is used in Figure 4–12 to illustrate the B-rep approach.

Figure 4–11 CSG Tree and
Code.

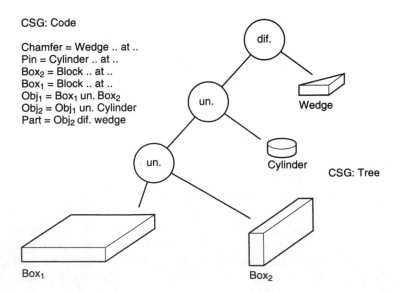

CSG: Code

Chamfer = Wedge .. at ..
Pin = Cylinder .. at ..
Box_2 = Block .. at ..
Box_1 = Block .. at ..
Obj_1 = Box_1 un. Box_2
Obj_2 = Obj_1 un. Cylinder
Part = Obj_2 dif. wedge

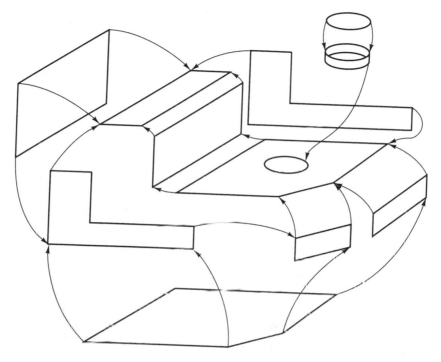

Figure 4-12   B-rep Solid Model.

The faces or surfaces are usually defined by using Bezier, B-spline, or NURBS surface-modeling techniques. (The NURBS technique is described in appendix 4–2.) As a result, the B-rep solid modeler can represent any object, regardless of surface contour complexity. However, it is difficult and complex to create a B-rep solid model, and the vector file for the completed model is usually quite large.

Sweeping.   The sweeping type of solid modeler is used for cutter path simulation. The two types of sweeping actions—translation and rotation—are good for certain types of geometry. The sweeping actions required to produce a rectangle and cylinder are illustrated in Figure 4–13.

Alternatives.   Solid-modeling software offers several alternative methods for creating and manipulating solid geometry, including *explicit, parametric, variational,* and *features-based* modeling. When applied to solid modeling, parametric analysis is associated with one- and two-way *associativity.*

Solid Modeling: Drawing Generation and Other Features.   Solid-model CAD systems offer significant advantages over traditional 2-D, 3-D, and surface-generating systems. After a part is modeled as a solid, the CAD software can automatically create other graphic renderings for the part, such as 2-D three-view

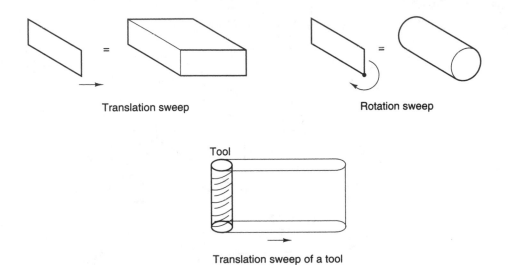

Translation sweep   Rotation sweep

Tool

Translation sweep of a tool

Figure 4–13   Sweeping Solid Modeler.

working drawings. When this conversion is performed, the new drawing sets have one- and two-way associativity with the original solid representation. *Associativity* means that a relationship is established between the solid model of a part and the 2-D working drawings prepared for the manufacturing area. If a dimension is changed on the solid model, all of the 2-D shop drawings are updated automatically. The opposite is also true: a change in some of the dimensions in the 2-D drawing would be reflected in the solid representation of the part. This reverse direction is important because changes in the solid model would normally have to be made by using the CSG or B-rep functions in the software. When associativity is present, the drivers controlling model shape can be changed on the 2-D drawing, and the solid model responds accordingly.

The *variational* characteristic is a subset of the parametric model. When variational capability is present, the user can add dimensional and geometric constraints without adhering to an exact sequential definition for the geometry and constraints. *Features-based* modeling is another subset of parametric analysis that permits the use of a command language, including features such as holes to build part geometry. This type of modeling is used most frequently for holes, bosses, and rounds. In practice, an operator can use features-based commands to build a countersunk hole by specifying the geometry desired and then providing the appropriate radii and depth. Features-based design currently focuses primarily on design requirements and does not support manufacturing-related operations.

The hybrid system that combines the CSG and B-rep techniques offers the best overall solid-modeling solution. For objects without contour surfaces, the CSG works well; when a contour surface is encountered, the B-rep portion of the modeler is used. Another type of hybrid system combines the surface-creation techniques found in surface modeling with solid-modeling products. Solid-modeling software provides many benefits, including calculation of the *mass properties, true cross sections, interference checking between models of parts, photo-realistic graphics of surfaces* and *finishes,* and *seamless exchange* of data to downstream applications such as *finite-element analysis, generation of numerical control code,* and *rapid prototyping.*

Solid-model software is required when the drawing database for the part must include the real-world characteristics of the object. Examples that require solid models include the following:

- Operational testing and simulation of parts and assemblies by using the physical, mechanical, electrical, and thermodynamic properties present in the part database
- Process planning of manufacturing operations by using computer software

The CAD Software Market

The CAD software market is very competitive; as a result, the software vendors are continuously improving the product and offering additional features. For example, Autodesk released over sixteen versions of the popular AutoCAD software in less than seventeen years. Each new release featured enhanced features plus a larger and more powerful set of drawing commands. In addition, new releases took advantage of the increased computing power of the latest Intel microprocessor in the PC hardware. While changes and continuous improvement in the CAD software are unsettling, they result in a steady decrease in the price-to-performance ratio. This increased performance at less cost permits many industries to use CAD technology as a first step toward CIM.

4–4 CAD: YESTERDAY, TODAY, AND TOMORROW

The product design and the manufacturing operations productivity provided by CAD is a requirement in this competitive global environment. However, as with most technology today, it is ever changing and demands careful planning since the effectiveness of the enterprise's shared database is affected by the choice of software selected. Therefore, a look at the past, present, and future in the design software area provides insight for these choices.

CAD Systems: Yesterday

In the early 1970s, the CAD software and the system hardware were typically developed by the same vendor and sold as a "turnkey" system. The system in Figure 4–14 illustrates a typical mainframe CAD system from the early 1970s. The vendor's hardware was a general-purpose mainframe computer or minicomputer connected to graphics displays and user-interaction devices, such as a mouse or a digitizing tablet for project development. The proprietary CAD software ran on a proprietary operating system and had vertically integrated add-on applications for specific industries, such as cutting code generation for computer-controlled machine tools. The system architecture was dictated by the computing power required for the CAD software. With one vendor supplying all parts of the system,

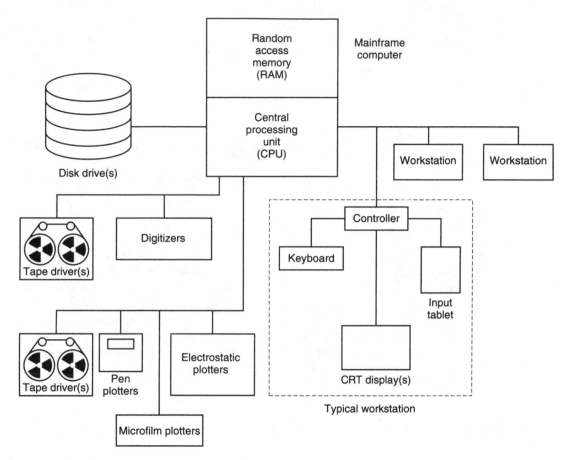

Figure 4–14   Mainframe Computer–Based CAD System.

integration and operation of the system was quite smooth. However, the cost for each CAD design station was the total cost of the system divided by the number of designers, called *CAD seats,* that were attached to the mainframe. This cost was typically $100,000 per designer.

CAD Software: Today

The migration from the closed and proprietary turnkey model to a more open model occurred in the mid 1980s. This change was primarily driven by the low-cost graphics hardware platforms entering the market. Lower-cost, general-purpose graphics systems started with UNIX vendors in the early 1980s, but by the late 1980s, both PC and Macintosh hardware had enhanced systems with the graphics capabilities required to run CAD applications.

The new CAD model that replaced the turnkey model was called the *Virtual Company.* Instead of CAD's being linked to a single proprietary system, it became available for all major hardware platforms. Also, third-party developers used the application program interfaces (APIs) in the PC Windows environment to link add-on applications to the basic CAD software. As a result, end users using software from a variety of companies developed robust CAD applications. In this Virtual Company model, the CAD vendor offered a basic functionality in CAD and worked with the hardware vendors to offer a portable environment. Third parties, with expertise in vertical niche areas, then developed software to port to the CAD vendor's platform for special applications. This technique had a significant impact on the cost of each CAD station. The Virtual Company is based on the synergy between the partners in the process—the application developers and the CAD vendor. Cost is reduced because each group works on the area it knows best. For example, an application developer such as ALGOR, Inc., can concentrate on development of finite-element analysis software using CAD drawing data without worrying about developing and maintaining the core graphics technology. The CAD vendor benefits by just focusing on the underlying CAD technology without the need to develop vertical market products for specific customers. The competition from competing application developers leads to lower-cost applications with greater potential for productivity gains. The number of 3-D and dimension-driven solid-modeling systems in use in the year 2000 is illustrated in Figures 4–15 and 4–16. The percentages of CAD seats by vendor are included.

The Virtual Company is the model in use today for mechanical CAD systems. While there are some problems associated with reliable data exchange between applications and interoperability, the problems are not severe.

CAD Hardware: Yesterday and Today

The turnkey mainframe computer systems used in early CAD implementations were produced by the top five giants of the era: Computervision, Applicon Ltd., Calma, Auto-trol Technology Corporation, and Intergraph. Today, Intergraph and

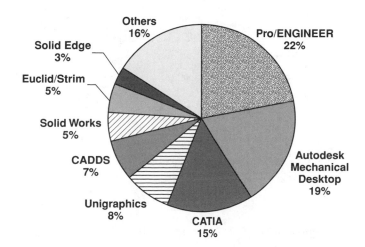

Figure 4–15 3-D Mechanical Design Software Sold by Brand (750,000 Systems Total).

Figure 4–16 Dimension-Driven Solid-Modeling Systems Sold by Brand (430,000 Systems Total).

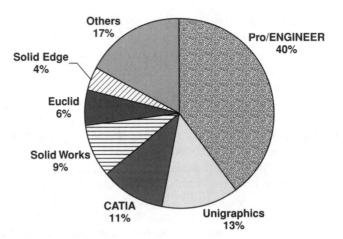

Auto-trol are mere shadows of their former selves, while more successful companies acquired the other three. Throughout the 1980s, mainframe CAD disappeared and CAD work shifted to single-user computers, each equipped with its own central processor, disk storage, and video card. An illustration of the single-user CAD workstation is provided in Figure 4–17. The single-user systems were faster and less costly than the shared mainframe systems. Prices for a CAD station dropped from nearly $100,000 in 1981 to less than $20,000 a decade later.

As CAD moved to stand-alone workstations, the industry started to segment into specialty niches. For example, hardware and software developers were in different companies, electrical and mechanical CAD separated into distinct industries,

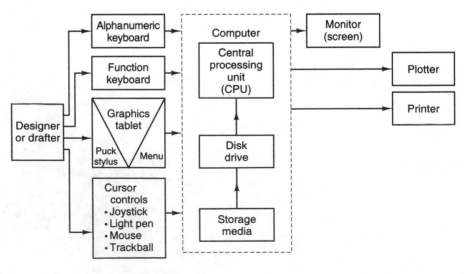

Figure 4–17 CAD System Block Diagram.

and the single-user computer industry divided into two groups: the IBM PC and other computer systems using Motorola microprocessors. The IBM PC used the Intel X microprocessor with the Microsoft DOS/Windows operating systems. The IBM PC was both the low-cost and the low-technology CAD solution. A host of other computers used Motorola computer chips and ran on AT&T's UNIX, Apple's Macintosh, and Apollo's DomainOS operating systems. The Motorola-based machines were better suited for higher-end CAD applications.

The separate domains continued to develop through the early 1990s. Intel increased the power and speed of X processors, and a new processor, called *RISC (reduced instruction set computer)*, was introduced to replace the Motorola microprocessor. The Intel X processor was a *CISC*, or *complex instruction set computer*. The RISC machines, with a specially selected instruction set, could perform the more complex mathematical routines demanded by CAD faster. As a result, most high-end CAD applications were developed to run on RISC-based computers called *engineering workstations*.

However, by the mid-1990s, advances in semiconductor manufacturing allowed Intel's processors, now called *Pentiums*, to match the speed and calculation capability of the RISC chips. As transistors became smaller, cheaper, and faster, Intel packed more of them into each microprocessor and raised the clock speeds. Intel chip sets were widely used by every PC builder. As a result, Intel was able to spread its higher design costs over hundreds of millions of units sold at commodity prices to numerous PC builders. By 1997, the PC became the platform of choice for most mechanical CAD work, and all the major CAD software vendors were porting their solutions over to the PC running Microsoft's NT operating system. The move to the PC platform has significantly reduced the entry cost for CAD users. The graphic in Figure 4–18 illustrates the changes in workstation cost during the period indicated. The first seven are for Sun Corporation RISC-based computers, the 1998 data point is for a Hewlett-Packard Kayak workstation with a 400-megahertz Pentium II processor, and the last two are Dell Computer Corporation machines with Pentium 3 and 4 processors, respectively. The CAD workstation cost has not changed significantly in the last five years.

Figure 4–18 3-D Workstation Prices, 1991–2000.

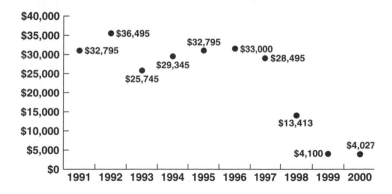

Today the Intel Pentium and the AMD equivalent are the processors that support most new installations of CAD from 2-D to solid modeling. A majority of the CAD software vendors have a version of their software that runs on a Microsoft operating system. While the RISC workstations remain strong in integrated circuit design applications and in transaction-server settings, Windows NT, 2000, and XP now dominate mechanical engineering, printed circuit design, and other graphical disciplines. The cost of CAD today makes it available to users in manufacturing regardless of company size and type of product produced. If CAD can improve productivity, then it is affordable.

CAD Hardware and Software: The Next Ten Years

Foreseeing the future is an inexact art, and it is not any easier today than it would have been in 1990. The technologies that would shape the 1990s were in place when that decade started. However, who could have predicted that new start-up Pro/ENGINEER would preempt industry leaders IBM and Computervision in mechanical design? Microsoft's Windows operating system was in version 1.0, but few people would have predicted that it would surpass UNIX and become the dominant platform for technical computing. Tim Berners-Lee had proposed the World Wide Web, but few persons envisioned the effect it would have on the world of computing in just ten short years. Future developments in software may use web-based tools that will enable the graphics to be available using any processor.

Future Technology Trends

The following seven trends should impact the direction of CAD in the next ten years:

First, computers will have multiple processors, and parallel processing will add new power and speed to the desktop.

How soon this is accomplished depends on the answers to a number of implementation questions now being resolved in corporate and university laboratories. Such questions include these: How can problems be divided across parallel machines? How many processors are optimal? How should workstations be configured versus the network server? How should application software be reorganized?

Second, the microprocessor transistor count and speed will continue to increase, and multiple processor chips will emerge.

Moore's law, named for Gordon Moore, the founder of Intel, states that the number of transistors in microprocessors doubles every two years. If Moore's law holds, then the Pentium 4, with 42 million transistors in 2000, must grow to at least 800 million by 2010 (Figure 4–19 illustrates this growth). With this device density, transistors will be just fifty silicon atoms wide. There is some question why applications would need that many devices or speeds in the 40-gigahertz range. There is no doubt that the device count and speed will increase, but the final numbers are still uncertain.

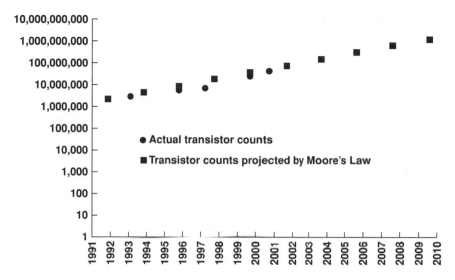

Figure 4–19 Transistor Counts in Intel Microprocessors.

This second advance—symmetric multiprocessing (SMP) workstations with two or more microprocessors sharing a single memory pool—will more than likely be employed in some format. It could also look like load-sharing software that distributes application tasks to computing farms with dozens of rack-mounted workstations. Intel may also choose to make highly parallel architectures such as Itanium or multiple Pentium-style chips on a single die or package. IBM currently has the POWER4, which is two processor chips on one module.

Third, the Internet and the World Wide Web will continue to impact business in general and product design using CAD and PDM in particular.

High-speed communications enables engineers to employ computing resources across the street or across the continent, and that trend will continue to expand. The Internet will provoke a growing demand for processing cycles and other computer resources, including storage and communication links. There is little doubt that some of that resource growth will occur in the Intel/Microsoft desktop computer system. The "extended PC" already is the natural hub for storing and processing data for devices such as PalmPilots, wireless telephones, portable computers, and digital cameras. The next evolution for CAD is a little more difficult to predict, but software technology has a way of consuming hardware resources. Given more re-sources, CAD software vendors will consume them. The resources will include both powerful stand-alone workstations and computing farms built from dozens or hun-dreds of inexpensive computers, or even computing farms of SMP systems. Since it will be difficult to predict demand levels, such systems have an advantage in being easier and cheaper to scale up and manage.

Another technology that will impact the design arena is Microsoft's .NET (pronounced "dot net") Internet programming environment. This innovation allows for the development of application programs that run on the Internet and allows multiple access to the target product. A good example is the CoCreate collaboration product OneSpace.net. This product is designed to allow project team members to manage projects, share ideas, and communicate among diverse product development teams and outsourcing partners. Built on Microsoft .NET technology, OneSpace.net would be a valuable asset for a concurrent engineering team.

Fourth, Linux, a UNIX-like operating system, will compete with Microsoft's latest release in a number of market areas.

Linux is an operating system kernel developed by Linus Torvalds as a student at the University of Helsinki, Finland, and is distributed as freeware. The Internet community of software developers has taken the original kernel to the level of a fully functional language with public, or "open," source code. It is estimated that there are over 10 million Linux installations worldwide, and Linux accounted for 17 percent of all server operating system shipments in 2000. All of the major hardware vendors are offering Linux as an operating system option on new machines, and many of the CAD software vendors are porting their software to Linux as well. If Linux avoids the fragmentation that hampered the growth of UNIX, it should be a player in the CAD operating system marketplace in the future.

Fifth, CAD software will support a "virtual mock-up" that enables a large number of component parts of a product to be assembled in a virtual world for analysis of the design.

Today, CAD systems do not permit several designers to easily share the same model. In the next ten years, virtual mock-ups will replace physical prototypes and permit part files from different-vendor CAD software programs to be integrated into the same model.

Sixth, data sharing between different CAD software programs will be possible.

In the next ten years, major CAD systems will be able to open models from most other major vendors. As a result, a designer using one type of CAD software could supply models to customers or part vendors for use in computer-based mock-ups, fixture design, or mold work, or for use in making engineering drawings with another CAD system. With this standardization, companies will offer CAD models of engineered components, such as motors, mechanical drives, and actuators. As a result, component manufacturers will supply models of half of the parts needed to create a product. CAD models or drawings of component parts, such as pneumatic cylinders, are now offered by pneumatic product vendors.

Seventh, continued consolidation will take place in the industry.

Tremendous consolidation took place in the 1990s in CAD and in the whole software industry. Examples include Microsoft in office automation and operating systems and Oracle, Microsoft, and IBM in relational database management software. In ten years, the cost of solid- and surface-modeling 3-D systems with integrated drafting should be about the same price as AutoCAD is today, $2,900 a seat. The likely

CAD/CAM survivors should include Dassault Systèmes (CATIA and SolidWorks) and Parametric Technology Corporation (PTC; Pro/ENGINEER). However, the major software companies today—Microsoft, Oracle, and SAP—may move into the CAD/CAM arena through acquisitions if any of the current leaders make a slip.

4-5 APPLICATION OF CAD TO MANUFACTURING SYSTEMS

A study of over 300 firms conducted by the Industrial Technology Institute in Ann Arbor, Michigan, indicated that CAD is used most often in two application areas: *concept and repetitive design* and *drafting*.

Concept and Repetitive Design

CAD software is used for a large percentage of the design function in manufacturing, including the design of the product plus the design of all the systems required to support the production process, such as fixtures for machining, assembly, or quality checking; gauges; and material-handling pallets. CAD and the design functions form a natural union because the part geometry is used frequently in the production process. For example, the curve surface data created on the CAD system during the design of a shampoo bottle can be sent to the mold manufacturer through electronic data interchange (EDI) for use in machining the injection molds needed to produce the bottles. The design of the plastic injection mold for the bottle is less expensive and faster because the surface geometry is directly available from the CAD database. The design data captured in CAD are reused often by many other departments and systems in the manufacturing process.

Drafting

The second major application for CAD is in the creation of all the working drawings required for product manufacturing. In most organizations, CAD was first introduced in this area as a replacement for manual drafting. In addition, CAD is used to create the documents that will be referenced during design verification and ordering of raw material and component parts. Manufacturing support departments require CAD as well; for example, facility planning, maintenance, and security make use of CAD to develop visual presentations of data that describe their operation.

4-6 SELECTING CAD SOFTWARE FOR AN ENTERPRISE

Criteria for Selecting CAD Software

Five criteria are frequently used to define the type(s) of CAD system(s) a company will use in an integrated solution. The selection process for a CIM-based CAD system becomes clear after each of these five criteria is examined. Every criterion affects both the hardware platform choice and the software solution.

Type and Complexity of Part Geometry and Drawings.   On the working drawing, a simple cross-sectional view of a complex geometry such as an aircraft wing rib can be represented easily by using a 2-D CAD system. However, a NURBS-based surface or solid modeler is necessary to draw the rib (Figure 4–8) as a pictorial. Factors that affect complexity are as follows:

- *Degree of axisymmetry.* Objects that are symmetric, or the same about one or more axes, can be represented with less-sophisticated CAD software.
- *Types of curved surfaces.* If the curves are circular or spherical, the object is less complex and only 2-D software is required. However, if conic or quadric surface data are present, the geometry is complex and the software must match.
- *Interior complexity.* An object with detailed interior surfaces represented with a 3-D wire frame is hopelessly confusing. Such a part would require software with a surface- or solid-modeling capability.
- *Parts in the assembly.* Objects with a large number of parts require analysis of fit, tolerance, and interference; therefore, the demand on the CAD software increases.

Concept Design Versus Repetitive Design.   The type of design practiced by a company is a major factor in the selection of a CAD system. Repetitive-design needs are often satisfied on a 2-D system, while concept design generally demands a surface or solid modeler with CSG and B-rep capability. In addition, the CAD software supporting the repetitive-design function should have a parametric-design capability.

Size and Complexity of the Product.   The size and complexity of the finished product affects the type of CAD system needed because big projects require several designers who each work on separate parts of the same design. One of the key considerations with this criterion is the system response time. In a study conducted by IBM, an improvement in the response time of a CAD system from one-half second to one-quarter second doubled the number of commands performed on the system and thus the productivity. Examples of complex projects include commercial aircraft, oil refineries, office buildings, bridges, cars, and microprocessors.

Enterprise Data Interfaces.   One major CIM axiom is "Create data once and use it many times." CAD is affected heavily by this precept because CAD is where most of the product geometry data originate. The CAD hardware and software must support file formats that permit the part geometry and specification data attached to the drawing to be used by (1) marketing for product brochures and manuals; (2) manufacturing engineering for CNC code generation, engineering change orders, routing, process planning, and quality analysis; and (3) production planning and control for product structure and part specification.

External Data Interfaces. Electronic data interchange (EDI), a technology that transfers data from one computer to another, is critical for vendors working with large companies. Two factors in EDI affect the CAD selection: (1) the type of EDI interface between the contractor's and vendor's computers, and (2) the transferability of the part geometry data file between the CAD software used by the contractor and that used by the vendor. Vendor certification with large companies or the federal government often dictates the EDI standard and CAD software the supplier must use.

The Varying Need for CAD Software

Figure 4–20 indicates how the need for the different types of CAD software varies in the different manufacturing systems and design processes introduced in chapter 2. The final comparison chart (Figure 4–21) indicates the type of CAD software recommended for different types of drawings and manufacturing interfaces. Take a few minutes and study these charts. Keep in mind that these are general recommendations and comparisons; in practice, there are always exceptions that increase or decrease the requirements for the CAD hardware and software used in

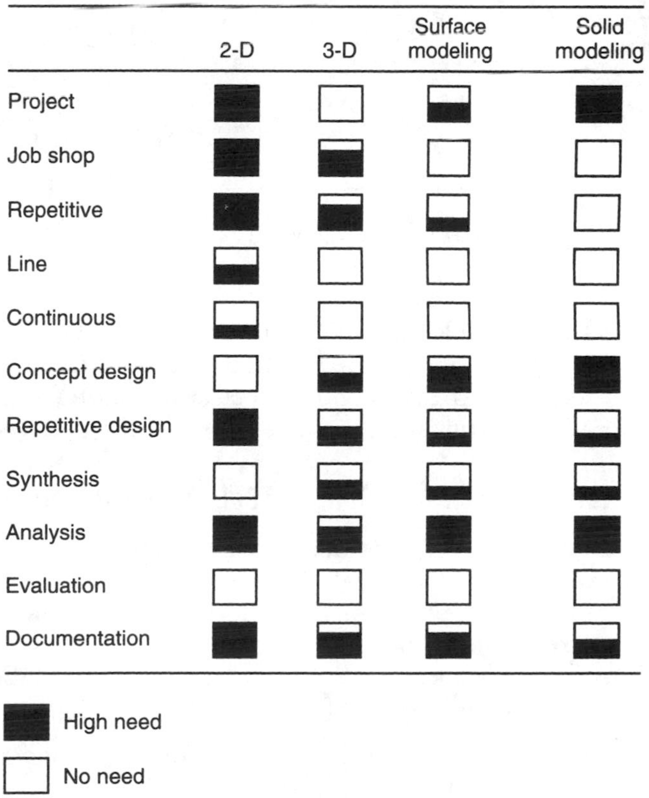

Figure 4–20 Need for CAD Stations in Various Types of Manufacturing and the Product Design Processes.

Part, drawing, and interface characteristics	Wire Frame		Surface and solid models			
	2-D	3-D	Ruled/revolution	UBS	NUBS	NURBS
Working drawings	•					
Assembly drawings	•			•	•	•
Pictorial drawings						
Normal surfaces		•	•			
Inclined surfaces		•	•	•		
Oblique surfaces		•	•			
Quadric surfaces						•
Simple sections			•	•	•	
Conic sections						•
Technical illustrations		•	•	•	•	•
Welding/drafting	•					
Surface development drafting	•	•	•	•	•	•
Architectural and structural drafting	•	•	•			
Map drafting	•	•				
Electrical and electronic drafting	•					
Aerospace drafting	•	•		•	•	•
CAM interface		•		•	•	•
DTP interface	•	•	•	•	•	•
MPC interface	•	•				
CMM interface		•				
CAE interface		•		•	•	•

Figure 4–21   Type of CAD Software for Modeling the Range of Part Geometries and Manufacturing Interfaces.

a specific application. Note that the boxes in Figure 4–20 indicate the level of need for CAD technology. A filled box indicates CAD would be highly necessary for product success. In contrast, a clear box indicates little need for CAD and a partially filled box indicates a CAD need falling between the two extremes.

Guidelines for CAD Software Selection and Training

Data collected by Anderson Consulting indicate that engineers and designers rarely devote more than fifteen hours per week (Figure 4–22) to the development of design graphics. As a result, they are part-time CAD users. CAD systems in general are designed for full-time users, especially the surface- and solid-model software required for many design applications. So that you use the CAD software efficiently, do the following:

1. Select the CAD system and software options that ensure the best fit available between the users and the CAD tools.

Figure 4–22 Time Study of a Design Engineer's Week.

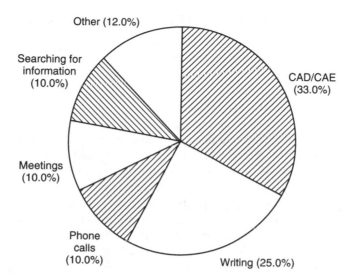

Other (12.0%)

Searching for information (10.0%)

CAD/CAE (33.0%)

Meetings (10.0%)

Phone calls (10.0%)

Writing (25.0%)

2. Customize the CAD software for each user or application area so that it meets specific needs. For example, a designer could be provided with macros that take care of the tedious and repetitive operations.

3. Select CAD software with a good history of reliability. In a survey of CAD users about the cause of lost productivity, 26 percent listed CAD software problems and defects, the second highest item listed. Only 2 percent picked computer failures, and 4 percent listed operating system failure.

4. Even with well-matched systems and users, customized support programs, and easier-to-use software, the CAD systems will remain complex. Therefore, ensure that all users obtain regular education.

In the productivity survey mentioned in item 3 in the preceding list, users indicated that the greatest cause (32 percent of responses) of productivity loss is operator ignorance or failure to follow procedures correctly. CAD continuing education must be a part of the job description for CAD designers if the expected CAD productivity is to be realized. Following are some guidelines:

- Ensure that every new user and users moving from 2-D to 3-D systems attend CAD software training.
- Hold a short training session on the new features of each new release of software.
- Schedule regular but brief productivity meetings on procedures, solutions to software defects, and CAD techniques for the company products.
- Develop online resources to help educate and inform the design group.
- Attend local and national user-group meetings for each CAD supplier.

■ Develop and document design automation tools specific to the company's processes or procedures.
■ Budget for training.

4–7 PRODUCT DATA MANAGEMENT

PDM systems are designed to organize and support the shared database so critical to a CIM enterprise. CAD vendors realized early that managing all of the CAD files associated with the parts of a product needed some type of automated system. Some vendors built the PDM function into the CAD system software, while others used software from other software vendors that interface with the CAD system data. Since the CAD data were not the only items that needed to be efficiently filed and retrieved during production, the large software companies developed PDM for the enterprise. The six largest PDM companies in 2001 are listed in Figure 4–23 along with the name of their data management software. Note that two of the companies, Dassault Systèmes and PTC, started out as major suppliers of high-end CAD systems.

PDM is sometimes called *product lifecycle management (PLM)* or *product development management (PDM)* in order to describe the function of the software. PDM systems store and organize CAD data, but this is not the only function performed. The primary purpose of PDM is to control a corporation's master engineering BOM. The BOM is a parts list that is organized by product with a structure that includes all major subsystems. The list for each of the subsystems contains the individual parts and materials used to make the product. For example, the subsystems for a computer would include the main processor board, power supply, memory boards, case, monitor, and drives. Each subsystem could be divided into individual parts. For instance, the power supply would include the transformer, the regulator board, the filter, and other necessary parts.

The PDM system stores and organizes information about each part or subassembly in a hierarchical system called a *product structure*. The technique used is called a *relational database* since part data are linked to numerous files related to the part, assembly, and product. The links could be to CAD models, CAD drawings,

Figure 4–23 Major Vendors Supplying PDM Software.

Company	Enterprise PDM product
Electronic Data Systems (EDS)	Teamcenter Enterprise
Dassault Systèmes	ENOVIA PM
Parametric Technology Corporation (PTC)	Windchill
SAP	SAP PLM
MatrixOne	eMatrix
Eigner	Eigner PLM

vendor specifications, or analyses results. On the larger scale, enterprise PDM vendors have links to enterprise resources planning (ERP) and many functions outside the design and manufacturing areas.

Data Interfaces

One of the terms used for PDM is *product development management* and it addresses the seamless sharing of CAD and other product and engineering data across the enterprise. The process includes (1) creating the product with CAD, (2) converting the CAD part geometry and attributes file to the format required by other departments, and (3) saving the different versions of the drawing files in the product structure of the PDM's relational database. For example, Figure 4–1 indicates the many areas and subsequent file formats that must be generated just in design. The enterprise units that frequently use part geometry and specifications created in CAD are production machines in manufacturing, manufacturing resource planning (MRP II) software systems in production planning and control, and text and graphics documentation software systems in marketing and other front office departments. In addition, most of the CAE software described in chapter 5 starts with CAD geometry and attributes.

Integrating product data within the extended enterprise will liberate an organization's overall design and manufacturing process. Studies indicate that for every CAD user in a manufacturing enterprise, nine additional nondesign personnel review, approve, and use the same product data. Enlarging the circle of CAD data users is hampered by the many different computer systems and software applications that must interface with the CAD product data. The solution taken by many CAD product leaders (see top ten CAD companies listed in Figure 4–24) is to use the Internet and the intranet as the common interface for all PDM files. Companies like PTC, IBM/Dassault Systèmes, SDRC, and Unigraphics have

1981	1991	2000
Computervision	IBM	Autodesk
Applicon Ltd.	Intergraph	Cadence
GE/Calma	Computervision	Dassault Systèmes
Auto-trol Technology Corporation	Mentor Graphics	Synopsys
Intergraph	Hewlett-Packard	PTC
McDonnell Douglas Automation Company (Unigraphics)	EDS/Unigraphics	Mentor Graphics
Gerber Scientific	Autodesk	Unigraphics Solutions
CalComp	Cadence	Avanti
Summagraphics	Schlumberger (Applicon)	SDRC
IBM	Valid Logic	MSC Software

Figure 4–24   Top Ten CAD and CAE Companies, Ranked by Sales.

all released CAD support products that give the corporate users with a standard Web browser a window into critical product data. PDM supports the need to "create data once and use it many times."

4–8 SUMMARY

Computer-aided design (CAD) is used at all levels in the design process, with the heaviest use in the design documentation area. The technology was started at MIT in 1963 during the Sage Project when the Sketchpad system was developed. The CAD software capability increased significantly in the 1970s with the introduction of the 3-D capability. The introduction of solid-modeling software, powerful engineering workstations, CAD hardware platforms, and powerful PC-based CAD software caused the use of CAD applications to increase exponentially in the 1980s.

The basic CAD system includes the items listed in the block diagram in Figure 4–17. Two components, a hardware platform and CAD software, form the basic CAD production station. The hardware platform includes a basic computer system with additional peripherals to expedite the entry of part geometry data. The computer systems used to produce the CAD database fall into three categories: mainframes, RISC-based engineering workstations, and microcomputers. The variation in CAD software is dictated by the type of computer platform chosen for the application. The capability of the software to handle complex part geometry increases with hardware power. For example, basic or entry-level CAD software running on a small microcomputer is restricted to 2-D drawing applications. In contrast, Pro/ENGINEER software running on a RISC/UNIX workstation could produce a surface or solid model of a complex shape such as the skin of a commercial aircraft. CAD software is divided into the following classifications: 2-D, 2½-D, 3-D, surface-model, and solid-modeling software.

A study of CAD applications indicates that the software is used in two industry application areas: design (concept and repetitive) and drafting. The selection of CAD software systems to support these two application areas requires analysis of five criteria: (1) type and complexity of part geometry and drawings, (2) concept design versus repetitive design, (3) size and complexity of the product, (4) enterprise data interfaces, and (5) external data interfaces. Using these criteria allows the CAD software to be matched to the needs of the CIM system, the department, the product, and the user.

The power of CAD systems offers tremendous productivity opportunities; however, using powerful graphics software is often difficult. The workplace must support the part-time CAD users at the upper end of the design process through better training and better design aids. The software, especially programs at the high end, must become more intuitive.

Product data management, or PDM, systems are designed to organize and support the shared database so critical to a CIM enterprise. PDM functionality is built into some CAD systems but may also be offered as a stand-alone software package. PDM is sometimes called *product lifecycle management (PLM)* or *product development management (PDM)* as a way to describe the function of the software. PDM systems store and organize CAD data, but their primary purpose is to control a corporation's master engineering bill of material (BOM). The BOM is a parts list that is organized by product with a product structure that includes all major subsystems and is stored in a relational database.

When the term *product development management* is used for PDM, it addresses the seamless sharing of CAD and other product and engineering data across the enterprise. The process includes (1) creating the product with CAD, (2) converting the CAD part geometry and attributes file to the format required by other departments, and (3) saving the different versions of the drawing files in the product structure of the PDM's relational database.

BIBLIOGRAPHY

ARNSDORF, D. R. *Technology Application Guide: CAD—Computer-Aided Design.* Ann Arbor, MI: Industrial Technology Institute, 1989.

CHALMERS, C. E. "Windows and the Web: Leveraging CAD Across the Enterprise." *Integrated Manufacturing Solutions,* March 1999, 26–31.

CHANG, T. C., R. A. WYSK, and H. P. WANG. *Computer-Aided Manufacturing.* Upper Saddle River, NJ: Prentice Hall, 1991.

GRAHAM, G. A. *Automation Encyclopedia.* Dearborn, MI: Society of Manufacturing Engineers, 1988.

KUTTNER, B. C., and M. A. LACHANCE. "NURBS: CAD by the Numbers." *Actionline,* January 1991, 20–23.

MITCHELL, F. H., Jr. *CIM Systems: An Introduction to Computer-Integrated Manufacturing.* Upper Saddle River, NJ: Prentice Hall, 1991.

SOBCZAK, T. V. *A Glossary of Terms for Computer-Integrated Manufacturing.* Dearborn, MI: CASA of SME, 1984.

VERSPRILLE, K. "Which CAD Is Right for You." *Integrated Manufacturing Solutions,* March 1999, 1–9.

QUESTIONS

1. Define *CAD.*
2. Describe the significant events in the development of CAD technology.
3. Describe the significance of the microcomputer in the growth of CAD applications.
4. List the basic components associated with a CAD system.
5. Describe some of the productivity gains provided by CAD applications.
6. Describe the evolution of CAD hardware platforms.
7. Describe a RISC-based CAD hardware platform.

8. Describe a microcomputer-based CAD hardware platform.
9. Compare and contrast the operation and capability of RISC systems with microcomputer CAD systems.
10. Describe the CAD software options.
11. Describe 2-D wire-frame CAD software.
12. Compare and contrast 3-D wire-frame CAD software with 2-D software.
13. What application areas are best suited for 3-D software?
14. Describe the three methods used to define surfaces in surface-modeling software.
15. What are the primary differences between geometry represented by a solid model and that represented by a surface model?
16. Describe the two commonly used techniques for creating a solid model.
17. Describe the criteria used in the selection of a CAD system.
18. Describe the function of PDM systems.
19. What are the three terms that refer to the management of products during their lifecycle?
20. How does CAD interface with PDM?

PROJECTS

1. Using the list of companies developed in project 1 in chapter 1, list the type of CAD hardware and software used in each company.
2. Use the CAD selection criteria and the data in Figures 4–20 and 4–21 to analyze the companies in project 1 that do not currently use CAD or do not use it effectively. On the basis of the analysis, list the CAD platform and software features that would increase productivity.
3. Select three companies with the largest CAD installations, one from each of the small, medium-size, and large groups, and describe how CAD is used across the enterprises.
4. Develop a table that illustrates the types of computer systems needed to run the software of the CAD vendors listed in appendix 4–1. Indicate the software that runs on Microsoft NT/2000/XP, UNIX, or Linux.
5. Develop a table that illustrates the major features of CAD software provided by the CAD vendors listed in appendix 4–1. The minimum features should include 2-D, 3-D, wire-frame, surface, and solid modeling; associativity; and parametric, variational, and features-based modeling.
6. Describe the product data management (PDM) options provided by some of the vendors listed in appendix 4–1.
7. Use the EDS Web site to determine what *virtual product development* and *predictive engineering* include.
8. Select a PC and RISC-based CAD solution from the CAD vendors listed in appendix 4–1. Then use the minimum machine specifications from the CAD

vendor Web site to build a computer from the computer vendors listed in appendix 4–3 that could be used for the CAD software.

The following projects are a continuation of projects 5 through 8 at the end of chapter 3.

9. A company that manufactures kitchen products wants to add a set of plastic storage bowls or containers to its product line. The company wants to distribute the product through Wal-Mart, Target, or a chain of stores in your region. Do the following: use a CAD system to document your product design developed in chapter 3.

10. A company that manufactures kitchen products wants to add a high-end manual can opener to its product line. The company wants to distribute the product through Wal-Mart, Target, or a chain of stores in your region. Do the following: use a CAD system to document your product design developed in chapter 3.

11. A company that manufactures kitchen products wants to add a manual handheld eggbeater to its product line. The company wants to distribute the product through Wal-Mart, Target, or a chain of stores in your region. Do the following: use a CAD system to document your product design developed in chapter 3.

12. A company that manufactures kitchen products wants to add a manual handheld garlic press to its product line. The company wants to distribute the product through Wal-Mart, Target, or a chain of stores in your region. Do the following: use a CAD system to document your product design developed in chapter 3.

APPENDIX 4–1: WEB SITES FOR CAD VENDORS

The universal resource locators (URLs) for suppliers of CAD systems that were discussed in this chapter are listed next. However, the URLs for supplier sites often change, so if any of the following URLs is not active, enter the vendor name into a search engine to find online material. When you visit the Web sites, look for links to images, demonstrations, case studies, or showcases so that you can see the capability of the product.

CAD Vendor	Web Address	Comments
Autodesk	http://www.autodesk.com	The creators of the popular AutoCAD software now have a long list of CAD-related products.
CADKEY Corporation	http://www.cadkey.com	Good links to customer examples.
Bentley Systems	http://www.bentley.com	Creator of the MicroStation CAD product with numerous applications packages.

CoCreate	http://www.cocreate.com	Creators of OneSpace.net and 2-D and solid-modeling products.
IBM	http://www.catia.ibm.com	Offers comprehensive CAD in the CATIA product plus ENOVIA, a PDM product.
Parametric Technology Corporation (PTC)	http://www.ptc.com	Creator of the popular Pro/ENGINEER CAD/CAM/CAE software, CADDS CAD software, and a number of CAD-related products.
EDS	http://www.eds.com	Supports the following CAD and PDM software: Unigraphics NX,
	http://www.sdrc.com	I-deas CAD/CAE, Solid Edge, and Teamcenter, plus other productivity software.
SolidWorks	www.solidworks.com	Creators of SolidWorks CAD solid-modeling software plus PDM packages.
IronCAD	www.ironcad.com	Creators of IronCAD software.

APPENDIX 4–2: B-SPLINES TO NURBS

The surface of a forming die for a car fender includes curves that cannot be represented by circular arcs. The CAD system uses line segments called *B-splines* (*BS*) to draw the complex contours of the surface. For more complex surfaces, such as conic sections, the system needs *nonuniform rational B-splines* (*NURBS*) to develop the surface. As a result, a basic understanding of BS and NURBS and their origin is essential.

This process started with the need to represent curves mathematically in the computer system. Circles and arcs of circles are relatively easy to present by using the equation for a circle. All other types of curves require polynomial equations, such as the quadratic $At^2 + Bt + C$, to describe the shape of the curve. The more complex the curves, the larger the power or degree of the polynomial equation. As a way to avoid higher-order polynomials, the complex curve is divided into

Figure 4–25   (a) UBS;
(b) NUBS.

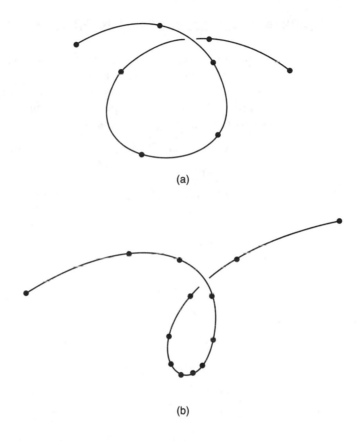

(a)

(b)

segments, with each segment represented by a much simpler polynomial. Five low-degree polynomials are used as an alternative to one equation of high degree. When these segments, called *splines,* are linked together smoothly and efficiently, the curve becomes a BS.

The *knots* along the curve mark the connections between splines and determine the BS's shape and the underlying properties. When splines of equal length are used, the knots are spaced uniformly along the BS and a curve like that in Figure 4–25a can be represented. The result is called a *uniform B-spline (UBS).* Representing curves with a shape like that described in Figure 4–25b requires nonuniformly spaced knots, and the curve becomes a *nonuniform B-spline (NUBS).* As these BS features are added to the CAD software, the system becomes capable of representing curves with greater complexity. Some curves have shapes that cannot be represented with BS formed from linking single polynomials, regardless of the spacing on the knots. In these applications the BS are created by forming a ratio of two polynomials so that *rational* polynomials are constructed. As a result, the new curve-generating software, called *nonuniform rational B-splines (NURBS),* can model the most complex curves and surfaces.

APPENDIX 4–3: WEB SITES FOR COMPUTER COMPANIES

The universal resource locators (URLs) for suppliers of computer systems that were discussed in this chapter are listed next. However, the URLs for supplier sites often change, so if any of the following URLs is not active, enter the vendor name into a search engine to find online material. When you visit the Web sites, look for links to products, systems, and/or custom configuration links.

Computer Vendor	Web Address
Acer	http://www.acer.com
Compaq	http://www.compaq.com
Dell	http://www.dell.com
Fujitsu	http://www.fujitsu.com
Gateway	http://www.gateway.com
Hewlett-Packard	http://www.hp.com
Hitachi	http://www.hitachi.com
IBM	http://www.ibm.com
Motorola	http://www.motorola.com
NEC	http://www.nec.com
SGI	http://www.sgi.com
Sun Microsystems	http://www.sun.com

Design Automation: CAE

OBJECTIVES

After completing this chapter, you should be able to do the following:

- Define *CAE*
- Describe how the CAE technology is integrated with CAD in the product design model and why it is critical to the design process
- Apply design for manufacturing and assembly (DFMA) and show how the manual and computer-based systems support efficient design and successful products
- List and describe the primary applications where finite-element analysis (FEA) is used
- List and describe the primary applications where mass-properties analysis is used
- List the three types of prototyping and compare and contrast their application in the design process
- List and describe six types of rapid prototyping systems
- Describe two methods for group technology (GT) and their application in production engineering
- Describe two types of computer-aided process planning (CAPP) and their application in the production planning process
- Describe the automation functions present in computer-aided manufacturing (CAM)
- Describe process modeling and simulation, production cost analysis, and maintenance automation
- Describe the topology used in the enterprise date network

In the last twenty years, the term *computer-aided engineering (CAE)* has had several meanings. To some people, it has implied software that works with computer-aided design (CAD) files to analyze geometric designs; other people have considered CAE as the umbrella covering all computer software used in design and manufacturing. More recently, the term *CAx* has been used to indicate that many operations in manufacturing are computer aided; you need only supply the applicable capital letter or initial for the *x*. Because of such varied usage of terminology, the study of CAE should start with a definition:

> *CAE is the analysis and evaluation of the engineering design by using computer-based techniques to calculate product operational, functional, and manufacturing parameters too complex for classic manual methods.*

Study of the form, fit, and function characteristics of products is covered by the terms *operational* and *functional* in the definition, while an examination of the match between the design requirements and the production capability of manufacturing is included in the phrase *manufacturing parameters*. The expression *too complex for classic manual methods* indicates that the process performed by CAE software could be performed manually, but the quantitative difficulty, number of computations, or time required for analysis would be prohibitive.

CAE fits into the *design process* (Figure 3–14) at the synthesis, analysis, and evaluation levels and is also consistent with the concurrent engineering (CE) principles described earlier. At the synthesis level, the primary CAE activity is focused on manufacturability using design for manufacturing and assembly principles. The output from the CAE operation at the analysis and evaluation levels is used by the CE team to determine the quality of the product design. On the basis of this CAE data, the product design is cycled through the top four steps in the design process until an optimum solution is generated.

CAE provides productivity tools to aid the production engineering area as well. Software to support *group technology (GT)*, *computer-aided process planning (CAPP)*, and *computer-aided manufacturing (CAM)* are grouped under the broad heading of CAE. As a result of these applications and the use of CAE in the *design process*, CAE is required in all five manufacturing system classifications and in the four production strategies described in chapter 2. The level and variety of CAE software used, however, depends on the amount of concept and repetitive design practiced in the company and the type of manufacturing systems present. In the sections that follow, we describe the type of CAE software commonly used throughout the *design process* and in the production engineering area.

5–1 DESIGN FOR MANUFACTURING AND ASSEMBLY

The synthesis stage in the *design process* enriches the product by adding basic geometric detail and reshapes the product by applying *design for manufacturing and assembly (DFMA)* guidelines and constraints. The DFMA process answers this question: Is the design optimal for manufacturing and assembly? *DFMA* is defined as follows:

> *DFMA is any procedure or design process that considers the production factors from the beginning of the product design.*

The definition states that every design activity, from conceptualization to evaluation, must focus on generation of a design that meets *market expectations* and can be *manufactured successfully*.

Brief History of DFMA

The DFMA effort started with two separate thrusts: *producibility engineering* and *design for assembly (DFA)*. Producibility engineering was an alternative to the older manufacturing reviews, which were typically held after the design was completed. Producibility guidelines focused on the producibility of individual parts rather than the total product with all parts assembled. Therefore, the major thrust was to produce simpler parts that were individually easier to manufacture. Producibility engineering guaranteed easily manufactured parts; as a result, the part count on the product often increased and caused a more complicated product structure. In many cases, the total product was more difficult and costly to manufacture and assemble.

The underlying principle of *DFA* was to reduce the number of parts by eliminating or combining them to achieve a simplified product structure. While the assembly efficiency was improved, the total cost of the product was not minimized because using the best manufacturing processes was not a high priority.

In 1977, DFMA started as an extension of the DFA process. The two methods (design for manufacturing [DFM] and DFA) must interact because the key to successful DFM is product simplification through DFA. DFMA is a holistic approach to design analysis because both the manufacture and the assembly of the finished product are considered simultaneously. This consideration of manufacturing and assembly during design results in lower total product costs.

Justification of DFMA

The use of DFMA early in the design process is necessary because 70 percent of the product costs are set by the time the product leaves the initial design team. The product costs are fixed because important decisions about material, processes, and assembly requirements are settled early in the design. Any effort to reduce cost after the design stage can influence only the remaining 30 percent of product cost. In addition to cost issues, 80 percent of quality problems are often a result of a poor design.

The greatest effect that DFMA has is in the *reduction* of total part count in a product. Reduction in part count leads to a reduction in total product cost and an increase in reliability of the product.

An analysis of 117 published case studies involving fifty-six companies indicated eleven reasons why DFMA was used to improve products. One reason, annual cost savings, was cited in 11 of the cases and averaged $1.4 million annually from DFMA practices. Study the results listed in Table 5–1 and note the significant impact that DFMA had on product design and cost.

Table 5–1 DFMA: Cited Results (117 Published Case Studies)

Benefit from DFMA	Times cited	Average change
Part count	100	−54%
Assembly time	65	−60%
Product cost	31	−50%
Assembly cost	20	−45%
Assembly operations	23	−53%
Weight	11	−22%
Labor costs	8	−42%
Manufacturing cycle time	7	−63%
Part costs	8	−52%
Assembly tools	6	−73%
Annual cost savings	11	$1.4 million

Note. DFMA = design for manufacturing and assembly.

The *interface* between separate parts is the major source of product failure and poor quality; therefore, a reduction in the part count results in fewer interfaces between parts and a better product. Also, with fewer parts there is a reduction in *dies, molds, stamping presses, assembly operations, inventory, material handling,* and *paperwork.*

DFMA: The Manual Process

The DFMA process is both *manual* and *computer based.* From the manual perspective, DFMA provides a step-by-step procedure to query the designer about part function, material limitations, and part access during assembly. The software used for DFMA calculates assembly time, product cost, and a benchmark theoretical minimum number of parts. In addition, a manufacturing element in the software assesses various material options (steel versus plastic) and manufacturing processes (machining versus die casting) for the most cost-efficient production of parts.

DFMA starts with effective DFA because assembly still remains a large part of the manufacturing effort and cost. When products are designed to make assembly easier and faster, costs are reduced and higher-quality products result. Ten design rules or guidelines, listed in appendix 5–1, are used by designers to create products with good assembly characteristics. Applying the guidelines to a production part will help explain how the rules are used. Study the spindle/housing assembly introduced in chapter 3 and repeated in Figure 5–1 so that you are familiar with the parts and assembly. Note that the U-shaped sheet metal bracket supports a stainless steel spindle. Two nylon bushings are inserted into holes in the sheet metal bracket and the spindle turns inside the bushings. Each bushing is held in place with three threaded fasteners screwed into tapped holes in the bracket.

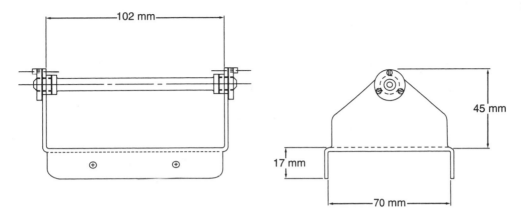

Figure 5–1   Spindle/Housing Assembly.

(Source: Boothroyd/Dewhurst, "Product Design for Manufacturing and Assembly," © April 1988, Manufacturing Engineering Magazine, p. 42. Reprint by permission of the Society of Manufacturing Engineers, Dearborn, MI.)

Exercise 5–1

Apply the ten design guidelines listed in appendix 5–1 to the spindle/housing assembly shown in Figure 5–1 and determine how many of the rules were violated.

Solution

- *Rule 1 (violated).* The total part count is ten, with only one moving part in the assembly. Any time that two mating parts, such as the bracket and the bushings, do not move relative to each other, every effort should be made to combine them into a single part.

- *Rule 2 (unknown).* With the information provided, it is not possible to determine if some of the surface area could be eliminated to reduce surface processing and finishing requirements.

- *Rule 3 (violated).* The part has a natural base, the sheet metal bracket, on which the assembly is constructed. The rule is violated, however, because the parts are assembled from the sides and not the top.

- *Rule 4 (satisfied).* Access to the parts during assembly is adequate, and standard tools could be used.

- *Rule 5 (violated).* There are no grooves or aligning guides for the bushings on the bracket to aid in aligning the holes for the screws.

- *Rule 6 (satisfied).* The parts are all symmetric.

- *Rule 7 (violated).* The screw size does not have adequate surface area for easy gripping.

- *Rule 8 (violated).* Six separate fasteners are used in the assembly.

Best ←

Assembly Method Scoring Chart

Approach / Connection	Weld	Solder	Stake	Adhesive	Pin	Nut	Tape	Screw	Snap ring	Snap fit
Assembly stack from top ■	10	20	30	40	50	60	70	80	90	100
Assembly stack from top ●		10	20	30	40	50	60	70	80	90
Assembly from side ■					30	40	50	60	70	80
Assembly from side ●					20	30	40	50	60	70
Assembly from bias ■						20	30	40	50	60
Assembly from bias ●						15	20	30	40	50
Rotated parts ■						10	10	20	30	40
Rotated parts ●						5	5	10	20	30
Assembly from bottom ■								5	10	20
Assembly from bottom ●										10
Connection	Weld	Solder	Stake	Adhesive	Pin	Nut	Tape	Screw	Snap ring	Snap fit
	Special tool or equipment required					Small tool required				Nothing

↑ **Good**

Legend:
■ Without hold-down
● With hold-down

Fastening or assembly method

Comments: Assign points to the open boxes; if the part you are placing in the assembly falls in one of the gray marked boxes, no points would be received. The upper right hand corner box would be assigned the highest points, and, as you go to the left or down, the points would decrease. After you evaluate your assembly, you add up the total points in boxes and divide the sum by the number of parts in your assembly; this gives you a design score.

Example: Design score = $\dfrac{\text{total points in boxes}}{\text{number of parts in assembly}} = \dfrac{750 \text{ points}}{10 \text{ parts}} = 75\%$

Figure 5-2 Assembly Method Scoring Chart.
(Source: Courtesy Xerox Corporation.)

- *Rule 9 (violated).* Self-locking features were not used in this design.
- *Rule 10 (violated).* This component is probably not standard.

The result of the analysis is that *seven* rules are violated, *two* are satisfied, and *one* is unknown. The value of using the ten design guidelines is that problem areas (all violations) are identified and attention is focused on satisfying the guideline in question. As a result of this analysis, the spindle/housing design could be improved in several areas.

Another approach used to analyze the quality of a design is illustrated in Figure 5–2. With the assembly method scoring chart, a score from 0 to 100 points is assigned on the basis of ease of assembly. Study the chart until you understand the general layout and the way a score is calculated. Note that there are five assembly motions or approaches (left side) from *assembly stack from top* to *assembly from bottom*. Each of these has two categories, with and without a hold-down, indicated by a black circle or square. In addition, the connection technique (across the bottom) can vary from *weld* to *snap fit*. The best score (100) is for an assembly from the top with no hold-downs and using a snap fit. If the block for an approach (left column) and connection (bottom row) is shaded, no score or a 0 score is recorded for that operation. For example, a weld from the top with no hold-down scores a 10, but all other weld operations score a 0. The application of the chart is illustrated in the following exercise.

Exercise 5–2

Calculate the assembly score using the assembly method scoring chart shown in Figure 5–2 for the spindle/housing assembly shown in Figure 5–1. Assume that the assembly starts with the insertion of the base into a holding fixture with a hold-down required. Use the following procedure for computing the assembly design score from the chart:

1. Start with the base of the assembly; next, assume that the base is placed in an assembly fixture with a snap fit from the top. Determine if hold-downs are required and enter either a 90 or a 100 on the score sheet.
2. Select the next logical part component to be added to the assembly and locate the appropriate *approach* row and *connection* column to assemble that part. The score in the box where the row and column intersect is entered on the score sheet for that part.
3. Repeat step 2 until every part, component, and subassembly has been added and the assembly is complete.
4. Add the part assembly scores to get a total score for the assembly.
5. Compute the assembly score by dividing the chart total by the total number of operations scored.

Solution

The following chart indicates the assembly score for the spindle/housing bracket shown in Figure 5–1.

Assembly part	Score	Comment
Base	90	
Spindle	60	From side, held by screws
Right bushing	80	From side, held by snap-in
Left bushing	80	From side, held by snap-in
Screw 1	60	From side, screw
Screw 2	60	From side, screw
Screw 3	60	From side, screw
Screw 4	60	From side, screw
Screw 5	60	From side, screw
Screw 6	60	From side, screw
Total	670	

The assembly score is 670, and there are ten operations, so the final score is 670/10, or 67. The score of 67 would indicate a below-average design (scores over 80 are considered good), which is what the analysis in exercise 5–1 indicated. The score could vary depending on how the chart is interpreted. Despite this possible variation, the assembly method scoring chart provides a good analysis tool for comparison of alternative designs. Also, a study of the chart indicates the type of assembly operations that would yield a good score.

DFMA: Computer Support

CAE software in the DFMA area aids the designer by calculating costs for alternative design solutions. With DFMA software, the designer enters the specifications for the part design, and the software provides a quantitative analysis of the alternative designs. An alternative solution for the spindle/housing assembly is illustrated in Figure 5–3. The requirement for bushings was eliminated by making the entire bracket or housing from nylon, so the part count is reduced from ten to two.

DFMA software from Boothroyd Dewhurst, Inc., was used to analyze the two spindle/housing assembly designs. Figure 5–4 shows the cost analysis of the two designs provided by the DFMA software. The cost analysis includes tooling, material, manufacturing, and assembly. Note that the total cost for a single part would be the sum of assembly, material, and manufacturing totals. Therefore, the sheet metal design would cost $5.69, while the two-piece design would cost only $1.61. The cost of tooling would be spread over the total number of parts produced. The DFMA software provides additional analysis of injection-molding cost, machining cost, and material alternatives.

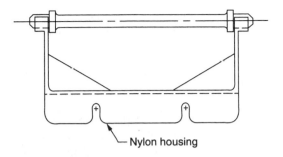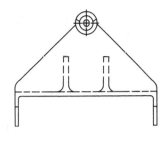

Nylon housing

Figure 5–3 Alternative Spindle/Housing Assembly.

(Source: Boothroyd/Dewhurst, "Product Design for Manufacturing and Assembly," © April 1988, Manufacturing Engineering Magazine, p. 42. Reprint by permission of the Society of Manufacturing Engineers, Dearborn, MI.)

In summary, DFMA encourages designers to break with old concepts in design and produce less costly products. DFMA provides the following typical advantages:

■ Reduction of design-to-product launch cycles by as much as 50 percent
■ Reduction of part counts by 30 to 70 percent
■ Reduction of assembly times by 50 to 80 percent

(a) Design using sheet metal housing cost

	Assembly	Material	Manufacturing	Tooling
Housing	0.02	1.74	1.56[a]	7,830[b]
Bush (2)	0.09	0.01	0.06[c]	9,030[d]
Screw (6)	0.35	0.72	—	—
Spindle	0.04	0.26	1.29	—
Total	0.05	2.73	2.91	16,860

[a] *Includes $1.35 for drilling and tapping screw holes.*
[b] *Three separate die sets for blanking, punching, and bending.*
[c] *Molded bushings have three-cored holes for screw clearance.*
[d] *Ten-cavity mold for least-cost manufacture.*

(b) Design using injection-molded housing cost

	Assembly	Material	Manufacturing	Tooling
Housing	0.02	0.14	0.24	10,051[a]
Spindle	0.02	0.26	1.29	—
Total	0.04	0.04	1.53	10,051

[a] *Two-cavity mold for least-cost manufacture.*
Note: A Comparison of the Two Spindle/Housing Assembly Designs Shows Significant Cost Reductions as a Benefit of DFA.

Figure 5–4 DFMA Cost Analysis.

(Source: Boothroyd/Dewhurst, "Product Design for Manufacturing and Assembly," © April 1988, Manufacturing Engineering Magazine, p. 43. Reprint by permission of the Society of Manufacturing Engineers, Dearborn, MI.)

Frequently, DFMA designs produce more complicated individual components but result in a simpler product structure and a lower total production cost.

5–2 CAE ANALYSIS

Testing a product, at the analysis step in the *design process*, requires a broad array of CAE software. The software selected is a function of the type of test desired. CAE and CAD are closely linked. In most applications, the data and information used as input for the CAE software are in the form of a drawing of the product created in CAD. The geometry produced in CAD is used by the CAE software to get the data needed for analysis.

CAE applications at the analysis step fall into two broad categories: *finite-element analysis* (FEA) and *mass-properties analysis.* Software in each of these categories is described in the following two sections. Special-purpose programs are described thereafter.

Finite-Element Analysis

The most frequently used CAE application is *FEA*, which is defined as follows:

> *FEA is a numerical program technique for analyzing and studying the functional performance of a structure or circuit by dividing the object into a number of small building blocks called* finite elements.

Most FEA applications fall into two categories: *mechanical systems* and *electronic circuits.* The FEA process for the mechanical system begins with the creation of the geometric model of a part or structure with CAD software. The three-dimensional (3-D) CAD model is divided into a finite number of small pieces (elements) that are connected to one another at points (nodes). The process is best illustrated by reviewing how heat-transfer FEA software from ALGOR was used by Delphi Corporation to design a wide-range oxygen sensor for the automotive industry. The sensor is pictured in Figure 5–5, and the two-dimensional (2-D) axisymmetric cross section used to model the part is pictured in Figure 5–6. The heat-transfer analysis, Figure 5–7, was performed with ALGOR heat-transfer-analysis software. The mesh and nodes on the part are visible in the left frame, and the predicted temperature variation across the part is illustrated in the right frame of Figure 5–7. Study the initial design configuration (Figure 5–8a) and the final design (Figure 5–8b). Using data generated from the heat-transfer analysis, Delphi engineers changed the configuration of the lower half of the sensor to reduce the operating temperature at the environment seal. The FEA software has mathematical equations that describe how the nodes respond when an external stimulus or force is applied. In the oxygen sensor example, the temperature at the base of the sensor was specified and the heat-transfer software used parameters from the model to predict how this temperature would be distributed across the

Figure 5–5 Wide-Range Oxygen Sensor for Automotive Engine Monitoring.

(Source: Image courtesy of ALGOR, Inc., and Delphi Corporation.)

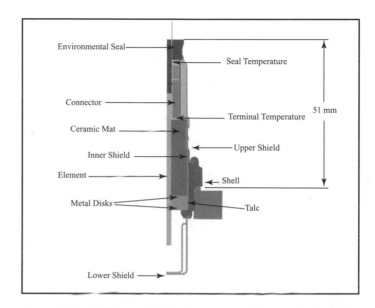

Figure 5–6 2-D Axisymmetric Cross-Section Model of Oxygen Sensor.

(Source: Image courtesy of ALGOR, Inc., and Delphi Corporation.)

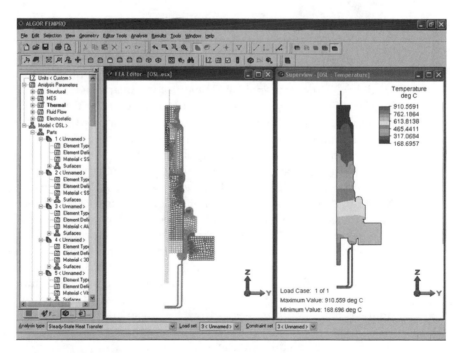

Figure 5–7 Heat-Transfer-Analysis Results for the Oxygen Sensor.

(Source: Image courtesy of ALGOR, Inc., and Delphi Corporation.)

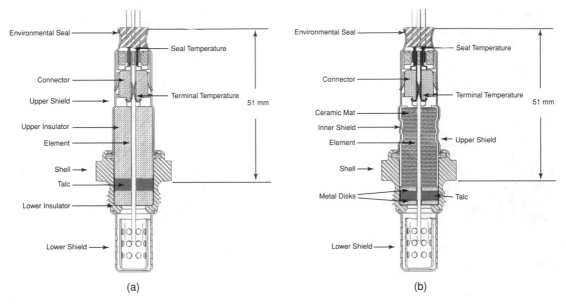

Figure 5–8   Oxygen Sensor: (a) Initial Design; (b) Final Design.
(Source: Image courtesy of ALGOR, Inc., and Delphi Corporation.)

length of the sensor. Delphi engineers adjusted the design until the analysis indicated that the desired temperature distribution was achieved.

The structural integrity of mechanical systems can also be analyzed by using FEA software. For example, a CAD file of a bridge element is used to create an FEA model for structural analysis. When a load is applied, the FEA equations predict how the finite elements throughout the structure will respond. The material, the composition, and other variables that describe the bridge element are represented in the equations. The overall response of a part or structure to the applied force is determined by the simultaneous solution of the equations representing each finite element in the model. An example best illustrates this process.

A Danly U4-1000-132-84 power press, shown in Figure 5–9a, was installed in a General Motors stamping plant in 1955 to trim and shape automotive body parts. Before some required cutouts were added to the base of the power press, FEA software was used to verify that the structural integrity of the 1,000-ton-capacity machine was not compromised. The Danly engineers built a solid model of the press bed, shown in Figure 5–9b, with added cutouts by using Solid Edge software by UGS. The CAD model was converted to an FEA model for static stress analysis with linear material models by using ALGOR's InCAD technology. The analysis model, Figure 5–9c, shows the mesh and the nodal displacement magnitude in inches. The calculated von Mises stresses were within Danly's standards for safe operation and fatigue.

In addition to the graphic solutions illustrated in the two preceding FEA examples, numerical solutions are available in tabular format.

(a)

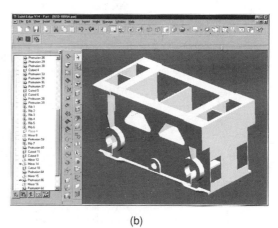

(b)

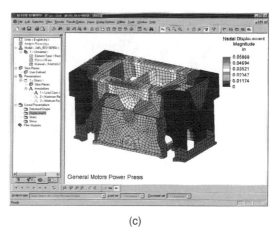

(c)

Figure 5–9   (a) Danly 1,000-Ton Press; (b) CAD Model of Press Bed; (c) FEA of Press Bed.
(Source: Image courtesy of ALGOR, Inc., and Enprotech Mechanical Services/Danly Division.)

FEA software covers a wide range of applications that include the following:

■ *Static analysis.* Static analysis software determines the deflections, strains, and stresses in a structure under a set of fixed loads. Typical structures include aircraft, bridges, buildings, cars, dams, and machine parts.

■ *Transient dynamic analysis.* Transient dynamic analysis software calculates the deflection and stress under changing load conditions by using the natural response time and frequency for the structure. Typical structures would be the same as those for static analysis.

■ *Natural frequency analysis.* Natural frequency analysis software computes the vibration of a structure at its natural frequency. For example, the destructive power of low-frequency vibrations created by earthquakes can be applied to structural models of buildings and bridges to determine the limits for catastrophic failure.

- *Heat-transfer analysis.* Heat-transfer-analysis software determines the steady-state and transient temperature distribution in a structure when thermal loads are applied and boundary conditions are known. Heat-transfer analysis is used on end products and production processes. For example, the efficiency of automotive cooling system designs for removing heat from engine components is tested with heat-transfer FEA. In the production area, mold-filling analysis using a heat-transfer FEA model is used to test plastic injection mold designs for optimum plastic flow, filling speed, and cooling parameters.

- *Motion analysis.* Motion analysis software, sometimes called *kinematic analysis*, computes the geometric properties (displacement, velocity, and acceleration) required for a mechanical mechanism to produce a desired motion. Few products designed today do not have moving parts, and motion FEA allows the designer to put the parts of the solid model into motion. Under these simulated motion conditions, parameters—such as limits of motion, interference, and the geometric properties of displacement, velocity, and acceleration—for any part can be analyzed.

- *Fluid analysis.* Fluid analysis software determines the flow, diffusion, dispersion, and consolidation characteristics of a fluid under different controlled conditions. Design of piping systems with complex turns for large volumes of fluid requires a model of the fluid flow.

- *Mechanical event simulation.* Mechanical event simulation (MES), a feature offered in ALGOR FEA software, allows engineers and designers to simulate the actual conditions that a mechanical component will experience in its application. MES is used to analyze component designs that include motion, from continuously moving parts to the event that precedes a static situation. Analysis of parts in motion is difficult because the loads are no longer constant and the magnitude and direction are not easily estimated. MES eliminates the need to estimate loads because it accounts for the interaction of the component with its surroundings and the inertial forces generated by the motion of the component itself. MES is a true dynamic analysis tool because it considers inertia and is not based on a linear static perspective. MES was performed on the Bourdon tube pressure gauge assembly illustrated in Figure 5–10 to determine motion and resulting stresses. The model utilized kinematic elements and surface-to-surface contact elements between the gauge's gears to simulate the motion of the indicator needle as pressure increases within the tube.

Use the Web sites provided in appendix 5–2 to find current examples of each of these applications.

The key to successful FEA is the selection of the *mesh,* which is composed of elements connected at nodes. The size and location of the elements is critical for

Figure 5–10 MES of a
Bourdon Tube Pressure Gauge.
(Source: Image courtesy of ALGOR, Inc.)

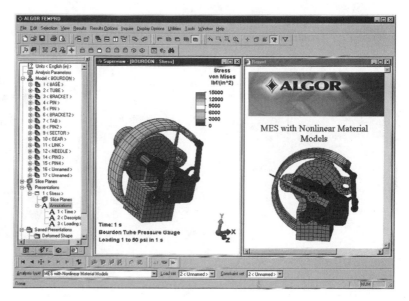

generation of useful results. Note in Figures 5–9c and 5–10 that the size of the elements varies. The elements are generally smaller at points on the geometry where the effects of the analysis are most pronounced. For example, in Figure 5–9c, the elements are smallest nearest the holes and circular features. Selecting appropriate elements is performed automatically by FEA software with mesh-generation features. The generated mesh can be modified by using mesh-refinement options or manual adjustment for optimum results.

In the past, special training was necessary to operate FEA software and interpret FEA results. For some analyses, FEA specialists had to perform the analysis work. However, the ease of use of current FEA software permits design engineers to perform analysis on their CAD designs.

Mass-Properties Analysis

Mass-properties analysis, a CAE function used frequently in the design process, is invoked through a command included in most CAD software packages. The mass-properties command signals the CAD system to calculate and return numerical values that describe properties of the selected drawing geometry. In the most basic application, the *area* of a 2-D CAD shape or the *volume* of a 3-D solid object is calculated and displayed. More complex mass-properties analysis produces the *mass, bounding box, centroid, moments of inertia, products of inertia, radii of gyration,* and *principal moments with x-y-z directions about centroids.* These complex parameters are important for analysis of part geometry that moves and rotates in the final application. Detailed explanations of these parameters are beyond the scope of this book, however.

An example from a company that produces extruded aluminum rails illustrates how important basic mass-properties data are for some manufacturers. A

Figure 5–11   Extruded
Aluminum Rail.

typical cross section of a rail used for building commercial aluminum windows is
pictured in Figure 5–11, and the CAD-drawing cross section used to produce the
extrusion die for production of the aluminum part is shown in Figure 5–12. CAD
could not be justified on the basis of drawing the simple outline required for the
extrusions; however, the extrusion area is important because it determines the
amount of metal used and the weight per linear foot of rail. The cross section of
the extrusion is drawn with CAD software, and the area is calculated by using the
mass-properties feature in CAD. As a result, the cross section for the rail and die
is designed to the desired weight standards, and the cost of the extrusion is accurately established. A secondary benefit is derived from electronic transfers of the

Figure 5–12   Die Drawing for
Extruded Aluminum Rail.

2-D CAD outline
of extruded rail

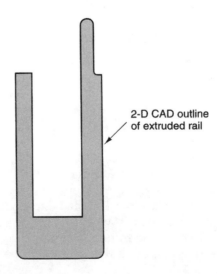

CAD data file. The CAD file of the cross section is sent electronically from the computer of the rail manufacturer to the computer of the company that produces extrusion dies. The die manufacturer processes the CAD shape file for the extrusion and sends the computer numerical control (CNC) file to a wire electrical discharge machine (EDM) on the shop floor to cut the dies. The single CAD database shared by the two companies guarantees that the die will produce the exact desired extrusion.

Mass-properties analysis of CAD solid models is also critical for companies that design plastic-injection-molded or metal-die-cast parts. DFMA software, used to evaluate trade-offs between assemblies in Figures 5–1 and 5–3, requires data from mass-properties analysis. The base for the spindle/housing assembly shown in Figure 5–3 is a nylon part produced on a plastic-injection-molding machine. Accurate values for the critical DFMA input parameters—*part volume* and *projected area*—are difficult to obtain without mass-properties software.

The DFMA software input screen illustrated in Figure 5–13 includes numerous mass-properties data. In this example, the part was drawn with solid-modeling CAD software, and the mass-properties feature provided the area and volume data. The output data from the DFMA analysis of the injection-molded housing illustrated in Figure 5–3 is provided in Figure 5–14. Compare the *total mold cost* ($10,051), *manufacturing cost* per part (23.9¢), and *polymer cost* per part (13.5¢) with the values used in Figure 5–4. The DFMA software analysis is for a production run of 100,000. Mass-properties analysis is important to every type of manufacturer where product design is performed.

Other CAE Design and Analysis Software

The expanding software market has many more examples of special-purpose programs for analysis of designs. Two additional types of analysis software used frequently include *circuit analysis* and *assembly interference, fit, and tolerance.*

Circuit Analysis Software.   Software in the CAE circuit analysis area supports a broad range of design activities across the electronics industry. In general, the software performs numerical analysis on electronic circuits to determine electrical performance and worst-case conditions. No circuit types are excluded; discrete, integrated, and hybrid circuit analyses are available. In addition, a host of design software is available to support integrated circuit chip design, surface-mount technology design, and printed circuit board design. Circuit analysis software such as the Electronic Circuit Analysis Program (ECAP) and the Simulation Program with Integrated Circuit Emphasis (SPICE) is used for analysis of electronic and power circuits.

Assembly Interference, Fit, and Tolerance Software.   Assembly interference, fit, and tolerance software focuses on design and analysis of mechanical parts and assemblies. A software package called *Valisys* runs on reduced instruction set computer (RISC) workstations using computer-integrated manufacturing (CIM)

Computer Input Screen for Nylon Housing

ESTIMATION OF INJECTION MOLDING COSTS FOR: HOUSING   THERMOPLASTIC: 6/6 NYLON

Dimensional Data		Part Complexity	
Part volume = 25.51 CM^3		Outer surface or cavity (0–5)?	2
Projected area = 71.25 CM^3 L = 95 MM W = 75 MM D = 57 MM		Inner surface or core (0–5)?	0
Thickness maximum = 2.5 MM Average = 2 MM			
		Mold Complexity	
Quality and Appearance		Standard two-plate mold?	Y
		Three-plate mold?	N
Tolerance factor (0–5)?	3	Multiplate stacked mold?	N
Appearance factor (0–5)?	1	Hot runner system?	N
Colored resin?	N	Number of side cores or pulls?	0
Textured surface?	N	Number of unscrewing devices?	0

At any time press: <H>elp, <V>olume, <A>rea, or <C>omplexity calculator.

Figure 5–13 DFMA Input Screen.

(Source: Boothroyd/Dewhurst, "Product Design for Manufacturing and Assembly," © April 1988, Manufacturing Engineering Magazine, p. 44. Reprint by permission of the Society of Manufacturing Engineers, Dearborn, MI.)

Results for Injection Molding of Nylon Housing

ESTIMATED INJECTION MOLDING COSTS FOR: HOUSING   THERMOPLASTIC: 6/6 NYLON

Total Production Volume (Thousands)	Number of Cavities	Total Mold Base Costs ($)	Cavity/Core Manufacturing Costs ($)	Total Mold Cost ($)	Mold Cost per Part (Cents)
100	2	3589	6462	10,051	10.1

Machine Size (kN)	Machine Rate ($/hr)	Cycle Time (Seconds)	Manufacturing Cost per Part (Cents)
1600	72	20.3	23.9

Part Volume (cm³)	Part Weight (Grams)	Polymer Cost ($/kg)	Polymer Cost per Part (Cents)
26	29	4.69	13.5

Total part cost (cents) = 47.5

Select required option:

1. Screen edit
2. Show mold cost/cycle elements
3. Print results and responses
4. Change basic cost data
5. Change responses/polymer
6. Exit

Figure 5–14   DFMA Mold Analysis for Alternative Design.

(Source: Boothroyd/Dewhurst, "Product Design for Manufacturing and Assembly," © April 1988, Manufacturing Engineering Magazine, p. 45. Reprint by permission of the Society of Manufacturing Engineers, Dearborn, MI.)

part files created with CATIA CAD software. Valisys features are worth describing because they illustrate the type of cross-discipline data generation and collection required in a CIM environment. In addition, a review of the software emphasizes the critical relationship among product design, manufacturing, and quality. The Valisys software has five separate modules, which are described next.

Design Verification.   Starting from the initial design, the Valisys software from Tecnomatix Technologies (see Web site in appendix 5–2) checks the dimensioning and tolerancing information against the geometric dimensioning and tolerancing (GD&T) standards, ANSI Y14.5 and ISO R1101. In addition, the software uses the existing part geometry to create *soft gauges. Hard gauges* are made by manufacturing to check the quality of a manufactured part. For example, if the size and separation distance between two holes is critical for an assembly, a metal gauge like the mating part is produced. This hard gauge is just a metal plate with two pins that can be inserted into the two holes to test the manufactured parts. The *soft gauge* is a 3-D wire-frame model of the hard gauge and is created by Valisys as part of the CATIA CAD geometry file. When a part is manufactured, the critical features of the finished part are measured on a coordinate measuring machine (CMM). These dimensional data are checked electronically by the soft gauge for a proper mating-part fit. The CAE Valisys soft-gauge technique provides better-quality monitoring with less investment in hard tooling.

Tolerance Analysis.   The tolerance analysis function in Valisys analyzes the fit between mating parts under worst-case tolerance conditions. Production tolerances on the part are set at the maximum value because this design feature of the software considers part fit as part of the design. In addition, optimum design geometry and drafting text for clearance and threaded holes for fasteners are generated automatically. An assembly option permits analysis of the tolerance buildup of assemblies.

Quality Engineering.   The quality engineering function in Valisys addresses inspection of the manufactured part by a CMM or a numerical control (NC) machine. Inspection paths and an inspection program for the measuring device are generated from the CAD geometry by Valisys to collect critical dimensional data on the manufactured part. An online graphical model of the "as-built" part is created with the data collected. A comparison of the as-built model with the soft gauge indicates whether the part passes or fails inspection. In addition, recommendations for rework are provided for parts that fail inspection. The process can be used for reverse engineering of a product, which requires going from a part to engineering drawings.

Inspection.   The inspection function controls the measuring machines and manages the acceptance and rejection of parts. Recommendations for process improvement are generated from the data collected during the quality measurement phase.

Process Control.   The process control module in Valisys collects quality data and creates control charts required for *statistical process control* (*SPC*). The data collected in this module are used to determine the accuracy and repeatability of the manufacturing process, to determine how closely the as-built part matches the design specifications, and to anticipate trends in the machine process so that corrections can be made before out-of-tolerance parts are produced.

5–3 CAE EVALUATION

The design analysis process provides ample data on the various design alternatives. The examination of those data to determine the degree of match between the actual design and the initial design goals and specifications is one part of the evaluation process. Examination of the data is performed by every member of the CE team, and recommended changes in the design are made. The reiterative nature of the *design process* makes it difficult to separate CAE activities in the analysis and evaluation functions. The important point is that CAE software to analyze and evaluate design quality is available. One CAE activity traditionally performed at the evaluation stage of the design is *prototyping.*

Prototyping

Building a prototype of a design is an age-old practice. The *prototype* is an functional model of the design, built to evaluate operational features before the start of full production of the product. The style of the prototype is dictated by the tests that are planned. For example, the automotive industry builds small-scale (about 1-foot-long) models of new cars from solid metal or wood for wind-tunnel tests of the aerodynamics of the body. The same industry builds full-sized fully working models of new engine designs and subjects them to operational tests.

The tools used for standard prototyping are conventional production machines. Frequently, prototype parts are machined from nonferrous metal or plastic; however, with the use of more complex plastic-injection-molded parts in products, the prototype process becomes difficult. Machining the complex shapes of injection-molded parts is difficult, expensive, and time consuming. While prototyping a design is still a critical evaluation process, the requirement to cut lead time to market requires faster prototyping techniques. Several different technologies, called *virtual prototyping* and *rapid prototyping,* are reducing the time required to evaluate a design.

Virtual Prototyping.   *Virtual prototyping,* also called *virtual mock-up,* is new on the technology scene because CAD technology does not support this technique today as it will in the future. Virtual prototyping is the process of building products or subassemblies of machines from components developed by a team of designers on different workstations. The virtual mock-up software that permits virtual prototyping is not ideal at present because synchronizing the CAD data requires costly integration by programmers very familiar with both the CAD software and the

Figure 5–15   Solid-Model
Assembly Prototype Using Parts
Designed with Pro/ENGINEER.

mock-up software. The process is especially stressful when multiple brands of
CAD systems attempt to supply components for the same prototyping project.

The mock-up software allows a large number of components to be integrated and seen in the prototype but not changed. CAD companies are addressing many of the integration issues by making their systems capable of displaying
larger assemblies. For example, CATIA, SolidWorks, Solid Edge, Pro/ENGINEER
(Pro/E), and Inventor can load lightweight faceted models of assemblies for display purposes only and not changes. Pro/E brings in objects devoid of internal details. The assembly of a car front end in Figure 5–15 illustrates a mock-up using
part components designed with Pro/E.

IX Design, from ImpactXoft, and Alibre Design, from Alibre, Inc., offer
mock-up software that more closely approaches the ideal. This improved process
lets large machinery models be controlled by a central server so that designers can
check out various subsystems and work on them while seeing all other details of
the machine. Other products currently available include Unigraphics Solutions'
VisMockUp, PTC's DIVISION MockUp, Prosolvia's Clarus DMU, and Tecoplan's
Digital Mockup.

Another technique for virtual prototyping uses software such as ALGOR's
Accupak/VE Mechanical Event Simulation. The software permits virtual prototyping with linear and nonlinear stress analysis on projects imported from a solid
model. The software allows virtual testing to replace physical testing by replicat-
ing mechanical events with a computer, helping to determine the behavior of a
product in its real-world, worst-case scenario. The capabilities include the ability

to predict motion and impact, perform stress analysis for each instant as the event unfolds, determine flexibility, intrinsically determine forces, handle complex shapes and nonlinear behavior, and test strength of materials.

Rapid Prototyping.   *Rapid prototyping,* a technique used to build a sample of a new design quickly, is another reliable tool in the evaluation process. These systems electronically divide a 3-D or solid CAD model of a part into thin horizontal cross sections and then transform the design, layer by layer, into a physical model or prototype.

Rapid prototyping systems are driven by RISC workstations or microcomputer platforms. Starting with a CAD 3-D fully closed surface- or solid-model file, the CAD software converts the geometry into a file format compatible with the rapid prototyping system. Most rapid prototyping systems use an STL (stereolithography) file format. The conversion software for the STL file format is available for all the major CAD software. Some of the rapid prototyping systems currently in use are discussed next. Visit the Web sites provided in appendix 5–3 to see examples of these systems. The best single source for rapid prototyping information is Castle Island's Worldwide Guide to Rapid Prototyping, listed in appendix 5–3.

Stereolithography.   The *stereolithography* process employs a tank of liquid photosensitive polymer with a vertically controlled table in the polymer and a servo-controlled laser focused on the surface of the polymer (Figure 5–16). The table is positioned with its top just below the level of the liquid in the polymer tank. The STL file consists of a closed mesh of triangles generated on the surfaces of a CAD model. The computer in the system reads the STL CAD file for the prototype part and cuts the part into cross sections from top to bottom. The cross sections are typically from 0.0015 to 0.005 inch thick. Thinner slices are possible to produce a smoother model, but the processing and fabrication time is increased significantly. The computer system stores the cross-section data as SLI (slicing) files and merges them to create files for controlling the laser and table elevator mechanism. The laser traces the area of the bottom cross section on the thin liquid layer of polymer

Figure 5–16
Stereolithography System.

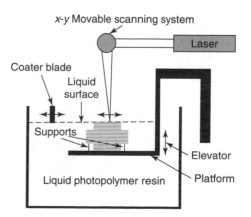

on the table; the laser light causes the polymer to harden. With the first cross section of the part created on the table, the table elevator is lowered by the cross-section thickness by the computer. Between layers, the top of the part is coated with a smooth layer of liquid resin by the coater blade that sweeps across the surface of the tank. The laser then solidifies the next cross section. This process continues until the entire part is created from liquid polymer.

Overhanging areas that are not directly above a previous layer must be connected with support structures either to the part below or directly to the bottom of the building platform. Both the part and any necessary supports are built from the same material. After a part is finished, the supports must be cut away and the surface sanded.

Processing time is a function of cross-section thickness and the size of the area traced. On average, the system processes about an inch of thickness per hour for a small part. After the part is produced in the system, it must be cleaned thoroughly with a solvent and water and cured to develop the full strength of the material. Curing is performed in a special chamber called the *postcuring apparatus* (*PCA*). There, the part is exposed to light from ultraviolet lamps and heat for certain resins. Postcuring takes between one and ten hours depending on the part size. Accuracy varies between 0.1 and 0.5 percent of the overall dimension from small to large parts.

Stereolithography is the best choice for prototypes where high accuracy and detail are critical but rigorous physical testing is not necessary. Figures 4–4 and 4–5 show 3-D CAD drawings of a plastic pencil sharpener used to produce the prototype in Figure 5–17. Study the drawings and compare the original pencil sharpener in Figure 5–17 with the model produced through stereolithography.

Selective Laser Sintering.   *Selective laser sintering* (*SLS*) employs a high-energy carbon dioxide laser to fuse, or sinter, powder into a solid object (Figure 5–18). Using a technique similar to stereolithography, the SLS laser traces the shape of each cross section, fusing the thin layer of powder. A mechanical roller then spreads more powder across the top of the finished layer, and the laser traces the next

Figure 5–17   (a) Original Pencil Sharpener; (b) Part Produced Through Stereolithography.

(a)   (b)

Figure 5–18 Selective Laser
Sintering.

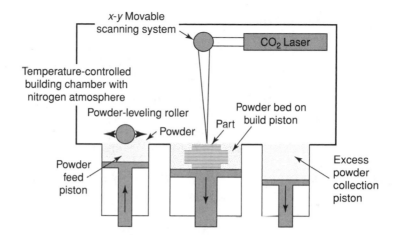

cross section. The building chamber is heated to just below the melting temperature of the powdered material so that the laser will sinter the powder grains with little additional energy. In addition, a heated chamber minimizes the thermal shrinkage of the layers during fabrication. The sintering takes place in the presence of nitrogen because the flammable powder could burn or explode if it is heated in air. Supports are not required because the unsintered powder surrounding previous layers supports separated or overhang areas of the part. A finished sintered part is encased in loose powder. Accuracy varies between 0.001 and 0.015 inch depending on the size of the part. Layer thickness varies from 0.004 to 0.006 inch.

SLS has the advantage of using a wider variety of materials with better mechanical properties than photopolymers and at lower cost. The most commonly used materials are nylon and glass-filled nylon powders, but other commercial materials include polycarbonate, polyamide, acrylic-based polymer, elastomeric polymer, zirconium and silica sands, and polymer-coated steel and metal-alloy powders. However, any material that can be powdered and melted could be used. SLS works better on some parts with complex internal shapes but does not hold tolerances as tightly as does stereolithography. Dimensional variation of ±5 mils per inch is normal for these systems. The SLS surface finish of PVC parts is quite smooth, but other materials exhibit a laminated appearance. Parts are ready for use as soon as they have cooled and the excess powder has been removed by using compressed air, brushes, or other tools. Sintered parts are porous and have a relatively rough surface finish as a result of the grains used in building.

Two vendors—EOS (Electro Optical Systems) and 3D Systems—have systems that can build metal parts with laser sintering. 3D Systems offers materials that consist of stainless steel powders coated with binder. The binder melts to hold the metal particles together. Later the binder is burned away and replaced by bronze to make a fully metal part. EOS uses a direct-metal sintering process with material that combines two metal powders with different melting temperatures.

During sintering, the components with lower melting temperatures melt and flow into the pores of the part, which creates a finished prototype.

Sintering machines have the same build time as that of stereolithography. The melting temperature of the material being used also determines building speed. As a result, materials with lower melting temperatures build parts faster. Laser sintering is best suited for parts requiring superior material properties or parts that must be exposed to weather, chemicals, or heat. Complex parts with inaccessible internal features can be built because sintered parts don't require supports that must be removed after building.

Three-Dimensional Printing (3DP).   *Three-dimensional printing* (3DP), the most popular concept modeling system, was developed at Massachusetts Institute of Technology (MIT) and is licensed for use by several companies. The process (Figure 5–19) utilizes a thin layer of powdered material spread one layer at a time. The powder can be any material, including ceramics, metals, polymers, and composites. An adhesive binder is then applied to an area that represents the cross section of a part or the mold for a part. The adhesive is dispensed in droplets through a device similar to an inkjet printer head.

3DP functions by building parts in layers in a process similar to SLS. Each layer begins with a thin distribution of powder spread over the surface of a powder bed. The binder material is selectively applied to form the object. A piston that supports the powder bed and the part-in-progress lowers so that the next powder layer can be spread. This layer-by-layer process repeats until the part is complete. The formed part is finished by infiltrating it with wax, cyanoacrylate, or two-part epoxy for added strength.

The advantages of this process include these: there is high geometric flexibility (Figure 5–20); no supports are required for overhangs, undercuts, and internal volumes; and any material in any color that can be produced in a powder form will work. In addition, 3DP is fast because the printheads sweep across the platform in a raster pattern and dispense binder from multiple jets simultaneously. In raster printing, the printheads sweep back and forth as in an inkjet printer rather than following the outline of the part. As a result, the process is more than forty times faster than other rapid prototyping technologies because every layer, regardless of complexity, takes the same amount of time, about thirty seconds, to build. According to Z Corporation, the primary supplier of 3DP products, the

Figure 5–19 Three-Dimensional Printing.

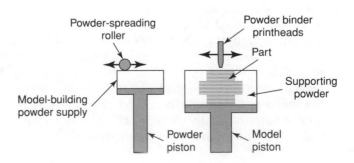

Powder-spreading roller

Powder binder printheads

Part

Model-building powder supply

Supporting powder

Powder piston

Model piston

Figure 5–20   Three-Dimensional Printing.
(Source: Courtesy of Z Corporation.)

system can build an automotive manifold in about four hours and a full-sized engine block in about sixteen hours. While the process is fast, it does not have the surface finish or accuracy provided by stereolithography.

A variation of 3DP, used by Soligen Corporation, is called *direct shell production casting (DSPC)*. In this variation, the powder is ceramic and the part produced is a casting. The engine manifold shown in Figure 5–21 illustrates this prototyping process. With this technique, castings for complex parts are produced much more quickly than the traditional model building process would permit. In addition to the speed advantage provided by DSPC, the process also permits parts too complex for standard model-making techniques to be manufactured efficiently and quickly.

Fused Deposition Modeling.   *Fused deposition modeling (FDM)*, illustrated in Figure 5–22, builds up each cross section by moving a thin extruded "wire" of plastic or

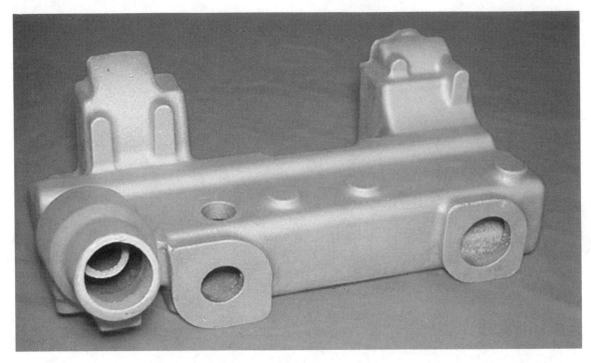

Figure 5–21   Engine Manifold Modeled with the DSPC System.
(Source: Courtesy of Soligen, Inc.)

wax just above the part location and heating it to its melting point. Again, the part is built one cross section at a time. The process is fast and employs relatively cheap materials. In addition, FDM can be used in an office setting, unlike stereolithography and laser sintering, which are not environmentally safe for an office. Some geometries produced by FDM have a grainy appearance, and materials are limited to investment casting wax and a "nylonlike" material.

Figure 5–22   Fused Deposition Modeling.

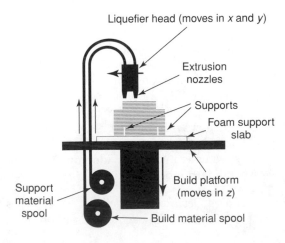

The chamber temperature is kept just below the melting temperature of the material, so the heat required to melt the filament sufficiently for extrusion is small. If the temperature is controlled correctly, the growing part won't sag or deform because the molten material solidifies on contact. FDM machines use a second nozzle located next to the primary nozzle to extrude *temporary* support material for overhangs and isolated features of the part. Stratasys has two types of supporting material: a waxlike material and a water-soluble material. The *wax* supports are used where removal is not blocked by finished-part geometry, and *water-soluble* supports are used for internal passages or small features. An agitated bath dissolves water-soluble supports. Finished FDM parts require no additional postprocessing.

An example of the process is illustrated in Figures 5–23 through 5–26. The Pro/E CAD model for a ratchet bottom and top is shown assembled in Figure 5–23 as a wire frame, in Figure 5–24 as a solid model of the bottom, and in Figure 5–25 as a solid model of the top. An STL file for each of the parts was created in Pro/E for use in an FDM machine. The model is built on a black foam base that is visible under the fully finished ratchet top and the half-finished ratchet bottom shown in Figure 5–26a and under the fully finished parts shown in Figure 5–26b. Note in Figure 5–26a the support material around the center ring of the part on the right. In Figure 5–26b the support material is visible through the hole of the part on the right and below the ratchets for the part on the left.

Figure 5–23   Assembly of Ratchet in Pro/ENGINEER CAD Wire Frame.
(Source: Courtesy of Stratasys, Inc.)

Figure 5–24 Ratchet Bottom in Pro/ENGINEER CAD Solid Model.

(Source: Courtesy of Stratasys, Inc.)

FDM systems build to an accuracy of ±0.005 inch (0.127 mm) and have a surface finish that is better than that of laser sintering but not equal to that of stereolithography. FDM systems build small or thin-shelled parts more quickly than large or solid parts because the material is put down in narrow beads called *roadways.* The speed is also affected by the nozzle size used in the application.

FDM parts built with ABS, polycarbonate, and polyphenyl sulfone materials are sufficiently resistant to heat, chemicals, and moisture that they are used for limited to extensive functional testing. For example, the development of the new model of Oreck vacuum cleaner was shortened by five months because thirty-one of the prototype machine's parts were built with FDM. All the parts that are dark

Figure 5–25 Ratchet Top in Pro/ENGINEER CAD Solid Model.

(Source: Courtesy of Stratasys, Inc.)

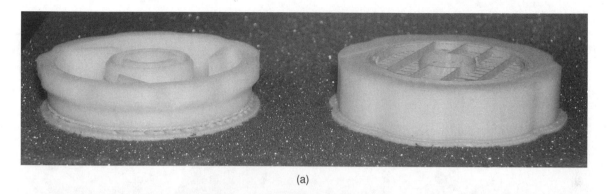

(a)

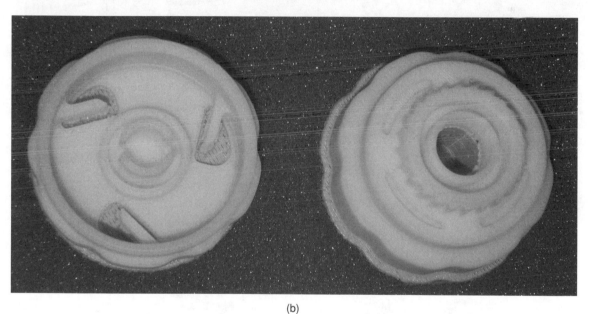

(b)

Figure 5–26   FDM Rapid Prototypes of (a) Fully Finished Top and Half-Finished Bottom and (b) Fully Finished Top and Bottom of a Ratchet.

in Figure 5–27 were built by using FDM from Stratasys, Inc. After the design was tested with FDM parts, five prototypes were built for life testing with regular production parts. The test stand for the life test is pictured in Figure 5–28.

Laminated Object Manufacturing.   In the *laminated object manufacturing* (*LOM*) system, parts are built up from sections cut from thin sheets of stock. The stock may be paper, plastic, or polyester composite. As the part is built (Figure 5–29), each sheet (0.002 to 0.02 inch thick) is glued with adhesive (which is activated by heat) to the already-constructed part, then trimmed with a laser. LOM is five to ten times faster than other rapid prototyping processes because the laser beam traces only the outline of each

COMMERCIAL VACUUM RUNS ON RAPID PROTOTYPE PARTS

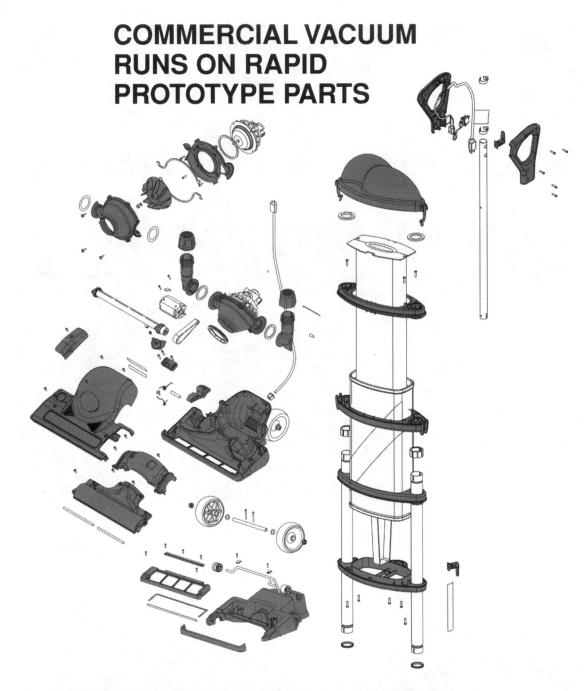

Figure 5–27   Oreck Vacuum with Thirty-One Parts (Dark) Produced with an FDM Machine.
(Source: Courtesy of Stratasys, Inc.)

Figure 5–28 Five Oreck Vacuum Prototypes in a Test Stand.

(Source: Courtesy of Stratasys, Inc.)

Figure 5–29   Laminated Object Manufacturing.

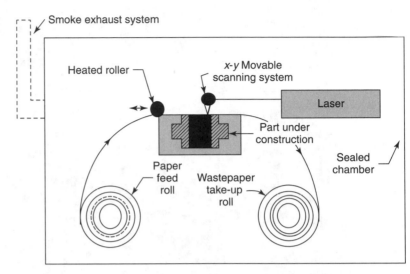

cross section, not the entire area. In addition, production of very large parts is possible with an accuracy of ±0.005 inch over the dimensions of the machine.

Multijet Modeling.   *Multijet modeling* (*MJM*) is in a class of rapid prototyping systems with 3DP called *concept modelers.* While stereolithography, laser sintering, and fused deposition modelers are designed to create accurate parts that can be used for limited functional testing, concept modelers produce relatively fragile parts cheaply and quickly for early evaluation of a product's design. In the concept modeling area, 3DP is better for larger parts and MJM is better for small, detailed parts.

The multijet modeler builds parts by spraying a molten waxlike material in layers onto a platform, using an 8-inch-wide printhead containing 352 jets (Figure 5–30). The droplets from the printhead, which measure just 0.003 inch (0.025 mm) in diameter, cool rapidly and harden when they strike the model surface. When a layer is complete, the build platform lowers by one layer thickness, typically 0.0017 inch (0.042 mm), and the next layer is sprayed. Thin, hairlike structures are used to support isolated or overhanging features of the part. The supports are brushed away after the part is built.

Figure 5–30   Multijet Modeling.

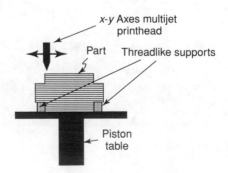

The real strength of MJM is its detail and surface-finish capability. This 3-D system has a resolution of 300 by 400 dots per inch along the horizontal x- and y-axes and 600 dots per inch in the z (vertical) direction. The smallest feature or wall can be as small as 0.003 inch (0.075 mm) across. The process is relatively fast, and the parts require no postprocessing, so they are available for use when built. Multijet modelers are best for applications where part detail is more important than durability, and they are ideal for building investment casting patterns.

The Present and the Future of Rapid Prototyping.   Rapid prototyping has made great strides in the last five years. The development of rapid prototyping departments by the larger manufacturers has driven the sales of systems: approximately 1,300 were sold in 2001. In a similar development, smaller producers moved their prototyping needs to service bureaus selling machine time. As a result, the number of service organizations selling machine time had grown to 397 by 2002.

An overview of the rapid prototyping industry technology is presented in Table 5–2. Study the table and review the data presented for each of the system types. The industry leader is 3D Systems because it has added two other technology types through acquisitions of other companies. This company is the sales leader, and 61 percent of the service bureaus use one of its machines. Stratasys is in second place, with 9 percent of its machines in service bureaus.

Predicting the future of any technology product is difficult, but several changes in rapid prototyping are clear. On the machine side, there should be improvements in accuracy and process speed. From a materials standpoint, improvement in the current materials and the introduction of new materials will increase model stability, strength, and durability. These changes should carry rapid prototyping forward into the realm of custom manufacturing. The manufacturing industry will continue to be the dominant user of the technology, but medical applications should start to gain momentum as well.

5–4 GROUP TECHNOLOGY

Group technology (GT) is not an automation strategy associated with either the design or the production engineering area, but the implementation of some form of GT is often necessary to achieve the order-winning criterion described in chapter 1. For example, GT is a critical first step for computer-aided process planning (CAPP) and many of the production engineering activities described in the next section.

Defining GT

GT is defined as follows:

> *GT is a manufacturing philosophy that justifies small and medium-size batch production by capitalizing on design and/or manufacturing similarities among component parts.*

Table 5-2 The Most Important Commercial Rapid Prototyping Technologies at a Glance

Technology type	Stereolithography	Wide area inkjet	Selective laser sintering	Fused deposition modeling	Single-jet inkjet	Three-dimensional printing	Laminated object manufacturing
Vendor	*3D Systems*	*3D Systems*	*3D Systems*	*Stratasys*	*Solidscape*	*Z Corp.*	*Cubic Technologies*
Maximum part size (in.)	20 × 20 × 24	10 × 8 × 8	15 × 13 × 18	24 × 20 × 24	12 × 6 × 9	20 × 24 × 16	32 × 22 × 20
Speed	Average	Good	Average to fair	Poor	Poor	Excellent	Good
Accuracy	Very good	Good	Good	Fair	Excellent	Fair	Fair
Surface finish	Very good	Fair	Fair	Fair	Excellent	Fair	Fair to poor (depending on application)
Strengths	Market leader, large part size, accuracy, wide product line	Market leader, office okay	Market leader, accuracy, materials	Office okay, price, materials	Accuracy, finish, office okay	Speed, office okay, price, color	Large part size, good for large castings, lower material cost
Weaknesses	Postprocessing, messy liquids	Size and weight, fragile parts, limited materials, part size	Size and weight, system price, surface finish	Speed	Speed, limited materials, part size	Limited materials, fragile parts, finish	Part stability, smoke, finish, accuracy
System price	$75–800K	$50K	$300K	$30–300K	$70–80K	$30–70K	$120–240K
Material costs ($/lb.)							
Plastics	75–110	100	30–60	115–185	100		9
Metal			25–30				
Other			5 (foundry sand)			Starch: 0.35/in^3. Plaster: 0.60/in^3. + infiltrant	5–8 (paper)

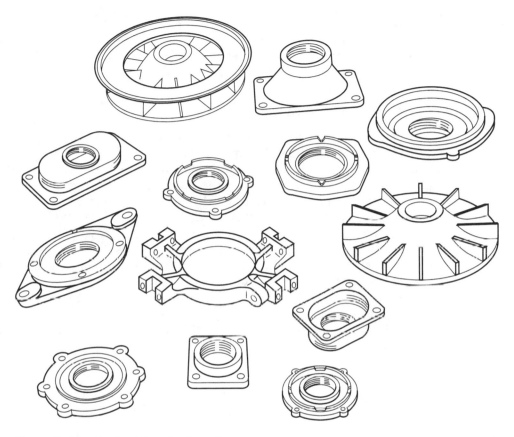

Figure 5–31   Family of Parts for Group Technology.

Under a GT implementation, dedicated production cells are created in which families of parts grouped by a selection code are produced on a set of production machines selected for the part group. For example, study the parts shown in Figure 5–31 and list all features common to all the parts. At first glance, the parts shown in the figure appear to have no common features. However, with further study the common features and machining operations become apparent: (1) all are similar in size and shape, (2) the raw material is a casting, (3) all require internal hole boring in a single direction, (4) all require face milling, and (5) most have drilled holes in a single direction. Common features that are not obvious from Figure 5–31 include required *dimensional tolerance*, *type of material*, and *surface-finish* demands. As a result of the similarity in process operations, a single production cell could be built to machine the part family shown in Figure 5–31.

GT offers a structured method of classifying parts on the basis of geometry and production characteristics. With GT coding on all parts and components, a company can sort all manufactured parts into part families suitable to a single production cell. Conversion to GT and cell manufacturing supports the following order-winning criteria: shorter lead and setup times, reduced work-in-process and finished-goods inventories, and less material handling. In addition, GT helps simplify production planning and control and is necessary for the successful implementation of other production engineering software applications.

Coding and Classification of Parts

Coding is a systematic process of establishing an alphanumeric value for parts on the basis of selected part features. *Classification* is the grouping of parts according to code values. Coding and classification in GT are highly interactive because the coding system must be designed to produce classified groups with the correct combination of common features. Relevant part features are used to place parts into groups by using either the *hierarchical, chain,* or *hybrid* code structures.

The *hierarchical code structure,* called a *monocode,* is based on the biological classification system established by Linnaeus. This type of coding, often called a *tree structure,* divides all parts of the total population into distinct subgroups of about equal size. Study the example shown in Figure 5-32, where the total population is all cylindrical parts. The group of parts is initially subdivided into two groups according to the ratio of the length divided by the diameter of the parts. The values on the conditions for the initial branch (>1.5 and <1.5) are selected so that each of the subgroups is about equal in size. Additional subgroups are selected

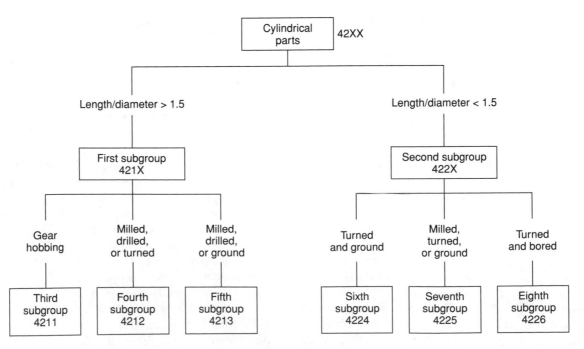

Figure 5–32 GT Monocode.

Digit	Class of feature	Possible values of digits			
		1	*2*	*3*	*4*
1	External shape	Shape$_1$	Shape$_2$	Shape$_3$	—
2	Internal shape	None	Shape$_1$	—	—
3	Number of holes	0	1–2	3–5	5–8
4	Type of holes	Axial	Cross	Axial and cross	
5	Flats	External	Internal	Both	
6	Gear teeth	Spur	Helical		

Figure 5–33   GT Polycode Structure.

according to the presence of gears and the type of machining performed. The number of digits in the code is determined by the number of levels in the tree. Another characteristic of the code numbers at each level is that only the least significant digit is different. Study the tree structure shown in Figure 5–32 until these last two concepts are clear. The advantage of the hierarchical structure is that a few code numbers can represent a large amount of information. The singular disadvantage is the complexity associated with defining all the branches.

The *chain structure*, called a *polycode*, is created from a code table or matrix like the example shown in Figure 5–33. Study the example matrix; note that the type of part feature and digit position are defined by the left vertical columns. The numerical value placed in the digit position is determined by the feature descriptions across each row.

Exercise 5–3
A part with the code 311412 has a 2 in position one, a 1 in position two, a 4 in position three, . . . , and a 3 in position six. Determine the features of the part using the polycode chart shown in Figure 5–33.

Solution
The features described by the code are no gear teeth (3 in position six), external flats (1 in position five), all axial holes (1 in position four), 5 to 8 holes (4 in position three), no internal shape or cutout (1 in position two), external shape in the Shape$_2$ category (2 in position one).

The major advantages of polycodes are that they are compact and easy to use and develop. The primary disadvantage is that, for comparable code size, a polycode lacks the detail present in a hierarchical structure.

A *hybrid code* captures the best features of the hierarchical and polycode structures. One of the best examples of a hybrid code is the Optiz code and classification system (Figure 5–34), developed in 1970. Note that the code starts and ends with a polycode and has a hierarchical code in the middle. Industry currently uses over

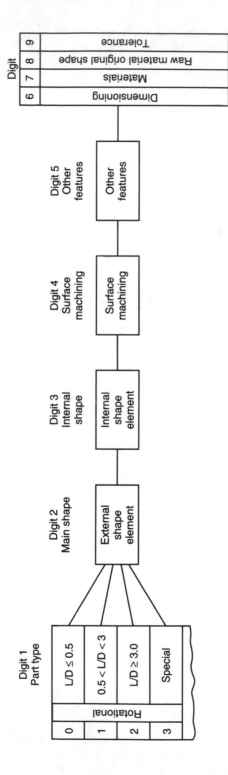

Figure 5–34 Adoption of an Optiz-Type Code That Combines Monocode and Polycode Structures. The Code Can Be Expanded for Nonrotational-Type Parts.

182

Figure 5–35 Traditional Process or Job Shop Layout.

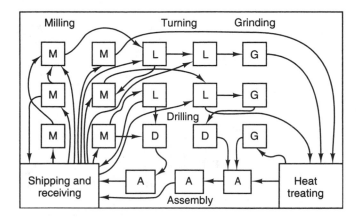

ninety GT coding systems; the code selected is dictated by the type of product, production system, and total mix of all parts and components. With the code and classification system selected, the development of a GT production cell follows.

GT Production Cells

Batch manufacturing has traditionally taken place in a functional layout where similar machines are grouped together (for example, all mills grouped together, all lathes grouped together, etc.) in the production facility. In the batch production process, parts are routed through these various work centers according to a specified sequence of operations. For example, the layout of a typical job shop is illustrated in Figure 5–35. Note that all similar machines are located in the same geographic area on the shop floor. Parts are routed to the machine for the specific operation required by the design. A partial list of the inefficiencies include *high in-process inventory, large number of parts handlers, longer lead times*, and *longer setup time*. The application of GT part families to batch production requires a physical rearrangement of the production facility. With GT applied to the production part mix shown in Figure 5–31, a family of parts is created and three linear production flow lines with cells are developed (Figure 5–36) to handle the same production. Production is no longer organized around machine similarity; instead, "groups" of different machines are identified according to their ability to produce families of parts.

Another method for identifying part families and the associated groups of machine tools is called *production flow analysis* (*PFA*). The primary grouping and classification data used in this process are the operation sequences and machine routings used for the parts under study. Parts with identical or similar routings are grouped together into families. From these family groups, logical machining cells and a GT layout are produced.

Regardless of the method used to code, classify, and eventually reorganize the shop floor, the primary benefit from the GT exercise is better organization, identification, and understanding of the products in manufacturing.

Figure 5–36   GT Layout.

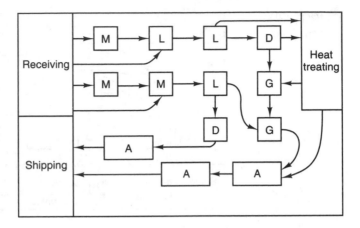

5–5 PRODUCTION ENGINEERING STRATEGIES

The production engineering function, highlighted in the organizational model in Figure 3–2, is closely linked to design. In the preceding chapters, we described the impact on the design function as a result of the computer and software revolution. The degree of change in the design function because of the development of computers is paralleled in production engineering. The term *computer-aided manufacturing (CAM)* was coined in the 1960s when early mainframe and special-purpose computers were interfaced to production machines to automate some of the production engineering and manufacturing functions. The introduction of computer automation into the production engineering areas continues to increase.

Production engineering has the responsibility for the development of a plan for the manufacture of goods and services for the enterprise. The elements in the plan, listed in chapter 3, include *process planning, production machine programming, tool and fixture engineering, work and production standards, plant engineering, analysis for manufacturability and assembly,* and *manufacturing cost estimating.* The analysis for manufacturability and assembly was described in section 5–1. A description of the CIM automation strategy associated with the rest of the elements in the manufacturing plan is provided in the following sections.

Computer-Aided Process Planning

Manual process planning includes the creation of all paperwork necessary to direct the flow of raw materials and parts through production and assembly. The process planner determines the sequence (Figure 5–37) and machines that will transform the raw material into a finished part. For example, in a typical job shop, the planner studies the drawing of the part, selects data from the machinability data handbook, checks the tooling and fixtures available, selects raw material stock, and then selects the metal-cutting operations available on the shop floor that are necessary to produce the part. The plan for the spindle in Figure 5–1

| | Process planner | | | |
	One	Two	Three	Four
1	Machine first face	Hole drilled in two steps: a. 20 mm dia b. 38 mm dia	Outside surface– 70 mm dia– turned	Hole drilled to finish in two steps: a. 30 mm dia b. 40 mm dia
2	Hole finished in three steps: a. Drill 10 mm b. Drill 38 mm c. Bore 40 mm	Machine first face	Hole drilled to finish in one step with drill of 40 mm dia	Outside surface– 70 mm dia– turned
3	Outside surface– 70 mm dia– turned	Cutoff	Machine first face	Machine first face
4	Cutoff	Machine second face	Cutoff	Cutoff
5	Machine second face	Outside surface– 70 mm dia– turned	Machine second face	Machine second face
6		Hole finished to 40 mm dia by boring		

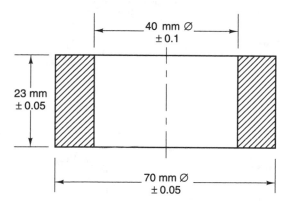

Figure 5–37   Comparison of Process Planning by Four Different People.
(Source: Schaffer, "Implementing CIM," © American Machinist, August 1981, p. 82, Reprint by permission of American Machinist, a Penton publication.)

would start with a cutoff saw operation (Figure 3–18) to cut a 90-mm length from a ¾-inch stainless steel bar. Following the deburring operation, the planner would route the material to a turning center, or lathe (Figure 3–18), and the contour would be cut. Following the lathe operation, the spindle would be routed to a centerless grinder to finish the bearing surfaces on each end. For a simple part like the spindle, the planning process is clear. However, for more complex parts, the order of operations depends on the planner's knowledge and experience.

For example, four planners were asked to plan a part; study the results, listed in Figure 5–37, from the four planners. The order of operations in each is different, and the method used to produce the 40-mm hole varies. Two of the planners decided they could get the required finish for the 40-mm hole from a drill, and two

Figure 5–38   Developing a Variant Process-Planning System.

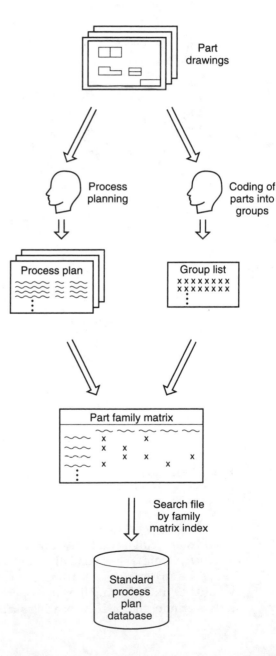

required a boring operation. The two calling for a boring operation were more experienced and correct. Consistent and correct planning requires both knowledge of the manufacturing processes and experience. Two automation techniques, *variant* and *generative* process planning, to improve the processing planning operation in the integrated environment are embodied in *computer-aided process planning (CAPP)*.

The CAPP *variant approach* uses a library of manually prepared process plans in a database and a retrieval system to match components on new parts to existing process plans of similar components. When the process plan is valid for a family of components, it is called a *standard* plan and is stored in the enterprise database with the family key number for identification. The retrieval method and the logic of the variant system are predicated on the grouping of parts into families, as in GT. In most situations, the standard plan must be modified to some extent before the plan can be used with the new component parts. After the preparatory stage (Figure 5–38), where the families of standard process plans are developed and saved in the variant database, the system is ready for production components. For a new part, the flow indicated in Figure 5–39 would be used for the variant process. A new production component is given a family code and then passed through a part-family search routine to find the family to which the component belongs. The standard plan for that family of components is retrieved and a human planner makes the adjustments necessary for the new component. Figure 5–39 illustrates the variant process-planning procedure. Study Figures 5–38 and 5–39 until you understand the variant process. The primary advantage provided by the variant technique is the reduction of process-planning time by almost 50 percent.

The CAPP *generative technique* for the creation of process plans is both more difficult to develop and more highly automated. In general, the generative process-planning system creates plans for new components without referring to existing plans or with the assistance of a human planner. Generative CAPP utilizes a process information knowledge base that includes the decision logic used by expert human planners. Frequently, the heuristic planning knowledge is captured in artificial

Figure 5–39 Variant Process-Planning System.

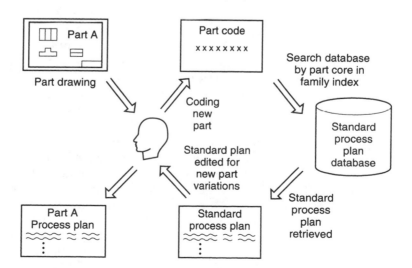

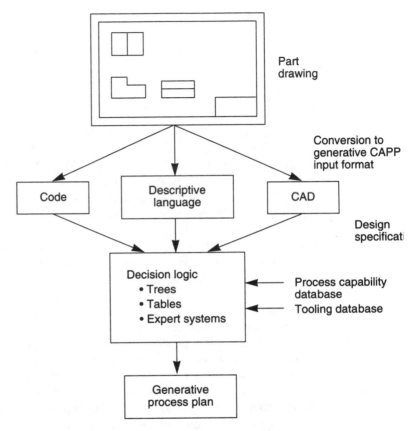

Figure 5–40 Generative Process-Planning System.

intelligence (AI) software, called *expert systems*. The AI software is designed to store and imitate the human decision-making process. The generalized system illustrated in Figure 5–40 shows the generative CAPP operation. A part drawing is received from design and needs a process plan created for manufacturing. The first step is to convert the design specifications into an input format compatible with the CAPP automation. The three techniques frequently used include code, a descriptive language, and CAD. In each technique, the complete design specification for the part is converted into a format compatible with the decision engine in the CAPP software. The decision logic portion of the CAPP system uses manufacturing database information, such as production machine capability, tooling, fixtures, and time standards, and the design specifications to arrive at an operational process plan. The three most commonly used decision logic algorithms are listed in Figure 5–40.

Leading CAPP software captures the knowledge and experience that has been developed on the shop floor, creates detailed process plans with accurate time standards, and then communicates this information to material requirements planning (MRP) and enterprise resources planning (ERP) databases. METCAPP is an example of CAPP software that provides these types of benefits in the planning process. METCAPP was developed by the Institute of Advanced Manufacturing Sciences in Cincinnati, Ohio, for machining operations. METCAPP uses cutting speeds and

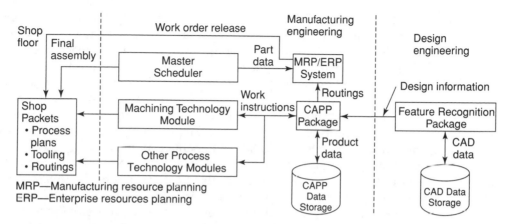

Figure 5–41 CAPP in the Enterprise System.

feeds from the industry standard handbooks and reference sources, which makes accurate cut time and cost calculations possible. METCAPP uses solid-part models to extract design features. The software separates part features into the appropriate setups and machining operations, then generates a complete process plan for the part production. Using the solid-part model from the CAD software, METCAPP imports the model and automatically generates a full-blown process plan with detailed work instructions for every step in the machining process. METCAPP is supplied with a database of 12,000 time standards, machining systems rules for forty-two common features, data on 18,000 cutting tools, and specifications for over 100 machine tool models. The interface between CAPP and the enterprise is illustrated in Figure 5–41. Note how the CAPP software forms the interface between design and the shop floor, thus providing critical integration of the product data from initial design to final part production.

The advantages provided by generative CAPP include the following:

- Process plans are created rapidly and consistently.
- Totally new process plans are created as quickly as plans similar to those for existing components.
- An interface to MRP and ERP software is possible.
- Documentation requirements of ISO or QS 9000 standards are completed and best practices are promoted, even without experienced workers.

Computer-Aided Manufacturing

The term *CAM* is used to describe a wide range of automation technologies. The reason for the confusion over the definition of *CAM* is explained by a brief review of the history of shop-floor automation. The emerging technologies chart shown in Figure 4–2 indicates that *numerical control* (NC) was the start of the shop-floor automation evolution. As U.S. industry emerged from World War II, the

NC technology and computers began to shape automation on the shop floor. The next significant jump in productivity was the development of *direct numerical control* (*DNC*) machine tools. Entire shop floors of CNC or NC machines were connected directly to a large mainframe computer that acted as a central program repository and control center. Later, NC tape-programmed production machines were replaced with *computer numerical control* (*CNC*) equipment that integrated a computer-type controller into the production machine. The CNC production equipment could now store and execute part programs without program tapes or a DNC connection to the mainframe computer and with less operator intervention. In the 1960s, other production systems, especially in the process control industries, incorporated computers into production processes. In the 1970s, computer hardware shrank in size, dropped in price, and increased in capability; as a result, "smart" production machines with powerful computers were available to every size industry. As a result of the rapid development of computer-driven shop-floor automation, the term *CAM* no longer applies just to automated machine tools. Understanding the broad new role that CAM plays in a fully integrated enterprise requires a clear definition:

> *CAM is the effective use of computer technology in the planning, management, and control of production for the enterprise.*

One of the major CAM applications, used by the discrete-part manufacturing industry, is the production of finished parts with information extracted directly from design-drawing data. In this application, often called *CAD/CAM,* the part geometry created with CAD in the design engineering area is used with CAM software to create machine code capable of machining the part on almost any CNC machine. The block diagram shown in Figure 5–42 illustrates the robust data interface between CAD drawing files and the CAM files required to machine the part. Study the diagram until you are familiar with all the names in the boxes.

The drawing of a machined part is created by using conventional CAD software and drawing techniques; no special commands or controls are embedded into the graphic file. However, it is customary to put all the geometry of the part on a separate drawing layer, with dimensions, notes, text, and other nongeometric information on other layers. This arrangement permits the part geometry, critical for the CAM process, to be stripped away from all the other drawing information by turning off all layers except the one with geometry. With the part geometry captured from the CAD drawing file, the geometry information is transferred to the CAM workstation.

The file format used to transfer the part geometry to the CAM workstation depends on several factors. Some CAM software has internal CAD capability to perform limited drawing of part geometry, and some of the CAD software systems have CAM capability as an option. For example, CAM software systems, such as SmartCAM and MasterCAM, have basic 2-D and 3-D CAD capability, while CAD software, like CATIA, offers true CAD/CAM, as the list of application modules indicates:

- *3-D design:* A 3-D graphics modeler
- *Drafting:* A basic drafting system

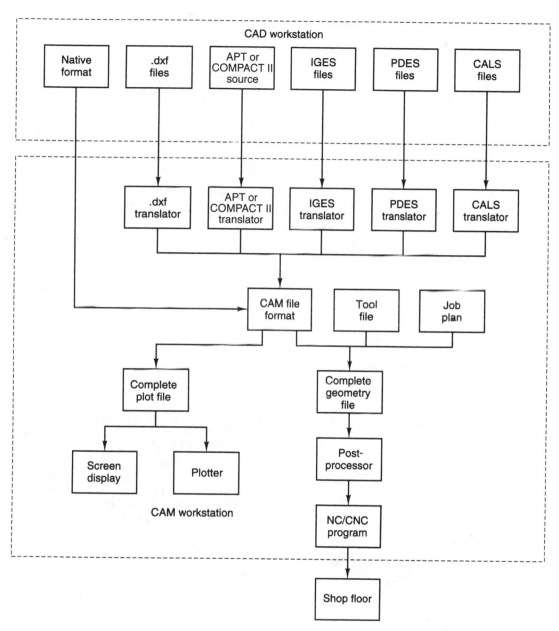

Figure 5–42 CAD/CAM Links.

- *Advanced surfaces:* A sculptured surface modeler
- *Solid geometry:* A solid geometry modeler
- *Building design:* An architectural design modeler
- *Library:* A library for custom symbols and objects
- *Numerical control:* An NC part programmer generator
- *Robotics:* A robot cell simulator and programmer

Most of the other CAD vendors offering workstation software support full CAM capability. Most of the CAD vendors who started in the UNIX environment have migrated their software to the PC with the NT/2000/XP operating systems. If CAD and CAM are integrated into the same software system, the CAD file in its *native* format can be transferred to the CAM application without the need for a format translation. Locate the native type of data transfer in Figure 5–42.

Another popular CAD file format used to translate drawing files between different brands of microcomputer CAD software is the .dxf (drawing interchange file) file format. Developed by the authors of AutoCAD, the .dxf format is accepted by most CAM software. For example, the AutoCAD drawing of the switch rotor displayed on the screen shown in Figure 5–43 was used with SmartCAM software to produce the plastic rotor pictured in Figure 5–44. The .dxf format is used most frequently with 2-D part geometry models. The 3-D features in the plastic rotor in Figure 5–45 were created by adding depth dimensions to the rotor geometry in the CAM software after the 2-D file was transferred by using the .dxf

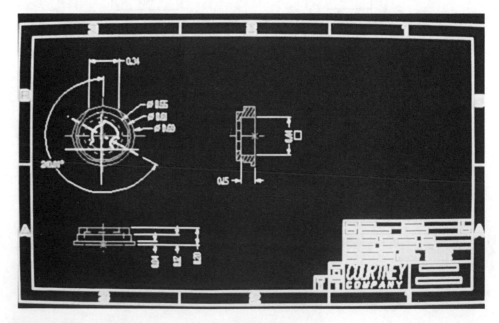

Figure 5–43   AutoCAD Drawing of Part for CAM System.

Figure 5–44 Milled Rotor.

format. The SmartCAM screen in Figure 5–46 shows an isometric view of the switch rotor after the z-axis dimensions were added. The .dxf format permits part geometry created on a microcomputer-based CAD software system to be sent to CAM software to prepare NC or CNC programs to cut the part. Inside the CAM software, a .dxf translator changes the .dxf file to a format that is native to the CAM system. Locate the .dxf file transfer in Figure 5–42.

The *automatically programmed tool (APT)* file format developed in 1956 at MIT and COMPACT II developed by Manufacturing Data Systems, Inc., in Michigan are two popular part programming languages in the United States. While .dxf is the common file format for transfer to CAM on microcomputer software, APT is used more frequently on RISC-based CAD/CAM solutions. APT and COMPACT II are NC programming languages with a format similar to that of the Fortran language. A sample APT program to mill a pocket is presented in Figure 5–47. After the program in Figure 5–47 is complete, a generalized solution in terms of a series of cutter location points, called a *CL* file in APT, is created. The cutter file is then

Figure 5–45 SmartCAM 2-D Representation of Rotor Part.

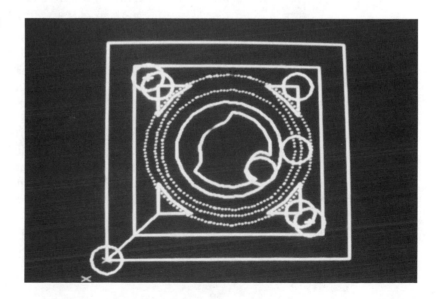

Figure 5–46 SmartCAM 3-D
Representation of Rotor Part.

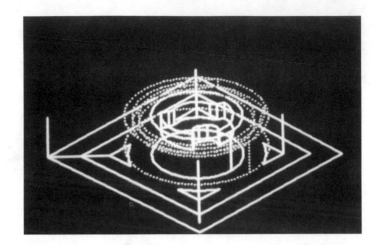

passed through a postprocessor, resident in the host computer or a CAM work-station, to create the NC program code. If the cutter file is sent to the CAM work-station, the cutter file is passed through an APT/COMPACT II processor to convert the file into a format compatible with the CAM software in the workstation. The CIMpro software from Intercim Corp. is an example of CAM workstation software that supports transfer of APT CL files and also offers a programming language called *Intercim APT*. Other NC programming languages in use include the following:

- *ADAPT and AUTOSPOT:* An IBM part programming language
- *UNIAPT:* Small computer version of APT
- *MAPT (Micro-APT):* A microcomputer version of APT

Before continuing, study Figure 5–42 and locate the APT file transfer path.

The *initial graphic exchange specification (IGES)*, jointly developed by industry and the National Institute for Standards and Technology in the 1970s, is the most widely used file format for CAD data exchange for RISC-based CAD software systems. For example, the CATIA software described earlier supports the IGES standard. The complex drawing files usually associated with this type of computer makes translation of drawing files between different software vendors difficult without several errors. While far from perfect, the IGES standard is the best-supported common format for 3-D CAD models at the present time. When IGES is used to bring part geometry into the CAM software, an IGES translator is required. The translator converts the standard IGES file format into a format compatible with the CAM software. Locate the IGES file exchange in Figure 5–42.

The last two file formats listed, the *product data exchange standard (PDES)* and *continuous acquisition and lifecycle support (CALS)*, are efforts to develop robust standard data part formats that extend beyond part geometry. Some projects have been undertaken to use these standards in the CAD/CAM interface, but wide-spread use in this application has not occurred. The CALS standard was adopted

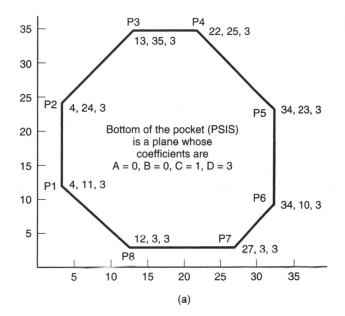

(a)

Pocket part program

```
 1  REMARK POCKET POLYGON COLLAPSE DEMONSTRATION TEST
 2  NOPOST $ $ NO POSTPROCESSING FOR THIS TEST CASE
 3  CLPRNT $ $ PRINT CUTTER CENTER DATA
 4  TOLER/.001 $ $ TOLERANCE BAND
 5  $ $ POINTS DEFINING POCKET PERIMETER
 6  P1 = POINT/4, 11, 3 $ $ STARTING POINT OF POCKET DEFINITION
 7  P2 = POINT/4, 24, 3
 8  P3 = POINT/13, 35, 3
 9  P4 = POINT/22, 35, 3
10  P5 = POINT/34, 23, 3
11  P6 = POINT/34, 10, 3
12  P7 = POINT/27, 3, 3
13  P8 = POINT/12, 3, 3 $ $ ENDING POINT OF POCKET DEFINITION
14  H = .01 $ $ SCALLOP HEIGHT MAXIMUM
15  D = .38 $ $ CONSTANT CUTTER DIAMETER
16  CR = .19 $ $ CUTTER CORNER RADIUS
17  D2 = SQRTF ((D * B) – B ** 2) $ $ A BALL END MILL EFFECTIVE CUTTER RADIUS
18  CV = (4 * D2)/D $ $ A MEASURE OF POCKETING CUT OFFSET
19  CUTTER/D, CR $ $ BALL END MILL
20  FEDRAT/50 $ $ MODAL FEED RATE
21  FROM/0, 0, $ $ STARTING CUTTER POSITION
22  GO TO/20, 20, 5 $ $ MOVE CUTTER TOWARD AND OVER CENTER OF POCKET
23  PSIS/(PLANE/0, 0, 1, 3) $ $ BOTTOM PLANE OF POCKET
24  POCKET/D2, CV, CV, 3, 10, 10, 1, 1, P1, P2, P3, P4, P5, P6, P7, P8, $ $ STATEMENT
25  GO DLTA/0, 0, 2 $ $ CLEARANCE POSITION OF CUTTER
26  GO TO/0, 0, $ $ END CUTTER POSITION
27  FINI $ $ END OF PART PROGRAM
```

(b)

Figure 5–47   (a) Part; (b) APT Program.

(Source: Chang/Wysk/Wang, Computer-Aided Manufacturing, *© 1991, p. 275. Reprinted by permission of Prentice Hall, Upper Saddle River, New Jersey.)*

for use in U.S. government Department of Defense purchasing applications and was used by companies working on government contracts. CALS has been replaced by *integrated data environments* (*IDE*). Again, a translator is necessary in the CAM system to convert these file formats to one acceptable to the CAM software.

PDES, Inc., is a consortium of members who are working together to accelerate the development and implementation of ISO 10303, which is commonly known as *Standard for the Exchange of Product Model Data* (*STEP*). This international standard exchanges data between different CAD/CAM and product data management (PDM) systems and is a proven way to ensure fast, reliable exchange and sharing of data among partners, customers, and suppliers. STEP enables interoperability, supply-chain integration, Web-based collaboration, and lifecycle management. If implemented, STEP allows manufacturers to reach new, higher levels of data quality and productivity while reducing costs and time to market. PDES, Inc., supports CIM and enterprise integration and interoperability for member companies by doing the following:

- Implementing STEP by using Web-based technologies
- Increasing modular STEP applications to support product data throughout the lifecycle
- Increasing interoperability among CAD/PDM, PDM/PDM, and electronic CAD/mechanical CAD (ECAD/MCAD) systems

Locate these last two formats—PDES and CALS—in Figure 5–42.

The technology just described used the CAD/CAM interface to move the part geometry required by the production equipment to the CAM *file format box* in Figure 5–42. This transferred part geometry file is used for two functions: (1) display of the cutting file for evaluation, and (2) preparation of a program to make the part on the production machine. In the first situation, the part geometry file is converted to a *plot file* that can be displayed on a computer screen or plotted on a paper plotter. Check the process shown in Figure 5–42 for the screen and plotter output. The computer screen from a SmartCAM program shown in Figure 5–48 indicates the tool paths required to produce the part pictured in Figure 5–44. Screen outputs are used to check the NC program code before it is sent to the production machine. Note that a single cutter size is used and all the cutter motion is represented. The example used in the last several figures started with a 2-D CAD file. CAM packages also support 3-D either through a .dxf file transfer or by creating the 3-D graphics directly in the CAM software. Figure 5–49 shows a 3-D mold created in SmartCAM.

The second function, preparation of the machine program, has several additional steps. Preparing a machine program requires the part geometry file to be merged with a *tool file* and a *job plan*. The *tool file* has a list of available tooling with the appropriate tool offsets; the *job plan* includes a recommended sequence of operations for a specific type of production machine. The completed part geometry file is presented to a *postprocessor* that converts the file to a sequence of machine codes. Each production machine vendor has machine-specific postprocessor software, so

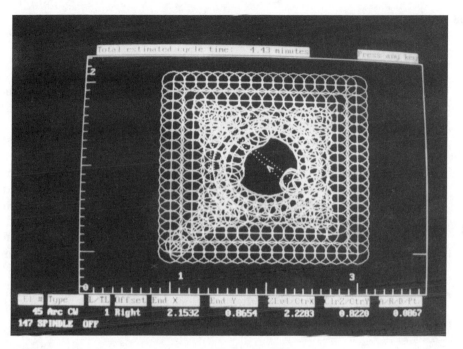

Figure 5–48   SmartCAM Screen of Rotor Part and Cutter Path.

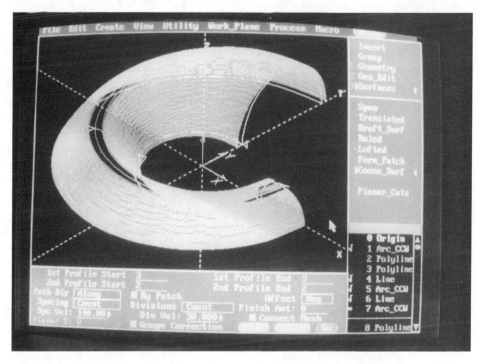

Figure 5–49   Mold Cavity Created by Using 3-D Graphics in SmartCAM Software.

machine code is generated to cut the part on a specific type of CNC production machine. The completed production program is sent to the machine on the shop floor over a local area network (LAN), through a direct link to the machine or on permanent storage media. In the past the permanent storage medium was paper or Mylar tape, but current systems use magnetic storage devices.

Programming discrete-part production machines is the most common application of CAM technology on the shop floor. The machines that are programmed to produce a product from CAD geometry include most of the metal-cutting machines: *mills, lathes, turning centers, punch presses,* and *nibblers;* and *boring, electrical discharge machining (EDM),* and *laser cutting machines.* In addition, CAM includes a host of other computer-controlled applications on the shop floor, too broad a range to list and describe. The linking of shop floor and continuous-process production areas to the enterprise computer system is easy to accomplish. Through this link, a two-way data exchange occurs between the machines and process and the departments controlling the production. The control of the many different systems and processes on the shop floor is possible only because of the rich selection of CAM software available to manufacturing.

Production and Process Modeling

The analytic models of parts, products, and manufacturing systems use mathematical equations that uniquely describe the behavior of the manufactured parts and production systems under study. The part or system is analyzed over a range of operating conditions by forcing the part or system variables in the equations to vary between their limits. When the variables reach worst-case conditions, the simulation provides a look at the part or manufacturing system under the most stressful conditions. The stress imposed by the worst-case environment may cause a catastrophic failure in the part, product, or manufacturing system. For example, consider the design of the aluminum alloy struts for the landing gear of a commercial jet aircraft. In the best-case scenario, the full weight of the aircraft must be supported by the struts when the plane is sitting at rest. However, the struts must be sized to handle the worst-case stress conditions that occur during landing and braking. The lightest design that can handle the worst-case condition with some safety margin is required. Subjecting an analytic computer model of the strut to the most vigorous landing conditions results in an optimum design and a significant reduction in engineering changes when the strut is sent to manufacturing. The techniques described earlier in the FEA section are additional examples of analytic models. Analytic models are applied to production problems as well.

Figure 5–50 illustrates an example of an analytic computer model used to analyze the design of a manufacturing cell including an industrial robot. The 3-D wireframe models of the cell and robot are tested by using kinematic motion analysis to determine the optimum position for the robot and associated cell machines and hardware. For some types of parts and for some elements of the production system, the equations required for the model are well defined and understood. However, the previous example illustrates two important points about analytic models: (1) the

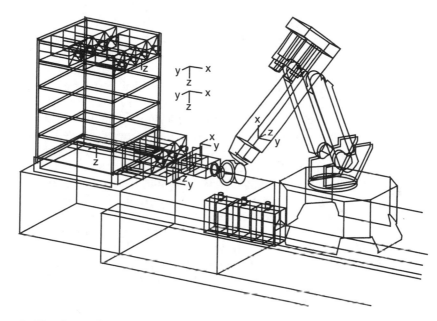

Figure 5–50   Screen Output for Work Cell Simulation.
(Source: James A. Rehg, Introduction to Robotics in CIM Systems, Fourth Edition, © 2000, p. 118. Reprint by permission of Prentice Hall, Upper Saddle River, New Jersey.)

use of models is a cost-effective and fast method of determining the limits of a design, and (2) the results obtained from the analytic model are valid only to the degree that the equations in the model truly describe the part, product, or manufacturing system. The last point is important to consider because manufacturing is a highly complex system of machines and operations; as a result, developing the equations to describe the operation accurately is often difficult. When a computer model is not possible, *simulations* offer a reliable alternative to analytic modeling of the more complex parts and manufacturing systems.

Production and Process Simulation

The construction of a physical model to analyze product and process behavior is called *prototyping*. To avoid the high cost of building this physical model, many manufacturers use computer-generated models called *simulations* to study the system. In manufacturing, *computer simulation* is defined as follows:

> *Computer simulation is the development of a theoretical or graphical model of a process or production system to evaluate behavior under varied conditions and in changing environments.*

Two computer simulation techniques are used to test manufacturing systems: discrete-event and continuous.

Discrete-event simulations use symbols to represent objects and resources, such as parts and machines, in the construction of a manufacturing model. The interaction between symbols is defined mathematically or with a logical relationship. Since the mathematical expression is for only a part of the manufacturing system, the complexity of the problem is reduced. The computer simulates the operation of the modeled process by capturing the operational data for objects and resources through time. The time for those intervals is established in the simulation by the computer clock. For example, if a machining work cell produces a finished part every five minutes, the output from the cell occurs at discrete intervals. The operational state of the objects and resources in the manufacturing system are saved in a series of variables called *state variables*. The manufacturing system is modeled by establishing the relationships between the symbols, defined in the simulation program, and an event calendar with the times for discrete events. Examples of events include job_arrival, begin_operation, end_of_operation, machine_breakdown, and machine_repair. For example, if the event calendar triggers a job_arrival event, production is initiated with the relationship between the part and the machine determining when the end_of_operation event is triggered. The time between the two events is the production time. Parameters embedded into the simulation software permit the production time to be affected by tool wear, setup problems, and other events, such as machine_breakdown. As the simulation executes, the statistical data associated with the objects and resources are captured in the state variables. Analysis of the state variable helps solve manufacturing problems plaguing the shop floor or identifies problems before the production system is built. For example, if the queue time for parts at machines is saved as a state variable, the queue variables with large values identify bottlenecks in the flow through the production system. In addition, many of the new manufacturing simulation programs provide a graphic overview of the production facility. The production flow is illustrated in color on the computer monitor, with problems highlighted for easier recognition.

Discrete-event simulation models use two programming methods: general-purpose computer languages, such as C and Pascal, or languages designed specifically for simulations. Commonly used simulation languages include the following:

- General Purpose Simulation Systems (GPSS)
- General Activity Simulation Program (GASP)
- Simulation Language for Alternative Modeling (SLAM)
- Research Queueing Package (RESQ2)

The simulation languages do not require the programming proficiency needed for languages such as C and Pascal; however, the special-purpose languages often fit one class of simulation better than others. Therefore, as the simulations become complex and sophisticated, the user is forced to adopt more powerful general-purpose languages.

Continuous processes require a different type of simulation strategy because the state of the system changes continuously with time. Examples of continuous systems include production of synthetic fiber, rubber, and many petroleum-based products. The continuous systems are modeled by using the mathematical and logical relationship between production components described in the last section. In addition, the model includes one or more differential mathematical equations that describe the rate of change in the state variables with respect to time. This last element in the model makes the continuous process different from the discrete-event technique. State variables are used to capture the critical production statistical data with the values plotted in real time or recorded at a specified sampling rate. In the past, all complex continuous-process simulations were performed on analog computers, but now most of the simulations are executed on high-speed digital machines.

The advantages of manufacturing simulation include the following:

- The optimum solutions for the manufacturing layout and production flow are identified.
- Alternatives (new products) and changes (quantity and mix of products) in manufacturing are evaluated rapidly.
- Production problems associated with product flow and material movement are identified.
- Manufacturing performance under various production rates is easily studied.
- Decisions that can change the manufacturing environment or product flow are analyzed before implementation.

The applications of discrete-event and continuous manufacturing simulations will increase as the price-to-performance ratio of computer platforms continues to fall.

Maintenance Automation

Machine and plant maintenance are major cost centers in manufacturing. The dollars spent in this area add no direct value to the products produced in the enterprise; however, if maintenance is ignored, a successful CIM implementation would be impossible to sustain. Automation in the maintenance area can include both hardware and software. The hardware includes better tools that are designed to shorten the time required to do plant and machine maintenance. Software designed to improve plant and machine maintenance has two components: (1) faster and more accurate identification of malfunctions in production hardware and manufacturing systems and (2) better management of the maintenance operations.

Artificial intelligence techniques in the form of *expert systems* permit the development of maintenance software to assist maintenance personnel in identifying machine problems more quickly and easily. An *expert system* is a computer program containing a series of rules that mimic the logical processes of an expert. For example, an expert system written for troubleshooting would use the same logic that a skilled troubleshooter with years of experience would follow in finding a faulty circuit in a production machine. When less-experienced technicians use this expert system software, they are helped by the experienced troubleshooter's logical set of measurements to identify the bad component or part. Expert systems or similar software is often embedded into the computer that controls a machine so that some internal faults can be diagnosed by the machine and reported to maintenance. In some cases, faulty machines are connected to the phone line through a modem, and a computer in an off-site location, sometimes another state, checks the machine and identifies the problem area.

The other area of maintenance automation focuses on improving the efficiency of the management of the maintenance operation. Software in this area performs functions that include scheduling of all planned and preventive maintenance, building databases of equipment and production machine part numbers, statically tracking machine maintenance cost, tracking the use of supplies and spare parts in maintenance inventory, supporting bar code input of machine numbers for annual capital equipment inventory checks, and producing a host of reports that help determine how maintenance operations can improve.

Production Cost Analysis

Establishing accurate product costs is critical in all five of the manufacturing systems (project, job shop, repetitive, line, and continuous) because price is an order-winning or order-qualifying criterion for many products. The accuracy of the cost estimate is especially vital in the first two manufacturing systems, project and job shop, because every job is a new product or design. In addition, a mistake in estimating fixed or variable cost for the product or job is not buffered by long production runs and large repeat orders. It is an equally important factor in the four production strategies, especially in the assemble-to-order and make-to-stock area. When customer lead time is so short that products must be available from stock or assembled from completed subassemblies, the work-in-process and finished goods inventory costs rise. The first step in controlling these inventory costs is an accurate determination of product cost in all phases of manufacturing. The information provided in this section focuses on hardware and software use in the analysis and evaluation phase of the product design model to estimate product cost.

The ambiguity present in the CIM model becomes more apparent with increased understanding of the technologies present. The benefits of specific types of hardware and software are a function of how effectively the organization uses the results across all enterprise departments. The integration of enterprise systems makes the lines that separate departmental functions and responsibilities less clear. For example, the DFMA analysis described earlier used computer software to force a good design for assembly and manufacturing. The product cost information generated from the DFMA software is critical for accurate cost estimating. Study Figures 5–4, 5–13, and 5–14 again and note the wealth of cost data present. The DFMA analysis of machining operations on the spindle/housing assembly (Figure 5–51) also provides cost data in the metal-working area. In the past, large manufacturing operations would have estimators working separately from the design area. The advent of concurrent engineering (CE) and the integrating effect of the CIM software force a desegregation of many enterprise areas. For example, the department responsible for loading the production manufacturing and control

Worksheet from Machining Software

			Load/unload time, etc. (seconds)		Set tool, engage cut, etc. (seconds)		Area (in.³)		Operation cost(s)
Operation		<S>teel <D>isposable <C>arbide <G>rind							
Operation number	<R>ough or <F>inish		Setup time (hours)	Number of operations		Volume (in.³)		Machine Time (seconds)	
1 Face	F C		1.41	29	1	9	—	0.15	0.66	0.30
2 Cylindrical turning	R C		0.22	—	1	9	0.03	0.28	2	0.08
3 Cylindrical turning	F C		—	—	1	9	—	0.28	0.50	0.07
4 Cylindrical turning	R C		—	—	1	9	0.28	2.98	16	0.19
5 Cylindrical turning	F C		—	—	1	9	—	2.98	4	0.10

Spindle

Material: stainless steel— ferritic free-machining   Machine: manual turret lathe

Batch 10,000   Totals   3.48   58   81   0.35   7.50   27   1.29

Cost/part($):   Material = 0.26   Setup = 0.01   Operations = 1.29   Total = 1.56

(Move indicator to required row/column/page using keypad functions)
Press <INS>ERT, <DEL>ETE, <C>HANGE, <M>ATERIAL, <H>ELP, OR <O>K

Figure 5–51   DFMA Cost Analysis for Spindle Machining.
(Source: Boothroyd/Dewhurst, "Product Design for Manufacturing and Assembly," © April 1988, Manufacturing Engineering Magazine, p. 46. Reprint by permission of the Society of Manufacturing Engineers, Dearborn, MI.)

software with manufacturing cost data must work closely with engineering to obtain reliable values.

Cost estimating in the job shop is especially critical for the following reasons:

- The number of quotes for new parts is high, with some shops quoting over 5,000 different parts a year.
- The response time on quotes must be fast, in some cases within twenty-four hours.
- The direct labor cost is often a significant part of the total manufactured part cost.
- Price is often the order-winning criterion.

The shortage of experienced estimators and the inconsistencies present in the manual system create many errors. Products that introduce consistency into estimating in the job shop setting are available. The systems automate the estimation process for metal-cutting operations by integrating a computer database with standard production times and costs with a data-entry device to get basic part shapes into the computer. On one of the systems, the human operator traces the outline of the part on a drawing and interacts with the software by answering questions about the part and production process. The software searches through extensive databases of production machines and a broad range of standards to extract manufacturing costs tailored to the specific machines and direct labor standards in the job shop. The geometric analysis and mass-properties commands present in CAD software also provide data important for accurate estimates of material and production cost.

5–6 DESIGN AND PRODUCTION ENGINEERING NETWORK

The CAD, CAE, and production engineering automation, described in chapter 4 and in this chapter, places two demands on the enterprise infrastructure: (1) easy, accurate, and instantaneous movement of part geometry files and product data between departments in the enterprise; and (2) a single, common database for all enterprise information, part files, and product data. To satisfy the last two conditions, the enterprise must have the automation systems and computers in all department areas linked to a product data management (PDM) system through an *information* and *data network*. Planning for electronic data communications in the enterprise must extend beyond internal divisions. The competitive nature of world markets demands data links with external vendors, equipment suppliers, and technology service bureaus. The most frequently used technology for external networks is electronic data interchange (EDI). The rapid development of the Internet has made links between companies fast, inexpensive, and reliable, which has resulted in significant changes in business models in many product sectors.

The Basic Enterprise Network

Figure 5–52 describes a typical enterprise network; be sure you are familiar with the layout and terms before continuing. Make frequent reference to the figure while the general operation of the enterprise network configuration is described. The best place to start is by defining an *enterprise network*:

> *An enterprise network is a nonpublic communications system that supports communications and the exchange of information and data among various devices connected to the network over distances from several feet to thousands of miles.*

The network in Figure 5–52 does not represent a unique solution for the flow of data or for the location of data storage devices because numerous options are available. The many network configuration options are driven by the variety of hardware present, the applications running on the hardware, and the performance requirements, especially speed, required by users on the network. Each of the separate networks, called *LANs*, serves a relatively small area or a specific group, or has a specific function. LANs are usually confined to a single building or group of buildings. However, LANs can be interconnected to span large distances by using telephone lines and/or radio waves. A system of interconnected LANs is called a *wide-area network (WAN)*.

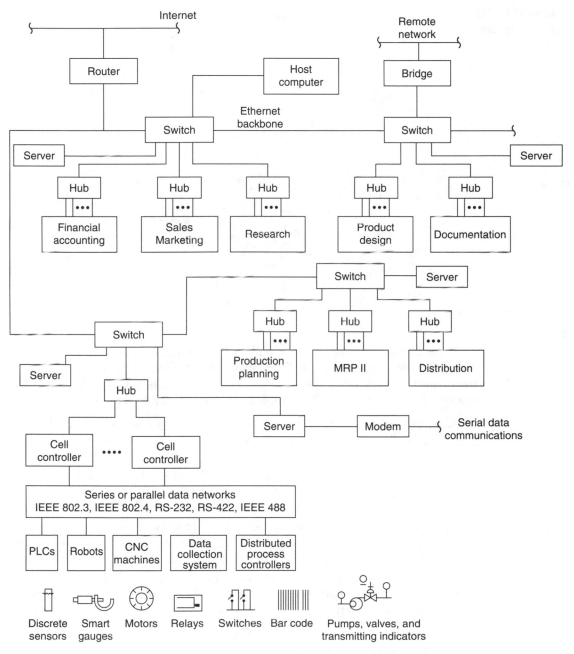

Figure 5–52   Enterprise Network.

Figure 5–53 Network Topologies: (a) Bus; (b) Ring; (c) Star.

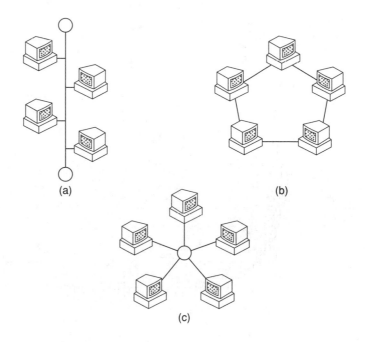

Three terms—*topology, protocols,* and *media*—are used to describe the structure and operation of a network and to differentiate one LAN from another. *Topology* refers to the shape of the network and is either *physical* (the way it is actually wired) or *logical* (the way it is made to work). There are four principal topologies used in LANs:

1. *Bus topology.* All devices are connected (Figure 5–53a) to a central cable, called the *bus* or *backbone.* Ethernet systems use a bus topology and are easy and inexpensive to implement.

2. *Ring topology.* All devices are connected (Figure 5–53b) to one another to form a ring or a closed loop. Ring topologies offer high bandwidth and cover large distances but are relatively expensive and difficult to install.

3. *Star topology.* All devices are connected (Figure 5–53c) to a central hub. While this topology is relatively easy to install and manage, data transfer can be slower because of the bottleneck created by all data passing through the hub.

4. *Tree topology.* A tree topology combines characteristics of linear bus and star topologies. The enterprise network illustrated in Figure 5–52 is a tree because groups of star-configured workstations are connected to a linear bus backbone.

The second term, *protocol,* refers to the rules and encoding specifications for sending data and whether the network uses a peer-to-peer or a client-server architecture. Two types of protocols are commonly used:

1. *Ethernet*—This is a LAN architecture developed in 1976 by Xerox Corporation that uses either a bus or a star topology, supports data rates of 10 megabits per second or higher, and is the basis for the IEEE 802.3 network standard. There are over twenty different designations for this protocol.

2. *Token ring*—This is a protocol used with a ring topology where a *token* (a special data pattern) moves around the ring to indicate the device or node on the network that can pass data. The IBM version of the protocol is the basis for IEEE standard 802.5. *Token passing* is a similar protocol that allows nonring topologies to use a passed token to indicate the node that can exchange data.

The final term, *media,* indicates how devices are interconnected. Examples include twisted-pair wire, coaxial cable, fiber-optic cable, and wireless networks using radio waves. With the structure of the network established, let's continue with some of the other terms and the function of network elements.

Internet IP Addressing

The network is populated with devices that are smart because they have an embedded microprocessor. The most common network devices are computers, but other devices, often grouped in a category called *peripherals*, include printers, plotters, and every conceivable production machine. Each attached device is called a *node* on the network and is assigned a unique network address called an *IP address.* The IP (Internet protocol) address is an identifier for a computer or peripheral on a network that uses TCP/IP (transmission control protocol/Internet protocol). There are a number of protocols; for example, IP handles the movement of data between host computers, and TCP manages the movement of data between applications. Networks use the TCP/IP protocol to route messages according to the IP address of the destination. Therefore, every computer that communicates over the Internet is assigned an IP address that uniquely identifies the device and distinguishes it from other computers on the Internet. An IP address is formatted as a 32-bit numeric address written as four numbers separated by periods. Each number can have a value from 0 to 255. An example of an IP address is 1.160.10.240.

Within an isolated network, like those illustrated for the departments in Figure 5–52, a device IP address can be assigned by the enterprise at random as long as each one is unique. However, when a private network, like the enterprise network in Figure 5–52, is connected to the Internet through a *router,* a registered IP address for the router is required to avoid duplicate nodes. Registered IP addresses are obtained for different regions of the world from four nonprofit registration organizations. The four regions and organizations are as follows:

1. *American Registry for Internet Numbers* (*ARIN*) for the United States
2. *Réseaux IP Européens Network Coordination Centre* (*RIPE NCC*) for Europe, the Middle East, and parts of Asia and Africa
3. *Asia Pacific Network Information Centre* (*APNIC*) for the Asia Pacific region
4. *Latin American and Caribbean Internet Addresses Registry* (*LACNIC*) for the Latin American and Caribbean region

The four numbers in an IP address are assigned according to three classes (A–C) and used in different ways to identify a particular network and a host on that network. However, the number of unassigned Internet addresses is running out, so a new classless scheme called *classless inter-domain routing* (*CIDR*) is gradually replacing the system based on classes A, B, and C. With CIDR, a single IP address can be used to designate many unique IP addresses by adding a slash followed by a number to the present IP process. An example CIDR IP address is this: 172.200.0.0/16. The number after the slash is not the number of IP addresses at the host site but indicates a number that can be addressed. An IP prefix of /12, for example, can be used to address 1,048,576 former class C addresses. The combination of TCP with IP defines one way that computers on a network can communicate by exchanging *packets* of data. It works well on a LAN, and works very well when you have networks of networks, which is what is present in manufacturing enterprises.

Network Structure and Elements

The network illustrated in Figure 5–52 uses the following network control devices:

- *Router*—A router is a device that forwards packets of data along a network. A router interconnects at least two networks, commonly LANs and WANs, and is often used to interface users with the external Internet. Routers present a unique IP address to the Internet and can assign IP addresses for users on its LANs. Routers are located at *gateways,* places where two or more networks connect. Routers use packet data and forwarding tables to determine the best path for forwarding the packets. The router is often responsible for the security of the enterprise network and contains software to build a *firewall,* or barrier to undesired access from the outside.
- *Gateway*—A gateway is an earlier term for a router or a node on a network that serves as an entrance to another network. In general, a gateway is any device that connects one type of LAN (for example, a wireless LAN) to another type of LAN (for example, a standard LAN using cable) within the enterprise.
- *Bridge*—A bridge is a device that connects two LANs or two segments of the same LAN that use the same protocols, such as Ethernet or tokenring.
- *Switch*—A switch is a device that filters and forwards data between LAN segments and provides very high-speed interconnections between individual LANs.
- *Hub*—A hub has multiple ports and is a common connection point for devices or nodes on a network. Data arriving at one port is sent to all the ports, so every node attached to the hub sees the data.

Study the network in Figure 5–52 and the definitions just presented until you can see how data are shared among enterprise departments. The computers attached to the network act as *workstations* for individuals or as *servers.* The operating system and software running on the computer determine its function on the network. Workstations run the software needed to support the function of the individual using the station. The servers play a number of roles, such as supplying Web pages, storing and providing files and data, or running software used by multiple departments, such as PDM. Data can be retrieved and stored on any server on the network for which access is provided.

Data Conversion

As enterprise data are created and shared across the enterprise, new problems arise. The PDM system becomes the central location for design and manufacturing data. The data are created by many of the functions shown in Figure 5–52 and used by many more of the areas listed on the enterprise network. There is no guarantee that users in manufacturing will have Pro/E CAD software on their machines so that they can open and view the product design. Putting design software on a manufacturing system to view the drawings is not a wise decision from an economic standpoint, so there needs to be another solution. Data conversion from one form to another is a critical function provided by a number of vendors. Trix Systems offers various conversion solutions, as illustrated in Figure 5–54.

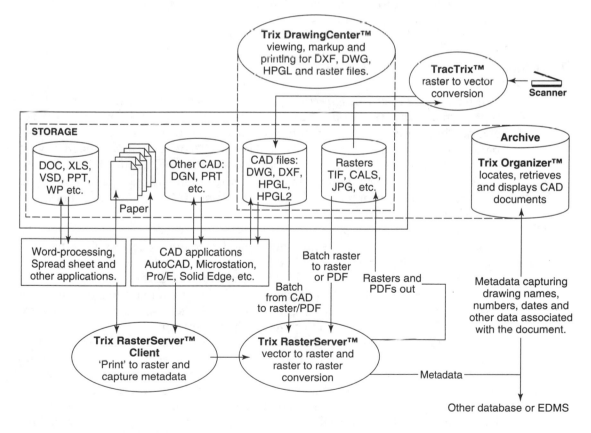

Figure 5–54   Enterprise File Conversion.
(Source: Courtesy of Trix Systems, Inc.)

Study the diagram and note the large number of options for conversion of CAD and other data. Note that CAD data can be converted to file formats readable by other CAD systems. In addition, CAD designs can be converted to raster/PDF format for viewing on computers with the free Adobe Reader software.

Data Beyond the Enterprise

The network is not limited to the confines of the enterprise, because the router connects nodes in the enterprise to servers anywhere in the world through the Internet. The Internet and EDI permit a company with corporate offices in South Carolina and plants in New York to look as if they are on the same network. Except for the speed of data exchange, the separation between the facilities would be transparent to the users; it would appear that the distant network was in the same office area. Another external interface illustrated in Figure 5–52 is a *serial data communications interface* that uses a *modem* to convert computer data into a standard serial data stream for transmission over standard telephone lines. The most popular standards for connecting between the host and the modem are the *RS-232* and *RS-422* interface definitions.

5–7 SUMMARY

The term *computer-aided engineering* (*CAE*) has had many definitions since its inception in the early 1970s, but today it clearly means the analysis and evaluation of engineering design by using computer-based software to calculate product parameters too complex for manual calculations. CAE is used in the design process from synthesis to evaluation and supports concurrent engineering efforts directly. CAE provides productivity tools to aid the production engineering area in group technology (GT), computer-aided process planning (CAPP), and computer-aided manufacturing (CAM).

The design for manufacturing and assembly (DFMA) process considers the production factors from the beginning of the design and is a major CAE technology capable of significant savings in production costs. At the analysis phase in the product design model, CAE uses the geometry file from CAD as the model for the analysis. The applications at this stage of the design include finite-element analysis and mass-properties analysis. Other CAE analysis tasks include software for circuit analysis and assembly interference, fit, and tolerance.

The CAE activity at the evaluation stage in the product design model includes examination of the data collected through all of the analysis processes and the building of prototypes of the design. The prototyping technique, called *rapid prototyping*, quickly constructs full-size models of products directly from the geometry stored in the CAD model. The two rapid prototyping techniques used most frequently are stereolithography and fused deposition modeling. *Virtual prototyping* is another alternative for evaluating a design before committing to full production. While the rapid prototype is a physical working model of the design, the virtual solution takes electronic images of product parts and assembles them into a finished part illustrated in the electronic format. The virtual prototype provides a host of useful information with no investment in material cost.

Group technology (GT), the grouping of parts according to design or manufacturing factors, is a technique used to support automation of the production engineering area. In GT, parts are coded and classified by one of the three following techniques: hierarchical, chain, or hybrid. With parts grouped by similar codes, creation of production cells to produce similar parts efficiently is possible. The major production engineering strategies used to increase productivity include CAPP, CAM, production modeling, maintenance automation, and automated product cost analysis. Of these technologies, CAM is the most widely used, with software implementations on all three computer platform types.

The enterprise must have the automation systems and computers in all department areas linked to common data storage through four principal LAN topologies (bus, ring, star, and tree) for information and data exchange. The CIM enterprise requires a central management facility to handle common product data and communication across all the internal and external entities in the organization. The primary devices used to build this enterprise network are routers, gateways, bridges, switches, and hubs. The two protocols used for the LANs are Ethernet and token-passing structures. Electronic data interchange (EDI) extends the network beyond the boundaries of the enterprise to external vendors, equipment suppliers, and technology service bureaus. Specifically, the product and process segment described in chapters 3 and 4, as well as in this chapter, must be linked through a network to achieve the productivity required for survival.

BIBLIOGRAPHY

BOOTHROYD, G., and P. DEWHURST. "Product Design for Manufacture and Assembly." *Manufacturing Engineering*, April 1988, 42–46.

CHANG, T. C., R. A. WYSK, and H. P. WANG. *Computer-Aided Manufacturing.* Upper Saddle River, NJ: Prentice Hall, 1991.

GETTLEMAN, K. M. "Stereolithography: Fast Model Making." *Modern Machine Shop,* November 1989.

GRAHAM, G. A. *Automation Encyclopedia.* Dearborn, MI: Society of Manufacturing Engineers, 1988.

GROOVER, M. P. *Automation, Production Systems, and Computer-Integrated Manufacturing.* 2nd ed. Upper Saddle River, NJ: Prentice Hall, 2001.

MITCHELL, F. H., Jr. *CIM Systems: An Introduction to Computer-Integrated Manufacturing.* Upper Saddle River, NJ: Prentice Hall, 1991.

SOBCZAK, T. V. *A Glossary of Terms for Computer-Integrated Manufacturing.* Dearborn, MI: CASA of SME, 1984.

WELTER, T. R. "Designing for Manufacture and Assembly." *Industry Week,* September 1989, 79–82.

WOHLERS, T. T. "Make Fiction Fact Fast." *Manufacturing Engineering,* March 1991, 44–49.

QUESTIONS

1. Define *CAE* in your own words.
2. Describe how CAE fits into the product design model developed in chapter 3.
3. Define *DFMA* in your own words.
4. Why is the combination of DFMA more effective than the use of DFM or DFA separately?
5. What is the major justification for using DFMA?
6. What is the difference between FEA and mass-properties analysis?
7. Define *FEA* in your own words.
8. What two categories of FEA software are used most frequently?
9. In mechanical applications of FEA, describe the process used to analyze a part for structural integrity.
10. Describe three of the seven FEA software application areas outlined in the chapter.
11. Describe the part of the FEA process most critical for valid analysis data.
12. Describe mass-properties analysis and list three typical parameters provided.
13. Describe the Valisys software and list the type of features that are useful to the analysis and evaluation process.
14. Describe the differences between prototyping, rapid prototyping, and virtual prototyping.
15. Compare and contrast the operation of the different types of rapid prototyping systems.
16. Define *group technology* in your own words.
17. Describe the hierarchical type code used to classify GT parts.
18. Describe the chain code used to classify GT parts.
19. What are the advantages and disadvantages of each of the GT code types?
20. Compare GT production cells with the machine production layout used in a conventional job shop. What advantages does GT offer over the conventional setup?
21. Describe the differences between the two techniques used to develop routings in CAPP.
22. What are expert systems and how can they improve the CAPP process?
23. Define *CAM* in your own words.
24. Describe the process used to manufacture a part on a CNC machine using CAM technology.

25. Compare the .dxf, IGES, PDES, and CALS file transfer protocols and describe the differences.
26. Describe the difference between process modeling and process simulation.
27. Compare the two simulation techniques: discrete event and continuous.
28. How can expert systems help in maintenance automation?
29. Why is cost estimating especially critical in a job shop manufacturing system?

PROBLEMS

1. List the major order-winning criteria, presented in chapter 1. Next to each criterion, list the CAE technologies presented in this chapter that support that order-winning criterion.
2. Select a group of three to five similar products that are easily disassembled and determine how many of the ten DFA guidelines listed in appendix 5–1 are violated. (The click-type ballpoint pen is an example of a product group to analyze.)
3. Select a moderately complex consumer product or a subassembly of a complex product. Apply the assembly method scoring chart shown in Figure 5–2 to the product and determine the total score.
4. Using the product from problem 3, apply the ten DFA guidelines and determine how many were violated. Using these results, suggest design improvements that would reduce the number of violations. Apply the assembly method scoring chart shown in Figure 5–2 to the improved design and compare this score with the original score from problem 3.
5. Select a group of products or parts and design a hierarchical code that could be used to group objects into common categories. (Use the parts from problem 2, if possible.)
6. The chain code for a part is 121322. Use the polycode structure shown in Figure 5–33 to determine and list the features present in the part.

PROJECTS

1. Using the list of companies developed in project 1 in chapter 1, create a matrix that compares the CAE technology in use at each of the companies.
2. Using the three companies selected in project 3 in chapter 4, determine what CAE technology could be installed that would improve design efficiency and productivity. List any changes that would be required, including changes in the design process and CAD system.
3. Select one company from the list in project 1 that includes assembly as part of the manufacturing process. Apply the ten DFA guidelines listed in appendix 5–1 and the assembly method scoring chart (Figure 5–2) to one of the assemblies. From the results, suggest changes that would improve the assembly process and product.

4. Using CAD software available at the college, draw several regular and irregularly shaped 2-D or 3-D objects. Use the mass-properties analysis command for the CAD software to generate mass-properties data for the objects.

5. From the list in project 1, select three companies that route parts on the shop floor. Describe the technique used by the process planners to route the parts to production machines. Using copies of sample routing sheets from each company, compare the various routing processes. Determine if CAPP would be useful for these companies.

6. From the list in project 1, select all the companies that use CNC machines, and identify the type of CAM software and system used.

7. Select three companies from the list in project 1 and prepare a report on how they generate product cost data.

8. Select several companies from the list in project 1 that have computer networks installed. Prepare a report describing how the networks are used and what departments are served by the network system.

9. View the Society of Manufacturing Engineers (SME) video *Design for Manufacturing* and prepare a report that describes the benefits of DFMA at one of the following companies: Storage Technology, Caterpillar, Xerox, or IBM.

10. View the SME video *Simulation* and prepare a report that describes the benefits of manufacturing simulation at one of the following companies: Rohr Industries, Intel, or General Electric.

11. Using data from the CAD Web sites listed in appendix 4–1, write a paper that describes how the CAD software from one of the CAD vendors integrates with the CAE functions covered in this chapter.

12. Identify two CAD companies listed in appendix 4–1 that have an integrated CAM solution and describe the CAM functions offered.

13. Develop a list of application software for CAD, CAM, and CAE using the Web sites listed in appendixes 4–1 and 5–2 that would support the design and manufacturing engineering departments of a small, medium-size, and large company.

14. Develop tables that compare the software capability provided by vendors for the following integration software: computer-aided manufacturing (CAM), finite-element analysis (FEA), and product data management (PDM).

The following projects are a continuation of project 9 through 12 at the end of chapter 4.

15. A company that manufactures kitchen products wants to add a set of plastic storage bowls or containers to its product line. The company wants to distribute the product through Wal-Mart, Target, or a chain of stores in your region. Use the CAD system drawing of the product to perform a mass-properties analysis for each of the bowls. Prepare a report on the results of the analysis and describe the parameters and their values.

16. A company that manufactures kitchen products wants to add a high-end manual can opener to its product line. The company wants to distribute the product through Wal-Mart, Target, or a chain of stores in your region. Do the following:

   a. Perform an FEA stress analysis using the CAD drawing and any FEA software available.
   b. Complete a DFMA analysis and recommend changes that would reduce the manufacturing cost for the product.

17. A company that manufactures kitchen products wants to add a manual hand-held eggbeater to its product line. The company wants to distribute the product through Wal-Mart, Target, or a chain of stores in your region. Do the following:

   c. Perform an FEA stress analysis using the CAD drawing and any FEA software available.
   d. Complete a DFMA analysis and recommend changes that would reduce the manufacturing cost for the product.

18. A company that manufactures kitchen products wants to add a manual hand-held garlic press to its product line. The company wants to distribute the product through Wal-Mart, Target, or a chain of stores in your region. Do the following:

   e. Perform an FEA stress analysis using the CAD drawing and any FEA software available.
   f. Complete a DFMA analysis and recommend changes that would reduce the manufacturing cost for the product.

APPENDIX 5–1: TEN GUIDELINES FOR DFA*

1. *Minimize the number of parts.* Combine or eliminate parts whenever possible. Combine parts in an assembly that do not move relative to each other unless there is strong justification otherwise.
2. *Minimize assembly surfaces.* Reduced surface processing results from reduced surface areas.
3. *Design for top-down assembly.* Establish a base part on which the assembly is built, and provide for assembly in layers from above the base. Insertion of parts from above takes advantage of gravity and usually results in less costly tooling and fewer clamps and fixtures. Use subassemblies to avoid violation of this rule.
4. *Improve assembly access.* Design for easy access, unobstructed vision, and adequate clearance for standard tooling. If possible, allow parts to be added to the assembly in layers.

*Adapted from "Designing for Manufacture and Assembly" by Welter (1989).

5. *Maximize part compliance.* Design with adequate grooves and guide surfaces for mating parts. Use the ANSI Y14.5M 1982 standard "Geometric Dimensions and Tolerancing" to ensure compliance between mated parts after processing.

6. *Maximize part symmetry.* Symmetric parts are the easiest to orient and handle. Where symmetry cannot be included, design in obvious asymmetry or alignment features.

7. *Optimize part handling.* For easier handling, design parts that are rigid rather than flexible, have adequate surfaces for mechanical gripping, and have barriers to prevent tangling, nesting, or interlocking.

8. *Avoid separate fasteners.* First, fasteners should be eliminated by using snapfits and other strategies. Second, the number of separate fasteners should be reduced to a minimum. Third, the fasteners used should be standardized to reduce variation and ensure availability.

9. *Provide parts with integral self-locking features.* Use tabs, indentations, or projections on mating parts to identify and maintain orientation through final assembly.

10. *Focus on modular design.* When parts have a common function, use a standard component or module; when parts must be interchangeable, use a common or standard interface.

APPENDIX 5–2: WEB SITES FOR CAE VENDORS

All the major CAE vendors have extensive Web sites that describe their products and offer other useful data for selecting CAE software. However, Web sites are dynamic and change frequently, so if a site does not link correctly, use a search engine to find the company site. Visit the following Web sites:

CAE Vendor	Web Address	Comments
ALGOR	http://www.algor.com	A broad line of FEA, analysis, and simulation software.
Ansys	http://www.ansys.com	Specialist in FEA plus thermal, computational fluid dynamics, magnetics, and electrical-field analysis.
Boothroyd Dewhurst, Inc.	http://www.dfma.com/ software/index.html	DFMA software.
Cosmos	http://www.cosmosm.com/	FEA solutions.
EDS	http://www.eds.com/ products/plm/femap/	Specialist in FEA.

MatrixOne	http://www. matrixone.com	Specializes in management of product data, documents, and configuration management tools.
MCS	http://www.mcsaz.com	CAD/CAM software.
MSC Software	http://www.mscsoftware.com	Simulation software for analysis.
SmarTeam Corporation Ltd.	http://www.smarteam.com	Specializes in product data management.
Cosmos	http://www.comosm.com	FEA solutions.
Tecnomatix Technologies	http://www.tecnomatix.com	Virtual quality control and analysis software.
Teksoft	http://www.teksoft.com	CAM software.

APPENDIX 5–3: WEB SITES FOR RAPID PROTOTYPING VENDORS

Rapid Prototyping Vendor	Web Address	Comments
Worldwide Guide to Rapid Prototyping	http://home.att.net/ncastleisland/	The best single source of rapid prototyping information.
Soligen	http://www.soligen.com	DSPC, a form of FDM and 3DP.
Stratasys, Inc.	http://www.stratasys.com	FDM.
3D Systems	http://www.3dsystems.com	Stereolithography.

Controlling the Enterprise Resources

PART GOALS

This part was written to introduce the reader to the management and control challenges faced by manufacturing organizations operating in today's global economy. A high-level look at manufacturing planning and control systems, and detailed discussions of material and capacity planning systems and advanced enterprise management systems are presented. Chapter 6 provides an introduction to the production-planning and control issues that affect all types of manufacturing operations. The focus of chapter 7 is on the computer-based systems for material and capacity planning, and the execution of the plan for procurement of materials and control of the shop floor. The key systems presented are material requirements planning (MRP) and manufacturing resource planning (MRP II). Enterprise resources planning (ERP), presented in chapter 8, extends the computer-based planning along the supply chain from suppliers to the customers. Chapter 9 presents a picture of the "lean production" revolution in manufacturing operations and the changing role of computer-based systems.

After you complete part 3, the following will be clear to you:

- The systems used to manage and control the production operations are part of computer-integrated manufacturing (CIM).
- The fundamentals of manufacturing planning and control are consistent with the fundamentals of CIM.
- Computer-based systems for material and capacity management are at the heart of CIM systems.
- A new generation of systems integrates the core elements of MRP and MRP II that support true enterprise-wide systems for operational control.
- The continuing emphasis on serving the customer is leading to new systems for electronic business and customer resource management.
- Modern methods and techniques such as just-in-time manufacturing and lean production systems have a place in the CIM environment.

■ Analysis of current enterprise operations and the elimination of unnecessary non-value-added operations must be completed early in the CIM implementation process.

CAREER INSIGHTS

Charles Kettering, a prominent industrialist, once said: "I am not sure what kind of future it is going to be. That is my reason for planning." Enterprises spend their life in a rapidly changing future; as a result, they share the same need for planning as that of Mr. Kettering.

In the past, the term *production* was often associated with factories, machines, and assembly lines. The scope of the management function was narrowly focused on problems related to manufacturing, with an emphasis on the procedures and skills necessary to run a factory. Since the early 1970s, however, evolutionary changes have occurred in enterprises across the world; as a result, production methods and techniques are now applied to a wide range of activity outside of manufacturing. For example, service areas such as health care, recreation, banking, finance, hotel management, retail sales, education, and transportation use management techniques derived from experiences in production management. In addition, many manufacturing organizations apply production management techniques to service functions such as information management and distribution. As a result, the traditional production management view was expanded and the term *operations management* coined to better describe the activities performed in both manufacturing and service organizations.

Graduates trained in the fundamentals of manufacturing planning and control and in the application of software to plan and control manufacturing resources have an opportunity for an interesting and rewarding career in the material management and production and inventory control segment of the enterprise wheel. If you find the material presented in part 3 interesting, then these are career areas that you should explore. Information technology and the application of computer- and network-based solutions to manufacturing problems is also a closely related area.

Introduction to Production and Operations Planning

OBJECTIVES

After completing this chapter, you should be able to do the following:

- Describe the basic issues that challenge production and operations planners
- Be familiar with the American Production and Inventory Control Society's definitions of key production and operations planning terms
- Draw and label a diagram of a planning system for a manufacturing operation
- Discuss the pros and cons of chase, level, and mixed production strategies
- Relate a master production schedule to the production strategy
- Understand the need for inventory and accurate inventory records
- Describe systems for material and capacity requirements planning
- Describe opportunities for automation and the application of computer-based systems for production and operations planning and the scheduling of work to be done

This chapter and the next three chapters address the operation and automation needs of the operations management area. The operations management activities utilize systems and shared knowledge to provide information needed for customer support, the definition of products and processes, and manufacturing operations. Refer to the Society of Manufacturing Engineers (SME) enterprise wheel shown in Figure 6–1. The manufacturing infrastructure is illustrated in Figure 6–2. The planning functions have formal interfaces with both the design and production departments and informal relationships with most of the enterprise. The operations management functions are a critical part of the computer-integrated manufacturing (CIM) implementation. The broad body of knowledge covering this important work is known as *manufacturing planning and control* (*MPC*). The American Production and Inventory Control Society (APICS) gives the following definition* of MPC:

*The *APICS Dictionary* terms and definitions in this chapter are published with the permission of APICS—The Educational Society for Resource Management, 10th edition, 2002.

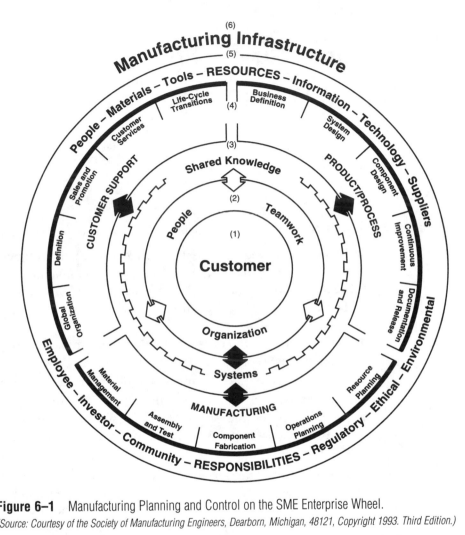

Figure 6–1   Manufacturing Planning and Control on the SME Enterprise Wheel.
(Source: Courtesy of the Society of Manufacturing Engineers, Dearborn, Michigan, 48121, Copyright 1993. Third Edition.)

> ***manufacturing planning and control system (MPC):*** *A closed-loop information system that includes the planning functions of production planning (sales and operations planning), master production scheduling, material requirements planning, and capacity requirements planning. Once the plan has been accepted as realistic, execution begins. The execution functions include input-output control, detailed scheduling, dispatching, anticipated delay reports (department and supplier), and supplier scheduling. A closed-loop MRP system is one example of a manufacturing planning and control system.*
>
> (Source: © *APICS Dictionary*, 10th edition, 2002)

6-1 OPERATIONS MANAGEMENT

The operations management area has responsibility for the administration of enterprise systems used to *create goods* or *provide services*. The health care industry serves as an example for each area. The activities performed by hospital management, a service

Figure 6–2 MPC in the Organizational Structure.

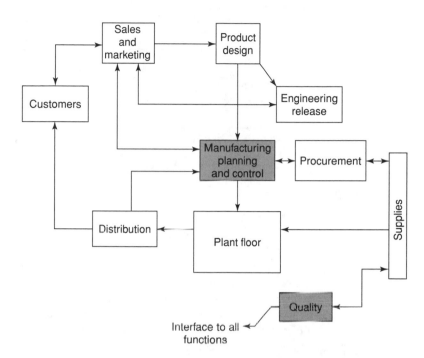

sector operation, include running the hospital, managing in- and outpatient medical services, providing food service, supervising and training staff, and housekeeping. Now compare this service side of health care to a manufacturer of health care equipment. The factory management must design new products, redesign current models, test designs, order raw material, determine the product mix and the quantity to produce, schedule the production machines, maintain production hardware and software, and adjust fixed and variable resources to meet changes in the market. The manufacturing operation has a *tangible output* that can be seen and touched.

This book focuses on manufacturing operations and the management of a *product-oriented* enterprise employing CIM concepts. However, changes in the global economy and enterprises over the last thirty years require the application of CIM principles to areas outside of manufacturing operations. The principles that provide an *order-winning* advantage to the manufacturing sector work in the service area as well. Clearly, the *computer-integrated enterprise* (*CIE*) needs an *operations management* emphasis for the service and manufacturing segments that must coexist. While this chapter and the next three chapters spotlight the management concepts required for a product-oriented operation, the principles are also applicable across other types of CIEs.

6–2 PLANNING FOR MANUFACTURING

All planning has a time horizon that dictates the number of days, months, or years into the future described by the plan. In general, enterprise planning is divided into the three levels illustrated in Figure 6–3.

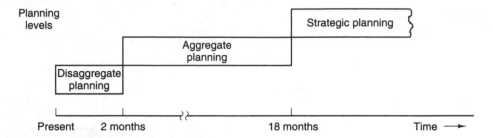

Figure 6–3 Enterprise Planning Levels.

The *strategic plan* is generally long range, one year to many years, and focuses, for example, on future capacity, products, and production plant locations. Planning for a new building or a production line takes many months. A strategic plan that covers two to four years should help identify long-range capacity needs while there is enough time to plan and implement them. It is fairly common for companies to consider a strategic view that looks ahead three, five, or even as many as ten years into the future. The strategic-planning process is performed at the highest management level in the enterprise and the plan is usually updated annually. Tracking the progress being made toward the planned goals and objectives is an ongoing effort.

The *aggregate plan* has an intermediate-length time horizon of two to eighteen months, with an emphasis on planning levels of employment, output, inventories, back orders, and subcontractors. The shortest time horizon, *disaggregate planning,* provides short-range planning with detailed plans that include machine loading, part routing, job sequencing, lot sizes, safety stock, and specific model order quantities. The disaggregate and aggregate planning processes have broad enterprise participation and are a major part of the MPC model illustrated in Figure 6–4. Study the figure until you are familiar with all the terms. An overview of each area in the MPC model is provided in the following sections, and a detailed study of some sections is provided in the following three chapters.

Introduction to Aggregate Planning

Aggregate planning, with a time horizon from two to eighteen months, is a "big picture" view of the production plan and includes all the functions in the top block of Figure 6–4. Aggregate planning is defined as follows:

> **aggregate planning**: *A process to develop tactical plans to support the organization's business plan. Aggregate planning usually includes the development, analysis, and maintenance of plans for total sales, total production, targeted inventory, and targeted customer backlog for families of products. The production plan is the result of the aggregate planning process. Two approaches to aggregate planning exist—production planning and sales and operations planning.*

(Source: © *APICS Dictionary,* 10th edition, 2002)

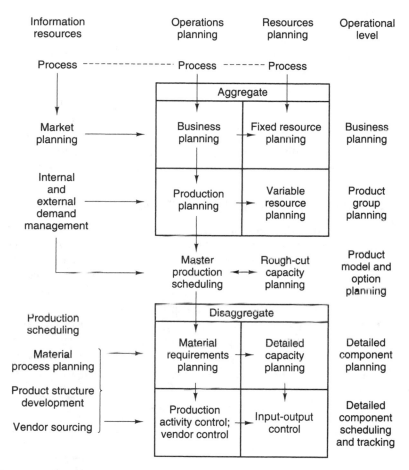

Figure 6–4 MPC Model.

(Source: Adapted from a chart from Dan Steele, University of South Carolina.)

Planners working at the aggregate level consider how output rates, inventory levels, and back orders affect the *variable* and *fixed* resources of the enterprise. Variable resources, such as full-time and temporary employees and subcontractors, are usually expressed in total labor hours per period. The fixed resources—facilities and machines—are often expressed as capacity in total machine hours per period. At this level, similar products or models with common characteristics are combined into product groups for planning purposes. For example, in the automotive industry, the aggregate number of cars or trucks planned would not specify two-door versus four-door models, four-cylinder versus six-cylinder engines, or other options that would not affect the results of the aggregate plan. With an aggregate plan established, three logical conditions relating the forecasted customer demand and enterprise production capacity are possible: (1) demand and capacity are about equal, (2) demand exceeds capacity, and (3) capacity exceeds demand.

When the first condition is present, the planners' activity shifts to matching existing capacity to forecasted demand in the most efficient manner. The last two conditions require additional action before allocation of production resources. For example, additional product promotion may be necessary to bring the lagging demand up to the current level of capacity. If demand exceeds capacity, however, expanding facilities or contracting with other manufacturers may be necessary. In the service sector, a similar process occurs, with the demand for services balanced against capacity limitations present. Aggregate planning forces management to address demand-capacity issues early in the production or service cycle so that the most cost-effective capacity solution is applied to meet customer demand.

Introduction to Disaggregate Planning

The term *disaggregate* means "to separate into component parts." At this planning level (Figure 6–4), a feasible aggregate plan is disaggregated into all the various models and options necessary to meet specific customer demand. The time horizon (Figure 6–3) for disaggregate planning often ranges from hours to months. The first step in disaggregation is the creation of a *master production schedule (MPS)* from the aggregate production plan. An example in Figure 6–5 from a manufacturer of fans illustrates the disaggregation process. The units listed under the aggregate production plan are similar models of exhaust fans with similar components, motors, and production processes; therefore, aggregation of these units is possible. The MPS in Figure 6–5, the first step in disaggregation, is required to

Figure 6–5 Disaggregation of Production Plan for Fans.

(a) Aggregate level, production plan

	Month				
	Jan.	Feb.	Mar.	Apr.	May
Planned output[a]	1,500	2,000	1,800	2,500	3,000

[a]*Aggregate units.*

(b) Disaggregate level, master production schedule

Planned output[a]	Month				
	Jan.	Feb.	Mar.	Apr.	May
24" 110 V	500	550	525	650	675
24" 220 V	150	450	450	500	775
36" 220 V	475	550	500	650	725
48" 220 V	175	200	150	325	475
54" 220 V	200	250	175	375	350

[a]*Fans by size and voltage.*

plan the production of the individual units at the model and option level. Note that the sum of all the disaggregate quantities in a given period should be equal to the original aggregate plan quantity.

The disaggregate strategies and actions included in the MPC model illustrated in Figure 6–4 describe a system for component planning at the disaggregate level. The MPC strategy used is based on the functional characteristics of the manufacturing system. Review the manufacturing characteristics for the five manufacturing systems listed in Figure 2–3 and note how *process speed* and *product complexity* vary under different manufacturing systems. Now compare those manufacturing characteristics in Figure 2–3 with the MPC disaggregate planning options included in Figure 6–6. The MPC method used to establish the MPS and to perform component planning at the disaggregate level is a function of manufacturing characteristics, such as process speed and product complexity.

Continuous-process systems use a flow-type MPC management system to disaggregate the production plan since the product has few component parts and tracking is straightforward. The management of raw materials and distribution of finished goods often is the most complex issue in continuous-process systems.

An assembly-type management system is used to disaggregate the production plan in the line and repetitive manufacturing systems, where typical products include watches, automobiles, microcomputers, and pharmaceuticals. Component complexity is higher than flow, and management of parts is required. However, the management issues in developing the MPS are less severe because components are coordinated through the assembly process.

The MPC strategies address the need for parts management of complex products and product mixes with relatively high rates of production. Figure 6–6 indicates that the MPC systems are used in some line, repetitive, and job shop settings.

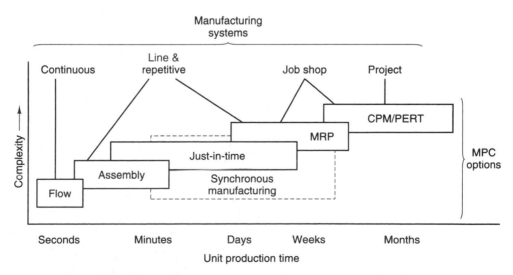

Figure 6–6   Comparison of MPC Disaggregate Options and Manufacturing Systems.

Products with hundreds of parts and production systems with thousands of purchased and manufactured items use MPC systems to disaggregate the production plan. The MPC model shown in Figure 6–4 supports the line, repetitive, and job shop production systems using material requirements planning (MRP) to disaggregate the production plan. A different model would be required to describe the operation of continuous and project manufacturing systems. Chapter 7 provides more detail on the MRP and manufacturing resource planning (MRP II) systems for MPC.

The just-in-time (JIT) and synchronous manufacturing management systems overlap the disaggregation options that support the line, repetitive, and job shop manufacturing systems. JIT and synchronous systems are described in detail in chapter 9; however, for the present overview, *JIT* is defined as a component planning process that encourages shorter cycle times, lower inventory, and reduced lead and setup times. *Synchronous manufacturing* or *drum-buffer-rope* (*DBR*) is a planning technique for components or products moving through multiple production stations. The DBR system, a combination of JIT and MRP, uses a pull-type production system at most work cells, but a push scheduling system to build buffers where bottlenecks occur at critical points in the production flow.* In general, moving the disaggregation strategy from right to left in Figure 6–6 promotes *order-winning criteria* because unit production time decreases. The practice repeated in many companies starts with implementation of MRP to gain control of the component planning activity. With MRP in place, specific parts or products are shifted to JIT and synchronous manufacturing to increase competitiveness.

The final disaggregation strategy shown in Figure 6–6, CPM/PERT, is used primarily for project work. While the project system has a large design component and often a large number of parts, the major problem is to schedule the delivery of parts at the required time during the extended assembly time. The commercial construction industry and manufacturers of ships and petrochemical plants would use this type of disaggregation from an aggregate plan.

6–3 MPC MODEL—MANUFACTURING RESOURCE PLANNING (MRP II)

The MPC model shown in Figure 6–4 indicates that all production planning is either global at the aggregate level or detailed in the disaggregate group. The *operational level column* emphasizes the change in planning scope, with intermediate-range business planning occurring at the top and detailed component planning, scheduling, and tracking located in the disaggregate group. The aggregate and disaggregate processes are divided into two sections: *operations planning* and *resources planning.*

A model for a formal planning system that links high-level business decisions, the planning of material and physical resources, and the detailed schedule for the shop floor was developed in the early 1980s. APICS (the American

---

*In a *push*-type manufacturing system, inventory is produced according to a schedule for future use. In a *pull*-type manufacturing system, parts are produced only when they are needed in the next level of the bill of materials.

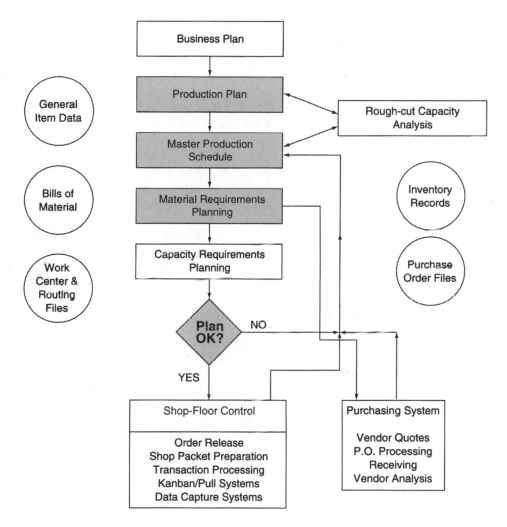

Figure 6–7   Model for a Manufacturing Resource Planning (MRP II) System.

Production and Inventory Control Society) has worked to develop and document this model. The model continues to be a solid depiction of the integrated central planning process. Several variations have emerged, and new methods and techniques such as JIT manufacturing and enterprise resources planning (ERP) have been developed since the original model, shown in Figure 6–7, was conceived. The model has been able to stand the test of time. It depicts clearly many of the functional relationships among plans, actions, databases, and priority schedules. High-level planning starts at the top. Plans for material and physical resources are in the middle of the diagram. Detailed output of the planning processes, such as the shop schedule, purchase orders, and transaction reporting, is found at the bottom of the model. The model does not describe the mechanics of how detailed plans are to be made or how the shop floor is to be managed and controlled. The focus is on the relationship of the critical planning and scheduling functions and the data that support them.

Note that the model does not try to show all the connecting relationships between the planning functions (the boxes in the model) and the primary databases (shown as circles in the model). Modern computer-based systems use database structures that provide the data elements to all the planning functions. An attempt to show all the links would make Figure 6–7 unnecessarily complicated.

Review the models presented in Figures 6–4 and 6–7 and become familiar with the information flows throughout these models. The discussion of the models focused on the general development of the aggregate and disaggregate plan of the production of finished goods. These models also provide a framework for the planning and production of assemblies, manufactured parts, and the related purchased material items. Chapter 7 provides additional information on the detailed planning that takes place in a formal system supporting the process of manufacturing.

High-Level Planning for the Business

Business planning starts with top management's definition of the business of the firm. The definition includes the products to be sold (in general terms), the markets to be served, and plans for market share, net sales, return on investment, profits, and the return on equity. The plans involve deciding how the firm will position itself in the marketplace and the strategies that will be used to compete (high volume, low cost, high service, specialty, etc.). Next come the documentation of measurable objectives, the development of plans to achieve the objectives, and the determination of realistic target dates for the implementation of the plans.

Marketing and financial plans are critical elements of the business planning process. This activity requires the participation of the true leaders of the enterprise. The true leaders control the available resources and determine the direction of the enterprise. The business plan tends to look several years into the future and is typically stated in dollars. A plan for producing and selling dollars is of little direct use to manufacturing, however. The general business plan must be translated into a language and form that the people in the operating areas can understand and use. Figure 6–8 shows an example of the planning progression (top to bottom).

Manufacturing's operating plan, the production plan, expands on the strategic decisions and marketing plans. It leads to more detailed plans and decisions concerning the production technology to be used, the type and quantity of facilities, the operating organization, quality requirements, operating policies, expected throughput times, and more.

Forecasting Future Demand

The MPC steps in planning the production of a product are detailed in Figure 6–8. Compare the blocks in the diagram in Figure 6–8 with the MPC model illustrated

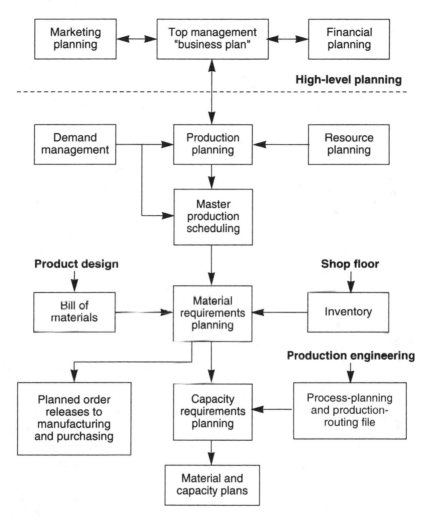

Figure 6–8 Production-Planning Function.

in Figure 6–4. At the highest level in the corporation, the marketing and financial areas pool customer and market data to produce the business plan. The business plan is a *forecast* of expected future demand for all the products produced by the enterprise:

> *forecast*: An estimate of future demand. A forecast can be constructed using quantitative methods, qualitative methods, or a combination of methods, and it can be based on extrinsic (external) or intrinsic (internal) factors. Various forecasting techniques attempt to predict one or more of the four components of demand: cyclical, random, seasonal, and trend.

(Source: © *APICS Dictionary,* 10th edition, 2002)

The data collected to produce the demand forecast that drives the production plan come from several sources. For example, historical sources provide sales information, quantity, and percentage change for product families from previous years. In addition to historical information, forecasting techniques use (1) the current economic condition of regional, national, and global markets; (2) the trends present in the current economic data and product markets; and (3) anticipated shifts in consumer demand. The forecast present in the business plan is management's "best guess" for the future demand for a product.

Computer-based systems can help process historical data and factor in adjustments for anticipated changes to develop a forecast of future demand. The system does the number crunching and produces information that can be used to make forecast decisions. Focus Forecasting is a computer-based system developed by Bernard T. Smith that is used by a number of well-known companies to improve the quality of their forecasts and reduce the work required to generate and maintain them. The system is based on three principles:

1. People will use only simple systems.
2. Simple systems are the only systems that work.
3. The only measure of any system is how well it works.

These principles have guided the design of the Focus Forecasting system. These ideas also make a lot of sense with respect to CIM systems.

The Focus Forecasting system works through a simulation process that evaluates over a dozen simple forecasting strategies and finds the one that is the best fit with the actual results of the recent past. The strategy that may fit the best could be as simple as "Whatever we sold in the last three months is probably what we are going to sell in the next three months." The system uses the identified strategy to predict the forecast for the next period, reports the forecast quantity, and identifies the strategy used to produce the forecast. The system tries to demystify the forecasting process.

So, does this use of simple strategies realle work? The customers of Focus Forecasting say it does. Of course, no forecast of future demand is going to be perfect. System users report that they can use 95 percent or more of the system-generated forecasts. The system takes care of the majority of the items and that frees time for people with more knowledge of the products, customers, and environment to concentrate on the remaining 5 percent or fewer items where the historically based strategies are not the best. Visit the Focus Forecasting site on the World Wide Web for more information (**http://www.focusforecasting.com**).

Planning for Production

Scheduling production to match the sales forecast is not realistic for many production operations. Therefore, the *production-planning process* is an important beginning step in the MPC effort. Production planning is defined as follows:

> **production planning**: *A process to develop tactical plans based on setting the overall level of manufacturing output (production plan) and other activities to best satisfy the*

current planned levels of sales (sales plan or forecasts), while meeting general business objectives of profitability, productivity, competitive customer lead times, and so on, as expressed in the overall business plan. The sales and production capabilities are compared, and a business strategy that includes a sales plan, a production plan, budgets, pro forma financial statements, and supporting plans for materials and workforce requirements, and so on, is developed. One of its primary purposes is to establish production rates that will achieve management's objective of satisfying customer demand by maintaining, raising, or lowering inventories or backlogs, while usually attempting to keep the workforce relatively stable. Because this plan affects many company functions, it is normally prepared with information from marketing and coordinated with the functions of manufacturing, sales, engineering, finance, materials, and so on.

(Source: © *APICS Dictionary*, 10th edition, 2002)

Production planning is not a new concept; the significance of this planning function was documented in the late 1970s. Berry, Vollmann, and Whybark described the production plan at a conference of production leaders in 1979: It is a top management responsibility. It is an agreement among marketing, finance, and manufacturing about what is to be produced and made available for sale. It is a clear link between the corporate plans and the MPS. This planning concept is still important today.

The aggregate forecast from the business plan is applied to the production plan (Figure 6–8) together with adjustable resource planning data and demand management information from marketing. The typical *production plan* shown in Figure 6–5 indicates that planned production is expressed in the total number of units, without reference to the mix of products. The aggregate data in the production plan are then broken down into individual products in the *MPS* example shown in Figure 6–5.

The production-planning block diagram shown in Figure 6–8 indicates that the MPS provides the data needed in *material requirements planning* (*MRP*) to produce the *time-phased* requirements for all manufactured and purchased parts. The term *time phased* implies that a purchasing and manufacturing schedule is generated that considers the lead time required to buy or manufacture each part needed for the finished products in the MPS. The time-phased schedule is generated by using the MRP process that is illustrated in Figure 6–19. The detail in the planning process has increased exponentially at this stage because every part in the finished product could require the generation of a separate MRP record. The MRP function in Figure 6–8 is supported with bill of materials (BOM) and inventory data. Note that critical links to other enterprise areas are present. BOM data are supplied by the design department, and inventory levels come from shop-floor data sources.

The diagram of the production-planning function shown in Figure 6–8 indicates that the MRP record also provides the data required to perform *capacity scheduling*. Note that capacity planning uses process-planning and production-routing data that come from another external source, production engineering. With the capacity and material plans defined, the planning process for production is complete. The next step is the implementation of the plans on the shop floor and the evaluation of production effectiveness.

The production-planning diagram shown in Figure 6–8 shows five sources of information outside the MPC function: *marketing, finance, product design, shop floor,* and *production engineering.* These links to other areas in the enterprise for production data underscore the need for a common database for production information and emphasize that integration of systems across the enterprise is required for an effective MPC system.

The overview of the production-planning function provided in this section is a preparation for the detailed analysis of the production-planning process in the following sections. Reread this section until the overall process is clear.

6–4 PRODUCTION PLANNING

The production-planning function, illustrated in Figure 6–8, provides the key communications link between top management and manufacturing. The production-planning function provides a framework for resolving conflict due to changes in product marketing and production resources. Suppose that marketing sees an opportunity to expand into a new market and requests production resources for the new product. With a specific production plan in place, resources for the new product cannot be allocated without reducing production of some other product. The production-planning process forces the business plan developed by top management to be consistent with the production capability required for the production plan. After marketing and financial issues are resolved and a production plan for the intermediate term is set, the manufacturing mission for the enterprise is clearly defined. The production plan provides manufacturing planning and the shop floor with the marching orders necessary to meet the objectives of the firm.

The production plan is usually stated in dollars or in aggregate units of output per month. The production plan is *not* a forecast of demand; rather, it is the planned production stated in aggregate terms for which everyone in the enterprise is responsible. For example, the forecast demand may exceed the aggregate units in the production plan. A decision by top management to produce less than the forecasted demand may be made for several reasons: a desire to have lower quantity and higher quality, a desire to defer investment in fixed or adjustable resources, or a desire to put available resources into other product areas. The production plan provides a vehicle for tough trade-off decisions in setting production goals at the aggregate level.

The Production-Planning Process

The production-planning process starts with a good sales forecast for the next year that discounts as many of the variables in the marketplace as possible. The demand management issues, such as interplant transfers, distribution requirements, and service parts must also be factored into the production plan. Changes in inventory or backlog levels that affect the overall production rate must also be considered.

Effective production-planning processes have reviews at regular intervals with a *time fence* for changes requested in the aggregate production levels. For example, successful firms often review the production plan monthly and make changes quarterly. The time fence frequently sets limits on how late in the planning cycle changes in the aggregate levels can be made. For example, the time fence may dictate that no changes can be made in the current or closest period and that no more than a 10 percent change can be made in the nearest future period. Routine reviews of the production plan keep the communication alive between top management and manufacturing.

Production Planning and Variable Resource Management

The production plan states the aggregate production goals for all products manufactured by the enterprise. For example, consider the production plan stated in aggregate dollars per month for ABC Manufacturing, illustrated in Figure 6–9. Converting this aggregate forecast into a production plan requires a decision on resources. A study of the chart reveals the following:

- Sales peak at $16 million in November.
- June has the minimum sales, $6.6 million.
- Two sales peaks occur, in spring and fall.
- Total sales for the year are $132 million.

Before variable resources can be allocated to meet this production plan, the sales in dollars per period must be converted to labor hours per period. The conversion is performed by using an estimate, obtained from company accounting records, that relates the dollar value of sales to hours of direct labor. In a low-technology production situation that relies on manual labor, for example, each hour of direct labor might equate to $30 of sales. In a production situation where high-technology production

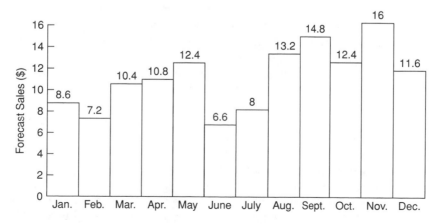

Figure 6–9   ABC Manufacturing Monthly Sales Forecast (Millions of Dollars).

Month	Sales (millions of dollars)	Labor (hours)	Days worked	Variable workforce (number of employees required)	Variable work week (hours per employee)
Jan.	8.6	86,000	21	512	29.72
Feb.	7.2	72,000	20	450	26.12
Mar.	10.4	104,000	23	565	32.81
Apr.	10.8	108,000	20	675	39.19
May	12.4	124,000	22	705	40.90
June	6.6	66,000	10	825	47.89
July	8	80,000	21	476	27.64
Aug.	13.2	132,000	22	750	43.54
Sept.	14.8	148,000	21	881	51.14
Oct.	12.4	124,000	22	705	40.90
Nov.	16	160,000	20	1,000	58.05
Dec.	11.6	116,000	20	725	42.09
Total	132	1,320,000	242		
Average			20	689	40.00

Sales dollar conversion estimated to be $100 per employee hour.

Figure 6–10   Chase Production Strategy Data.

equipment is used, the figure might be $100 of sales resulting from every hour of direct labor. Using the $100 conversion value, the table in Figure 6–10 was developed for the production plan in Figure 6–9. Study the headings across the top of the table. The Sales column comes directly from the forecast. The Labor column is calculated by dividing the sales dollars by the labor conversion rate ($8,600,000 divided by an estimated $100 per labor hour equals 86,000 direct labor hours). The Days Worked column indicates the number of working days in each period or month. Note that the plant is closed in June for 10 working days for employee vacations. A study of Figure 6–10 indicates that 160,000 direct labor hours are required in November to satisfy the maximum sales value of $16 million, and only 66,000 labor hours are required in June to meet the demand of the lowest sales period. This wide variation in human resource requirements is the basis for planning in the variable resource area. Three different production-planning strategies, called *chase, level,* and *mixed,* are used to address this variation in direct labor hours required by the production plan. A description of each of these strategies is provided in the following sections.

Chase Production Strategy.   The *chase production strategy* requires that the production in each period equals the planned production for that period, which implies that the product inventory level at the beginning of each period would be zero because all the production planned for the period would be produced during the period. A pure chase strategy requires that either the number of employees or

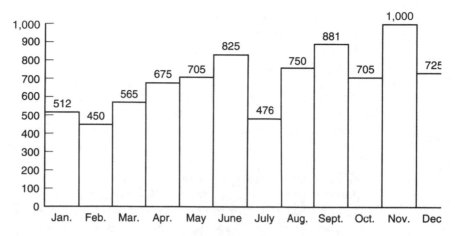

Figure 6–11   Monthly Number of Full-Time Employees in the Chase Strategy.

the hours worked per week by each employee must change to meet the production planned for the period.

Chase data for the ABC Company, for example, are provided in the last two columns in Figure 6–10. Study the data in Figure 6–10 and the graph of the monthly employment in Figure 6–11. The column in Figure 6–10 labeled Variable Workforce indicates the number of full-time employees required each month to meet the planned production exactly. The number of employees is found by first calculating the total hours worked by each employee in the period (21 days worked times 8 hours per day equals 168 hours per period per employee). The number of employees required per period is then calculated by dividing the total labor hours for the period (86,000) by the employee hours (168) per period (86,000 labor hours divided by 168 hours per employee equals 512 employees). Stop and calculate the total employees required for the months of February and March to verify that you understand the calculations.

Analysis of this strategy indicates that employee levels vary from a low of 450 in February to a maximum of 1,000 in November. A change of this magnitude in the level of full-time employees during twelve months would be difficult to support. The only industries that can use this type of variable resource strategy successfully usually require only a low skill level in the workforce.

Another approach used in the chase strategy keeps the number of employees in the workforce constant and varies the hours worked per week. The change in the hours worked per week for the ABC Company example is listed in Figure 6–10 under Variable Work Week. The average number of employees required for a year's production (689) is indicated at the bottom of the Variable Workforce column in Figure 6–10. This strategy keeps the workforce at 689 and varies the hours worked per week to meet exactly the production planned. Note that the weekly workload varies from a low of 26.12 in February to a high of 58.05 in November. The hours required per work week for the chase strategy are calculated by first finding the hours per month per employee (86,000 labor hours per month divided by 689 employees equals 124.8 labor hours per employee per month). Dividing the previous value by the working days in the month yields the hours worked per day per employee (124.8 divided by 21 days equals 5.943 hours worked per day per employee). Finally, multiplying the daily hours by 5 days provides the total weekly hours for each employee. Stop and perform the calculation for February and March to verify that you understand how to calculate the total weekly hours per employee in this chase strategy. The swings in hours worked per week is wide, but this strategy is frequently used to chase a production schedule with wide variations.

Level Production Strategy. The *level production strategy* requires that the production in each period equals the monthly average production calculated from the total production value for the year. With this strategy, the workforce and weekly work hours are constant, so that production is roughly the same each month. As a result, in some months products produced are not sold, so that an inventory of parts develops to handle the months where market demand is greater than the production. The monthly production totals are listed in Figure 6–12 in the columns labeled Monthly Production and Inventory Balance. Note that the workforce is held at 689 employees and the work week is 40 hours. The monthly production is calculated by multiplying the number of employees (689) times 8 hours per day times days worked (21) times the sales dollar conversion factor of $100 per employee hour worked established in the variable resource section. The

Month	Sales (millions of dollars)	Labor (hours)	Days worked	Level workforce (number of employees required)	Monthly production (millions of dollars)	Inventory balance (millions of dollars)
Jan.	8.6	86,000	21	689	11.58	2.98
Feb.	7.2	72,000	20	689	11.02	6.80
Mar.	10.4	104,000	23	689	12.68	9.08
Apr.	10.8	108,000	20	689	11.02	9.30
May	12.4	124,000	22	689	12.13	9.03
June	6.6	66,000	10	689	5.51	7.94
July	8	80,000	21	689	11.58	11.51
Aug.	13.2	132,000	22	689	12.13	10.44
Sept.	14.8	148,000	21	689	11.58	7.22
Oct.	12.4	124,000	22	689	12.13	6.94
Nov.	16	160,000	20	689	11.02	1.97
Dec.	11.6	116,000	20	689	11.02	1.39
Total	132	1,320,000				
Average			20	689	11.12	7.05

Planned work week is 40 hours long.

Figure 6–12 Level Production Strategy Data.

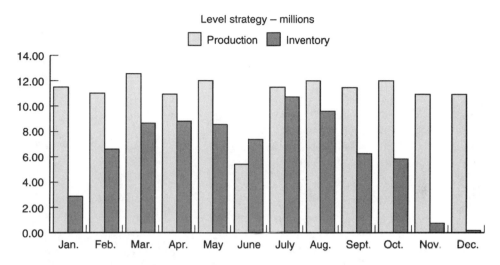

Figure 6–13 Level Production Strategy Graph.

dollar inventory balance is the difference between sales dollars and production dollars. The level strategy is illustrated graphically in Figure 6–13 with the production level and inventory balance plotted. The production is not exactly level because the number of days worked per month varies slightly. The inventory balance increased early and reached a peak in July in preparation for the high product demand in the fall.

Each method, chase and level, has advantages and disadvantages. The advantage of the chase strategy is that no excess inventory is carried; however, the disadvantage of the variable workforce implementation of chase is the high cost of hiring and firing employees. In the variable work week implementation of chase, the disadvantage is the cost of overtime and in implementing shortened work weeks. The advantage of the level strategy is the constant workforce and work week, but that constancy creates a disadvantage in the cost of carrying the excess inventory created. As a result, many companies use a combination of the chase and level strategies in implementing the production-planning requirements.

Mixed Production Strategy.   The *mixed production strategy* uses the best parts of the chase and level strategies. Study Figure 6–14, which illustrates the process. Note that the workforce is held level at 610 employees for the first seven months and then increases to 809 employees for the last five months. The numbers in the chart assume a constant 40-hour work week during all periods. With this method, the average inventory is reduced from $7.05 million for the level strategy (Figure 6–12) to $3 million for the mixed approach. The savings in carrying cost for the inventory must offset the cost of changing the employment levels twice a year. The mixed strategy often includes a varying work week, which might reduce the number of employees changed. In most manufacturing situations, the mixed strategy is used to adjust the variable resources to meet the production-planning goals.

Month	Sales (millions of dollars)	Labor (hours)	Days worked	Workforce (employees required)	Monthly production (millions of dollars)	Inventory balance (millions of dollars)
Jan.	8.6	86,000	21	610	10.25	1.65
Feb.	7.2	72,000	20	610	9.76	4.21
Mar.	10.4	104,000	23	610	11.22	5.03
Apr.	10.8	108,000	20	610	9.76	3.99
May	12.4	124,000	22	610	10.74	2.33
June	6.6	66,000	10	610	4.88	0.61
July	8	80,000	21	610	10.25	2.86
Aug.	13.2	132,000	22	809	14.24	3.89
Sept.	14.8	148,000	21	809	13.59	2.69
Oct.	12.4	124,000	22	809	14.24	4.52
Nov.	16	160,000	20	809	12.94	1.47
Dec.	11.6	116,000	20	809	12.94	2.81
Total	132	1,320,000				
Average			20		11.23	3.00

Planned work week is 40 hours long.
January starting inventory balance is zero.

Figure 6–14   Mixed Production Strategy Data.

The aggregate values in the production plan drive the MPS (Figure 6–8), where the first level of disaggregation of the product plan occurs. The development and operation of the MPS function is described in section 6–5.

Resource Capacity Considerations.   The production plan must be a realistic statement of what is expected to be available to supply the demands of customers. The plan must consider the capacity of identified critical resources. A technique known as *rough-cut capacity planning* is often used at the production plan level to evaluate expected critical resources. These key resources may include machines, labor, floor space, vendor, or even the dollars required. The purpose of the rough-cut analysis is to evaluate the feasibility of the plan before attempting additional detailed planning and implementation activities. The idea is to identify problem areas early in the planning process and resolve conflicts and issues early by taking a planning approach.

6–5 MASTER PRODUCTION SCHEDULE

The *MPS* is the basic communications link between the business "game plan" and manufacturing:

> **master production schedule (MPS):** *The master production schedule is a line on the master schedule grid that reflects the anticipated build schedule for those items*

assigned to the master scheduler. The master scheduler maintains this schedule, and in turn, it becomes a set of planning numbers that drives material requirements planning. It represents what the company plans to produce expressed in specific configurations, quantities, and dates. The master production schedule is not a sales item forecast that represents a statement of demand. The master production schedule must take into account the forecast, the production plan, and other important considerations such as backlog, availability of material, availability of capacity, and management policies and goals. Syn: master schedule.

(Source: © *APICS Dictionary*, 10th edition, 2002)

The output or result of the MPS process is called the *master schedule*. The master schedule provides a specific statement of the products that are going to be produced and made available to fill the demand. This statement includes the identification of the items that will be made, the quantities that will be produced, and when that production is scheduled to be complete and ready for shipment. The MPS must be realistic, attainable, and reasonable. An overstated master schedule becomes a wish list, not a true production schedule. Once the operating departments determine that the master schedule is unrealistic, they are forced to develop alternative informal systems for planning what is really produced. The informal system plans will lead to confusion and chaos.

Note in Figure 6–8 how the MPS is positioned strategically between manufacturing and business or production planning. The MPS is an anticipated build schedule for end products; as a result, the MPS states production plans, not market demand. The MPS is not a forecast. In most cases, the MPS and the forecast have different values because the MPS represents what can be built given the enterprise resources, whereas the forecast portrays anticipated market demand.

The MPS and Production Strategies

The MPS must support the production strategies—engineer to order (ETO), make to order (MTO), assemble to order (ATO), and make to stock (MTS)—described in chapter 2 and illustrated in Figures 2–4 and 2–5. Study these figures and review the characteristics of each strategy. The ETO and MTO strategies use a similar process in the MPS implementation. The ATO and MTS processes are quite different. The primary impact on the MPS for all of these production approaches is in the choice of units for the MPS.

In the ETO and MTO company, where no finished-goods inventory exists, all products are built to customer orders. This type of production is used when, because of a large number of possible product configurations and knowledge, the customer will wait for complete manufacture and assembly of the specific end item. In the case of the ETO, the customer will also allow the time required for basic product engineering to be included in the delivery time. The units used for these two strategies are specific end items or sets of products that compose a customer order. Since production often starts before the end item is completely designed or specified, development of the MPS is difficult. Frequent adjustments to the MPS in this type of production setting are necessary and expected.

The ATO production strategy is used when there is a wide variety of end-item configurations built from a broad range of basic components and subassemblies. In this case, the required customer delivery time is shorter than the production lead times for components, so production is started in anticipation of customer orders. The large number of end-item configurations makes forecasting of specific end-item configurations difficult and stocking of finished goods risky. As a result, assembly of common base components and building of subassemblies are started into production before the customer order is received. The final assembly to a specific end configuration is started only when the customer order is recorded. Therefore, an ATO company does not include specific end items in the MPS. The MPS units reflect an average product requirement based on percentage estimates of typical market demand. For example, in the automotive industry, the MPS schedules into production the average number of four-door versus two-door vehicles with the average number of four-cylinder versus six-cylinder engines. The percentages of different options are also averaged, so that subassemblies of the options are available to build into the car during final assembly. The averages and percentages, used to determine the number of units in the MPS, come from a *planning bill of materials*. The planning bill of materials has common parts and options as its components because the development of a specific BOM for all possible combinations of finished products is impractical.

The units for the MPS in the MTS operation are all end-item catalog numbers, which is reasonable because MTS companies usually use batch production techniques and carry finished-good inventories for most, if not all, of the end items. In this case, the MPS is a statement of how much of each of the standard products to make and when to produce them. In some companies, the specific products are aggregated into model groupings so that planning of common components achieves a better economy of scale. The items are then broken into specific products at the latest possible time. For example, in furniture manufacturing a *consolidated item number* is used in the MPS for furniture that is identical except for finished color. Then a separate system allocates lot size in the MPS for each possible color at the last step in the production process.

MPS Techniques

The principal method used to represent MPS data is the *time-phased record*, illustrated in Figure 6–15. The record is used to show the relationship among the *rate of output, the sales forecast,* and the expected *inventory balance.* The number of periods indicated on the record is a function of the individual industry and product. The Forecast row represents the number of end-item units that the business anticipates will be sold in each period. The end items or numerical values put in each period in the Forecast row represent a disaggregation of the product data in the production plan. As the information flow diagram in Figure 6–8 indicates, the MPS time-phased record is developed directly from the production plan. Depending on the production strategy, the units listed in the Forecast row will be either actual catalog product numbers for MTS or information from planning bills of materials for ATO.

	Period number									
	1	2	3	4	5	6	7	8	9	10
Forecast	5	5	5	5	5	5	20	20	20	20
Available	26	32	38	44	50	56	47	38	29	20
MPS	11	11	11	11	11	11	11	11	11	11
On hand	20									

Figure 6–15   MPS Time-Phased Record.

The Available row represents the inventory balance at the *end* of the period or the number of units available for sale in the next period. The MPS row indicates the number of units scheduled for production during the period and available to satisfy the forecast for the period. The On Hand value is the number of units present in inventory at the start of the first period. To avoid confusion between the On Hand and Available values for period 1, remember that On Hand is the inventory coming into the first period, and Available is the inventory leaving the first period. The time-phased record is a method for visualizing how the MPS relates to the forecast and the inventory.

The time-phased record in Figure 6–15 shows all the data for 10 periods of production. Study the data in the figure and try to determine if the production process is level or chase. Note that the forecast calls for the sale of five units per period through period 6 and twenty units per period through period 10. The production process is determined from the entries in the MPS row. Since the plan calls for the production of eleven units in every period, a level production process is present.

The master production scheduler starts with a time-phased record that has the Forecast row and On Hand value present. On the basis of company policy and available resources, a decision to use a level, chase, or mixed production strategy is reached. For the record in Figure 6–15, a level production rate was chosen. The number of MPS units per period was determined by adding the forecast for 10 periods (110 units) and dividing by the number of periods (10), the result of which was a period production rate of 11. Note that inventory is built up in the first 6 periods to cover the higher sales rate in the last 4 periods. With the MPS row now inserted into the record, the scheduler calculates the Available row, starting with period 1. The Available value at the end of period 1 equals the On Hand balance plus the units produced in period 1 minus the forecast sales for the period $(20 + 11 - 5 = 26)$. The same process is used for period 2, except that the Available value in period 1 becomes the On Hand value for period 2 $(26 + 11 - 5 = 32)$. Stop and make the calculations for periods 3 through 10 to verify that you know how to complete an MPS time-phased record.

In many production situations, the products and parts are produced in *lot sizes*. In the MPS record shown in Figure 6–15, the assumption was made that eleven units was an economical lot size for the product. If a different lot size is

	Period number									
	1	2	3	4	5	6	7	8	9	10
Forecast	5	5	5	5	5	5	20	20	20	20
Available	45	40	35	60	55	50	60	40	20	0
MPS	30			30			30			
On hand	20									

Figure 6–16   MPS Time-Phased Record with Lot Sizing.

required to manufacture the product economically, that requirement must be reflected in the time-phased record. To demonstrate this concept, assume that a lot size of thirty units is the smallest production level used. The new time-phased record is illustrated in Figure 6–16. Note that the MPS would call for production of thirty units in periods 1, 4, 7, and 10. Compare the Available row of the records in Figures 6–15 and 6–16. Note that inventory levels are higher when larger lot sizes are necessary. Also, no MPS value is present in period 10 because the twenty available in period 9 satisfies the forecast for twenty in period 10. However, that means the inventory for period 11 is zero. If there is a minimum level of inventory, or *safety stock*, required by manufacturing, an MPS order for thirty units in period 10 would need to be established.

If no lot size is required, the production system produces *lot for lot*, or just what is required; however, because of production machine setup time and several of the other cost-added operations present in manufacturing, many manufacturing operations set a minimum lot size for production.

Computer-based production-scheduling systems used in CIM implementations capture the same data presented in the time-phased record illustrated in Figure 6–15 and make all the calculations for future periods. However, the computer database displays the information in report formats different from those used for hand calculations.

Order Promising

The time-phased records illustrated in Figures 6–15 and 6–16 do not allow for production situations where customers place orders for future delivery. The earlier record format assumed that sales planned for each period would occur only in that period, when in reality sales planned for future periods may occur as an order from a customer in an earlier period. One way to handle this additional variation in planning the master schedule is to expand the time-phased record to include two more rows: *ATP* (available to promise) and *Orders*. The record in Figure 6–17 illustrates this concept. The Orders row indicates all orders received for future products that will be consuming the forecast. The quantity listed in the Forecast row is expected to be sold and the values in the Order row represent firm sales of the product. The calculation for the Available row is unchanged.

	Period number									
	1	*2*	*3*	*4*	*5*	*6*	*7*	*8*	*9*	*10*
Forecast	10	10	10	10	10	15	15	15	15	15
Orders	6	5	3							
Available	30	20	10	30	20	5	20	5	20	5
ATP	26			30			30		30	
MPS	30			30			30		30	
On hand	10									

Figure 6–17   MPS Time-Phased Record with Order Promising.

The ATP value for period 1 is calculated by adding the On Hand and MPS values and subtracting the total orders received up to period 3 (10 + 30 − 6 − 5 − 3 = 26). Space does not permit coverage of the many variations on the ATP concept practiced by companies using a time-phased MPS system, but it is important to know that order promising is an integrated part of the MPS process.

6–6 INVENTORY MANAGEMENT

The American Production and Inventory Control Society (APICS) defines *inventory* as follows:

> ***inventory***: *1) Those stocks or items used to support production (raw materials and work-in-process items), supporting activities (maintenance, repair, and operating supplies), and customer service (finished goods and spare parts). Demand for inventory may be dependent or independent. Inventory functions are anticipation, hedge, cycle (lot size), fluctuation (safety, buffer, or reserve), transportation (pipeline), and service parts. 2) In the theory of constraints, inventory is defined as those items purchased for resale and includes finished goods, work in process, and raw materials. Inventory is always valued at purchase price and includes no value-added costs, as opposed to the traditional cost accounting practice of adding direct labor and allocating overhead as work in process progresses through the production process.*
> (Source: © *APICS Dictionary*, 10th edition, 2002)

Inventories consist of *raw materials, component parts, work-in-process (WIP)*, or *finished products and goods*. The singular reason for carrying a finished-goods inventory is to satisfy a customer lead time that is less than the production lead time. In many situations, it is not possible to eliminate the finished-goods inventory without losing market share. As long as the need for the inventory is an order-qualifying criterion for every competitive manufacturer, the cost of the inventory is included in each vendor's product prices. In that case, the battle to gain new customers and increase market share is fought in the production area.

Two fundamental reasons exist for carrying the raw material, component parts, and WIP inventory items listed earlier. The primary justification is the elimination or reduction of disturbances in the production cycle. These disturbances have many causes, including unreliable part supplier deliveries, poor quality, poor scheduling, and undependable production operations. The secondary reason to carry production inventory is to take advantage of the economies of larger order sizes from suppliers. Data from world-class companies indicate that *order-winning* advantages are possible for the industry willing to change how manufacturing functions. Therefore, the focus for the rest of this section is on better management of the production inventory.

The level of production inventory is a function of the complexity of the product and the layout of the manufacturing process. Flow-oriented production systems that use effective group technology require less inventory in both raw materials and WIP. Regardless of the system used, however, many manufacturing operations require some level of production inventory. Therefore, the critical issue in minimizing the cost-added effects of inventory is to know how much inventory is present and where it is located.

Benefits of Inventory Accuracy

The benefits of an accurate inventory include fewer missed manufacturing schedules and delays in shipping because incorrect inventory counts created shortages of critical parts. In addition, safety stock can be eliminated or reduced significantly because inventory records indicate exactly the parts on hand. From a financial viewpoint, inventory accuracy means correctly stated inventory cost reports, less costly material expediting, and reduced losses due to obsolete and excessive inventory in stockrooms. Although the benefits of an accurate inventory are numerous, the most important reason to solve the inventory accuracy problem is that MRP and the other MPC software modules fail to function if the inventory values are less than 98 percent accurate.

Inventory accuracy is most often measured by one of two methods: inventory dollar basis or a physical count of the parts. The dollar method compares the dollar value of the parts in inventory with the dollar value stated in the financial records. The degree to which the two totals agree is a measure of the accuracy of the inventory part count. This method has two serious problems. First, the ability to identify poor inventory practices is reduced because the records often have offsetting errors. For example, consider an inventory error on two parts, one that has a $100 value and a second that has a $10 value. If the count were 10 high on the $10 part and 1 low on the $100 part, the dollar value of the inventory would be unchanged. The second serious problem is related to the first. As a result of offsetting errors, manufacturing believes that the inventory stated in dollars means that the parts are available. In fact, when manufacturing obtains the last of the $100 parts from the stock area to finish an assembly, no part is found because the records were wrong. As a result, a shipment date is missed and a customer is not satisfied.

The only method that guarantees parts will be there for manufacturing is the *physical count method.* Inventory management is successful only if the inventory records and the physical count of parts in the stockroom have more than

98 percent agreement. For all parts tracked, the records and count must agree on both total parts present and location in the warehouse or plant. Therefore, the important questions for inventory management are as follows: What needs to be counted? And how often does it need to be done?

Cycle Counting the ABC Part Classification

An accurate inventory record of every part in a product is not necessary as long as a sufficient quantity is available for final assembly. For example, it is not necessary to keep accurate on-hand balances of nuts, bolts, and washers that cost less than a few cents each. In contrast, the part count and location of high-cost components (e.g., a $150 part) must have an accurate inventory record. The 80-20 rule applies in this situation. Experience shows that 20 percent of the parts account for 80 percent of the inventory value. Figure 6–18 illustrates this concept and indicates that maintaining accurate records for all of the A parts and probably some of the B parts is a good inventory management process.

   For the balance of the parts in classifications B and C, a *two-bin* inventory system satisfies the production requirements. In such a system, parts are divided into two bins, the size of each being determined by the production rate and reorder lead time of the part. Parts are drawn from the production inventory bin to satisfy manufacturing; when this bin is empty, the purchasing department orders

Figure 6–18   Inventory
Classification.

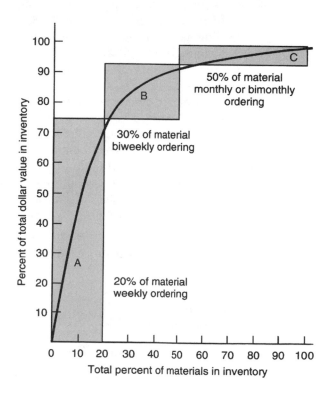

more parts. Production uses the second bin, which holds a sufficient number of parts to last through the lead time on the replacement parts order. The cost of parts in category C is low, so the burden of carrying the excess inventory is negligible.

Managing the more costly parts effectively requires a physical count and verification of the warehouse location. The method preferred is called *cycle counting*. In cycle counting, every item in classification A is checked regularly to verify that the quantity and location on the inventory record agree perfectly with the number of parts at a specific location in the warehouse. The cycle-counting activity is assigned to specific persons with training in accurate cycle-counting techniques. For example, some companies have cycle-counting teams check a different group of part numbers each day. This practice ensures that all parts in classification A get counted every two or three production periods. In this type of system, accuracy is determined as follows:

$$\text{Inventory accuracy} = \frac{\text{total hits}}{\text{total counts}}$$

where the term *total hits* refers to every part number checked that had the inventory present within the required quantity range and all the parts in the correct location. For some parts, there is an allowed range in the count number, or error, where the count still falls within the definition of a *hit,* or an accurate count. These parts usually are lower-cost items or parts where the count is determined through a weight scale reading. *Total counts* is the total number of parts checked through cycle counting. All discrepancies between the inventory record and the count are investigated thoroughly to improve continuously the reliability of the inventory management system. Companies that implement cycle counting often achieve an inventory accuracy of 0.99. Under these ideal production conditions, 99 percent of the time when a part is checked, the count value falls within the range on the inventory record and the location is correct. In these situations, manufacturing knows that parts are available if the inventory record indicates parts are on hand.

The *annual physical inventory* is intended to prove that the dollar value of the inventory is correct as stated in the financial records. However, MRP and the MPC software system require an accurate inventory status on a day-to-day basis, not just once per year. The goal of cycle counting is to demonstrate a level of inventory accuracy that makes MRP effective and eliminates the need for the costly annual physical inventory.

6–7 PLANNING FOR MATERIAL AND CAPACITY RESOURCES

The MRP techniques and logic were created to plan for the material required to support the production of the master schedule. How much to order and when to order was logically planned to minimize the investment in raw materials, parts, and subassemblies. The details of MRP are presented in chapter 7. Think through the concept of MRP before jumping into the details of the logic and calculations. Pay particular attention to the inputs and outputs of this planning system as shown in Figure 6–19.

Figure 6–19 MRP Operational Model.

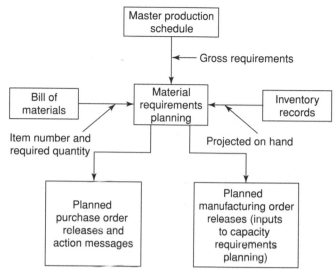

Planning Inputs and Outputs

The MRP process starts with the MPS, which provides the quantity of each model or product required in future periods (Figure 6–5). This required production quantity becomes the driver of plans for all the material items and subassemblies that make up the product.

Two additional inputs, *BOM* and *current inventory,* provide critical information for an effective MRP system. The inputs from these two sources must be *timely* and *accurate* for the formal MRP system to work. Updates of the inventory control system for changes in inventory due to part movement in manufacturing or purchasing must be continuous. For example, in some manufacturing operations, parts from vendors arrive daily, so when the parts arrive, the inventory control system is updated to provide timely information for planners.

A *product structure diagram* graphically represents the BOM for the end product in terms of all required component parts. For example, in Figure 6–20, product A is produced by assembling one of part B, one of part C, and one of part E. Part C is made by assembling one of part D and one of part E. The items in the product structure diagram are identified by levels, with the zero level representing the finished product. The product structure diagram illustrates clearly the sequence required to build the product. For example, the first step in manufacturing product A is the production and assembly of parts D and E. The uncomplicated product structure shown in Figure 6–20 has three levels and five parts; however, a product structure for a riding lawn mower could have many levels and hundreds of parts.

In most MRP installations, a BOM accuracy of 95 percent or better is required, and a location and count accuracy of 98 percent is necessary for specific

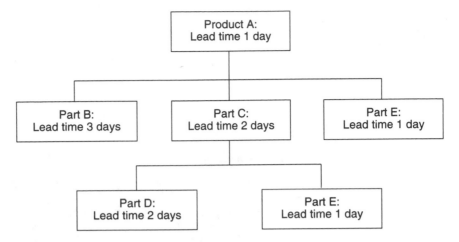

Figure 6–20 Product Structure Diagram with Planned Lead Times (nearest whole day).

parts in the inventory system. The BOM (Figure 3–10) provides the MRP system with the *part number* and *quantity* of all parts required to build and assemble the product. The inventory control system supplies the MRP system with the *projected on-hand balance* of all parts and materials listed on the BOM.

The planning system, whether manual or computer aided, is driven by the data supplied to it. The critical data needed by the MRP system include manufacturing order information that incorporates the critical information in the master schedule: the specific quantities of products to be made and when they are due, and common item-related data such as the item number, description, unit of measure, group technology code, planning lead times, costs, and who is responsible for the item. Product structure information describes the relationships of items and the products that the company makes. Inventory records tell how much material is on hand, where it is stored, and the status of the material. Purchase order data provide information on what material is currently on order and the status of open purchase orders.

The MRP logic is almost always executed by a computer-based system. Many number-crunching calculations are required to develop the detailed ordering plans for all the materials and subassemblies that go into most products. Even simple products can require several subassemblies and have more than 10 parts or components. Most companies have to manage hundreds of products and thousands of material items.

The results or output of a computer-based MRP system can be generalized into the two information streams shown in Figure 6–19. The primary output of MRP is the system-generated plan for future purchase orders for the material needed to produce the quantities of products shown in the master schedule. Planned purchase orders are evaluated and finalized by the purchasing function and are then released to material suppliers. In addition, systems often provide action and exception messages that help guide the approval and review actions of the purchasing department.

Action messages identify orders that are recommended for release to material vendors on a particular day. *Exception messages* alert the materials planner or buyer that a planned order may need special attention or expediting to satisfy the master schedule. A more complete discussion of the MRP logic is provided in chapter 7.

A good material plan has little value if manufacturing does not have the needed resource capacity to convert the material into the required products to satisfy the master schedule. Insufficient capacity leads to poor delivery performance, high WIP inventories, and frustrated manufacturing personnel. Lack of the needed capacity makes the master schedule meaningless, along with all the plans that have been made to support it. Conversely, excess capacity is inefficient and can lead to overproduction of products and growing WIP and finished-goods inventories.

The primary output of the MRP process for manufacturing is the production order schedule for the subassemblies that go into the finished products. This schedule of orders for manufactured items becomes the plan that drives production on the shop floor. *Capacity requirements planning (CRP)* is a computer-based extension of the MRP process that uses the results of MRP along with detailed production operation information and labor information to calculate planned workloads. A sample of the detailed routing information for a part is shown in Figure 6–21.

Manufacturing planners can evaluate the projected workload and assess the capability of manufacturing to complete the planned work. The results of the CRP process are used to determine short-term capacity needs for equipment and labor skills. More discussion of CRP follows that of MRP in chapter 7.

Part D routing (hours)

Operation	Work center	Run time	Setup time	Move time	Queue time	Total time	Rounded time
1	201	1.6	0.5	0.4	2.6	5.1	5.0
2	208	1.5	0.3	0.2	2.8	4.8	5.0
3	204	0.1	0.1	0.3	0.6	1.1	1.0
4	209	1.2	0.8	0.3	2.3	4.6	5.0

Total lead time 16 hr (2 days)

Part E routing (hours)

Operation	Work center	Run time	Setup time	Move time	Queue time	Total time	Rounded time
1	201	1.1	0.4	0.3	1.8	3.6	4.0
2	204	0.2	0.3	0.2	0.5	1.2	1.0
3	205	1.2	0.1	0.4	1.5	3.2	3.0

Total lead time 8 hr (1 day)

Figure 6–21 Part Routing.

6–8 INTRODUCTION TO PRODUCTION ACTIVITY CONTROL

Production activity control (PAC), formerly called *shop-floor control,* manages the detailed flow of materials inside the production facility. *PAC* is defined as follows:

> **production activity control (PAC):** *The function of routing and dispatching the work to be accomplished through the production facility and of performing supplier control. PAC encompasses the principles, approaches, and techniques needed to schedule, control, measure, and evaluate the effectiveness of production operations.*
> (Source: © *APICS Dictionary,* 10th edition, 2002)

PAC supports *order-winning criteria* by seeking a balance among the following three goals: (1) to minimize inventory investment, (2) to maximize customer service, and (3) to maximize manufacturing efficiency. The MPC model (Figure 6–4) illustrates the links between the material plans generated by the MRP system and PAC. Three different processes, *Gantt charts, priority rules for sequencing jobs at a work center,* and *finite loading,* are used for scheduling production in manufacturing. The first two are discussed next. The third is discussed in section 6–9.

Gantt Charts and Graphic Schedule Techniques

Gantt, or *bar, charts* are basic shop-floor control tools used by small- and medium-size manufacturers. The process starts with the preparation of a *setback chart* (Figure 6–22), which shows the manufacturing start and finish dates based on routing sheet (Figure 6–21) or MRP lead times for all parts in the product (Figure 6–20). On the basis of the lead times, production must start on May 3, which is 5 days before the product ship date. Study the setback chart and the routing sheet (Figure 6–21) for parts D and E and note anything common between the two parts.

Figure 6–22 Operation Setback, or Back-Scheduling, Chart.

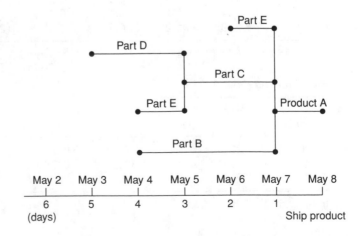

Figure 6–23 Gantt Chart for Work-Center 201: Parts D and E.

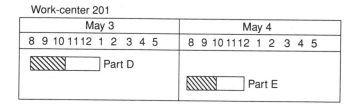

Two events critical to shop-floor control are apparent: common work centers and overlap in the production schedule. Both parts use work-center 201 as the first operation and work-center 204 for a later operation. In addition, the production of part D and that of part E overlap on May 4; also part E is scheduled for production on two different days. Using the setback, or back-scheduling, chart (Figure 6–22) and data from the routing sheet for each part, the Gantt chart in Figure 6–23 for work-center 201 was developed. The chart includes total lead time (run, setup, move, and queue) and indicates that work-center 201 can handle both parts. No conflict occurs between the schedule for part D and that for part E. The cross-hatched area in the bar indicates the queue-time component in the total lead time. If a conflict develops in the work center as other parts are scheduled, the work-center schedulers can use queue time to adjust part schedules by shortening the waiting time and starting work on the parts ahead of the original schedule. Frequently, the move and queue times are not included in the work-center Gantt charts, so that the chart reflects work-center productivity more accurately.

In some applications, companies put the Gantt chart on a large scheduling board where the planned activity for every work center is easy to check. Scheduling work with Gantt charts is well understood and effective for a limited number of work centers and part variations. However, keeping the chart current with the latest schedule change is difficult when the number of parts and work centers increases. In addition, communicating the changes to the shop floor in real time is a demanding requirement. Some companies have addressed this problem by using personal computers on the shop floor and software that creates a scheduling board.

Priority Control and Dispatching Techniques

The second technique used to control activity on the shop floor includes a set of *priority sequencing rules*. Effective application of the rules requires an MPC system with an *integrated database, a dispatching mechanism,* and *priority rules.*

First, the system requires an integrated database for all manufacturing scheduled receipts with open shop orders that includes part identification data, routing information for all operations, scheduled operation dates, and the scheduled completion date. In addition, the database must accumulate the following

Figure 6–24 Dispatch List
with Look-Ahead.

Plant: 10   Department: 25   Work center: 15
Capacity: 100 hr
Date: 9-5

Plant number	Order number	Quantity	Hr/unit	Total hours	Due date
12-9201	SO 73421	300	0.2	60	9-20
12-4510	SO 73107	100	0.3	30	9-22
18-2009	SO 73560	150	0.2	30	9-28
				120	

Arrival on 9-6

12-7210	SO 73416	100	0.5	50	10-4
15-0379	SO 73601	100	0.2	20	10-7

data on WIP: actual completion date, part and material move dates, labor utilization, and scrap.

A second requirement for effective shop-floor control is a reliable *dispatching mechanism*, defined as follows:

> *A dispatching mechanism is a process for the selection and sequencing of manufacturing jobs at individual work centers and a process to assign jobs to workers.*

The dispatching system performs four functions: (1) determines the *relative* priority ranking through priority rules for all jobs released to manufacturing, (2) communicates job priority rankings to the shop floor through the use of a *dispatch list*, (3) tracks the movement of jobs on a real-time basis across the shop floor, and (4) monitors and projects the contents of work-center queues. The dispatch list, a primary component of the dispatching system, provides a daily listing of manufacturing orders in priority sequence, oriented by work center. Study the dispatch list example illustrated in Figure 6–24 and note the type of data present.

A set of rules for establishing the priority of jobs at work centers is the last requirement for a *priority rule* system. The primary requirement for an effective rule-based system is consistency in the priority rankings. The rule-based processes use several different routines to rank manufacturing jobs in priority order. Some examples of routines include the following: *shortest operation next, first come–first served, earliest due date, critical ratio, order slack, order slack time per remaining operations, queue ratio rule,* and *operation start and operation due date.* A description and example of each of these rules is provided in appendix 6–1.

6–9 SHOP LOADING

Loading is defined as the process of committing capacity and implies a scheduling process for work centers and machines. Shop loading is either *infinite* or *finite*.

Infinite Loading

Infinite loading results generated by MRP software establish a work-center schedule that does not balance the planned work order resource needs with the capacity of the work center. Study the dispatch list in Figure 6–24 and compare the capacity of work-center 15 with the total hours scheduled. The work-center capacity of 100 hours is insufficient to handle the 120 hours scheduled. The concepts of loading are illustrated in the load profile diagrams of Figure 6–25. Figure 6–25a shows the concept of infinite loading, with a work center with a capacity of 80 hours per period. The infinite loading illustrated in this figure was generated by CRP software and MRP data. Note that the WIP, called *open shop orders,* and the new jobs scheduled for the work center, called *planned orders,* exceed the 80-hour capacity during the first three periods; in addition, there is a large number of past-due open shop orders, and the open shop orders exceed the capacity in the first two. All the open shop orders

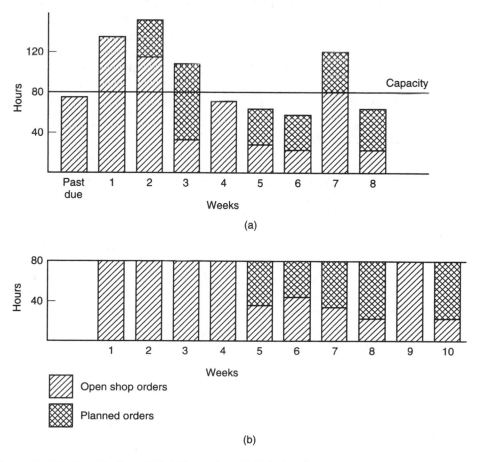

Figure 6–25   Load Profiles: (a) Infinite Loading; (b) Finite Loading.

scheduled for period 1 will not be completed; therefore, deliveries may be late and shop work will be pushed to the following period in a past-due category. The priority for work at the infinitely loaded center is established by production control personnel or supervisors who take control of the work-center schedule and set the job schedule by using one of the priority rules described previously. Without human intervention, the multiple jobs scheduled into the work center at the same time would cause chaos.

Finite Loading

Finite loading systems produce a detailed schedule for each part and work center through simulated shop order start and stop dates. The result of finite loading (Figure 6–25b) is a work center scheduled at capacity for the first four periods, with open orders and no past-due WIP. One output of the finite loading system is a simulation of how each machine and work center will operate during the planned time horizon. The schedule is built from the following information: jobs currently waiting in the queue, the completion time for operations scheduled at the work center ahead of the current work center, and the priority on current WIP. On the basis of that information, a decision could be made to let the work center remain idle until the next job scheduled for the center is finished on the previous machine. Two approaches are used to fill a work center: *vertical loading* and *horizontal loading.*

Vertical and Horizontal Loading

In the vertical loading approach, the jobs currently in the work center plus those that will be finished at a previous operation are evaluated from a priority standpoint and the highest priority is loaded. In the horizontal method, an entire shop order or job with the highest priority is completed at each work center in the sequence, with the shop order taking priority over other parts. For example, in Figure 6–21, part E starts in work-center 201 and the center would schedule the run time for all the parts plus the setup time as the highest priority for the work center. If the previous job in work-center 204 finished before part E was ready to move to the work center, 204 would sit idle, waiting for the high-priority job to arrive. The horizontal process often produces "holes" on idle times in the capacity profile for the work center.

In addition, a choice must be made between a forward or a backward scheduling process. The *forward scheduling* process (Figure 6–26a) starts with the current date and builds through the total manufacturing lead time (LT) to the completion date. The alternative approach, *backward scheduling* (Figure 6–26b), starts with the order due date and subtracts the manufacturing LT to arrive at the shop order release date. In both cases, the manufacturing LT may not be consistent with the time remaining to complete the job; as a result, corrective action is necessary.

Figure 6–26 (a) Forward Scheduling; (b) Backward Scheduling.

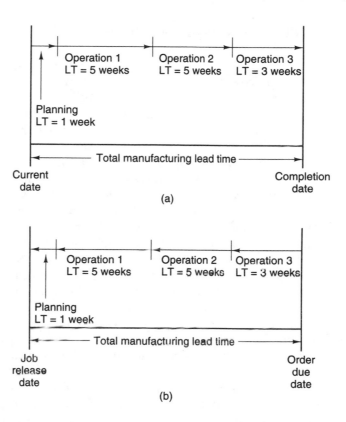

6–10 INPUT-OUTPUT CONTROL

The function of *input-output control* in the MPC model (Figure 6–4) is to evaluate the number of production orders that were released to manufacturing versus the work completed. Input-output control provides a check on shop-floor efficiency and is a valuable feedback tool for production planning. For example, if the input-output control report shows that back orders of some products develop whenever orders are released to the shop, that is a reliable indication of some type of production problem. Input-output control is defined as follows:

> *input/output control: A technique for capacity control where planned and actual inputs and planned and actual outputs of a work center are monitored. Planned inputs and outputs for each work center are developed by capacity requirements planning and approved by manufacturing management. Actual input is compared to planned input to identify when work center output might vary from the plan because work is not available at the work center. Actual output is also compared to planned output to identify problems within the work center.*

(Source: © *APICS Dictionary*, 10th edition, 2002)

6–11 AUTOMATING THE PLANNING AND CONTROL FUNCTIONS

The graphic model of MPC shown in Figure 6–4 describes the overall planning and control process. For a small organization with a few simple products, a fairly manual system could be implemented. Most manufacturing operations would soon bog down in a manual attempt at this level of planning. Companies may start out small and be successful with manual systems before they outgrow the manual system capabilities.

Computer-based systems offer real opportunities and support for MPC systems. Various parts of the process have been automated as "islands," much the way production processes have been automated. A CIM approach offers real opportunities for companies to develop integrated systems. Integrated systems share and reuse critical data, and thereby eliminate the cost and waste associated with redundant data in the system. The sharing and exchange of information between the functional areas shown in Figure 6–27 is critical. New computer-based MPC systems provide a framework that can enable timely communication and effective exchange of data and information among the functional groups of the company.

Inventory management and control is an area that can benefit from an automated system. Many functional areas use inventory data for various calculations, for reports, and for decision making. Automated systems provide program modules or screens that support the entry and maintenance of inventory data. Data in the inventory database are often used for cost analysis and inventory management decisions. Systems can also support the analysis of inventory value by classification, location, age (receipt date), and the producer's lot number. Support for inventory cycle counting is often a feature of automated inventory systems. Automated data capture technologies integrated into the inventory control and manufacturing planning systems increase the accuracy and timeliness of the data in the system.

The MPS function is often supported by integrated computer-based software that aids in the preparation and "what-if" analysis of the master schedule. Programs can assist in the development of the aggregate plans and in the work to disaggregate the family plans into the specific production items that are shown in

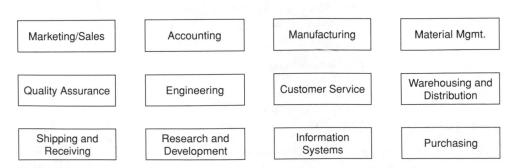

Figure 6–27   Functional Areas in the Enterprise.

the master schedule. "What-if" analysis lets the master scheduler test different versions of the production plan and the related master schedule. Rough-cut capacity analysis tools are often included in automated systems. The computer-based support of this function helps provide the best inputs possible to the material and capacity planning processes that follow.

MRP and CRP are techniques that have been enabled by computer-based systems. Manual planning using the MRP logic is possible only on a limited scale. CRP is driven by the results of the MRP process and requires a computer-based system to be effective and timely.

6–12 SUMMARY

Welcome to the exciting world of manufacturing planning and control (MPC) systems. The MPC function is an important part of the Society of Manufacturing Engineers (SME) enterprise wheel. The primary activities of MPC are planning for the product and enterprise resources at both the aggregate level and the disaggregate level. The MPC function is responsible for generating production schedules, planning for needed material and resource requirements, and tracking and control of work through the production processes. The MPC model describes a process and flow of production-planning and -scheduling information. In addition, the model indicates areas of responsibility and decision points for the critical steps in the planning and scheduling of production.

Companies face a real challenge when they are designing the systems they will use to plan and schedule production and to ensure the availability of the material and production resources needed. Huge investments have been made in software and systems to better schedule production operations and ensure effective production and high customer service levels while keeping inventory costs to a minimum.

This chapter sets the stage for the discussions that follow in the next three chapters. Chapter 7 covers, in more detail, the problem of getting materials and resources together at the right time to meet the customer's requirements. Chapter 8 takes a broader look at the manufacturing enterprise and the linking of information and systems needed to be competitive. Chapter 9 looks at the application of the methods and techniques of just-in-time (JIT) manufacturing and lean production. These methods not only offer the promise of significant gains for companies that implement them, but also challenge nearly all the traditional beliefs and methods of production.

BIBLIOGRAPHY

BERRY, W. L., T. E. VOLLMANN, and D. C. WHYBARK. *Master Production Scheduling Principles and Practice*. Falls Church, VA: American Production and Inventory Control Society, 1979.

CLARK, P. A. *Technology Application Guide: MRP II Manufacturing Resource Planning.* Ann Arbor, MI: Industrial Technology Institute, 1989.

COX, J. F., and J. H. BLACKSTONE, eds. *APICS Dictionary.* 10th ed. Alexandria, VA: American Production and Inventory Control Society—The Educational Society for Resource Management, 2002.

GROOVER, M. P. *Automation, Production Systems, and Computer-Integrated Manufacturing.* 2nd ed. Upper Saddle River, NJ: Prentice Hall, 2001.

SHRENSKER, W. L. *CIM: Computer-Integrated Manufacturing: A Working Definition.* Dearborn, MI: CASA of SME, 1990.

SMITH, B. T. *Focus Forecasting—Computer Techniques for Inventory Control: Revised for the Twenty-First Century.* Butler, PA: B. T. Smith and Associates, 1997.

SOBCZAK, T. V. *A Glossary of Terms for Computer-Integrated Manufacturing.* Dearborn, MI: CASA of SME, 1984.

STEVENSON, W. J. *Production/Operations Management.* 3rd ed. Homewood, IL: Richard D. Irwin, 1990.

VOLLMANN, T. E., W. L. BERRY, and D. C. WHYBARK. *Manufacturing Planning and Control Systems.* 4th ed. Homewood, IL: Richard D. Irwin, 1997.

QUESTIONS

1. Compare production management with operations management.
2. How are service operations and manufacturing operations different, and how are they similar?
3. What is the goal of MPC?
4. What is *aggregate planning,* and how does it fit into MPC/MRP II?
5. Describe disaggregate planning and its function in MPC/MRP II?
6. Compare the three disaggregate planning options.
7. Describe the MPC/MRP II model presented in the chapter.
8. Describe the MRP system input and output data.
9. What is the relationship between MRP and CRP?
10. Define *production activity control.*
11. Describe the three processes used in PAC.
12. What is the function of a product structure diagram?
13. What is the function of a routing sheet, and what manufacturing information is included on the sheet?
14. Describe how Gantt charts and scheduling boards are used for PAC.
15. What are *priority sequencing rules?*
16. Define *dispatching mechanism* as used in a priority rule system.
17. Describe the four functions performed by a dispatching system.
18. List the priority rule routines described in appendix 6–1.

19. Discuss the concept of infinite versus finite shop loading.
20. Compare vertical and horizontal shop loading.
21. Discuss forward and backward scheduling for shop orders.
22. Where does the MPC function reside in a manufacturing operation?
23. Describe the production-planning process, using the model in Figure 6–8.
24. How does MPC link to the CIM common database and other departments in the enterprise?
25. Describe the resources used to forecast production requirements.
26. Describe the production-planning process.
27. Describe the three different production-planning strategies: chase, level, and mixed.
28. Describe the two options available when the chase strategy is used.
29. Compare the advantages and disadvantages present when the chase and level production strategies are adopted.
30. What is the *master production schedule (MPS)*?
31. Compare and contrast the MPS with the production plan and the forecast and describe similarities and differences.
32. What changes occur in the MPS for the various production strategies listed in Figure 2–5?
33. Describe the function of the MPS time-phased record and include a description of the terms.
34. What impact does the introduction of lot sizes have on an MPS record?
35. Compare lot-for-lot and lot-size production and describe the differences.
36. What is *order promising*, and how does it affect the MPS?
37. How is the MPS function automated?
38. Define *inventory* and list the sources of inventory.
39. Describe the causes of inventory inaccuracy and discuss the positive and negative aspects of inventory balances in each of the sources.
40. Describe the two methods used to measure inventory accuracy, and discuss the advantages and disadvantages of each.
41. What is *cycle counting*?
42. Describe the ABC parts classification method of inventory control.
43. How do cycle counting and the ABC method provide accurate inventory control?
44. How is the inventory function automated?

PROBLEMS

The shop order information that follows provides the data necessary for the following priority scheduling problems. The future due dates and current date are expressed as the number of days from the first day of the year, and all times are in days. The current M-day is 110.

Shop order number	1B	2B	3B
Date arrived	107	110	109
Date due	118	116	119
Lead time remaining (LTR)	11	5	8
Processing time remaining (PTR)	9	2	6
Operations remaining (OR)	4	1	3
Standard scheduled queue (SSQ)	4	0.5	2
Next operation time	1.5	2	1

1. Calculate the priority schedule for the shop orders from the preceding data, using the following methods:

   a. Shortest operation next

   b. First come–first served

   c. Critical ratio rule

   d. Order slack time rule

   e. Slack time per remaining operations rule

   f. Queue ratio rule

   g. Earliest due date rule

2. Determine the impact on shop order priorities if the due date on the 3B shop order is changed to 115.

3. An engineering change causes the OR and PTR for the 2B shop order to change as follows: OR = 3 and PTR = 5. How do these changes affect the priority schedule for the shop orders?

4. Compare the first come–first served and earliest due date rule with the remaining methods. How well does each method account for process time, number of operations, and lead time?

5. Develop an electronic spreadsheet template to generate the chase production strategy data shown in Figure 6–10. Input the Sales and Days Worked values from the figure and use $100 of sales per direct labor hour to verify the values in the columns for Labor, Variable Workforce, and Variable Work Week.

6. Develop an electronic spreadsheet template to generate the level production strategy data shown in Figure 6–12. Input the Sales and Days Worked values from the figure and use $100 of sales per direct labor hour to verify the values in the columns for Labor, Level Workforce, Monthly Production, and Inventory Balance.

7. The demand and production rates for a company that permits back orders is illustrated next. With a beginning inventory of 150 units at the start of period 1, calculate the back orders per period, average inventory, and ending inventory.

Period	Demand	Production
1	5,500	5,350
2	8,200	5,800
3	3,100	6,000
4	4,500	4,000

8. A company using a level production policy has the following demand for six periods:

Period	Demand
1	8
2	6
3	14
4	6
5	9
6	5

With a zero starting inventory balance, determine the following:

a. The level production rate that will result in a zero inventory balance after period 6.

b. The inventory and back orders per period with the production rate in period 4.

c. The level production rate that would eliminate back orders.

9. Atlas Furniture Company has the following annual sales (times 100,000) forecast: January, $6; February, $7.5; March, $8.2; April, $9.5; May, $13.8; June, $10.5; July, $7; August, $5.8; September, $6.7; October, $7.6; November, $9.6; December, $12.3. Analyze the chase and level production strategies, using the spreadsheets developed in problems 5 and 6. Use the same values for days worked and $50 of sales per direct labor hour in the calculations. Discuss the strengths and weaknesses of each strategy.

10. Use the graphing function of the spreadsheet to generate graphs to support the conclusions from the analysis in problem 9.

11. Complete the following MPS record, assuming a level production strategy and a safety stock of 10.

Period	1	2	3	4	5	6	7	8	9	10
Forecast	5	5	5	5	10	10	20	20	20	20
Available										
MPS										
On Hand	15									

12. Complete the MPS record in problem 11 while assuming a minimum lot size of 25. Would the resulting MPS require a level, chase, or mixed production strategy?

13. A company inventory accuracy standard has the following tolerance on cycle counts (percentage variation from recorded inventory value): class A items, ±0.5 percent; class B items, ±1 percent; class C items, ±3 percent. All class A items are hand counted, and the class B and C items, which are weight counted, have an additional 1 percent of error permitted. Determine which of the following items have cycle-count data within specifications.

Item	Class	Count basis	Recorded inventory	Counted inventory
1	A	Hand	56	55
2	C	Weight	10,425	10,950
3	B	Weight	2,850	2,825
4	A	Hand	430	431
5	B	Weight	8,530	8,745

PROJECTS

1. Using the list of companies developed in project 1 in chapter 1, determine the type of MPC model used by the companies.

2. Compare the MPC model presented in this chapter with the model used by one company from the small, medium-size, and large groups, and describe how their operations differ from that of the model.

3. Determine the type of shop order priority scheduling method used most often by the companies listed in project 1.

4. Identifying the problem of manufacturing: In part 1 of this problem, you are the manager of a manufacturing company. Brainstorm a list of the critical elements of your operation (the essential actions or activities that must happen or are required for you to be able to serve your customers effectively and stay in business). In part 2, prioritize the list by identifying your top five elements, then draw a circle around the element you believe is most critical. (This activity is best done in small groups but may be done individually.) Be prepared to share your ideas with others.

5. Visit the APICS Educational and Research (E&R) Foundation site on the World Wide Web (**http://www.apics.org**) and find the requirements for the Donald W. Fogarty International Student Paper Competition. Select a topic of interest and write a research paper, following the contest guidelines. Enter your paper in the contest through your local APICS chapter. Categories are open for part-time and full-time students at the undergraduate and graduate levels.

6. Using the list of companies developed in project 1 in chapter 1, determine the type of production method (chase, level, or mixed) used for each of the products they produce.

7. Compare the production-planning flowchart shown in Figure 6–8 with the process used by one company from the small, medium-size, and large groups, and describe how their operations differ from that of the model.

8. Compare the bill of materials and inventory accuracy of the companies selected in project 7 with the standards established in the chapter. Describe the inventory counting method used by each company and discuss how that method contributes to the accuracy of inventory data.

9. Consider the problem of manufacturing from the materials management perspective: getting the right materials to the right place, at the right time, to satisfy the customer's requirements.

   a. Who are the people (functional job titles) involved?
   b. What data related to a specific item or part need to be documented in the manufacturing database?
   c. What information needs to be included in the product structure?
   d. What information related to inventory is needed, and what level of accuracy is needed?
   e. What additional information is needed for effective material planning and the issuance of purchase orders?

10. Consider the problem of manufacturing from the manufacturing management perspective: getting the right products delivered to the right place, at the right time, to satisfy the customer's requirements.

   a. Who are the people (functional job titles) involved?
   b. What information is required to define clearly the manufacturing resources (capacity) needed for production?
   c. What is the critical information that should be included on the manufacturing order to be released to the shop floor?

APPENDIX 6–1: PRIORITY RULE SYSTEMS

The information shown in Figure 6–28 provides the data used for the following priority scheduling examples. The future due dates and current date are expressed as the number of days from the first day of the year. For example, a due date of February 7 would be listed as a due date of 38 (31 plus 7) for these calculations.

Exercise 6–1
Calculate the priority schedule for the shop orders shown in Figure 6–28, using the *shortest operation next* rule. The order with the shortest operation time has the highest priority, and the one with the longest operation time has the lowest priority.

Figure 6–28 Shop Order Data
for Priority Calculations.

Shop order number	1A	2A	3A
Date arrived	117	115	116
Date due	125	120	130
Lead time remaining (LTR)	8	5	10
Processing time remaining (PTR)	6	4	5
Operations remaining (OR)	1	2	8
Standard scheduled queue (SSQ)	2	1	5
Next operation time	6	3	1

Note: All Times in Days; Today's Date, M-Day 117.

Priority	Order number	Next operation time
1	3A	1
2	2A	3
3	1A	6

Solution

The rule ignores the due date and processing-time-remaining (PTR) information because it processes orders on the basis of speed of completion. This rule maximizes the number of shop orders processed and minimizes the number waiting in queue.

Exercise 6–2

Calculate the priority schedule for the shop orders shown in Figure 6–28, using the *first come–first served* rule. The priority is set by the arrival date of the shop order in the work center.

Solution

Priority	Order number	Arrival date
1	2A	115
2	3A	116
3	1A	117

The rule schedules on the basis of the arrival time of the part in the work center and will result in the shortest average queue times for parts.

Exercise 6–3
Calculate the priority schedule for the shop orders in Figure 6–28, using the *critical ratio (CR)* rule. The rule is based on the ratio of *time remaining* to *work remaining*. If the ratio is 1, the job is on schedule; if the result is greater than 1, the job is ahead of schedule; and if the result is less than 1, the job is behind schedule. The priority is set on the basis of the ratio values. The *CR formula* is as follows:

$$CR = \frac{\text{due date} - \text{today's date}}{\text{lead time remaining}}$$

Order 1A: $CR = \dfrac{125 - 117}{8} = \dfrac{8}{8} = 1$ (on schedule)

Order 2A: $CR = \dfrac{120 - 117}{5} = \dfrac{3}{5} = 0.6$ (behind schedule)

Order 3A: $CR = \dfrac{130 - 117}{10} = \dfrac{13}{10} = 1.3$ (ahead of schedule)

Solution

Priority	Order number	Critical ratio
1	2A	0.6
2	1A	1.0
3	3A	1.3

The rule identifies the job that is most behind schedule (the one with the lowest ratio) and the job most ahead of schedule (the one with the highest ratio). The rank order of the jobs by ratio makes sure that those behind schedule are performed first.

Exercise 6–4
Calculate the priority schedule for the shop orders shown in Figure 6–28, using the *order slack* rule. The rule is based on *slack time,* which is the difference between the remaining production time (due date minus current date) and the sum of setup plus run time. The highest priority is assigned to the part with the lowest slack time. A positive slack value indicates a part ahead of schedule; negative slack, a part behind schedule; and zero slack, a part on schedule. The highest priority is assigned to the part with the lowest slack time. The *slack formula* is as follows:

Slack = due date − today's date − processing time remaining

Order 1A:

   Slack = 125 − 117 − 6 = 2   (ahead of schedule)

Order 2A:

   Slack = 120 − 117 − 4 = −1   (behind schedule)

Order 3A:

   Slack = 130 − 117 − 5 = 8   (ahead of schedule)

Solution

Priority	Order number	Slack
1	2A	−1
2	1A	2
3	3A	8

The rule identifies the job that is most behind schedule (the smallest or most negative value) and the job most ahead of schedule (the largest or most positive value). The rank order of the jobs by slack value (smallest to largest) makes sure that the jobs with too little slack time are performed first. This priority rule addresses the sequencing of jobs on the basis of the value of work remaining.

Exercise 6–5

Calculate the priority schedule for the shop orders shown in Figure 6–28, using the *order slack time per remaining operations* rule. The rule is based on the ratio of *slack time* to *total number of operations remaining*. The highest priority is assigned to the part with the lowest slack time for each operation remaining. A positive value indicates a part with slack in the operations, a negative value indicates a part with insufficient operations time to meet the schedule, and a value equal to 1 indicates that the operation times are on schedule. The highest priority is assigned to the part with the lowest value. The formula is as follows:

$$\text{Slack per operation (S/O)} = \frac{\text{slack time}}{\text{operations remaining}}$$

Note: Values for slack time are obtained from exercise 6–4.

Order 1A:

   $S/O = \dfrac{2}{1} = 2.0$ (ahead of schedule)

Order 2A:

   $S/O = \dfrac{-1}{2} = -0.5$ (behind schedule)

Order 3A:

   $S/O = \dfrac{8}{8} = 1.0$ (on schedule)

Solution

Priority	Order number	Slack per operation
1	2A	−0.5
2	3A	1.0
3	1A	2.0

The rule identifies the job that is most behind schedule (the one with the lowest ratio) and the job most ahead of schedule (the one with the highest ratio). The rank order of the jobs by ratio makes sure that the jobs with insufficient operations slack are performed first.

Exercise 6–6

Calculate the priority schedule for the shop orders shown in Figure 6–28, using the *queue ratio (QR) rule*. This rule is based on the ratio of *slack time* to *standard scheduled queue time*. The highest priority is assigned to the part with the lowest ratio value. A positive value indicates slack in the queue time, a negative value indicates a part for which the standard queue time must be shortened for an on-time delivery, and a value equal to 1 indicates that the part is on schedule. The highest priority is assigned to the part with the lowest value. The *QR formula* is as follows:

$$QR = \frac{\text{slack time}}{\text{standard scheduled queue}}$$

Note: Values for slack time are obtained from exercise 6–4.

Order 1A:

$$QR = \frac{2}{2} = 1.0 \qquad \text{(on schedule)}$$

Order 2A:

$$QR = \frac{-1}{1} = -1.0 \qquad \text{(behind schedule)}$$

Order 3A:

$$QR = \frac{8}{5} = 1.6 \qquad \text{(ahead of schedule)}$$

Solution

The rule identifies the job with the least sufficient queue time (the one with the lowest ratio) and the job with the most queue time to use (the one with the highest ratio). The rank order of the jobs by ratio makes sure that those with insufficient queue slack time are performed first.

Priority	Order number	Slack per operation
1	2A	−1.0
2	1A	1.0
3	3A	1.6

To simplify priority sequencing rules, many companies have adopted a process based on *operation start and operation due date*. The due date is generated from the MRP output, and the due dates for all shop orders are updated daily through a *dispatch list* (Figure 6–24) or work-center schedule. In this type of system, the due dates would not appear on any shop paperwork that travels with the WIP inventory. The shop paperwork would show just the static information, such as work standards and routings. This process permits the due dates to be revised frequently because the only reference to the due date is on the daily dispatch list. Confusion in the work center is avoided because the job priority is set from the due date on the dispatch list.

Exercise 6–7

Determine the priority schedule for the shop orders on the dispatch list shown in Figure 6–24, using the *earliest due date* rule. The highest priority is assigned to the part with the earliest due date, and the lowest is assigned to the part with the latest due date.

Solution

Priority	Order number
1	SO 73421
2	SO 73107
3	SO 73560

The rule schedules order priority according to the order due date; as a result, the process works to maximize the number of on-time deliveries.

The specific priority rule selected by the production automation control group depends on several business and production factors.

Detailed Planning and Production-Scheduling Systems

OBJECTIVES

After completing this chapter, you should be able to do the following:

- Follow the progression of systems for material planning and scheduling from the economic order quantity (EOQ) to material requirements planning (MRP) to manufacturing resource planning (MRP II)
- Understand the logic of MRP calculations and complete a partial calculation grid
- Describe the expected benefits of material planning using MRP logic
- Extend the output of the MRP process to plan for resource capacity by using capacity requirements planning (CRP)
- Describe the benefits of closed-loop MRP
- Identify several features found in modern manufacturing planning and control systems
- Prepare an outline of the critical steps in the implementation process for an MRP II system

The manufacturing planning and control (MPC) process in the computer-integrated manufacturing (CIM) enterprise is responsible for the aggregate and disaggregate planning of production and scheduling of manufacturing resources. The aggregate planning process starts with a production plan stated in broad product specifications. The first disaggregate plan, broken into specific product models, is called the *master production schedule* (*MPS*). The MPS states the production plan for each model for several production periods in an MPS record. The output of the MPS record provides the data for the *material requirements planning* (*MRP*) scheduling system.

7-1 FROM REORDER-POINT SYSTEMS TO MANUFACTURING RESOURCE PLANNING (MRP II)

A study of the scheduling and planning systems used in MPC during the last forty years illustrates the impact of computer technology on manufacturing in the global market. The evolutionary change in the production and planning process was fueled by advances in technology that permitted manual planning processes to use the computational power of the computer for increased control. Initially, order timing decisions were based on a *reorder-point system.*

Reorder-Point Systems

The timing of replenishment orders under the *reorder-point rule* was determined by the use of trigger levels. The inventory level was monitored continuously and a replenishment order for a fixed quantity was issued when the part count dropped to a specified level. The graph of inventory levels shown in Figure 7–1 illustrates the reorder-point concept. The sawtoothed solid lines in the figure represent the inventory level at any time during production. The inventory starts at the highest level, M, on the graph and drops at a uniform rate as parts are used in production. When the inventory level reaches the reorder level, marked by the dashed line labeled R, an order is placed for the quantity of parts indicated in the figure. The lead time required to produce the parts is highlighted in the figure between points 1 and 2. Note that the trigger point is selected so that a new order arrives as the current inventory reaches the safety stock level, S. Usually some safety stock is carried for critical parts to prevent stock shortages.

The selection of the reorder point is influenced by four factors: the production demand rate for parts, the lead time required for replenishment inventory, the degree of uncertainty in the demand rate and lead time, and the management policy concerning inventory shortages. If the demand rate and lead time have a high degree of certainty, there is no need for safety stock and the reorder point is easily set.

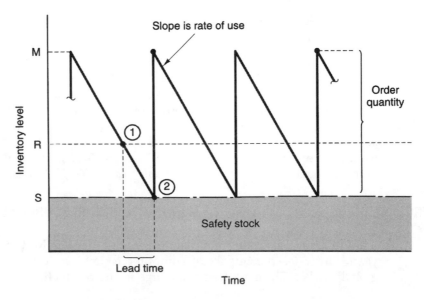

Figure 7–1   Reorder-Point System.

7-2 MATERIAL REQUIREMENTS PLANNING

The American Production and Inventory Control Society (APICS) defines* MRP as follows:

> **material requirements planning (MRP)**: *A set of techniques that uses bill of material data, inventory data, and the master production schedule to calculate requirements for materials. It makes recommendations to release replenishment orders for material. Further, because it is time-phased, it makes recommendations to reschedule open orders when due dates and need dates are not in phase. Time-phased MRP begins with the items listed on the MPS and determines (1) the quantity of all components and materials required to fabricate those items and (2) the date that the components and material are required. Time-phased MRP is accomplished by exploding the bill of material, adjusting for inventory quantities on hand or on order, and offsetting the net requirements by the appropriate lead times.*
>
> (Source: © APICS Dictionary, 10th edition, 2002)

The description of the MRP operational model was given in chapter 6 and is supported by Figure 6–19. Study this figure, and verify that the model is compatible with the MRP definition.

The relationship of MRP to the bill of materials (BOM) and the use of the MRP record to calculate the time-phased release of orders for manufacturing are fundamental to the operation of the MPC system. Therefore, the function and creation of the MRP record is an appropriate place to start a study of MRP.

The MRP Record

The record shown in Figure 7–2 represents the production plans for the part number under study. The accepted convention considers period 1 the current period and periods 2 through 8 to be in the future. The period is a specified production time used for planning. For example, some companies equate each period to five working days, with the period starting on Monday and ending on Friday. When this period convention is used, all the entries in the record have the value listed at either the start of the period, the beginning of the workday on Monday, or at the end of the period, the end of the workday on Friday. An understanding of the MRP process starts with a working knowledge of all the words used in the record. Study the following definitions until you understand the terminology for MRP record data:

- *Period number.* The period number is the time duration used in the MRP process. Usually, one period represents a day, week, or month; however, it could be any number of days or hours.
- *Part number.* The part number identifies the specific part being planned with the MRP record.

---

*The *APICS Dictionary* terms and definitions in this chapter are published with the permission of APICS—The Educational Society for Resource Management, 10th edition, 2002.

Item: tube steel #246784		Period number							
		1	2	3	4	5	6	7	8
Gross requirements					100				
Scheduled receipts									
Projected on hand	140	140	140	140	40	40	40	40	40
Planned order receipts									
Planned order releases									

Order Policy = Fixed lot   Order Quantity = 200   Lead time = 5

Figure 7–2   MRP Record.

■ *Gross requirements.* Gross requirements equal the anticipated future demand, both independent and dependent, for an item, stated period by period, not aggregated or averaged. The gross requirements consist of a statement of the exact number of parts needed in each of the periods covered by the MRP record.

■ *Scheduled receipts.* The scheduled receipts represent all orders released to manufacturing (production, manufacturing, or shop orders) or to suppliers through purchase orders. Another term used for scheduled receipts is *open orders.* The quantity listed as a scheduled receipt arrives at the *start* of the period in which the item quantity number appears. Remember, scheduled receipts represent orders that have been placed with either manufacturing or a vendor and *do not* include orders for parts or raw material that will be placed in the future.

■ *Projected on hand.* The projected available balance represents the calculated inventory for the item projected through all the periods on the record. The projected available balance is the running sum of on-hand inventory minus gross requirements plus scheduled receipts and future planned order releases. The value for the projected available balance represents the inventory at the *end* of the period in which it appears. Therefore, the projected available balance is the inventory available at the start of the next period. A special value of the inventory is provided in the box immediately after the term "Projected on hand." This value is the inventory at the end of the previous period or the on-hand inventory balance ready for use in period 1.

■ *Planned order receipts.* The planned order receipts entry in the record indicates when a planned order will be received if the planned order release date is exercised. As soon as the order is placed, the value in planned order receipts moves up to a scheduled receipt at the beginning of the period.

■ *Planned order releases.* The planned order releases are the suggested order quantity, release date, and due date generated with MRP software. Orders are released at the beginning of a period. These suggested planned orders are calculated by the MRP software according to the gross requirements for the period and the inventory balance available to satisfy the gross requirements. When a planned order is finally released, it converts into a scheduled receipt.

- *Lead time.* The lead time is the time between the release of an order and the completion or delivery of the order. For a manufactured item, lead time represents the time required to produce the quantity of parts in the order. In a subassembly, lead time represents the time required to complete the assembly.

- *Lot size.* The lot size, when a specific value is given, is the required minimum order quantity determined by the economics of the production process. Usually, if the net part requirements exceed the lot size, the planned order is specified in multiples of the minimum lot size. If the lot size is listed as *lot for lot*, the planned order quantity is equal to the *net requirements* for the period. Lot-for-lot capability indicates that production efficiencies permit any quantity to be manufactured economically.

- *Safety stock.* Safety stock is the lowest level of inventory allowed in the projected on-hand line. Safety stock protects against variations in delivery from manufacturing or from vendors due to production or quality problems.

Under normal operation, seven of the entries—part number, gross requirements, scheduled receipts, projected on-hand inventory available for period 1, lead time, lot size, and safety stock—are known quantities in the MRP record. Calculation of the remaining data in an MRP record with computer software is performed in an MRP run. After the MRP run, the projected on-hand balance and planned order releases are calculated for all periods covered by the record. The MPC system uses the MRP records to generate a time-phased production plan for each part in the product structure.

MRP Calculations

The MRP record in Figure 7–3 illustrates the values present in the record before the start of MRP calculations. The *gross requirements* (16 units) come from either the next-highest level in the product structure diagram or from the MPS. The *scheduled receipts* value (three times the lot size of 5, or 15 units) represents planned order releases that became firm orders when the orders were placed with either manufacturing or a vendor in the last period. The *on-hand* inventory for the start of the first period (4 units listed in the box) is determined from the MRP record from the last period and is often verified by a cycle count of part inventory. The *lead time, lot size,* and *safety stock* are values set by purchasing and manufacturing.

The calculations for the record start with the first period and proceed to the last. The calculations determine the *projected on-hand balance* and the need for a *planned order release.* If the projected inventory balance is positive and above the

Part number	Period number 1	2	3	4	5	6	7	8	9	10
Gross requirements	16		8	15	21		12	15		28
Scheduled receipts	15									
Projected on hand [4]										
Planned order receipts										
Planned order releases Lead time = 1 Lot size = 5 Safety stock = 0										

Figure 7–3 Basic MRP Record with Starting Values.

safety stock level, no action is required for that period. However, if the projected inventory balance is negative or less than the required safety stock, a planned order release is required and must be included in the inventory balance calculations. The calculations are easy to understand when illustrated by an example using the starting data in Figure 7–3. Check the fully completed MRP record in Figure 7–4 as you read through the calculations that follow.

Period 1 Calculations:

$$\begin{array}{c}\text{Starting}\\\text{inventory}\end{array} + \begin{array}{c}\text{scheduled}\\\text{receipts}\end{array} - \begin{array}{c}\text{gross}\\\text{requirements}\end{array} = \begin{array}{c}\text{projected available}\\\text{balance}\end{array}$$

$$4 + 15 - 16 = 3 \text{ units}$$

The projected available balance of 3 units would be available at the start of the second period.

Part number	Period number 1	2	3	4	5	6	7	8	9	10
Gross requirements	16		8	15	21		12	15		28
Scheduled receipts	15									
Projected on hand [4]	3	3	0	0	4	4	2	2	2	4
Planned order receipts			5	15	25		10	15		30
Planned order releases Lead time = 1 Lot size = 5 Safety stock = 0		5	15	25		10	15		30	

Supplies gross requirements for next record in product structure

Figure 7–4 Completed MRP Record.

Period 2 Calculations:

$$\underset{\text{inventory}}{\text{Starting}} + \underset{\substack{\text{receipts} \\ \text{(or planned} \\ \text{order} \\ \text{receipts)}}}{\text{scheduled}} - \underset{\text{requirements}}{\text{gross}} = \underset{\text{balance}}{\text{projected available}}$$

$3 + 0 - 0 = 3$ units

The starting inventory for period 2 is the ending inventory for period 1. Period 2 could have either a scheduled receipt or a planned order receipt, depending on the production needs and the lead time. In this situation, neither is present. The equations for all subsequent periods will be the same as the equation given for period 2.

Period 3 Calculations:

$3 + 5 - 8 = 0$ units

The inventory balance at the end of the period is 0 units because the gross requirement for 8 units consumes the inventory balance of 3 plus the planned delivery of 5 additional units.

Period 4 Calculations:

$0 + 15 - 15 = 0$ units

Period 5 Calculations:

$0 + 25 - 21 = 4$ units

Continue to use the formula for the last five periods and verify that you can understand the calculations.

Appendix 7–1 presents a version of a classic MRP logic example based on the production of bicycles. Work through this example from the schedule for finished units down to the material that is transformed into handlebars.

Data Used in MRP

Management of the product data is critical for the MRP system to be effective because the accuracy and the reliability of the output information are related directly to the quality of the data going into the system. The BOM and inventory levels are critical items in the management of product data. For example, BOM accuracy of more than 97 percent and inventory information with more than 98 percent accuracy are critical for dependable MRP system generation of planned orders for manufacturing and purchasing. A technique called *cycle counting*, used to achieve highly reliable and accurate inventory information, was described in chapter 6. Bills of materials are discussed next.

Bills of Materials.   World-class manufacturing organizations maintain just one BOM for a product; however, the BOM is stated in different terms for different departments in the organization and has several representations. The BOM originates in product design when the parts to produce the product are either designed or purchased from vendors. The representation of the BOM in design engineering is often just a list of parts and subassemblies necessary to build the product. The information associated with each part includes quantity, part number, and specifications necessary for manufacturing or purchase. Figures 3–9 and 3–10 illustrate a part design and the BOM.

Production engineering, responsible for planning the total manufacturing and shipment of the product, frequently adds to the design bill boxing and packaging items along with raw material requirements. In MPC, the bill is represented as either a *product structure diagram* or an *indented BOM*. The product structure diagram and indented BOM for a simple product, a table, are provided in Figure 7–5. These two representations of the BOM are important in MPC because they describe how the product must be manufactured and assembled. For example, the legs, long rail, and short rail must be manufactured before a leg assembly can be produced. In addition, the product structure diagram indicates that every leg assembly requires *two* short rails, *two* long rails, and *four* legs. While both the product structure and the indented bill contain the same information, the representation as an indented bill is much easier to capture in MPC computer software. However, the product structure diagram helps to explain how the time-phased MRP records are used to plan the production of the table and its components.

In other types of production operations, *planning* BOMs are generated to represent product families with large numbers of end-item configurations. The planning bill is not a different BOM, just a different representation of the design bill. BOM accuracy in the 97-plus percent range is possible when three conditions are satisfied: (1) responsibility for maintenance of the bill rests in one department, (2) a formal process for engineering change approval is established and religiously followed, and (3) a single image of all product data and BOM information is maintained in a central CIM database.

The Product Structure and the MRP Record.   Product structures provide a critical input to the MRP process. The structure guides the level-by-level planning process. Structures describe the path that the MRP logic will follow from the zero level to the lowest-level material items.

Study the product structure diagram and the indented BOM illustrated in Figure 7–5 and determine how many different assemblies, subassemblies, and

Figure 7–5 Product Structure Diagram and Indented BOM.

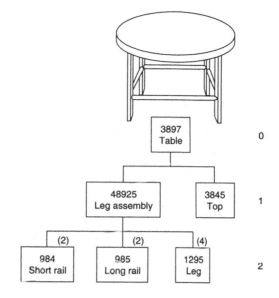

Indented bill of materials (BOM)

3987 Table (1 required)

    48925 Leg assembly (1 required)
        984 Short rail (2 required)
        985 Long rail (2 required)
        1295 Leg (4 required)

    3845 Top (1 required)

Item	Low-level code
Table	0
Leg assembly	1
Top	1
Short rail	2
Long rail	2
Leg	2

parts are represented. The table assembly in the product structure diagram, part number 3897, is constructed from five components and subassemblies. Every box in the product structure diagram is covered by an MRP record; as a result, the MRP records are linked, as Figure 7–6 indicates. A study of the links between the MRP records in the figure indicates that the gross requirements for records at level 2 are generated from the planned order releases from level 1. For example, the gross requirements for the short rail, number 984, came from the planned order releases from the 48925 leg assembly. Figure 7–7 shows the planned order release for the 48925 leg assembly flowing into the MRP record for the 984 short rail. The planned order releases for the leg assembly in period 1 is 30 units; therefore, the gross requirements for the short rails in the same period is 60 units. The gross requirements for the short rails is twice the number of required leg assemblies because the product structure (Figure 7–5) indicates that it takes two short rails for every leg assembly. Check periods 4 and 7 and verify that you understand how the planned

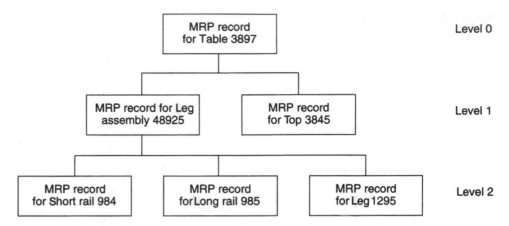

Figure 7–6   MRP Records for Table Product Structure.

order releases from one record flow into the gross requirements of the record at the next-lower level.

The final assembly MRP record, part 3897, does not have a higher-level record to dictate the gross requirements for finished parts. Therefore, the part number at the zero level in the product structure diagram (Figures 7–5 and 7–6) uses the MPS

Product 48925 Leg assembly		Period number									
		1	2	3	4	5	6	7	8	9	10
Forecast		10	10	10	20	10	10	20	20	20	20
Available		25	15	5	30	20	10	65	45	25	5
MPS		⃝30			⃝45			⃝75			
On hand		5									

| | | ◀2 x | | ◀2 x | | ◀2 x | |

Part number 984 Short rail		1	2	3	4	5	6	7	8	9	10
Gross requirements		⃝60			⃝90			⃝150			
Scheduled receipts		40									
Projected on hand	23	3	3	3	3	3	3	3	3	3	3
Planned order receipts					90			150			
Planned order releases			90			150					
Lead time = 2											
Lot size = 5		Note: 2 of the 984 short rails are required for each									
Safety stock = 0						48925 leg assembly					

Figure 7–7   MRP Record Interface.

Product Table		Period number									
		1	2	3	4	5	6	7	8	9	10
Forecast Available		10	10	10	20	10	10	20	20	20	20
MPS		25	15	5	30	20	10	65	45	25	5
On hand		㉚			㊺			㊲			
		5									
Part number 3897 Table		1	2	3	4	5	6	7	8	9	10
Gross requirements		㉚			㊺			㊲			
Scheduled receipts		40									
Projected on hand	5	15	15	15	10	10	10	15	15	15	15
Planned order receipts					40			80			
Planned order releases			40			80					
Lead time = 3											
Lot size = 20											
Safety stock = 0											

Figure 7–8   MPS and MRP Record Interface.

to determine the quantities for the gross requirements entry. The MPS and MRP records in Figure 7–8 show how the MPS quantities flow down to the final assembly MRP record. The MPS is used to determine the MRP gross requirement quantities in each period for the table, part number 3897. With production information for the table available, the MRP record at the highest level (Figure 7–6) in the product structure is completed. With the highest level planned, the MRP records at level 1 and then at level 2 are completed, and planning is complete for all subassemblies and components of the table. These calculations for all items in the product structure are performed by MRP software when an MRP run is executed.

Transient Items in the Product Structure.   *Transient items* are often used to represent items in the product structure and help keep engineering and manufacturing synchronized. Traditionally, the BOM structure shows how various parts come together to make subassemblies, and how the subassemblies come together with other parts to make a final product. Manufacturing may determine that it is more efficient to combine steps and operations in a way that appears to change the product structure. The coding of levels of the structure that are to be combined at the request of manufacturing as "transient items" allows the structure to meet the needs of engineering and manufacturing. Transient items exist on the shop floor for a short time before they are consumed or converted into a higher-level product. This type of item is also known as a *phantom item* or a *blow-through item.* Transient items exist for such a short time that there is essentially no need to track them closely and report on them. They allow the structure diagram to show the various levels of assemblies and sub-assemblies, but they do not create additional work for manufacturing. Transient items have two important characteristics: they have no manufacturing lead time, and manufacturing orders are not created by the system for these items. The use of transients leads to the reduction of the effective levels in the BOM. When transients are used in an MRP system, the results are more efficient planning, less processing time, fewer planned orders for subassemblies, and much less reporting and transaction processing. In addition, the lower number of transactions leads to fewer possible reporting errors and improved system accuracy. A great source of additional information about the use of transients and how to improve BOMs and product structures can be found in the 1997 book *Bills of Material: Structured for Excellence,* by Dave Garwood.

Computer-Assisted MRP

MRP is often a program module or a defined function in a formal system software package. MRP software is somewhat mysterious. The software typically follows the model of the system shown in Figure 6–8. The MRP function pulls schedule data from the master schedule, production forecast, or customer orders. The structure or recipe is pulled in from the BOM. The on-hand balances are found in the inventory data files, and general item data are found in the part master files. The module then works to complete the following:

1. The calculation of the gross material requirements for each production order
2. The consideration of on-hand inventory and open purchase orders for the items needed to complete the production orders
3. The calculation of the remaining material that must be obtained to complete the production orders
4. Time phasing or back scheduling from the due date of the scheduled production orders to determine the critical dates for the planning, release, and scheduled receipt of new materials and intermediate subassemblies
5. Identification of problems and conflicts and the generation of exception messages and reports to the responsible people working with the system
6. Generation of dependent-item subassembly schedule dates that become input to the capacity requirements planning (CRP) function

Figure 7–9 describes the general data flows and system interfaces related to the MRP system.

The table example presented earlier uses a simple product with few subassemblies and components; as a result, the calculations could be performed manually. However, imagine the product structure diagram for a riding lawn mower, an automobile, a videocassette recorder, or a computer. There are often over 30 levels in the bill; hundreds of different parts and subassemblies, each requiring an

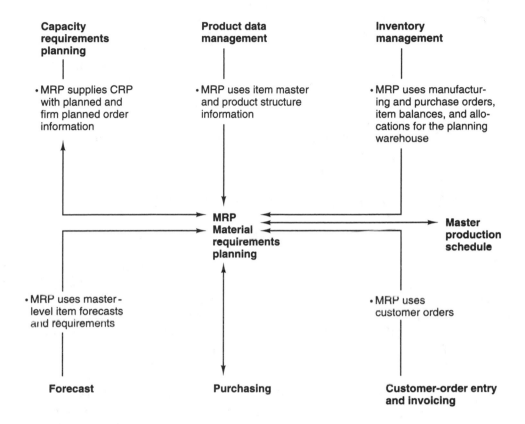

Figure 7–9   MRP System Interface.

Note: MRP's net change planning reflects activity by any of the interfacing applications.

MRP record; and thousands of individual items. In addition, there are usually multiple models of the final assembly, which share common parts and subassemblies. As a result, the MRP records for these common items have gross requirements coming from different sources that must be combined before the final production plan is completed. As an added burden, the MRP plan is never static; gross requirements, lead time, and on-hand balances change frequently. It is obvious that manual MRP calculations on final assemblies of this size are impossible; as a result, computer software is used.

Even with computer assistance, the MRP run often takes hours for the computer to complete. As a result, MRP runs on products with large parts counts are often performed monthly or at best once a week. Changes in the gross requirements, lead times, and on-hand inventory, which can occur daily, make the monthly or weekly MRP recommendations for planned order releases inaccurate. To overcome this problem, MRP software designers included a feature called *net change*. When executed, the net change option recalculates the MRP records affected

Part number		Period number									
		1	2	3	4	5	6	7	8	9	10
Gross requirements		15		8	15	21		12	15		28
Scheduled receipts	9	15									
Projected on hand	4	4 9	4 9	1	1	0	0	2	2	2	4
Planned order receipts				5	15	20		15	15		30
Planned order releases			5	15	20		15	15		30	
Lead time = 1						▼					
Lot size = 5		Supplies gross requirements for next record in product structure									

Figure 7–10   Impact of Change on an MRP Record.

by the changes and produces a revised set of recommendations for the planned order releases. For example, if an inventory cycle count changed the starting on-hand balance for the top (part number 3845 in Figure 7–5), the net change calculations would require recalculation of the MRP record for the top. However, if the starting inventory balance changed for the leg assemblies (part number 48925 in the figure), the net change option would need to recalculate the MRP record for the leg assemblies and possibly for all three components at level 2. An example of the change in an MRP record due to a change in information is illustrated in Figure 7–10. The MRP calculations from the previous period indicated that an on-hand balance of 4 units was expected for period 1. However, an inventory cycle count turned up 9 units instead of 4. The impact on the MRP record is illustrated in the figure. Notice that the net result is the cancellation of a planned order release for 5 units in period 2.

The Benefits of MRP

MRP helps planners understand and manage their operations. The correction of the on-hand inventory from cycle-count data is common in manufacturing. The use of the MRP system allows production control personnel to view the effect of the change in future periods. They can see that a planned order recommended by MRP for period 2 could now be canceled. The effect of that cancellation on parts lower in the product structure is also easily analyzed. The primary benefit is that problems in manufacturing due to disturbances in the production system are solved early when a greater number of alternatives are available to the planner.

Another substantial benefit from implementing MRP results from the preparation for the installation. As stated earlier, an accurate BOM and a cycle-count process to guarantee reliable inventory records are minimum conditions for successful MRP operation. The self-study used to improve the BOM and inventory tracking uncovers other operations that do not add value to the product. The correction of these problems and the improvements in inventory and the BOM add substantially to the profitability and quality of the products. The following list of improvements in the operation of the enterprise are frequently attributed to implementing MRP:

- Improved customer service
- Reduction in past-due orders
- Better understanding of capacity constraints
- Significant increases in productivity
- Reduction in lead time
- Reduction in the inventory for finished goods, raw materials, component parts, and safety stock
- Reduction in work-in-process (WIP)
- Elimination of annual inventory
- Significant drops in annual accounting adjustment for inventory problems
- Usually, a doubling of inventory turns

7–3 CAPACITY REQUIREMENTS PLANNING

A review of capacity management and control in the MPC model shown in Figure 6–4 reveals the fact that capacity planning occurs at three different points in the production control process: (1) variable resource planning supports the production-planning process, (2) rough-cut capacity planning aids in the development of a valid MPS, and (3) *capacity requirements planning (CRP)* supports the MRP. Priority schedules for work cells are set through the production activity control process discussed in chapter 6. The effectiveness of work-cell scheduling in utilizing capacity is monitored by the capacity control function using input-output control reports. Capacity planning and CRP are defined as follows:

> *capacity planning: The process of determining the amount of capacity required to produce in the future. This process may be performed at an aggregate or product-line level (resource requirements planning), at the master-scheduling level (rough-cut capacity planning), and at the material requirements planning level (capacity requirements planning).*
>
> (Source: © APICS Dictionary, 10th edition, 2002)

> *capacity requirements planning (CRP): The function of establishing, measuring, and adjusting limits or levels of capacity. The term capacity requirements planning in this context refers to the process of determining in detail the amount of labor and machine resources required to accomplish the tasks of production. Open shop orders and planned orders in the MRP system are input to CRP, which through the use of parts routings and time standards translates these orders into hours of work by work center by time period. Even though rough-cut capacity planning may indicate that sufficient capacity exists to execute the MPS, CRP may show that capacity is insufficient during specific time periods.*
>
> (Source: © APICS Dictionary, 10th edition, 2002)

The CRP process produces the infinite shop scheduling and loading discussed in chapter 6. The shop load, usually expressed in hours of work per work center per period, is produced by the CRP software from open shop orders and planned order releases generated in MRP. Infinite loading occurs because CRP

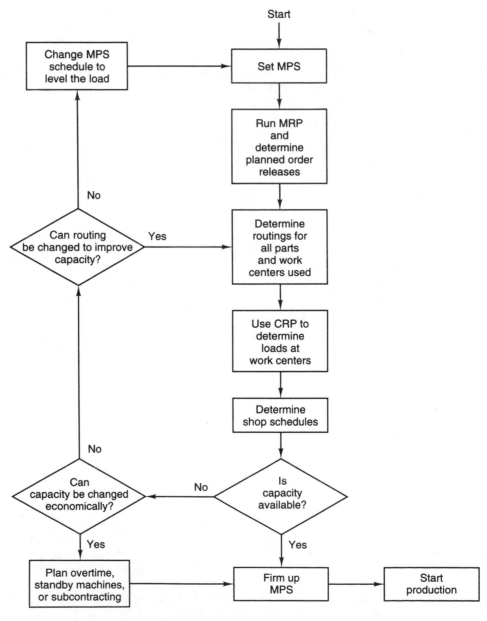

Figure 7–11 Process to Establish MPS for Available Capacity.

schedules capacity for each part independently and then sums the planned capacity for each work center for every part processed. If a work center is used by many of the parts, there is a good likelihood that the planned capacity exceeds the hours available in the period. Study the CRP process described by the flowchart shown in Figure 7–11. The process starts with a tentative MPS that is converted into planned order releases through the MRP process. After the routings are used to identify the work centers required, the material requirements are converted into labor and/or machine loads at the work centers. The machine loads are scheduled through CRP and a check on planned capacity versus available capacity is made. Work centers with sufficient hours to produce the parts have the orders released. However, work centers with loads that exceed capacity require a change in either the routings or the MPS to meet the existing capacity, or an increase in capacity through overtime, additional shifts, use of additional machines, or subcontracting of the work. Study the flowchart until you understand the interaction among the MPS, MRP, and CRP functions in achieving a workable schedule for the shop floor.

Automating the CRP Process

CRP is often a program module or a defined function in a formal system software package. The CRP function can be found in Figures 6–7 and 6–8. CRP is a logical process for calculating the requirements for specific resources. It works to plan physical resource requirements in much the same way that MRP works to plan material resource requirements. The CRP function receives schedule data from the master schedule and the subassembly schedules generated by MRP. Data files are established to provide work centers with specific information on labor requirements, scheduling, and the expected capacity available for use. The resource requirements for the items to be produced are determined from the routing documentation of the specific work operations to be completed and the work center where that work is to be done. General item data are pulled from the part master files. The CRP module then works to complete the following:

1. The calculation of the workload for each operation required to complete the production orders using the time standards stored in the item master files.
2. The generation of load profile information showing the calculated amount of work to be done at each workstation. This generation is done most often without restraints on the scheduled load in any time period, which creates a so-called infinite load situation.
3. Identification of problems, schedule conflicts, and resource shortages, and the generation of exception messages and reports to the responsible people working with the system.
4. Generation of dependent-item subassembly schedule dates that become input to the CRP function.

Figure 7–12 describes the general data flows and system interfaces related to the CRP system.

Figure 7–12 CRP System
Interfaces.

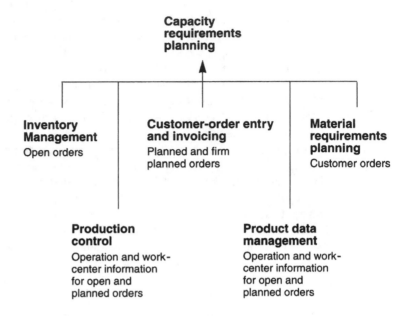

7–4 MANUFACTURING RESOURCE PLANNING

The evolution in the planning and scheduling processes used for dependent demand items has been continuous since the early 1960s. However, a major change occurred in the 1970s that expanded the interface between enterprise operations significantly. The revised process, called *manufacturing resource planning*, or *MRP II*, was described and promoted by Oliver Wight, a manufacturing consultant. MRP II links a broader range of enterprise departments into the production-planning process.

Open-Loop MRP

The open-loop MRP process described earlier in the chapter provided an alternative planning mechanism for parts and subassemblies in the product structure. These *dependent demand* items are identified by exploding the schedule for the finished product through the BOM and accounting for orders due and inventory on hand. The demand for these items is related to the product structure; for example, the quantity of parts in level 2 of the product structure shown in Figure 7–5 is set by the production quantities identified in level 1. Therefore, the MRP process was used to schedule items that had a dependent demand, and the reorder-point system was used to manage finished-goods inventory where the demand was *independent* and set by historical data and forecasting techniques.

Closed-Loop MRP

The open-loop MRP system answered the *when* and *what* aspects of the make-or-buy questions. In addition, the open-loop system had well-defined links between MPS, MRP, and CRP. However, the system was static and not tuned to the dynamic nature of most manufacturing operations. Problems with production machines, labor, quality, and late vendor deliveries made the MRP-generated schedules irrelevant, and manual updating of order status was difficult and error prone.

The problem was solved by interfacing the purchasing and production activity control modules with the MRP module. The purchasing interface provided MRP with a dynamically updated order status from which new planned order releases and suggested new schedules could be generated. The links to the shop floor were critical for collecting daily production and operator time-sheet data so that the production status could be determined. The use of automatic tracking devices such as bar codes permitted a near-real-time production activity control interface to MRP and a responsive closed-loop planning system. Study Figure 7–13, which illustrates the feedback process in the *closed-loop MRP* system. The loop includes shop production data and the planning data present in MPS, MRP, and CRP. Closed-loop MRP is defined as follows:

> *closed-loop MRP: A system built around material requirements planning that includes the additional planning processes of production planning (sales and operations planning), master production scheduling, and capacity requirements planning. Once this planning phase is complete and the plans have been accepted as realistic and attainable, the execution processes come into play. These processes include the manufacturing control processes of input-output (capacity) measurement, detailed scheduling and dispatching, as well as anticipated delay reports from both the plant and suppliers, supplier scheduling, and so on. The term closed loop implies not only that each of these processes is included in the overall system, but also that feedback is provided by the execution processes so that the planning can be kept valid at all times.*
> (Source: © APICS Dictionary, 10th edition, 2002)

Companies must be able to take prompt corrective action when conditions require a change. There must be feedback throughout the organization to communicate the exceptions and plan changes to everyone concerned. The discipline to keep the information in the system correct and up to date must be instilled in the company's employees at all levels. Confidence that the information in the formal system is correct must be developed and maintained. A formal system that lacks employee confidence is of little value to any company.

Figure 7–14 compares the reorder-point system and the MRP system for several of the processes present in manufacturing. Study Figures 7–13 and 7–14 until the differences among reorder point, open-loop MRP, and closed-loop MRP are clear.

The closed-loop MRP system was expanded to include interfaces to all the financial areas, including purchasing with accounts payable, sales order processing with accounts receivable, and inventory and WIP with general ledger. The MRP system is illustrated in Figure 7–13. Note that MRP, material requirements planning, is just one part of MRP II. As a result, material requirements planning is often called *little mrp* and manufacturing resource planning is called *big MRP*.

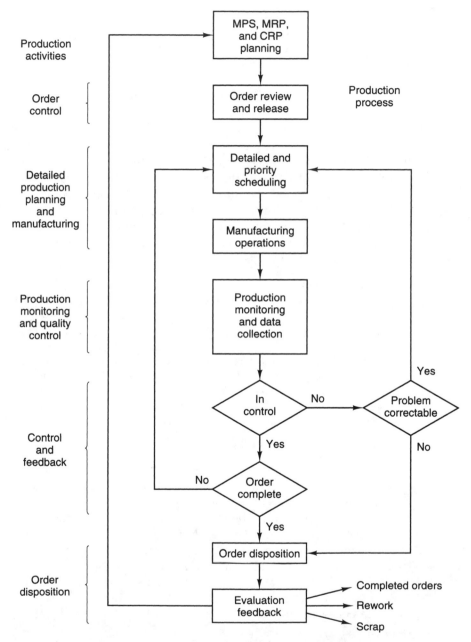

Figure 7–13 Closed-Loop MRP System.

	MRP	*Reorder point*
Demand	Dependent	Independent
Order philosophy	Requirements	Replenishment
Forecast	Based on master schedule	Based on past demand
Control concept	Control all items	ABC analysis
Objectives	Meet manufacturing needs	Meet customer needs
Lot sizing	Discrete	Economic order quantity (EOQ)
Demand pattern	Lumpy but predictable	Random
Types of inventory	Work-in-process and raw materials	Finished goods and spare parts

Figure 7–14   Comparison of MRP and Reorder Point.

Together with the development of MRP II, Oliver Wight produced a set of evaluation questions to measure the degree of compliance between an enterprise's operation and the ideals defined in MRP II. The first set of evaluation questions was published in 1977 and revisions to the questions have occurred regularly. The current evaluation process, published in 1993, divides the questions into the following five basic business functions: strategic planning processes, people and team processes, total quality and continuous-improvement processes, new product development processes, and planning and control processes. The overview questions for each of the five business functions are listed in appendix 7–2. Read through these questions so you can understand the level of operation required for a world-class manufacturing operation. The complete checklist has more detailed questions for some of the overview questions to help determine the best answer for the overview questions. The ABCD checklist process requires a company to answer each of the overview questions on the basis of a rating scale. If a company is perfectly executing the activity described by the overview question, then an "excellent" rating is chosen and a 4-point value is earned for that question. In contrast, if the company has not yet started to work in the area addressed by the overview question, a score of "not doing" is assigned and the point value is 0. After all sixty-eight questions are answered, the average for the numeric score is found and a letter grade is assigned. The operational descriptions for companies working at the four grade levels are also listed in appendix 7–2.

7–5 FEATURES OF MODERN MANUFACTURING PLANNING, AND CONTROL SYSTEMS

This section introduces an example of several of the important features of the modern enterprise resources planning (ERP) systems. The examples were originally developed by Joe Wilkinson, creator of the WinMan software system, and enhanced for presentation at the Society of Manufacturing Engineers (SME) Education in Manufacturing Conference in 1998. WinMan is a Windows-based package that takes advantage of relational database technology and gives a fresh look at how a system can support manufacturing. Details of the example are explained in appendix 7–3.

A central database for item-specific information is a common feature of modern systems. The creation of items in the system forces the system users to consider the type of item being created. These items include the item numbering system, the effective use of descriptive text, the unit of measure, the person responsible for planning and buying the item, the cost of the item, and procurement lead time information. Lots of questions need to be answered before the system can do any planning. All material items are not the same and must be carefully considered. A partial list of the key material issues that must be addressed includes the following:

- What is the unit of measure?
- What planning parameters do we want the system to use?
- How much inventory do we want to keep?
- What items may require safety stock?
- How will the warehouse be controlled?
- How will locations be named and identified?

Remember, this list is partial.

Current versions of business systems come with special controls and field edits to protect the integrity of the system data. New users must first sign on to the system and will be granted access to screens and transactions according to their assigned security level. Some of the information accumulated in the system is accessible to people throughout the organization. Users who have been given the appropriate authorization can access security-controlled information and data.

Typically there are several standard reports and screens that come ready to use. System users soon learn where to look in the system to find the data and answers needed to solve problems. Most systems now utilize online updates for most transactions. Additions of items to the system, reporting, and maintenance transactions are typically processed online. The processing of most transactions is completed quickly, with little user waiting time.

There is a growing trend toward system designs that support a nearly paperless operation. Most often, the information needed to solve a problem is available for online inquiry and action. Occasionally, a printed report is required to find the needed information or for special documentation. It is recommended to view information on the screen whenever possible or to route report output to a screen window rather than to generate a printed page. Printed reports are available to users on a request basis. Report writers often provide filters that can be applied to limit the size of the print output or to focus on a particular issue. In addition to standard reports, a report writer function may be available to system users. Report-writing features often allow more experienced users to select the fields they want from the database and to develop customized reports. Print output can be directed to a screen for viewing or to printers on a network.

Describing Items and Products in a Formal System

The work with a formal system for manufacturing really begins with the entry of item-specific information into the database. Nothing can happen in the system until the specific information that defines an item has been entered and processed. Once the item is created in the system, the information that has been loaded can be viewed and used by people throughout the company. For example, the product designer or engineer may select the item for use in a product structure. The purchasing department may place orders for items and actually schedule delivery of items needed to complete production. Manufacturing may issue some of the item out of stock and use it to produce a product.

Modern systems try to make the entry of item information into the system as easy as possible. However, many decisions are required as part of the item setup process. The current generation of systems with graphical user interfaces often leads the user step by step through the data-entry process. In many cases, systems will provide the user with a prompt in a dialog box or with a pull-down menu of choices. Often the data-entry process is as easy as pointing and clicking on a desired entry with the mouse. Selecting from the choices provided and minimizing the actual entry from the keyboard helps reduce data-entry errors. Editing or correcting fields that were entered incorrectly is easier in the modern systems.

Accurate data entry is a critical part of the formal system process. The infamous "central database" is only as good as the data entered into it. People from all around the company will be using these data. The system cannot do a good job planning for you if the data you input are not the best you have. A sample view of the product setup screen for a new item is shown in Figure 7–15.

Maintenance of the data is also important. Many of the data entered into the item master files are fairly static and will not change often. Several fields are more dynamic, such as the procurement lead time (in days) and the order policy information. The new generation of formal systems has made changing these fields as easy as clicking on the field and following the instructions in the resulting dialog box. Additional discussion of planning with WinMan is included in appendix 7–3.

This edition of *Computer-Integrated Manufacturing* includes a CD-ROM with a single-user demonstration version of the WinMan software and related documentation. The WinMan manufacturing system is used by over 100 companies to plan and control their operations.

Numerous vendors provide MRP II software support for all three types of computer platforms. These systems now often run on a minicomputer and can support small to very large enterprise production control databases. A formal system provides management with a tool that can help link the corporate strategy to the operational objectives of the firm. Management's plans and systems to achieve desired objectives address the following:

Setting goals that describe management's vision for the company

Developing plans and objectives to achieve the goals

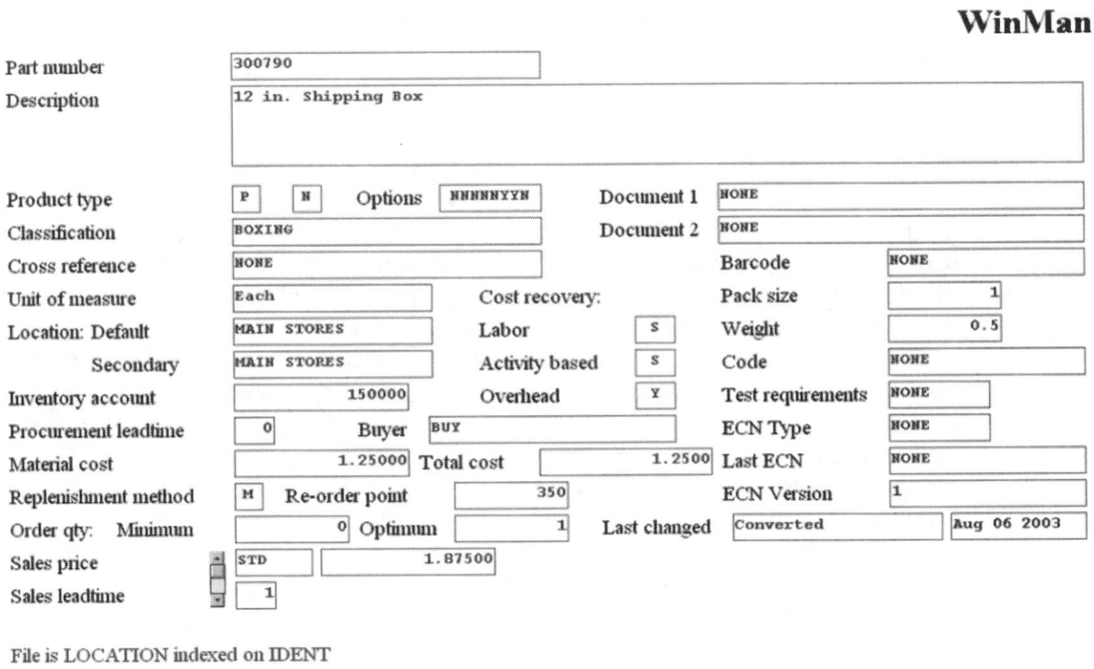

WinMan

Part number	300790
Description	12 in. Shipping Box

Product type	P	N	Options	NNNNNYYN	Document 1	NONE
Classification	BOXING				Document 2	NONE
Cross reference	NONE				Barcode	NONE
Unit of measure	Each		Cost recovery:		Pack size	1
Location: Default	MAIN STORES		Labor	S	Weight	0.5
Secondary	MAIN STORES		Activity based	S	Code	NONE
Inventory account	150000		Overhead	Y	Test requirements	NONE
Procurement leadtime	0	Buyer	BUY		ECN Type	NONE
Material cost	1.25000	Total cost		1.2500	Last ECN	NONE
Replenishment method	M	Re-order point	350		ECN Version	1
Order qty: Minimum	0	Optimum	1	Last changed	Converted	Aug 06 2003
Sales price	STD	1.87500				
Sales leadtime	1					

File is LOCATION indexed on IDENT

Figure 7–15   Part Master File Screen from the WinMan System.
(*Source: Used with permission of TTW Services, LTD.*)

> Executing and implementing the approved plans
>
> Providing feedback and communication of the results and outcomes

The planning system must be understood and supported by the functional area managers. The plan must be future oriented, with checks of internal and external resources. It must integrate the expertise of the functional areas. The higher-level formal plans become the basis for preparing detailed functional operating plans. A formal planning approach is an ongoing effort. It cannot be done correctly as a one-shot effort that is reviewed only annually.

System packages of software for MRP II computer applications for controlling material and inventory began to emerge in the 1960s with the release of communications-oriented production information and control system (COPICS) by IBM Corporation. COPICS initially was not a complete MRP II system, but over the years the package was modified and expanded beyond the initial material management focus. Advances in technology have led to the design of systems that go well beyond the initial material planning applications to integrate all the functional areas of the company. The new systems help link the strategic plans of the firm to specific operational objectives.

Recent publications by APICS now provide information on over 100 software packages for MRP/MRP II and ERP. Packages are available for companies of nearly any size and market type. The evaluation and selection of a package that best fits a company is a major part of the system implementation process. The selection process requires a thorough understanding of the business and the operations required to support it.

Benefits from MRP II

A formal manufacturing planning system can be a powerful tool that helps all the functional areas of the operation work more effectively. MRP II systems have proven to be valuable to companies and have produced real, "tangible" cost savings in several areas. Reductions in inventory have reduced the amount of money spent on investments for material and the cost of keeping that material on the shelf until it is needed. Better planning in communications often leads to reductions in the amount of inventory declared obsolete and written off as a loss. Manufacturing effectiveness is improved and savings result from having better schedules with fewer delays, which results in reduced direct labor costs and reductions in unplanned overtime. Better scheduling provides the purchasing function with better visibility of future requirements and allows better planning with suppliers and lower purchasing costs. Component waste is reduced through better scheduling of similar items, planning of changeovers, and improved control of material issues to the shop floor. Customer service levels are improved by shipping more complete orders on time and with improved responsiveness to customer requirements.

In addition, companies have found other intangible savings that cannot easily be assigned a dollar value. Although difficult to quantify, these intangible savings are still considered significant. Changing requirements and business conditions are realities in manufacturing. The formal plan becomes a basic reference point supporting the planning for change. Companies skilled in planning can often respond better to changing conditions. It is difficult to put a value on the benefit of an increased ability to develop and execute plans for the operation and to deal with change. Additional intangible benefits often include improved customer relations, reductions in expediting costs, improved financial planning, and improved employee morale.

Achieving class A status through the implementation of MRP II requires a commitment by management to an eighteen- to twenty-four-month plan, a great deal of hard work, and the investment in computer hardware and software. As with any investment in business, the payoff must justify this level of effort and cost. A study performed by the Oliver Wight companies asked participating companies to indicate their MRP II rating level and the percentage improvement for operations in four areas: customer service, productivity, purchase cost, and inventory turnover. The results are listed in Figure 7–16. Note that class A companies are three times better than class D, twice as good as class C, and almost 50 percent better than class B. As a result, companies that successfully implement MRP II achieve the world-class metrics described in Figure 1–6, have the order-winning criteria necessary to expand market share, and receive tangible dollar benefits ranging from 10 to 25 percent of gross sales. Review the case study in chapter 2, which describes a company moving from class D to class A through a structured improvement program. Note

Figure 7–16   Percentage
Improvement by MRP II Class
Ratings.

	Percentage improvement for class:			
	A	B	C	D
Customer service	26	18	13	8
Productivity	20	13	9	5
Purchase cost	13	9	6	4
Inventory turnover	30	21	13	8

the benefits obtained by reaching the class A status. Success in using MRP II concepts is frequently accompanied by work in another production process, *just-in-time* manufacturing, which is covered in chapter 9.

Cost Justification

Companies and their planning teams must take time to consider carefully the expected benefits and the costs associated with a formal system implementation. There are no simple solutions and turning-point implementations. Planning and implementation decisions will have an ongoing impact on the cost-effectiveness of the new system. A meaningful justification requires careful analysis and a conservative look at the benefit and cost figures. The savings have proven to justify the high cost of implementation for many companies. Many companies have approached the implementation of a formal planning system with a "quick fix" approach that has produced few tangible benefits. Management commitment must be real and demonstrated for a successful implementation.

Implementing MRP II

The implementation process for an MRP II system (see Figure 7–17) requires a highly structured approach that involves every employee, with a minimum of 90 percent of the workforce trained, including management. A team made up of representatives from all the functional areas often leads a successful implementation of formal systems. Some team members may be full time; others may participate only as needed. The project team typically reports to a high-level steering committee made up of the executive managers who control the business functions. Figure 7–18 identifies many of the planning considerations that must be addressed by the project team and the steering committee. Decisions that cannot be made at the project team level are taken to the steering committee. The steering committee ultimately resolves any conflicts.

The process used in many successful implementations, illustrated in Figure 7–19, is called the *proven path*. Note that education is the first step in the process and is continuous throughout the implementation. Top management commitment and involvement in the MRP II program is critical. Management must understand the MRP II process and comprehend the costs and effort required to install and operate the process fully. In addition, management must know how MRP II will affect every department and the benefits that will result from successful implementation.

Figure 7–17 MRP II System.

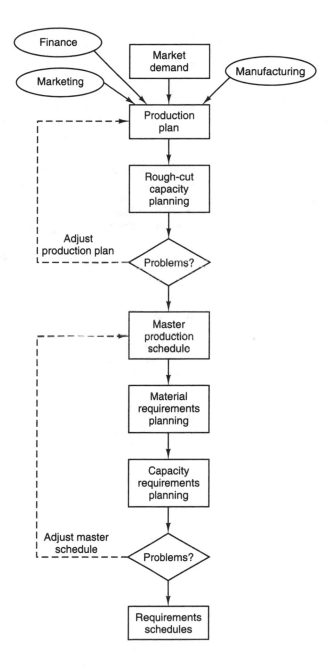

The installation process is divided into four phases: *initial, preparation, implementation,* and *operation.* After the initial phase, the project team is in place with a full-time project director, and work on problem analysis begins. Many times the project team will identify between 50 and 500 problems that must be addressed before the system is installed. Problems are divided into functional areas and prioritized, then teams of employees from the areas start working on solutions. For

Essential prerequisites
- Knowing what systems are currently in place and how they work.
- Identification of features required of future systems.

Organizational issues
- Who are the key players, and what are their roles?
    - —The members of the steering committee?
    - —The members of the implementation project team?

Project management plans
- Preparing a realistic plan of action.
- Plotting milestones and tracking progress toward them.
- Initial measurements (who, what, when, and how).
- Auditing progress along the way and following completion of the project.

Developing the project justification and implementation plans
- What are the expected results?
    - —Tangible benefits.
    - —Intangible benefits.
- What are the expected system costs?
    - —Software and hardware.
    - —Implementation costs.
    - —Education costs.
    - —Any additional personnel costs related to the implementation.

Education plan details for all employees
- Who, what, when and why it is needed.
- How the education will be delivered.
- Identification of the education resources needed and currently available.

Plans for a pilot test of the new system and software
- System requirements and timing of availability.
- Setup and organization of test data and exercises.
- Location and accessibility.
- Assessment and follow-up.

Figure 7–18 System Implementation Planning Considerations.

example, the inventory group could be assigned to work on inventory count accuracy, location accuracy, and damaged-goods problems. As Figure 7–19 indicates, MRP II implementation can take eighteen months in a medium-size company, with the first eight to twelve months used to get the current manufacturing system in proper order for the implementation of the hardware and software. Study the proven path model until you are familiar with all the activities.

The cost of the implementation is directly proportional to the size and type of company. The costs are usually divided into four categories: (1) consulting (10 percent), (2) education and problem analysis (40 percent), (3) hardware (20 percent), and

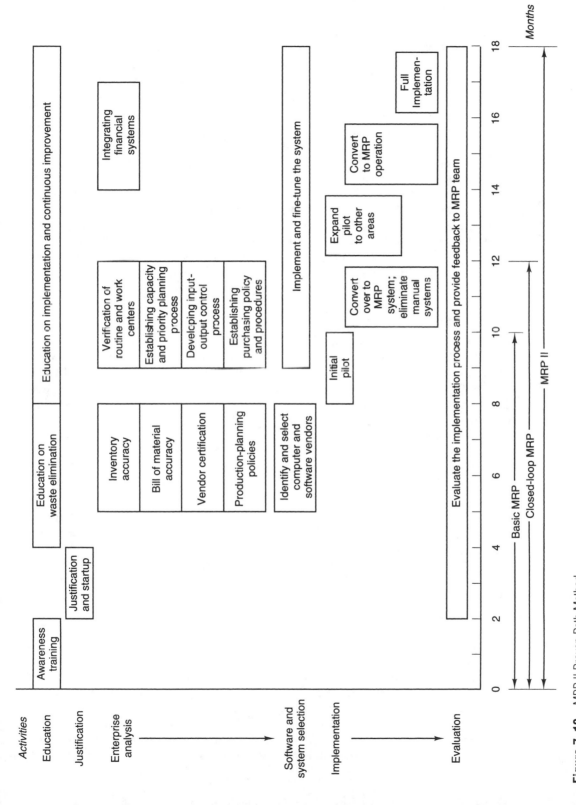

Figure 7-19 MRP II Proven Path Method.
(*Source: Courtesy of Industrial Technology Institute.*)

(4) software (30 percent). The cost of the software is a function of the computer hardware. For example, the cost of base and scheduling software for a job shop operation that runs on a microcomputer is in the range of $20,000 to $30,000. MRP II software for a microcomputer-based system would be less than $50,000, while software for a mini- or mainframe computer is usually over $100,000. While roughly half of the cost for MRP II is in hardware and software, the resources spent on education and problem analysis are the most critical for successful system implementation.

Implementation costs for a formal system are significant. The costs extend well beyond the cost of software and hardware that will run it. An MRP implementation is not a small task, even for companies that already have an MRP system installed. A mainframe-based system implementation, including software, hardware, and the human resources needed to implement them in a company, can easily cost more than $1 million. Modern systems that use the currently available technologies for minicomputers and personal computer networks may reduce the cost of hardware and software by 60 to 70 percent, but the cost of a new formal system remains high.

Manufacturing System Education

Employee education is the "glue" that holds the implementation project together. It is more than the training of people in the operation of the new system. Education prepares people to deal with the changes that come with the implementation of new tools and new ways to perform their jobs. The need for employee education does not end with the implementation of a new system. The project team should develop a detailed education plan that meets the needs of system users at all levels of the organization.

An effective way to develop and display the education plan is to build a spreadsheet. The spreadsheet lists all employees by functional group and all the topics to be covered in training sessions. An example is shown in Figure 7–20. The spreadsheet shows the plan for education by the topic area that the project team recommends for each employee. It becomes the basis for planning detailed and overview education, time requirements, class size, the number of sessions, and more. The spreadsheet is a powerful tool for education planning.

7–6 SUMMARY

The MPC model described in chapter 6 presented a process and flow for production-planning and -scheduling information. This chapter provided a deeper look into the core systems used by companies to plan for material and production resources.

The process of material requirements planning (MRP) provides a very logical, engineered approach to the problem of how much material to order and when to order. The success of this process depends on the quality of the data inputs to the process. The bill of materials (BOM) must provide an accurate and complete list of the materials required to produce each product and subassembly. Inventory records must provide correct information on the amount of material on hand.

Education Plan Matrix	Overview of Mfg. Systems	Bills of Material	MRP logic	Planning Using MRP	Product Data Management	Master Scheduling	Rough-Cut Capacity Planning	Other topics to be determined . . .
Planners:								
Bill Jones	×	×	×	×	×	×	×	
Mary Davis	×	×	×	×	×	×	×	
Karen Hopping	×	×	×	×	×	×	×	
Accounting:								
Mary Johns	×	×						
Larry Captain	×							
Engineering:								
Bob Cramer	×	×			×		×	
Jane Williams	×	×			×		×	
Sales/Marketing:								
Robert Keyes	×				×	×	×	
Joe Smith	×	×			×	×	×	
Manufacturing:								
Johnny King	×	×		×	×	×	×	
Mike Mitchell	×	×		×	×	×	×	
Sue Brock	×	×		×				

Figure 7–20 Education Plan Spreadsheet.

Procurement records must also be accurate and correctly show information on vendor lead times, open purchase orders, and the planning information to be used to drive the ordering process. The master production schedule (MPS) drives the MRP process. The MPS provides the specific statement of what production is to supply to satisfy the demand for products. The master schedule must have the support of all the functional areas and be a realistic statement of what production can do. The MPS cannot become an overloaded wish list.

The detailed material plans produced by the MRP process become the drivers of the detailed capacity requirements plan (CRP) for the operation. Production order schedules planned by the MRP process become one of the critical inputs to the capacity planning process. Detailed capacity planning goes beyond the rough-cut analysis of key resources that is a part of the production planning and master scheduling. CRP brings in detailed operation data from the routing details documented by the engineering function. The routing details include the production operation sequence that is to be followed to transform material into parts, subassemblies, and products. Routings tell where work is to be done, how long it is expected to take, and the type and quantity of labor required.

Detailed material and capacity plans provide information for people in the various functional groups to consider before they release purchase orders to vendors and work orders to the shop floor. The detailed plans are a guide. The planning exercise often identifies problems and conflicts that people in the functional groups will need to resolve. The plans are all designed to push the production described by the master schedule through the manufacturing process and make the stated quantities of products available for customers by the planned due dates.

MRP, CRP, and manufacturing resource planning (MRP II) are logical, data-driven systems that can be extremely effective. These systems are almost always computer driven. The results can be very good, or very bad, depending on the quality of the data inputs and the skills and knowledge of the people interpreting the system outputs. Companies have invested heavily in these systems. Many companies license software for MRP, CRP, and MRP II. Some companies have developed their own systems and software internally. In all of the proven systems, the basic logic that drives MRP and CRP follows the conventions established and supported by the American Production and Inventory Control Society (APICS). Successful systems require a great deal of attention and care. Companies that lack the organization and discipline needed to ensure timely and accurate data inputs to the system have not found success with their systems. The class A system users have had great success and have developed internal processes that meet the needs of the planning system and support accurate planning outputs.

The environment of manufacturing is continually changing. The next chapters cover the generation of systems that have followed MRP II and a new approach to planning and operations that is challenging some of the traditional thinking behind MRP II systems.

BIBLIOGRAPHY

CLARK, P. A. *Technology Application Guide: MRP II Manufacturing Resource Planning*. Ann Arbor, MI: Industrial Technology Institute, 1989.

Cox, J. F., and J. H. Blackstone, eds. *APICS Dictionary.* 10th ed. Alexandria, VA: American Production and Inventory Control Society (APICS)—The Educational Society for Resource Management, 2002.

Cox, J. F., J. H. Blackstone, and M. S. Spencer, eds. *APICS Dictionary.* 7th ed. Falls Church, VA: American Production and Inventory Control Society, 1992.

Garwood, D. *Bills of Material: Structured for Excellence.* Rev. 5th ed. Atlanta, GA: Dogwood Publishing, 1997.

Goddard, W. E. "ABCD Rankings and Your Bottom Line." *Modern Materials Handling,* March 1989, 41.

Kraebber, H. W. "Teaching Manufacturing Systems." In *Manufacturing Education for the 21st Century. Volume V, Manufacturing Education for Excellence in the Global Economy.* Dearborn, MI: Society of Manufacturing Engineers, 1998, 259–264.

Lunn, T., and S. Neff. *MRP: Integrating Material Requirements Planning and Modern Business.* Chicago: Irwin Professional Publishing, 1992.

Sobczak, T. V. *A Glossary of Terms for Computer-Integrated Manufacturing.* Dearborn, MI: CASA of SME, 1984.

Vollmann, T. E., W. L. Berry, and D. C. Whybark. *Manufacturing Planning and Control Systems.* 4th ed. Homewood, IL: Richard D. Irwin, 1997.

Wight Co. *Survey Results: MRP/MRP II JIT.* Essex Junction, VT: The Oliver Wight Companies, 1990.

Wight, O. *Production and Inventory Management in the Computer Age.* Boston: CBI Publishing, 1974.

QUESTIONS

1. Define *material requirements planning.*
2. Define *MRP record,* describe all the terms present, and identify those that are normally given at the start of an MRP run.
3. Clearly define the difference between planned order receipts and scheduled receipts.
4. What is the relationship between the product structure diagram and the MRP records for parts and assemblies?
5. Use the product structure diagram shown in Figure 7–6 to describe how the gross requirements for the MRP records of each component are generated.
6. List five benefits achieved as a result of an effective MRP system.
7. Compare the level of capacity planning detail associated with the three places in MPC where capacity planning is performed.
8. Describe the interaction between the MPS and the shop-floor capacity scheduling process that is used to achieve a reliable MPS.
9. Define *reorder-point system.*

10. Describe the four factors that influence the reorder point.

11. Describe why the reorder point is used for independent demand items and why MRP is used for dependent demand items.

12. Compare and contrast open-loop MRP, closed-loop MRP, and MRP II.

13. What are the key elements in the implementation of MRP II using the proven path method?

14. What distinguishes a class A company from the rest of the classes, and what benefits does a class A company have over companies in lower classes?

15. How do BOMs differ in different departments in the enterprise?

16. Describe the relationship among BOMs, product structure diagrams, indented BOMs, and planning bills.

17. What conditions must be met to achieve BOM accuracy in the 97-plus percent range?

PROBLEMS

1. Using the production information for a product in the following table, complete an MRP record to determine planned order releases and inventory balances. On-hand inventory is 15 units, lead time is 2 periods, lot size is 25, and safety stock is 0.

Period	Gross requirements	Scheduled receipts
1	20	25
2	10	0
3	5	0
4	35	0
5	15	0
6	25	0

Problems 2 through 6 use the production data in problem 1 with a single value changed. Change only the data required and do not carry the changes forward to other problems.

2. A cycle count determines that the on-hand balance in problem 1 is incorrect and should be 5 units instead of 15. How does that affect the MRP record, and what changes in the planned orders are required?

3. During the first period, marketing adjusts the gross requirements for the data in problem 1 from 35 units in period 4 to 15. Recalculate the MRP record for data in problem 1. How were planned orders affected?

4. Recalculate the MRP record for problem 1 with a lot size of 30. How are planned order releases affected? How does the average inventory for the two lot sizes compare?

5. Recalculate the MRP record for problem 1 with a lot-for-lot production capability. How are planned order releases affected? How is the average inventory affected?

6. Recalculate the MRP record for problem 1 with a safety stock of 15 units per period. How are planned order releases affected? How is the average inventory affected?

7. The MPC system at a company performs a weekly update of the MPS and MRP production-planning data. At the start of week 1, the MPS for products A and B are as listed next.

Period (weeks)	Product A	Product B
1	5	15
2	15	0
3	0	20
4	25	5
5	0	10
6	15	0

One unit of component C is required to manufacture products A and B. The lead time for C is 1 week, the safety stock is 0, the lot size is 10 units, the on-hand balance of C at the start of week 1 is 15 units, and a scheduled receipt of 10 units of C is due in week 1. Complete the MRP record for component C for the 6 periods, and determine the planned orders, inventory balance, and average inventory level.

8. Develop a spreadsheet for an MRP record and repeat problems 1 through 7 using a computer to generate inventory balances.

9. The following inventory data apply to the product structure for the table assembly shown in Figure 7–5. There are no scheduled receipts for any parts, and the lead time for each is 1 week.

   For each of the following situations, determine how many tables could be delivered to the customer at the end of the period.

   a. No problems; the inventory data are accurate and all parts meet quality specifications.

   b. Of the 985 part numbers, 10 are defective.

   c. A cycle count determines that the inventory for part 984 is 35 parts and not 40.

PROJECTS

1. Using the list of companies developed in project 1 in chapter 1, determine which companies use open-loop MRP, closed-loop MRP, or MRP II manufacturing systems.

2. Select one of the companies in project 1 that uses MRP II and compare the world-class standards achieved by that company with the standards presented in chapter 1.

3. Select one of the companies in project 1 that uses MRP II and describe the process used to achieve continuous improvement.

4. Develop a spreadsheet to calculate the MRP requirements for all the parts in the product structure of the table assembly shown in Figure 7–5. Link the MRP records so that MPS requirements for tables for 6 periods are entered along with beginning on-hand inventories and scheduled receipts. Make assumptions, as necessary, for other production parameters.

APPENDIX 7–1: WIGHT'S BICYCLE EXAMPLE

Oliver Wight is recognized as one of the innovators of the tool that has become known as *material requirements planning (MRP)*. He presented an example of the MRP logic using a bicycle as the product. Almost everyone can relate to this basic product. This exercise keeps the MRP calculation logic fairly simple. Later examples add to the complexity of the calculation, but the logic behind the planning stays the same.

The bicycle used in the example is a simple, single-rider device. It has two wheels, two tires, a frame, a seat, and one handlebar. The handlebar in this example is manufactured from tube steel that is cut and formed. The BOM for the handlebar is shown in Figure 7–21. One handlebar is required for each bicycle produced.

Figure 7–22 details the MRP records that link the finished bicycles to the handlebars and to the tube steel. Gross requirements for the finished bicycles have been predetermined and are not part of this example. The lead time for assembly of bicycles is 1 time period, and there is no stock on hand of finished bicycles. Note the 1-period offset that links a requirement for 40 finished bicycles in period 2 and a planned order release for 40 units into assembly at the start of period 1.

Planned order releases into final assembly drive lower-level requirements for material. Note that the planned order release for 40 in period 1 becomes a gross requirement for 40 handlebars in period 1. The MRP logic considers the 60 handlebars on hand before period 1, a lead time of 4 periods to convert raw material into finished handlebars, and a planning requirement for lot quantities of 120 units. The 120 handlebars shown as "scheduled receipts" in period 3 are the result

Figure 7–21 BOM for the Handlebar.

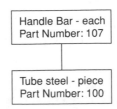

Item: bicycles		1	2	3	Period 4	5	6	7	8	9
Gross requirements (from MPS)			40		50			60		60
Scheduled receipts										
Projected available balance	0									
Planned order release		40		50			60		60	

Order policy = lot for lot; order quantity = variable; lead time = 1.

(a)

Item: handlebars		1	2	3	Period 4	5	6	7	8
Gross requirements		40		50			60		60
Scheduled receipts				120					
Projected available balance	60	20	20	90	90	90	30	30	90
Planned order release				120					

Order policy = fixed lot; order quantity − 120; lead time = 4.

(b)

Item: tube steel		1	2	3	Period 4	5	6	7	8
Gross requirements					120				
Scheduled receipts									
Projected available balance	140	140	140	140	20	20	20	20	20
Planned order release									

Order policy = fixed lot; order quantity = 200; lead time = 5.

(c)

Figure 7–22   MRP Records.

of the release of a past order to the shop, with 120 expected to be available for use at the start of the period. Work through the projected available balance line from left to right. The projected available balance considers the quantity on hand at the start of the period, plus any new receipts expected, minus any gross requirements. Note that in week 8 not enough stock is on hand and there are no scheduled receipts. This condition forces the planning of a new order release for 120 units at the start of period 4.

Tube steel is a purchased item. The planned order release for handlebars at the start of period 4 drives the gross requirement for 120 pieces of tube steel in period 4. In this example, there is little activity at the tube steel level because there are 140 pieces of steel on hand at the start of period 1 and only one requirement for 120 pieces in period 4.

*APPENDIX 7–2: ABCD CHECKLIST**

1. Strategic Planning Processes

Qualitative Characteristics

Class A   Strategic planning is an ongoing process and carries an intense customer focus. The strategic plan drives decisions and actions. Employees at all levels can articulate the company's mission, its vision for the future, and its overall strategic direction.

Class B   Strategic planning is a formal process, performed by line executives and managers at least once per year. Major decisions are tested first against the strategic plan. The mission and/or vision statements are widely shared.

Class C   Strategic planning is done infrequently but provides some direction to how the business is run.

Class D   Strategic planning is nonexistent, or totally removed from the ongoing operation of the business.

Overview Items

1–1 COMMITMENT TO EXCELLENCE
The company has an obsession with excellence; there is dissatisfaction with the status quo. Executives provide the leadership necessary for change. They articulate the motivations for positive change and other core values, and communicate them widely throughout the organization—by actions as well as by words.

1–2 BUSINESS STRATEGY/VISION
There is an explicit written business strategy that includes a vision and/or mission statement. This strategy articulates the commitment to excellence and the overriding importance of customer satisfaction.

1–3 BENCHMARKING
The company continuously measures its products, services, and practices against the toughest competitors, within and outside the industry. This information is used to identify best practices and establish performance benchmarks.

1–4 SUSTAINABLE COMPETITIVE ADVANTAGE
The business strategies recognize the principle of sustainable competitive advantage: those items not directly under the company's control may not yield competitive advantage over the long run.

---

*Courtesy Oliver Wight Publications.

1–5 ONGOING FORMAL STRATEGIC PLANNING
There is an ongoing formal strategic planning process in place, in which all senior executives have active, visible leadership roles.

1–6 CONGRUENCE TO STRATEGY
Requests for capital expenditure are tested first for congruence to the business strategy and appropriate functional strategies (i.e., does this proposal fit the strategy?).

1–7 BUSINESS PLANNING
A business process is used to develop and communicate annual financial plans that incorporate input from all operating departments of the company.

1–8 GENERATION OF PRODUCT COSTS
Executives and managers believe the accounting system generates valid product costs, reflecting the true costs involved in producing and delivering the company's products. Activity-based costing as well as other costing methodologies are understood, and the most suitable costing method is used.

2. People/Team Processes

Qualitative Characteristics

Class A   Trust, teamwork, mutual respect, open communication, and a high degree of job security are hallmarks of the employee/company relationship. Employees are very pleased with the company and proud to be part of it.

Class B   Employees have confidence in the company's management and consider the company a good place to work. Effective use is made of small work groups.

Class C   Traditional employment practices are largely being used. Management considers the company's people to be an important, but not vital, resource of the business.

Class D   The employee/employer relationship is neutral at best, sometimes negative.

Overview Items

2–1 COMMITMENT TO EXCELLENCE
All levels of management have a commitment to treating people with trust, openness, and honesty. Teams are used to multiply the strength of the organization. People are empowered to take direct action, make decisions, and initiate changes.

2–2 CULTURE
A comprehensive culture exists to support and enhance effective people and team processes.

2–3 TRUST
Openness, honesty, and constructive feedback are highly valued and demonstrated organizational traits.

2–4 TEAMWORK
Clearly identifiable teams are used as the primary means to organize the work, as opposed to individual job functions or independent workstations.

2–5 EMPLOYMENT CONTINUITY
Employment continuity is an important company goal as long as the employee exceeds the minimum acceptable job requirements and the level of business is viable.

2–6 EDUCATION AND TRAINING
An active education and training process for all employees is in place and is focused on business and customer issues and improvements. Its objectives include continuous improvement, enhancing the empowered worker, flexibility, employment stability, and meeting future needs.

2–7 WORK DESIGN*
Jobs are designed to reinforce the company goal of a team-based, empowered workforce.

2–8 CONGRUENCE
People policies, organizational development, and educational and training maintain consistency with the company vision and business strategies.

3. Total Quality and Continuous Improvement Processes

Qualitative Characteristics

Class A   Continuous improvement has become a way of life for employees, suppliers, and customers. Improved quality, reduced costs, and increased velocity contribute to a competitive advantage. There is a targeted strategy for innovation.

Class B   Most departments participate in these processes; they have active involvement with suppliers and customers. Substantial improvements have been made in many areas.

Class C   Processes are used in limited areas; some departmental improvements have been achieved.

Class D   Processes are not established, or processes are established but static.

---

* Same item included in more than one business function.

Overview Items

3–1 COMMITMENT TO EXCELLENCE

There is a commitment to total quality in all areas of the business and to continuous improvements in customer satisfaction, employee development, delivery, and cost.

3–2 TOP MANAGEMENT LEADERSHIP FOR QUALITY AND CONTINUOUS IMPROVEMENT

Top executives are actively involved in establishing and communicating the organization's vision, goals, plans, and values for quality and continuous improvement.

3–3 FOCUS ON CUSTOMER

Various effective techniques are used to ensure that customer needs are identified, prioritized, and satisfied. Customers are identified both internally and externally, and all functions participate. External customers include users, other external links in the chain to the end user, shareholders, stakeholders, and the community.

3–4 CUSTOMER PARTNERSHIPS

Strong "partnership" relationships that are mutually beneficial are established with customers.

3–5 CONTINUOUS ELIMINATION OF WASTE

There is a companywide commitment to the continuous and relentless elimination of waste. A formal program is used to expose, prioritize, and stimulate the elimination of non-value-adding activities.

3–6 ROUTINE USE OF TOTAL QUALITY CONTROL TOOLS

Routine use of the basic tools of total quality control and the practice of mistake proofing has become a way of life in almost all areas of the company.

3–7 RESOURCES AND FACILITIES—FLEXIBILITY, COST, QUALITY

Resources and facilities required to receive, produce, and ship the product economically are continuously made more flexible, cost effective, and capable of producing higher quality.

3–8 PRODUCE TO CUSTOMER ORDERS

The time required to manufacture products has been reduced such that the planning and control system uses forecasts to project material and capacity needs, but production of finished products is based on actual customer orders or distribution demands (except where strategic or seasonal inventories are being built).

3–9 SUPPLIER PARTNERSHIPS

Strong "partnership" relationships that are mutually beneficial are being established with fewer but better suppliers to facilitate improvements in quality, cost, and overall responsiveness.

3–10 PROCUREMENT—QUALITY, RESPONSIVENESS, COST
The procurement process is continuously being improved and simplified to improve quality and responsiveness while simultaneously reducing the total procurement costs.

3–11 KANBAN*
Kanban is used effectively to control production where its use will provide significant benefit.

3–12 VELOCITY
The velocity and linearity of flow is continuously measured and improved.

3–13 ACCOUNTING SIMPLIFICATION
Accounting procedures and paperwork are being simplified, eliminating non-value-adding activities, while at the same time providing the ability to generate product costs sufficiently accurate to use in decision making and satisfy audit requirements.

3–14 USE OF TOTAL QUALITY CONTROL AND JUST-IN-TIME
A minimum of 80 percent of the plant output is produced using the tools and techniques of TQC and JIT.

3–15 TEAMWORK
Clearly identifiable teams are used as the primary means to organize the work, as opposed to individual job functions or independent workstations.

3–16 EDUCATION AND TRAINING
An active education and training process for all employees is in place and is focused on business and customer issues and improvement. Its objectives include continuous improvement, enhancing the empowered worker, flexibility, employment stability, and meeting future needs.

3–17 WORK DESIGN
Jobs are designed to reinforce the company goal of a team-based, empowered workforce.

3–18 EMPLOYMENT CONTINUITY
Employment continuity is an important company goal as long as the employee exceeds the minimum acceptable job requirements and the level of business is viable.

3–19 COMPANY PERFORMANCE—QUALITY, DELIVERY, COST
Company performance measurements emphasize quality, delivery, and cost. Performance measures are communicated to all through visible displays that

---

*See chapter 9 for a more detailed description of *kanban*.

show progress and point the way to improvement (e.g., run charts coupled with Pareto charts).

3–20 SETTING AND ATTAINING QUALITY GOALS
Short- and long-term quality goals that cause the organization to stretch are established, regularly reviewed, and monitored. These goals are targeted on improvements in total cost, cycle time (or response time), and customer quality requirements.

4. New Product Development Processes

Qualitative Characteristics

Class A   All functions in the organization are involved with and actively support the product development process. Product requirements are derived from customer needs. Products are developed in significantly shorter time periods, meet these requirements, and require little or no support. Internal and external suppliers are involved and are an active part of the development process. The resulting revenue and margins satisfy the projections of the original business plan proposals.

Class B   Design engineering (or R&D) and other functions are involved in the development process. Product requirements are derived from customer needs. Product development times have been reduced. A low to medium level of support is required. Few design changes are required for products to meet the requirements.

Class C   The product development process is primarily an engineering or R&D activity. Products are introduced close to schedule but contain traditional problems in manufacturing and the marketplace. Products require significant support to meet performance, quality, or operating objectives. The manufacturing process is not optimized for internal or external suppliers. Some improvement in reducing development time has been achieved.

Class D   The products developed consistently do not meet schedule dates or performance, cost, quality, or reliability goals. They require high levels of support. There is little or no internal or external supplier involvement.

Overview Items

4–1 COMMITMENT TO EXCELLENCE
An intense commitment to excel in the innovation, effectiveness, and speed of new product development is broadly shared by all levels of management throughout the organization.

4–2 MULTIFUNCTIONAL PRODUCT DEVELOPMENT TEAMS

Multifunctional product development teams—including manufacturing, marketing, finance, quality assurance, purchasing, suppliers, and, where appropriate, customers—are used during the design process for new product development.

4–3 EARLY TEAM INVOLVEMENT

Design for manufacturability/concurrent engineering processes are used at the beginning and throughout the product development processes.

4–4 CUSTOMER REQUIREMENTS USED TO
DEVELOP PRODUCT SPECIFICATIONS

Customer requirements are determined utilizing processes such as quality function deployment (QFD) and competitive benchmarking. These customer requirements are used to develop the specifications for the product.

4–5 DECREASE TIME-TO-MARKET

An ongoing effort to decrease the time-to-market—the elapsed time between the start of the design and the first shipment of the product—is viewed as an important competitive weapon, as highly visible, and as generating improvements.

4–6 PREFERRED COMPONENTS, MATERIALS, AND PROCESSES

There is an agreed-upon product development and manufacturing policy on commonality and use of preferred components and preferred processes in new designs.

4–7 EDUCATION AND TRAINING

An active education and training process for all employees is in place and is focused on business and customer issues and improvements. Its objectives include continuous improvement, enhancing the empowered worker, flexibility, employment stability, and meeting future needs.

4–8 NEW PRODUCT DEVELOPMENT INTEGRATED
WITH THE PLANNING AND CONTROL SYSTEM

All phases of new product development are integrated with the planning and control system.

4–9 PRODUCT DEVELOPMENT ACTIVITIES INTEGRATED
WITH THE PLANNING AND CONTROL SYSTEM

Where applicable, product development activities in support of a customer order are integrated with the planning and control system.

4–10 CONTROLLING CHANGES

There is an effective process for evaluating, planning, and controlling changes to the existing products.

5. Planning and Control Processes

Qualitative Characteristics

Class A   Planning and control processes are effectively used company-wide, from top to bottom. Their use generates significant improvements in customer service, productivity, inventory, and costs.

Class B   These processes are supported by top management and used by middle management to achieve measurable company improvements.

Class C   The planning and control system is operated primarily as a better method for ordering materials and contribution to better inventory management.

Class D   Information provided by the planning and control system is inaccurate and poorly understood by users, providing little help in running the business.

Overview Items

5-1 COMMITMENT TO EXCELLENCE

There is a commitment by top management and throughout the company to use effective planning and control techniques—providing a single set of numbers used by all members of the organization. These numbers represent valid schedules that people believe and use to run the business.

5-2 SALES AND OPERATIONS PLANNING

There is a sales and operations planning process in place that maintains a valid, current operating plan in support of customer requirements and the business plan. This process includes a formal meeting each month run by the general manager and covers a planning horizon adequate to plan resources effectively.

5-3 FINANCIAL PLANNING, REPORTING, AND MEASUREMENT

There is a single set of numbers used by all functions within the operating system that provides the source data used for financial planning, reporting, and measurement.

5-4 WHAT-IF SIMULATIONS

What-if simulations are used to evaluate alternative operating plans and develop contingency plans for materials, people, equipment, and finances.

5-5 ACCOUNTABLE FORECASTING PROCESS

There is a process for forecasting all anticipated demands with sufficient detail and an adequate planning horizon to support business planning, sales and operations planning, and master production scheduling. Forecast accuracy is measured to improve the process continuously.

5–6 SALES PLANS

There is a formal sales planning process in place, with the sales force responsible and accountable for developing and executing the resulting sales plan. Differences between the sales plan and the forecast are reconciled.

5–7 INTEGRATED CUSTOMER ORDER ENTRY AND PROMISING

Customer order entry and promising are integrated with the master production scheduling system and inventory data. There are mechanisms for matching incoming orders to forecasts and for handling abnormal demands.

5–8 MASTER PRODUCTION SCHEDULING

The master production scheduling process is perpetually managed to ensure a balance of stability and responsiveness. The master production schedule is reconciled with the production plan resulting from the sales and operations planning process.

5–9 MATERIAL PLANNING AND CONTROL

There is a material planning process that maintains valid schedules and a material control process that communicates priorities through a manufacturing schedule, dispatch list, supplier schedule, and/or a kanban mechanism.

5–10 SUPPLIER PLANNING AND CONTROL

A supplier planning and scheduling process provides visibility for key items covering an adequate planning horizon.

5–11 CAPACITY PLANNING AND CONTROL

There is a capacity planning process using rough-cut capacity planning and, where applicable, capacity requirements planning in which planned capacity, based on demonstrated output, is balanced with required capacity. A capacity control process is used to measure and manage factory throughput and queues.

5–12 CUSTOMER SERVICE

An objective for on-time deliveries exists, and the customers are in agreement with it. Performance against the objective is measured.

5–13 SALES PLAN PERFORMANCE

Accountability for performance to the sales plan has been established, and the method of measurement and the goal has been agreed upon.

5–14 PRODUCTION PLAN PERFORMANCE

Accountability for production plan performance has been established, and the method of measurement and the goal has been agreed upon. Production plan performance is more than ±2 percent of the monthly plan, except in cases where mid-month changes have been authorized by top management.

5–15 MASTER PRODUCTION SCHEDULE PERFORMANCE
Accountability for master production schedule performance has been established, and the method of measurement and the goal has been agreed upon. Master production schedule performance is 95 to 100 percent of the plan.

5–16 MANUFACTURING SCHEDULE PERFORMANCE
Accountability for manufacturing schedule performance has been established, and the method of measurement and the goal has been agreed upon. Manufacturing schedule performance is 95 to 100 percent of the plan.

5–17 SUPPLIER DELIVERY PERFORMANCE
Accountability for supplier delivery performance has been established, and the method of measurement and the goal has been agreed upon. Supplier delivery performance is 95 to 100 percent of the plan.

5–18 BILL OF MATERIAL STRUCTURE AND ACCURACY
The planning and control process is supported by a properly structured, accurate, and integrated set of bills of material (formulas, recipes) and related data. Bill of material accuracy is in the 98 to 100 percent range.

5–19 INVENTORY RECORD ACCURACY
There is an inventory control process in place that provides accurate warehouse, stockroom, and work-in-process inventory data. At least 95 percent of all item inventory records match the physical counts, within the counting tolerance.

5–20 ROUTING ACCURACY
When routings are applicable, a development and maintenance process is in place that provides accurate routing information. Routing accuracy is in the 95 to 100 percent range.

5–21 EDUCATION AND TRAINING
An active education and training process for all employees is in place and is focused on business and customer issues and improvements. Its objectives include continuous improvement, enhancing the empowered worker, flexibility, employment stability, and meeting future needs.

5–22 DISTRIBUTION RESOURCE PLANNING (DRP)
Distribution resource planning, where applicable, is used to manage the logistics of distribution. DRP information is used for sales and operations planning, master

production scheduling, supplier scheduling, transportation planning, and the scheduling of shipping.

APPENDIX 7–3: AN ERP EXAMPLE USING WINMAN*

The product structure is a series of parent-item to component-item relationships. The way the product is structured can have a great impact on the operation of the planning system. The type of product, the structure of the product, and the planned routing of the product through its required operations each have a significant impact on the planning done by the WinMan system. Excessive product structure levels create unnecessary planning activities and reporting requirements for manufacturing. Decisions on the level of control required will also affect the number and complexity of reporting transactions. It is easy to make the product structure much more complicated than necessary; however, the ocean liner *Queen Mary* has been described in fourteen structure levels. The space shuttle is reported to have been described in only eight structure levels. Most manufactured products can be described in five levels or less. The use of fewer levels greatly simplifies manufacturing and the material planning process.

This example begins with a simple product. A graphical representation of the structure of product A and its related subassemblies and component items is shown in Figure 7–23. The structure diagram for product A shows how the required parts are put together in the finished product. Each step in the manufacturing process can take some time. The actual operating time to complete the work is critical for determining costs, but at this level, a formal planning system is looking for something a little different. The system must know how to plan for the total time needed to manufacture the item. This total time includes collecting the needed materials, the setup of the required workstations, the time to do the work, any quality control checks or inspections, and any material-handling time. The expected total production time is the sum of all these inputs and any others that are relevant to this production work. The time is calculated and entered into the planning system to the nearest whole day. This time allowance is fairly coarse, but for most material planning applications, it works quite well. The detailed capacity planning and shop-floor execution schedules require more detailed time standards.

The diagram in Figure 7–24 adds the dimension of the processing location (AS1 or AS2 in this example) to the given product structure information. It is assumed in this example that a special piece of equipment is needed to assemble parts D and E and this equipment is in assembly location AS2. A future improvement to this process might lead to the redesign of the two workstations into one. Product C could then be coded as a transient item, which would simplify the planning, assembly process, and reporting of work completed.

---

*This description of the operation of part of the WinMan ERP system has been adapted from documents prepared for TTW, Inc., and a paper presented at the 1998 SME Education in Manufacturing conference held in San Diego. Additional information about WinMan may be found at its Internet site: **http://www.winman.com**.

Figure 7–23 A Simple Product Structure "Tree Diagram" for Product A.

(Source: Used by permission of the Society of Manufacturing Engineers. Copyright 1998.)

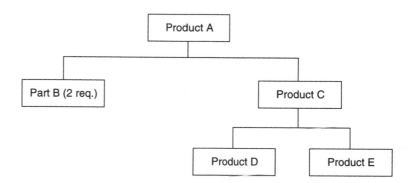

Requirements for product A may come from a sales order, the master production schedule, a specific job, a higher-level product or assembly, a what-if order, or a reorder-point signal. The system logic considers requirements for product A, the quantity needed, and the date it is required. The system works backward from the required date to determine the appropriate material plan for all the items in the product structure. Plans are developed by the system on the basis of the planning parameters defined for each item. During the planning cycle, the demands for lower-level items are driven from the top-level item in the structure. The planning process works through all the lower-level assemblies and components that make up the manufactured item. The formal system uses the product structure, stated lead times, planning parameters, and current stock (inventory levels) of the required items to generate a time-phased material plan.

An MRP record for product A can now be considered. Product A is a top-level finished-goods item. The replenishment requirements for product A are a function of demand, current stocks, and the planning parameters. In Figure 7–25, the current stock on hand for product A is 2 units.

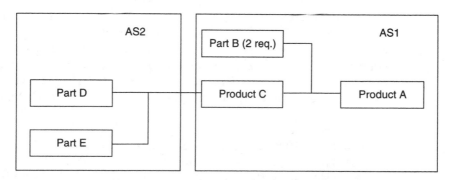

Figure 7–24 Linking the Product Structure with Workstations on the Shop Floor.

(Source: Used by permission of the Society of Manufacturing Engineers. Copyright 1998.)

Item: A   Period Number

		1	2	3	4	5	6	7	8	9	10	11	12
Gross Requirements						25					25		3
Scheduled Receipts													
Projected On-Hand	2	2	2	2	2	17	17	17	17	17	0	0	2
Planned Order Receipts						40					8		5
Planned Order Releases		40					8		5				

Lead time = 4   Order Policy = MRP   Optimum Order Quantity = 20

Minimum Stock = 0   Minimum Order Quantity = 5

Figure 7–25   MRP Record for Product A.

(Source: Used by permission of the Society of Manufacturing Engineers. Copyright 1998.)

Some discussion of this record is needed. During the execution of the preceding plan, the quantities stated as planned order releases and planned order receipts will be reviewed and, if approved, will be replaced with firm scheduled receipts. For example, if the planned order release quantity of 40 on day 1 is confirmed, that quantity will drop out of the planned order release and planned order receipt lines and appear as a scheduled receipt quantity of 40 on day 5. The scheduled receipt will be available at the start of day 5. In reality, the formal system must schedule receipt to occur by the end of work on day 4.

Note also how the system applies the planning parameters. The first new order release for product A is for a quantity of 40 units, or two times the stated optimum quantity. The second release is for the exact number of units required on the basis of the MRP logic because the required quantity is between the stated minimum and optimum. The last scheduled release for 5 units was made because of the stated minimum requirement. These quantities may seem a little confusing at first, but they do provide planning flexibility for the system users.

The formal system will review and calculate the replenishment requirements for all the assemblies and component items that go into product A. Consider the proposed manufacturing order for a quantity of 8 units of product A scheduled for completion on day 10. Requirements for lower-level items are driven by the planned order release for 8 units at the start of day 6 based on the 4-day lead time for product A. The manufacturing order for 8 units of product A might look like Figure 7–26.

Material moving and handling is often one of the highest costs of manufacturing that does not add any value to the product. The developers of WinMan have designed a system of locations and priority codes that allows users to tailor the system to fit their operation and reduce the associated costs of reporting and system transactions. WinMan encourages the definition and use of storage locations for materials at the location where the work is to be done. Parts will be

---

Information on the Manufacturing Order for Product A

Description: Assembled Product A   MO: 23987

Start date: x/6/xx   Due date: x/10/xx   Order Qty. = 8

Oprn # Operation description

0010   Assemble part D to part E as shown in the drawing "Product C/100"

0020   Test in accordance with Proc 456/89, then deliver to location AS1

0030   Assemble with two part B as shown in the drawing "Product A/001", then deliver to location "FINGDS"

Figure 7–26   Information on the Manufacturing Order for Product A.
(Source: Used by permission of the Society of Manufacturing Engineers. Copyright 1998.)

issued automatically from the defined storage locations if the parts are shown by the system to be on hand. The material on hand the longest in approved locations will be issued first, which will create a dynamic first-in, first-out system. If the parts are not in the expected location when they are needed, the system will look for other available parts in approved locations before generating new planned orders for material. Manual issues can be made to an open manufacturing order at any time from any inventory holding. WinMan will issue the parts required on the planned issue date from the location designated in the product setup record.

Consider the issue of part B to workstation AS1. Sixteen of these parts are required to complete the order for product A. Stock of part B can be found in several locations, as shown in Figure 7–27. Following the priority scheme in WinMan,

---

INVENTORY INFORMATION FOR "Part B":

Location	Qty.	Age	Availability
AS1	4	5	Y (only issue to manufacturing in AS1)
AS2	8	30	Y (only issue to manufacturing in AS2)
FINGDS	40	30	S (only shipped via a sales order)
QNTN	10	25	N (not to be issued or shipped)
MNST	10	25	B (both manufacturing and sales issues)
MNST	20	12	M (for manufacturing issue only)

Figure 7–27   Inventory Information for Part B.
(Source: Used by permission of the Society of Manufacturing Engineers. Copyright 1998.)

the 4 items in the designated manufacturing location AS1 will be issued first because the work is done in that location and material already there will be used first. The system then looks for material in other locations to complete the issue of the remaining 12 units of part B. WinMan will look at each available batch and issue the material on hand the longest first, which means that 10 units from the 25-day-old batch in the main stockroom (MNST) and 2 units from the 12-day-old batch in MNST will be issued to the order. Note that parts in a location such as the main stockroom have an additional code letter that allows more control of material issues. Parts with the B code may be used for sales or manufacturing, while the M code allows parts to be used only for manufacturing.

Note that some older pieces of part B in stock were not issued. The parts in assembly area 2 (AS2) are not issued because they are expected to be issued to manufacturing orders assigned to location AS2. Parts in the finished-goods warehouse location (FINGDS) that are coded with an S are issued only to sales orders. Parts in the quarantine location (QNTN) may not be issued or shipped without approval. In many companies, a material review board will evaluate quarantined parts and make a decision on their status. Only authorized individuals with access to the system may change the availability indicator of a batch of material.

Enterprise Resources Planning, and Beyond*

OBJECTIVES

After completing this chapter, you should be able to do the following:

- Be aware of the ongoing changes to the traditional material requirements planning (MRP) and manufacturing resource planning (MRP II) systems with the development of new software and systems technologies
- Describe key issues facing the information technology (IT) departments of companies using computer-based management systems
- Understand several of the critical issues that face companies trying to integrate their business and engineering systems to better manage design data
- Define the new data management initiatives known as *product data management* and *product lifecycle management*
- Develop an awareness of the complexity and rapid change taking place in data management for the enterprise

Since the late 1990s, the pace of change in manufacturing systems has been increasing exponentially as information technology (IT) tools grow and advance. Keeping up to date with all the improvements is not an easy task, but it is a mandatory one. The intention of this chapter is to highlight the technology being offered in the marketplace today.

The objective of this chapter is to convey to the reader a basic view of the expanded enterprise-wide system tools that build on the foundation of material requirements planning (MRP) and manufacturing resource planning (MRP II). Consideration of the new *enterprise resources planning* (ERP) tools opens up related system topics and the operations issues that face manufacturing companies. If

---

*Much of the supporting material for this chapter comes from Patrick Delaney, President of SIBC Corp., a manufacturing industry educator whose clients include many Fortune 500 companies. Mr. Delaney's consulting work keeps him up to date with the changes taking place in the application of information technology to manufacturing problems. This chapter deals with topic areas that are driven by rapid changes in computer and information systems technologies and the competitive pressures of the global business environment.

there ever was an area in manufacturing that will require lifelong learning, the study of the evolving systems for enterprise control is it. This is where the action is. ERP and the ERP system are defined* as follows:

> *enterprise resources planning:* A method for the effective planning and control of all resources needed to take, make, ship, and account for customer orders in a manufacturing, distribution, or service company.
>
> (Source: © APICS Dictionary, 10th edition, 2002)

> *enterprise resources planning (ERP) system:* 1) An accounting-oriented information system for identifying and planning the enterprise wide resources needed to take, make, ship, and account for customer orders. An ERP system differs from the typical MRP II system in technical requirements such as graphical user interface, relational database, use of fourth-generation language, and computer-assisted software engineering tools in development, client/server architecture, and open-system portability. 2) More generally, a method for the effective planning and control of all resources needed to take, make, ship, and account for customer orders in a manufacturing, distribution, or service company.
>
> (Source: APICS Dictionary, 10th edition, 2002)

ERP is one of the newer system concepts that focuses on the integration of business systems. These integrated systems support all of the functional departments in the enterprise: sales and order entry, engineering, manufacturing, finance and accounting, distribution, order planning and execution, and the supply chain flows. It is a logical extension of traditional MRP II, which takes advantage of increased computer power and new client-server architectures to bring heavy-duty data analysis tools to the user's desktop.

Figure 8–1 traces the development of manufacturing software, beginning with the early computer-based systems for inventory control and moving toward the modern systems popular at the start of the millennium. During this evolution, software companies have moved toward increasing the integration of the disconnected enterprise systems with the more mature manufacturing and financial systems. The primary path runs through the center of the figure, beginning at the left with inventory control and working across to ERP/enterprise resource optimization (ERO) at the right. System capabilities were developed and expanded as computers and software became more powerful and cost efficient. System users came to appreciate and develop relationships with the functional areas that provided them with the critical data and information needed to operate manufacturing efficiently. For example, MRP is a more-effective process if the higher-level plans that feed into it are more accurate and timely. Improving the inputs to MRP required more participation and inputs from functional areas outside of the traditional manufacturing organization. The development of detailed plans for material resources became the logical driver of detailed capacity plans for physical resources and the evolution of closed-loop systems for MRP II. The customer focus evident in the Society of Manu-

---

*The *APICS Dictionary* terms and definitions in this chapter are published with the permission of APICS—The Educational Society for Resource Management, 10th edition, 2002.

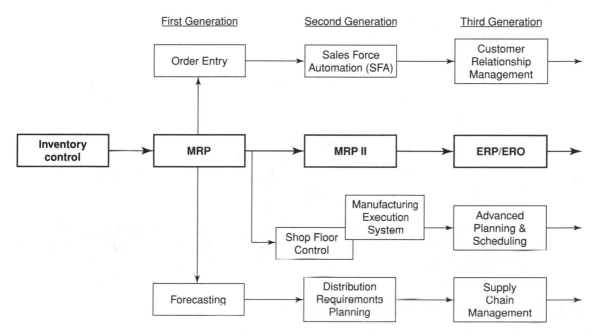

Figure 8–1 Manufacturing Software Development Path.
(Source: SIBC Corp. Technology Course Series 2003.)

facturing Engineers (SME) enterprise wheel provided an additional driving force to develop advanced systems that considered the resources of the enterprise and all the functional areas that it encompasses. The experience of the past twenty years has proven that manufacturing cannot function effectively by itself. The ERP architecture forces each functional area, including manufacturing, to join the program. Examples in this chapter show that obtaining full cooperation is not an easy task. Besides ERP, several other system solutions have evolved to address specific planning problems.

Companies are paying more attention to the information that can help them better serve their customers. A better understanding of the customer can produce significant benefits. This is where *customer relationship management (CRM)* comes into play. CRM is defined as follows:

> ***customer relationship management (CRM):*** *A marketing philosophy based on putting the customer first. The collection and analysis of information designed for sales and marketing decision support (as contrasted to enterprise resources planning information) to understand and support existing and potential customer needs. It includes account management, catalog and order entry, payment processing, credits and adjustments, and other functions.*
> (Source: © *APICS Dictionary*, 10th edition, 2002)

CRM systems are being developed to help companies manage the information they have about their customers, the products these customers buy, and the way the customers prefer to do business. Effective use of this information can lead to improved customer service and increased sales.

Managing data and information to better serve the customer and make the operation more competitive is also leading companies to consider their suppliers and the systems they use to communicate and work with them. Interest in computer-based systems for *supply chain management* (*SCM*) is growing. SCM is defined as follows:

> **supply chain management:** *The design, planning, execution, control, and monitoring of supply chain activities with the objective of creating net value, building a competitive infrastructure, leveraging worldwide logistics, synchronizing supply with demand, and measuring performance globally.*
>
> (Source: © APICS Dictionary, 10th edition, 2002)

The need for effective management of data and information throughout the supply chain is becoming critical. This is a growing area for systems development and integration.

8–1 MRP II: A DRIVER OF EFFECTIVE ERP SYSTEMS

MRP II is an essential part of ERP. Figure 8–2 depicts the core modules of the standard MRP II model and highlights the new modules offering expanded functioning for ERP. The core system logic enabling the MRP explosion process remains intact. Significant improvements have been made to planning and analysis modules such as finance and accounting, demand management, engineering, and global planning.

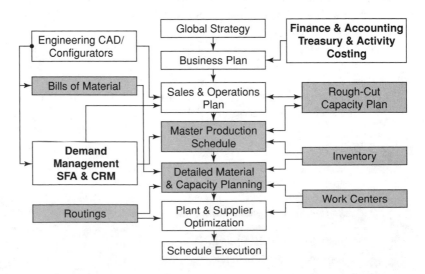

Figure 8–2   The MRP II Core Inside ERP.
(Source: SIBC Corp. Technology Course Series 2003.)

These changes are covered in detail in this chapter. At the other end of the model, the plant and supplier optimization module and the schedule execution module have been restructured through capabilities such as supply chain planning and optimized scheduling software. Again, these tools have been in place for many years, but they have been structured for broader applications across industries in ERP systems. The schedule execution module has received significant attention as companies look toward just-in-time flows of material and short-cycle manufacturing with *kanban*-type* pull signals. The close-in schedule is much more focused on finite execution and broadcast capabilities than on a frozen master schedule with multiple time fences for accepting customer orders. Today's approach is to produce to satisfy the demand schedule and keep the order board open around the clock.

8-2 INFORMATION TECHNOLOGY

The synopsis presented in Figure 8–3 was prepared by a consultant familiar with the backroom operations taking place in the IT department of many companies. His recent experiences depict the challenges IT teams face as they respond to various system issues. The IT team may have to address upgrades or replacement of software code, the implementation of new systems, conversion of data from legacy subsystems, the introduction of new architectures (for example, client-server), and debugging of code. Other issues may include the training of system users, process reengineering, documentation of new procedures, upgrading the staff skills of the department, and searching for new employees. All of this takes place while IT placates sales and marketing and engineering functions that are not getting critical software systems implemented in their areas. It is clear that companies will have to face many difficult systems issues and have to make many difficult systems and operations decisions in the short and long terms.

Many schedule execution problems are caused by the slow transmission of critical data throughout the organization. Software companies around the world have responded to this problem and now offer several features in a transaction processing and control system known as the *ERP system*. Such enterprise-wide information systems attempt to connect all business units in a company and facilitate data acquisition and sharing across all departments and functional areas. Companies using ERP can now share data much more efficiently and make major advances toward sharply decreasing data redundancy. Design engineering can now gain access to the customer information gathered by the sales and marketing departments to make product design decisions. Sales and marketing personnel can electronically communicate product configuration data, cost estimates, price quotations, and forecasting information with manufacturing and engineering. This ability provides a significant improvement in the ability to schedule the shop floor finitely. In this view, ERP is much like MRP II but with some new features that make it more attractive for all the functional areas.

---

*See chapter 9 for a description of *kanban*.

Figure 8–3 Today's Typical Systems Environment.
(Source: SIBC Corp. Technology Course Series 2003.)

- IT strategies have been replaced with survival tactics.
- Many incomplete projects are showing no paybacks.
- Needed technologies are only partially implemented or interfaced.
- Budget constraints constantly force redirection of resources.
- Business acquisitions and divestments cause major steps backward.
- Y2K time constraints caused ERP implementations to be only partially implemented—then budgets were cut.
- IT backroom operations personnel are forced to run a patchwork of systems on multiple platforms.
- Short-term needs interfere with obtaining quality deliverables.
- Sales/engineering/marketing/purchasing/field service feel they have been abandoned by cutbacks in e-commerce applications.
- ERP systems are undergoing "second-wave" implementations.
- The user/customer/client isn't well trained.
- Outsourcing is frequently disappointing to users and IT management.
- Hiring, projects, new software, and integration are all on hold until the conflict in Iraq is over and the economy recovers.

Poorly integrated IT infrastructures can make a company's information systems seem like "islands within the enterprise." Consultants from Grant Thornton report,

The systems implemented to control vital manufacturing functions often cannot communicate with one another. Data and information do not flow where they are needed, but must be shared through manual processes and/or by rekeying data into multiple systems within the company. Poor integration can lead to critical omissions, erroneous data, production delays, and ultimately reduced profit.

Data Collection and Control Issues

ERP systems have helped solve many of the data acquisition and consistency problems encountered by medium-size to large manufacturers. Kurt Freimuth, president of Factory Floor Solutions, has found that problems may persist even after the implementation of an ERP system. Personnel often have to ask, "How did we make that part the last time? Sam always did it this way, but what do we do now?" The Deloitte Touche Tohmatsu consulting firm has diagnosed some of the problems with ERP implementations in its tool kit titled "Second-Wave." This firm's assessment of many ERP implementations is that they are only partially completed and lack the data accuracy and integration that helped to sell the systems to company officers in the first place. This firm now has a large contingent of consultants performing "second-wave" implementations and cleanup tasks at these ERP sites.

MRP II is an important component of ERP systems that provides critical information to all areas of an organization, from design engineering to sales and marketing, accounting, inventory management, and shipping. While important to the daily operations of a manufacturing concern, MRP II is more or less an accounting and scheduling system oriented to ship on time, every time. In many midsize companies, it is an information system that runs daily or weekly and is not real time. Variations on the shop floor between the MRP II system and actual practice may not be discovered or corrected until the next time the system is run. The biggest problem with using MRP II alone is that it is based on two major assumptions: (1) the based data are accurate, and (2) the lead times are unchanging. The key to accurate data on the shop floor lies in the base files that support the entire planning hierarchy within an organization.

In practice, there are several fairly static base files used by MRP II, such as the bill of materials, part and item master, inventory location master, work-center file, routing files, vendor master, lead times, and employee master file. These files constitute the foundation for all planning activity and provide the reference foundation for planning; however, they are not updated regularly in many systems. The information contained in the base files may be gathered only once every one to five years. The resulting information that supports the entire ERP system may be based largely on static historical data rather than on dynamic current data.

Sophisticated and highly integrated systems provide the greatest value when the transaction data are timely and highly accurate. Transaction data include material movement, labor reporting, work-in-process changes, quality control data, and hundreds of other transactions that affect productivity and accounting records and controls. The currently available technologies for data collection can eliminate inaccurate warehouse records, invoicing errors, and similar transaction-related problems. Manufacturing environments that utilize modern data-collection technologies make life a lot easier for workers, material planners, and operations managers in these facilities. Bar-coding technology can improve transaction data accuracy significantly. Wireless technologies such as radio frequency identification (RFID) are being used to help to improve the overall control of goods throughout the manufacturing operations and in transit.

The Information System Integration Nightmare

Companies often operate with systems that have evolved throughout many years. Many times, companies are operating several different software packages to satisfy the need for information and to help them plan and operate the business. Likewise, systems are found to operate on different computer hardware platforms (mainframe, midrange, and more recently networked PCs). The information systems group in many companies has custom designed and programmed elaborate and complex support systems to link the software programs, databases, and systems used in the different functional areas. Figure 8–4 provides a system map depicting various modules on different platforms with varying levels of interconnection at an actual case study company.

The challenge facing many companies with systems at all levels of integration and installed across various computer platforms is how to pull them together and how to keep them running. These two tasks are not easy for the information

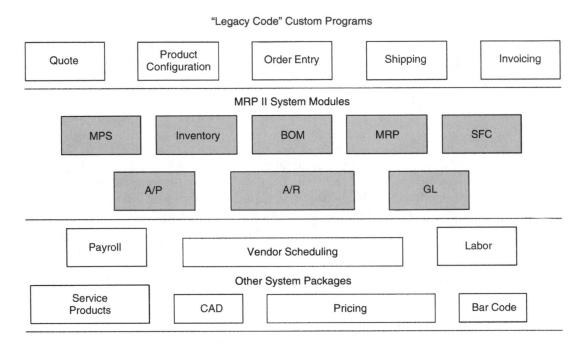

Figure 8–4 A Multiplatform System Environment Map.
(Source: SIBC Corp. Technology Course Series 2003.)

systems department, and they can become more difficult. The technology advances are relentless and powerful, yet many companies do not take advantage of the new systems because of problems that plague their current operations. Companies have shortchanged the investments that should have been made in platform consolidation and software upgrades. Company executives complain about their large investment in systems that do not show any improvement to the bottom line of the corporation. The risk to those progressive companies that have implemented ERP is that they too may not achieve the original payback projections if the systems are implemented only partially and lack the data accuracy and integration necessary to be classified as a class A system.

8–3 THE DECISION TO IMPLEMENT AN ERP SYSTEM

Companies that choose to implement an ERP system face several critical decisions. There is much more to this process than selecting a software package. The rapid growth of information systems technology and the explosion of the Internet have made this process more complex. Patrick Delaney, of SIBC Corp., identified several of the key questions that must be addressed:

What are the hardware and software requirements?
Will the system architecture be client-server?

Will there be an extranet Web site?

What bolt-on software modules might be required?

How will data be warehoused and mined?

What are the e-commerce applications of the system?

What business entity partitions will be needed in the system?

What network bandwidth will be needed to support a responsive system?

What is the migration path from where we are now to the future system?

Many of these questions are new to today's executives, but they must be answered correctly and with an eye toward the future. These are critical questions and the way they are answered can have a significant impact on the potential for success and survival of the company.

The decisions made in the past ten years related to systems and technology have a direct impact on companies now. Decisions made today will have far-reaching effects on the company well into the future. For many companies, the picture today is not optimistic. Bankrupt software companies have left system users with no support, system maintenance, or software upgrades. Complex interfaces written to link software packages on different systems often complicate migration to a new package. Decisions made in the past have limited the network capability to expand for e-business, Internet applications, and mobile access. Discontinued or obsolete hardware may make migration to a new system expensive; examples are reported weekly as companies move to decentralized computing models that require "fat clients" on the desktop. In many corporations, the IT staff members do not have the skill set needed to work in the networked PC environment evolving rapidly toward full e-commerce capability.

Regardless of the difficulty and complexity, many companies will invest in new ERP systems. The systems can help companies sustain a competitive advantage in their industry. ERP systems can help companies manage information, a critical resource, more effectively. Figure 8–5 provides a list of some of the typical information needs of ERP systems. Notice that a lot of information in the list is outside the scope of the traditional manufacturing functions. The information requirements outlined in the figure address the enterprise, not just manufacturing.

Figure 8–5 Information Requirements of an ERP System.

- Sales, customer, and order demand–related information
- Manufacturing resource data
- Inventory status data
- Manufacturing process information
- Internal control and security access tables for client-server
- Cost collection: standard-actual-activity costs
- Performance measurement extracts
- Customer information
- Customer satisfaction information
- Stockholder and treasury information
- Vendor and supply chain detailed data
- Employee human resources data

8–4 IDENTIFYING ERP SYSTEM SUPPLIERS

ERP systems are now available from several software suppliers. These systems run on various computer platforms. Costs for ERP systems vary greatly. Note that specific system implementation cost figures are not typically provided by software vendors. When comparing costs, make sure you know what is included and what is not. Interested readers can find some of the most current information about ERP and available cost estimates for ERP systems by visiting sites on the Internet.

The Internet has quickly become an important source of information about manufacturing systems and the companies that supply them. The sites listed next are expected to remain active; however, in the fast-paced world of the Internet, Web addresses often change. Finding current active sites may require the use of an Internet search engine that will list the addresses of sites of possible interest on the basis of key words you enter.

- *APICS* (*American Production and Inventory Control Society—The Educational Society for Resource Management*). APICS has been a leader in developing the body of knowledge of manufacturing systems such as MRP and ERP: **http://www.apics.org.**
- *The Gartner Group*. This special consulting company studies trends in manufacturing and operations and makes predictions about the future; it is credited with the abbreviation *ERP*: **http://www.gartnergroup.com.**

Many of the new generation of manufacturing systems that were formerly known as MRP or MRP II systems now promote themselves as ERP systems. They still try to utilize a single database. New networking and open computer architectures have made the communication and integration of companywide systems possible. New links that were not possible just a few years ago can now be made to allow links to other systems through application interface programs. The continuing growth of the Windows operating systems and open system architectures have made communications among many of the systems used in manufacturing companies much easier. Interface programs are written to connect the ERP core software and the central database with detailed systems that support the operations throughout the company. This connection allows, for example, data collected by the sales force to be incorporated seamlessly into the central database and to be used for improving planning in ERP. The advancements in the available technology are making meaningful and efficient new links between the functional area programs and the central planning programs possible. Figure 8–6 provides an illustration of these relationships.

Modern ERP systems such as WinMan support the concepts of lean production. Lantech, Inc., a maker of shrink-wrapping equipment in Louisville, Kentucky, worked for nearly two years to become a lean manufacturing company. Lantech redesigned its manufacturing processes to eliminate waste and move to lean production methods. The traditional push-style MRP II system was in place before the lean transformation was replaced with the WinMan ERP system and an effective *kanban* system. The improved work flow and details of the operation have been

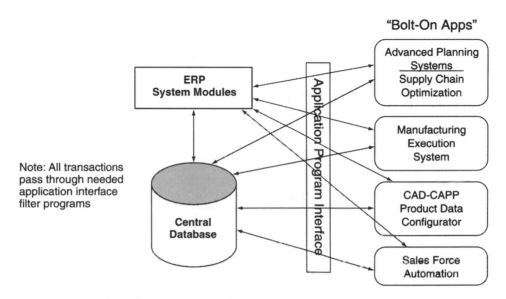

Figure 8–6   Advanced System Architecture "Bolt-On" Applications That Supplement the ERP Modules. *(Source: SIBC Corp. Technology Course Series 2003.)*

described in the new planning system. Lantech used the data fields and planning parameter codes provided in WinMan to match the system to what is actually done on the shop floor. (See appendix 7–3 for a more specific example of planning using WinMan.) WinMan has helped provide system users at Lantech with the data they need and helped to minimize manual reporting to the system. The cost savings realized by Lantech as a result of running an effective formal system have been significant. The work to streamline the processes and eliminate waste throughout the operation also helped Lantech complete the WinMan implementation in only two months, which saved additional time and money. The case story of Lantech can be found in the book *Lean Thinking* (Womack and Jones, 1996).

Refer again to Figure 8–1 and note that there is another three-letter abbreviation associated with ERP: ERO. This subtle change to *enterprise resource optimization (ERO)* is descriptive of the work being done to enhance and develop the concept. It is also a contributor to the jargon problem in manufacturing terminology that can often confuse matters. More abbreviations will be coming and the systems will continue to grow more complex, but do not let ERP/ERO confuse you.

And what can we predict about the future? We can be sure that our systems will continue to change. Computer systems will continue to increase in speed and become more cost effective. At the same time, the systems will become more complex and sophisticated. SIBC Corp. presents a possible scenario for the future in Figure 8–7. ERP systems may grow to absorb the advanced production scheduling (APS) functions that are now considered "bolt-ons" to current packages. The functions now handled by third-party software packages will become part of the true functioning of the ERP system. These changes could take us into a systems

Figure 8–7 Linking Asset Optimization Software to ERP.

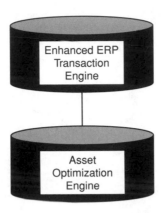

Module functions that were formerly achieved by using "bolt-on" packages such as advanced planning and execution systems, integrated design and process links, configuration management tools, and sales force automation tools are incorporated into the ERP system.

New optimization software with seamless links to ERP provides software tools to help companies get the best utilization of materials, equipment, and human resources.

environment with fewer software providers involved, and they should make the information systems function better able to manage and control the system. The traditional interface programs that have to be written and maintained by the information systems function today will be absorbed into the primary ERP system. This move would really bring ERP up to the level that was intended for it.

With ERP functioning at the highest level and traditional bolt-on applications absorbed and working seamlessly across all the functional areas, it may be possible to move to a higher level of system use and performance. Traditional systems have provided tools for companies to make better decisions about their resources. The tools documented plans and scheduling changes, and the system tools helped companies determine how to change the plans in light of the new conditions. Perhaps we may see new system engines that will help companies optimize their operations by going beyond the current responses to change. Future systems should make it easier for companies to test multiple scenarios and play the "what-if" game more effectively. Future systems may provide users with more options to consider and do it in such a way that considering the options becomes a timely reality.

One of the core benefits of computer-integrated manufacturing (CIM) is an accurate and valid cost accounting system with clean interfaces to the accounting and financial reporting systems. Fully integrated manufacturing systems can be a deterrent to fraudulent financial reporting. Well-managed operations that are controlled by fully integrated systems, regardless of software, are much more likely to be easy to audit and to track costs from one process or operation to another. Cost and transaction data should flow directly from the shipping dock and the shop floor to the accounting system and the general ledger without requiring off-line spreadsheets, manual adjustments, and outright plugs at the end of each reporting period.

Companies can get into trouble and frequently end up having to adjust earnings reports when they get into a pattern of running operations on off-line spreadsheets. Companies that operate with historical cost averages and historical work-center labor costs leave themselves open to off-line tracking of actual data and off-line development of accounting data to adjust the computer system outputs at the end of each period. In some companies, this has become accepted practice, but it results in officer terminations and public embarrassment when surprise announcements are made in the fourth quarter regarding revenue, earnings, costs, inventory, and other key variables. Once this process of adjusting system output reports is accepted as valid practice, then the outputs of the entire computer system can be questioned. If the system output reports for the general ledger system are in error, then the output reports are probably not correct for the cost accounting system, the material system, the quality system, or any other system used to run the operations.

It is clear that computer-based systems are here to stay in manufacturing. They bring great opportunities but will place great demands on the people in the organizations who use them and maintain them. The next several years should be exciting in manufacturing.

8–5 DEVELOPING TECHNOLOGIES: CONVERGING AND ENABLING

The advancements in technology and the development of modern pull systems for manufacturing are forcing several systems to be combined. Islands of technology that once separated the finance, engineering, manufacturing, and sales and marketing functions are being networked together. Figure 8–8 lists several of the primary system groups. No single system now covers all the areas and systems.

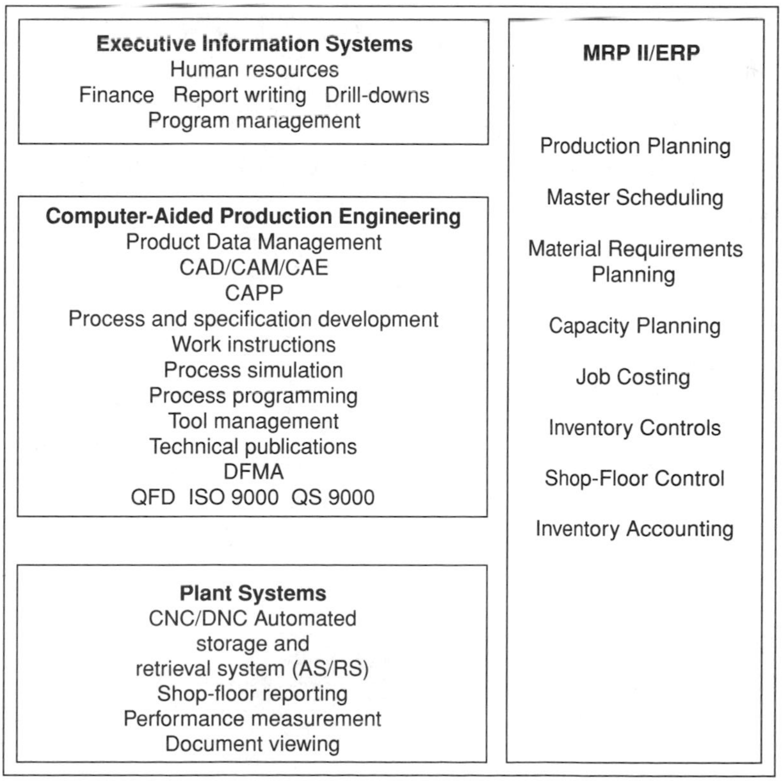

Executive Information Systems Human resources Finance Report writing Drill-downs Program management	**MRP II/ERP**
Computer-Aided Production Engineering Product Data Management CAD/CAM/CAE CAPP Process and specification development Work instructions Process simulation Process programming Tool management Technical publications DFMA QFD ISO 9000 QS 9000	Production Planning Master Scheduling Material Requirements Planning Capacity Planning Job Costing Inventory Controls Shop-Floor Control Inventory Accounting
Plant Systems CNC/DNC Automated storage and retrieval system (AS/RS) Shop-floor reporting Performance measurement Document viewing	

Figure 8–8   Converging Systems for Manufacturing Planning and Control Systems.

The ongoing development of computer operating systems and hardware platforms has made sharing data and improved communications possible. The Windows NT operating system, for example, allows programs to share data, pictures, text, spreadsheet and word processor documents, and even sound and video files. Report-writing features built into modern systems allow users to prepare database queries that in the past required special attention from someone in data processing. Drill-down features allow users to work through system fields to find the real drivers of schedules and costs. Production engineering functions related to design and documentation of parts and processes are much more efficient when data sharing is possible. Modern quality programs such as ISO 9000 (the international quality standard) and QS 9000 (the automotive industry quality standard) require extensive documentation and controls that are enhanced through the use of new computer-based systems. Systems are designed to take advantage of the emerging technologies to improve communication and break down barriers that cause confusion and add cost to the operation, and these systems help companies become more competitive.

Information will be a significant driver of the factory of the future. The newly linked programs, systems, and the central database will provide a framework for even more computer-based integration. The key systems issues will continue to address the improvement of the company's data accuracy and its ability to communicate internally and externally. The management and control of the company's data cannot be left to chance. The data and information contained in the company's

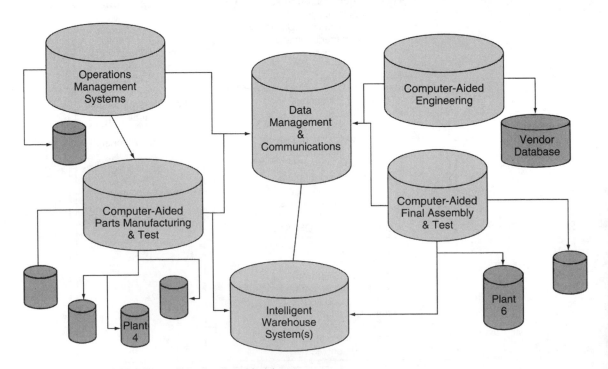

Figure 8–9   Factory-of-the-Future Database Model.
(Source: Adapted from Fundamentals of Computer-Integrated Manufacturing, *Prentice Hall, 1991.)*

systems are real assets that have value and must be protected. Figure 8–9 provides a factory-of-the-future vision according to General Electric. Each site is seamlessly connected to the major databases that provide data to all of the plant sites. Data and information flow throughout the organization in a fashion similar to the way the human brain and nervous system send signals and commands throughout the body. The environment is paperless, intranet enabled, extranet dependent, client-server structured, online, and real-time enabled with fully staffed teams of "knowledge workers" maximizing the flow of the information pipeline.

Many companies have seen their systems grow and evolve during the past fifteen to twenty years. The result is often a messy system that has been patched, modified, bridged, enhanced, customized, and so forth. This type of system is extremely difficult for the information systems groups to work with and maintain. In these environments, the selection of a single new software package is also difficult. Software packages have often been developed to solve a particular issue or support a particular functional area. Many of the leading packages are recognized for their functioning in a particular area. In a growing effort to provide a total solution for a company, software packagers have added features that fall outside their niche so they can round out their approach to the marketplace. The result is a package that is strong for some functions but very weak for others.

Some major corporations have elected to implement systems that utilize the best features from several different commercial packages. Figure 8–10 provides an

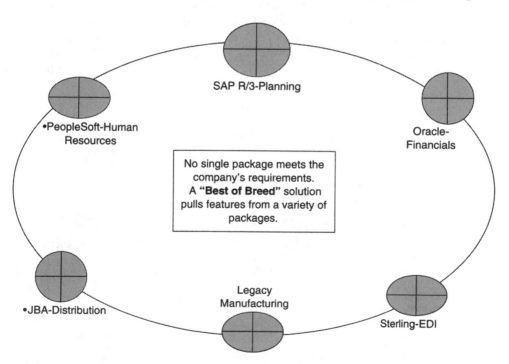

Figure 8–10 Single Corporation Selecting Diverse Applications.
(Source: SIBC Corp. Technology Course Series 2003.)

illustration of this approach. In the figure, the company could not find a single system that covered all its needs. It did recognize, however, that several packages are known for their strengths in particular areas and that by pulling in the best of the available systems, it could achieve the best system. It will be interesting to see if this approach works effectively, or if the company has defaulted to a politically complex situation in the short term. Working with only one central package and its related interfaces is a huge challenge. Trade-offs must be considered. Companies must not lose sight of their competitive and strategic reasons for investing in information systems.

The system users in the functional areas may become closely attached to their systems. A change in how the work gets done, even if it represents an improvement, can turn people against it. The system implementation team will need to develop carefully a plan for communicating the need for change to the people involved. It will be critical to get the end users to accept the new system and thus ensure its success.

The systems that will drive the factory and enterprises of the future are continuing to be developed. Figure 8–11 presents some of the emerging applications that will become part of the integration challenge of the future. Nearly instantaneous communication from one point to any other around the world is becoming a reality. The growth in communication and the information it will make available to companies will highlight the challenge. Communications and systems will no longer have an internal company focus. Much of the growth will come as companies attempt to reach out to customers and suppliers along the length of the supply chain. The growing customer focus will drive companies to use the emerging technology for becoming more responsive and for making business easier to conduct. The amazing growth of the Internet has only begun for industrial applications. The specific applications are emerging slowly, but the direction toward doing more business online, for example, is clear. The technology is already allowing

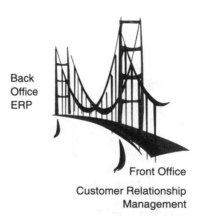

Back Office ERP

Front Office
Customer Relationship Management

- Sales Force Automation (SFA)
- Laptop Configurators
- QA-Remote Diagnostics
- Global Dealer Pipelines
- Web-based Procurement
- Supply Chain Management
- In-transit GPS Tracking
- Legacy Data Mining
- E-Commerce Ready Applications
- Electronic Catalogs
- Supply Chain Optimizers
- APS "Bolt-On" Applications
- Finite Execution Systems
- RF Data Collection
- Mobile Access

Figure 8–11   The Integration Challenge of the Future.
(Source: SIBC Corp. Technology Course Series 2003.)

- **Distributors and retail stores may no longer add value.**

- **Internet purchases and EC links between end customers and OEMs may eliminate the need for a midlevel buffer.**

- **Combining "Quick Response Manufacturing" with electronic commerce may eliminate a major linkage in the supply chain, and reduce costs.**

Figure 8–12   E-Commerce Issues.
(Source: SIBC Corp. Technology Course Series 2003.)

companies to do more and do it faster, with virtual plant tours of remote sites, transmission of complex engineering graphics, direct customer access to product configuration systems, and order tracking through each link in the supply chain.

Figure 8–12 highlights several of the critical issues related to e commerce. E-commerce is already changing the way we communicate and do business outside and inside the firm. We may no longer have retail stores and displays if the value that they provide to the customer can be provided more economically using the Internet. The traditional distribution channels that linked suppliers with distributors, re-sellers, and the customers may be drastically shortened. A new computer-based warehouse and distribution system may emerge for many companies. The pressure to respond quickly in a competitive environment will compel companies to explore and implement new communications technologies. These technologies will directly affect each company's information and planning systems.

The emerging technology of "portals" on the World Wide Web supports e-commerce trading exchanges. Web portal marketplaces are being established to link groups of buyers and sellers of particular products or families of products. The portal is more than just another site on the Web. E-commerce portals provide easy and secure access to the marketplace and support tools designed for the customers and suppliers being served. A map of the critical Web portal functions supporting the e-commerce procurement process is presented in Figure 8–13.

"Virtual manufacturing" may take on new meaning as companies link to-gether to share information electronically and respond to manufacturing orders initiated over the Internet. Imagine that a company needs to have a particular part produced with the highest quality and at the lowest cost. The specifications and order details could be posted to the Internet at a site where approved suppliers can find the demand. Bids for producing this part could be accepted in a defined bid window. Proposals from suppliers could be presented and managed electron-ically. The parts made to the company's specifications could arrive at a specified assembly operation at a given time, and the finished products can be completed and shipped. Thus, traditional orders and paperwork could be replaced by inter-active systems using the Internet.

Figure 8–13 E-Procurement Web Portal.

(Source: SIBC Corp. Technology Course Series 2003.)

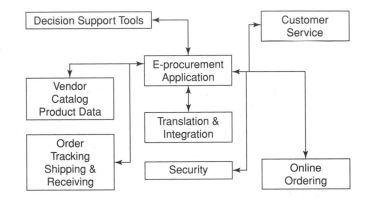

Connecting the field sales and service personnel with the main corporate system is an exciting possibility with the advanced information systems and communications technologies. These systems can dramatically change the way a company does business. Imagine the power and potential competitive advantages that can result from the services listed in Figure 8–14. Customer questions can be handled in real time with real answers. Order-entry people can have access to a customer's order history, which will make it easier for them to reorder. Configuration issues and option selections can be managed better before production begins and errors can be

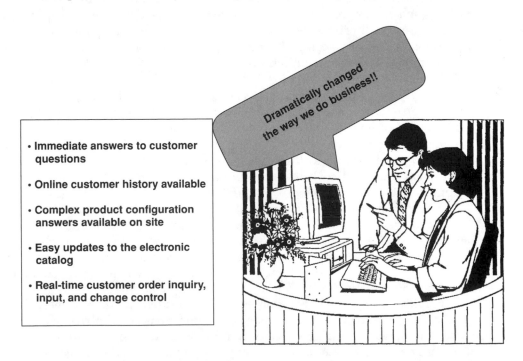

Figure 8–14 Sales and Field Force Automation.

(Source: SIBC Corp. Technology Course Series 2003.)

found on the shop floor. Printed catalogs that become outdated can be replaced by electronic versions maintained on the Internet. Changes to customer orders can be managed better, and the implication of changes on the production process identified sooner. Companies with this improved communication capability will have the potential to become the preferred supplier and thus enhance their competitive position.

8–6 INTEGRATING SYSTEMS TO MANAGE DESIGN DATA

A manufacturing company's product is the source of its profits; a key factor to success is being able to manage all relevant product data. The company must be able to harness its intellectual property and put the product at the center of all processes. By doing so correctly, it will have control over its products, be able to manage the work flow throughout the entire enterprise, build an intelligent data structure, and secure these data. Product data management (PDM) solutions, correctly implemented, aim to do just that, with the ultimate purpose of increasing net profitability. *Product lifecycle management* (PLM), defined by writers from IBM Corporation in their trade paper titled "What Is PLM?" is a business strategy that puts the product, the source of a company's profit, at the center of everything the company does. PLM addresses the broad scope of these data decisions from concept design to the delivery of products to the customer.

The study of PDM and PLM systems began at Purdue University with support from the IBM Corporation and Dassault Systèmes, S. A., and the creation of the Digital Enterprise Center in the School of Technology. Graduate students have worked with advanced software to develop a working model of a manufacturing enterprise in a complete e-business environment. Donald Lucas and Hugo Ramos, graduate students at Purdue University in the School of Technology, have had the opportunity to work in the Digital Enterprise Center and study PDM and PLM applications for manufacturing. This section presents information from their research that provides insight into the data management needs of industry and software applications for PDM, and the broader application of PLM.

Manufacturers are placing the customer first and doing whatever is necessary to meet the customer's needs. Staying competitive requires companies to find ways to be more efficient from the concept stage through the delivery of the finished good. Suppliers are seeking to remove redundant or lost-time operations. The decisions made early in the development stage have far-reaching effects. Large percentages of the costs incurred across the life of the product may be the result of decisions made during the product development stage. New cost-reducing approaches are evolving in PDM, PLM, and ERP areas.

In the twentieth century, engineers issued a blueprint drawing to communicate design details, but in the twenty-first century, the approach most often uses computer-aided design (CAD) digital models. In its short history, CAD has evolved into a powerful communications tool that promotes concurrent design, manufacturing, and service applications with the ease and speed of a keystroke. CAD programs now feed design information into computer-aided engineering (CAE) and computer-aided manufacturing (CAM) software packages.

The development of a product creates new data that are needed by many different functional groups throughout the enterprise. The data have typically included drawings and paper records kept in various engineering, quality, and service storage files. With the wide acceptance of computers and the growing number of software packages available, various departments now have their own digital information files. These data, however, are frequently available to only a limited number of people and functional areas.

A Central Data Repository—A Foundation of CIM

A controlled central source of information for use jointly by all team members has been a foundation concept of CIM from its beginnings. The centralized data concept should help eliminate delays and data errors and facilitate the sharing and reuse of data and information. A central data repository should produce cost savings for the enterprise. Information gleaned from the product development cycle transformed into knowledge can benefit the entire enterprise. Advanced communications systems should distribute information to multiple locations and enable global collaboration. Accurate, timely, and accessible data should speed design decisions, help optimize production, improve quality, reduce the time to market, help control costs, and use the system information to gain a competitive edge.

The following expectations drive the requirements for a central data resource repository (CDRR):

- Provide a complete repository for all planning data, which gives a consistent description of the entire planning process, and a common planning environment
- Support new planning projects and modifications of existing production systems
- Help manage and control changes to data relevant for planning
- Provide a knowledge base for subsequent projects
- Enable the creation and evaluation of planning alternatives
- Permit analysis of production from different manufacturing processes

"Knowledge is information that is organized, evaluated, valuable, and available to the appropriate people at the appropriate time" (IBM, n.d.). The manufacturing companies that accept and use this new technology will have a competitive edge over those that do not.

Product Data Management

PDM, as first conceived in the late 1980s, was a design-centered package to manage product design data. Originally, the focus of PDM systems was on engineering and manufacturing and the generic theme of capturing product data from the product's original release to the data's obsolescence. The key concerns were managing the initial release of data to manufacturing and managing the engineering change order process initiated by manufacturing. A basic PDM system provides secure information, stores and organizes files, controls revisions, manages communications, and aids in application integration.

PDM is built around the following five functions and characteristics (Lui and Xu, 2001):

1. Data vault management, which provides information storage and retrieval
2. Process and procedures management, which provides product data-handling procedures
3. Product structure management, which handles the engineering bill of materials (BOM) and product configurations
4. Part classification management, which provides information on reusing standard designs
5. Project management, which provides structure and allows coordination among process, resource, and project tracking

The central data repository (vault) maintains the product data, controlled by the current acting engineering authority. If possible, data should be stored in a neutral format that can be accessible to a broad spectrum of users. Authority for the data transfers from the engineering disciplines to other change authorities to make any necessary modifications as the product lifecycle progresses from a design to a mature product.

Computer-based PDM solutions help engineers become more efficient by providing software tools to help organize, control, administer, and map vital data. Modern PDM issues deal with collaboration to expedite time to market and enable concurrent engineering. Access and retrieval of information are now possible by using the infrastructure of the Internet. Internet-based technology allows companies to collaborate remotely and concurrently, not only among corporate departments but also with the entire supply chain from the suppliers of materials to the customer. Valente (2001) presents PDM as a system by describing three key functions in addition to the data storage vault. These functions include configuration management, process management, and structure control.

Configuration management provides control over products by using procedures, guidelines, and technology. This function of PDM enables synchronization among parts, documents, and change data. *Process management* functionality deals with the method of data exchange from one group or participant to another for action according to a set of procedural rules. The *structure control* function of PDM addresses part-specific relationships. The system links every part listed on a BOM with data related to that part. PDM systems also provide the storage vault function used to secure the data. The integrity of the data is secured and monitored, and all changes are documented.

PDM systems address the creation, flow, control, and use of critical product and process data elements. The diagram in Figure 8–15 provides a graphic representation of the central role of a PDM system. A computer-based system using advanced software that provides the needed linkages and controls is now available in the marketplace.

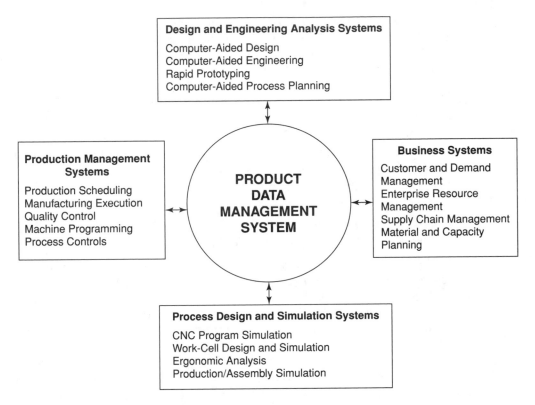

Figure 8–15   PDM System Interface Model.
(Source: Developed by the Digital Enterprise Center at Purdue University.)

Web-Centric PDM Functionality

The process by which a company communicates and shares information across the extended enterprise, not just with other corporate departments but also suppliers, resellers, and ultimately the customer, is known as *collaborative product commerce*. The expanded functionality of Web-enabled PDM (also known as *Web-centric PDM*) makes true collaboration more of a possibility now than ever before. Web-enabled PDM promises a new facet in manufacturing. Seamless business process integration and supply chain collaboration will link the PDM system to the ERP system to produce a more empowered virtual enterprise.

The Internet provides an infrastructure that is relatively easy to use, reliable, and readily available. There has been a growing need to expand PDM functionalities to the Web because product definition and development has been extended beyond the enterprise's geographic boundaries, a growing percentage of manufacturing is being outsourced, and supply chains are global and more complex.

In generic terms, Web-centric PDM is an expanded PDM application that uses standard Web browser and portal technologies to allow users to have access to product information remotely. The user will interface with the enterprise and be able to control and manage product data over the Web. This technology allows storage of files on the Web for use by remote team members in synchronicity. Team members utilize standard Web browser protocols to gain remote data access and search and find files and attributes such as the drawing name, a revision, a part number, and so on. The Web-based system software can monitor the history of design document revisions and file downloads. Web-enabled PDM allows the user to share engineering change processes with the members of the supply chain in real time. A business can benefit because Web-enabled PDM helps streamline processes throughout the product lifecycle, enables access from virtually anywhere with just a Web browser, and saves valuable search time when the user is looking for documents and drawings.

The Product Lifecycle and the Virtual Enterprise

The initial focus of PDM has grown to include the product's entire lifecycle, from concept definition through production, sales and service, and finally product disposal. The model of the product lifecycle shown in Figure 8–16 considers the relationships of concept definition, production, and customer service. *Concept definition* includes all the processes that bring the product from being an idea to becoming a visual design, usually a three-dimensional (3-D) graphic model. This includes planning, concept engineering, product design, and design prototyping. *Production* refers to all the planning and processes that are required to make the product, including manufacturing engineering, manufacturing planning, product testing, and quality control. *Customer service* refers to every business process needed to deliver the product to the customer and to support the customer until the product has no further use; this includes marketing, sales and distribution, maintenance and repair operations, and finally disposal and recycling of the obsolete product.

At every stage of the product's lifecycle, information regarding the product is being used and new information is being produced. During the first stage, CAD files are being created, along with assembly drawings, the BOM, specifications, engineering analysis data, and so on. During the second stage, all these data need to be readily available for use by production engineering and operations. CAD files and the BOM are key inputs to manufacturing engineering needed to plan processes and production. Process plans and customer requirements drive production operations. Data and information important to the enterprise continue to develop even after the product has shipped to the customer.

The PLM Solution

One of the challenges a company faces is closing the communication gap between all of its components and partners to allow for concurrency, integration, and collaboration. Linking all of an enterprise's business processes seamlessly has been a dream of CIM from the beginning. In practice, however, this has proved to be a daunting task. A PLM system can provide a framework that helps a company address what products it needs to build, how the products will be managed and collaboration facilitated, and the plans and processes that will drive production.

Dassault Systèmes is a world leader in the development of PLM systems. Dassault's vision for an integrated system is to enable people to invent and create

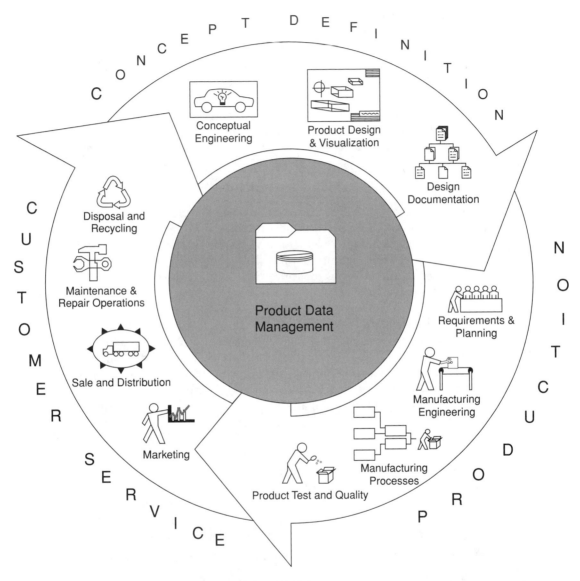

Figure 8–16   Product Lifecycle Management and the PDM Core.
(Source: Used with permission of H. Ramos, 2003.)

innovative products and to simulate the whole product lifecycle. Dassault's primary business focus is PLM and its mission is "to provide companies with e-Business solutions to enable people to invent and create innovative products and simulate the entire product lifecycle." Dassault achieves these objectives and seamless future objectives through combined methodologies, software solutions, and services to design, simulate, optimize, control, and monitor all production means.

Dassault's PLM system is design-, process-, and market-centered through the integration of three software entities: CATIA, ENOVIA, and DELMIA. The design system is configured around CATIA, high-end 3-D CAD software. CATIA is a tool for authoring the product design. ENOVIA provides access to data and enables collaboration, visualization, and analysis. DELMIA supports process design and evaluation through digital manufacturing and graphic simulation. Digital manufacturing addresses the authoring of the manufacturing process. The CATIA, ENOVIA, and DELMIA systems are all based on an open system architecture that enables the sharing and reuse of data across the enterprise and throughout the product lifecycle. The software functionality is developing continually. Visit the ENOVIA Corp. Web site for the latest information (**http://www.enovia.com**).

The central "product/process/resource hub" (PPR hub) developed by ENOVIA provides a unified and open model of the products, manufacturing processes, and resources of the enterprise across the product lifecycle. The PPR hub supports integration with Supply Chain Management (SCM), Customer Relationship Management (CRM), and ERP systems. Figure 8–17 provides a

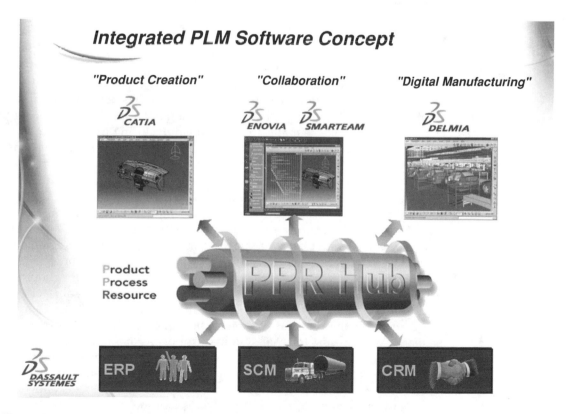

Figure 8–17   Integrated PLM Software Concept.
(Source: Used with permission of ENOVIA Corp. 2003.)

graphic representation of the PLM system concept and the critical PPR hub. Note that in this system model, the PPR hub is dynamic; it is intended to provide links to data and information that flow through the product development and production pipeline. The integrated systems from Dassault support virtual product modeling that builds on industry-recognized best practices. This exciting new technology enables collaboration, visualization, simulation, and evaluation that lead to better and faster business decisions throughout the life of a product.

PLM supports a collaborative environment where team members, even from remote locations, can tap into the creative process by getting the critical information fast and being able to share ideas and knowledge in real time. PDM-PLM systems address the issues of harnessing and organizing the mass amounts of data a company has to manage. Modern systems, however, respond to many more problems than just managing engineering data and processes. A properly implemented PDM-PLM system is expected to provide better control of projects, reduced time to market, seamless product information sharing and improved collaboration, security for critical data, enabling support for decentralized operations, and an increased capability for handling product and process complexity.

8–7 SUMMARY

Systems for manufacturing planning and control can be expected to evolve and grow as new computer and information systems technologies develop. Enterprise resources planning (ERP) systems represent the leading edge of applied technology. The new generation of systems that is following manufacturing resource planning (MRP II) is helping companies realize the unfulfilled promises of earlier systems. Material requirements planning (MRP) could not be effective without reliable schedule inputs. The expected benefits from uniting the functional areas with a formal system may have been lost as a result of the apparent manufacturing focus of MRP II. The ERP systems have the potential to really integrate the functional areas of manufacturing, the supporting areas that make up the enterprise, the customers who are served, and the suppliers who provide the materials. The emerging systems for product data management (PDM) and product lifecycle management (PLM) offer significant opportunities for sharing and reusing data throughout the enterprise and the life of products from design concept through disposal.

Computer-based information systems for manufacturing will continue to develop and increase the demand for integration and sharing of information. Systems emerge from needs. Internal systems supporting manufacturing operations now include ERP, product data management (PDM), and a host of others. Customer-focused systems address the issues of sales, order management, customer relationship management, and more. Supplier-directed systems allow for the management of vendor resources, advanced planning and scheduling, and more. SIBC Corp. presents a picture of the SCM integration challenge in Figure 8–18. The integration and effective linking of these software systems across the

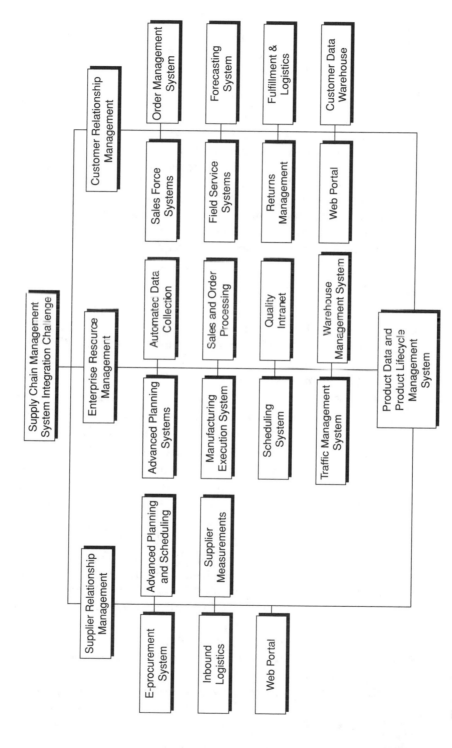

Figure 8–18 Supply Chain Management Software Integration Tree.
(Source: SIBC Corp. Technology Course Series 2003.)

supply chain will challenge the information technology (IT) function and the overall computer-integrated manufacturing (CIM) system.

This chapter gave the reader a look at the leading edge of technology and provided some speculation about the future direction of computer-based systems for CIM in manufacturing. The one thing that is certain is that all of the systems presented in this chapter will soon be replaced by a newer, faster, and better set of systems. Companies will continue to access and use information about their operations to sustain their competitive advantage. Careful management and planning will be needed as older systems are upgraded or replaced as the technology evolves. The decisions that are made by the leaders of the operations and information systems areas will have significant impacts on the overall performance and profitability of the company. Manufacturing professionals will be challenged as they try to keep up with the continuing flood of changes and enhancements to available manufacturing planning and control systems.

BIBLIOGRAPHY

Cox, J. F., and J. H. Blackstone, eds. *APICS Dictionary.* 10th ed. Alexandria, VA: American Production and Inventory Control Society (APICS)—The Educational Society for Resource Management, 2002.

Enovia Corp. **http://www.enovia.com.**

Factory Floor Solutions, Inc. **http://www.factoryfloorsolutions.com.**

Foston, A. L., C. L. Smith, and T. Au. *Fundamentals of Computer-Integrated Manufacturing.* Upper Saddle River, NJ: Prentice Hall, 1991.

IBM. Retrieved February 26, 2002, from **http://www-3.ibm.com/solutions/plm/.**

Lucas, D. "Product Data Management." Unpublished project report for TECH 590 independent study course, Purdue University, School of Technology, 2002.

Lui, T., and X. Xu. "A Review of Web-Based Product Data Management Systems." *Computers in Industry 44,* no. 3 (2001): 251–62.

Pond, K., A. Bond, and A. Dailey. *On-line Research Note, 1998 EP Large User Magic Quadrant.* Boston: The Gartner Group, 1998.

"Powering Product Information Collaboration from Concept Through Realization." *IBM Product Lifecycle Management.* Retrieved July 10, 2002, from **http://www-3. ibm.com/solutions/plm/pub1/05256965005be62f/8/9b5f111c734b05bc87256a6a0001 4ac8.jsp.** *

Ramos, H. X. "PDM: The Core of Collaborative Commerce and Its Link to ERP." Unpublished project report for TECH 598 directed project, Purdue University, School of Technology, 2003.

SIBC Corp. *Technology Course Series.* Evanston, IL: SIBC Corp., 2003.

---

*This white paper is no longer available at the IBM Web site. Please refer to all other IBM PLM white papers at **http://www-3.ibm.com/software/applications/plm/library/.**

TTW, INC. *WinMan.* Reston, VA: TTW, Inc., February 2003.

Valente, T. "Product Data Management: It's More than Just the Latest Fad." *Printed Circuit Design 18,* no. 1 (January 2001): 22–26.

"What Is PLM?" *IBM Product Lifecycle Management.* Retrieved July 10, 2002, from **http://www-1.ibm.com/solutions/plm/doc/content/resource/thought/ 758034113.html.**

WOMACK, J. P., and D. T. JONES. *Lean Thinking.* New York: Simon & Schuster, 1996.

QUESTIONS

1. What is *product data management* (PDM), and how do the accuracies of the bill of materials (BOM) and inventory affect PDM?
2. How can the PDM function be automated?
3. What is *product lifecycle management*, and how does it integrate with other computer-based systems supporting a manufacturing enterprise?

PROJECTS

1. Enterprise resources planning and related planning and control systems are changing rapidly. You can find some of the most current information possible on the Internet. Complete a key-word search on the Internet for one or more of the new systems for enterprise planning and control. For example, select a search engine such as AltaVista and search for information on sales force automation or advanced planning systems. You should find links to some of the most current publications available. Write a short summary about your findings.
2. Visit the Internet sites listed in section 8–4. Write a summary outline of the key issues discussed and presented at the sites.
3. Visit the APICS site on the World Wide Web and research the current information on three commercial systems for MRP II-ERP. Choose your systems first on the basis of the price of the software and include systems that fall into the high-, mid-, and low-price ranges. Remember that MRP-ERP systems are expensive. Even a low-priced system may have a software cost of $2,000 or more per seat.
4. Think of the functions required within the manufacturing enterprise and along the supply chain. Prepare a graphic diagram that shows the relationships of the functions and the computer-based systems used to manage and control them.

The Revolution in Manufacturing

OBJECTIVES

After completing this chapter, you should be able to do the following:

- Describe the revolution that is taking place in manufacturing in response to the increased competition of products made in Japan
- Describe the features that can be expected to be found in effective just-in-time (JIT) and lean production systems
- Give examples of waste in manufacturing operations and make recommendations for change that can lead to the elimination of waste
- Describe the way a manufacturing resource planning (MRP II) or an enterprise resources planning (ERP) system can work in conjunction with JIT and lean operations
- Discuss ways that the concepts of lean manufacturing can be applied to make manufacturing support systems for purchasing and accounting more effective
- Know where to look for additional information on this expanding topic

Computer-integrated manufacturing (CIM) is in many ways a response to changes that have rocked the foundations of manufacturing operations. This chapter covers the big picture of manufacturing. The discussion may seem to be a long way from the shop floor, automated machine tools, and computer-aided systems for manufacturing. In fact, some critics may say that this chapter does not even belong in a book about CIM. Nevertheless, the growing interest of company executives in just-in-time (JIT) manufacturing, the production system of Toyota, and lean production is part of the manufacturing revolution. In many cases, these new systems seem to be replacing computer-based systems. We believe that CIM is a vital part of the revolution taking place in manufacturing and that computer-based systems (when properly planned and implemented) can be a complement to other improvement strategies.

Competition from Japan

It was great to be ranked number one. The manufacturing efforts of companies in the United States were number one for more than twenty-five years following the end of World War II. The United States was the undisputed manufacturing leader, and the world knew it. Some writers believe that there should be an asterisk by all of this in the official manufacturing record book. For most of the record period, the United States had no significant competition, so perhaps we were number one in the world by default. Still, the number one ranking was good. Competitive goods from companies in Japan and others in the Far East were considered by many consumers to be little more than cheap copies of U.S. goods. Cars, for example, produced in Japan by Toyota and Honda in the mid-1970s were considered by many U.S. consumers to be a joke.

The joke, people learned, was not really funny. Once consumers took a test drive, they had a surprise. The Japanese cars had surprising performance, great fuel economy, new standard features, and styling that made consumers reconsider buying a car from one of the "Big 3" U.S. automakers. Buyers of the Japanese cars found new levels of quality and overall satisfaction that pulled them out of the market for a traditional American car. The game had suddenly changed, and manufacturers in the United States were not ready to play. It took the U.S. auto industry nearly ten years to make a serious effort to respond to the new competition. The book *The Machine That Changed the World* (Womack, Jones, and Roos, 1990) describes in great detail how the auto industry was changed as a result of the new competition.

9–1 JUST-IN-TIME MANUFACTURING

The term *just-in-time (JIT)* is defined in the American Production and Inventory Control Society (APICS) dictionary* as follows:

> **Just-in-Time (JIT)**: *A philosophy of manufacturing based on planned elimination of all waste and on continuous improvement of productivity. It encompasses the successful execution of all manufacturing activities required to produce a final product, from design engineering to delivery, and includes all stages of conversion from raw material onward. The primary elements of Just-in-Time are to have only the required inventory when needed; to improve quality to zero defects; to reduce lead times by reducing setup times, queue lengths, and lot sizes; to incrementally revise the operations themselves; and to accomplish these activities at minimum cost. In the broad sense, it applies to all forms of manufacturing—job shop, process, and repetitive—and to many service industries as well.*

(Source: © *APICS Dictionary*, 10th edition, 2002)

JIT encompasses every aspect of manufacturing, from design engineering to delivery of the finished goods, and includes all stages in the processing of raw

---

*The *APICS Dictionary* terms and definitions in this chapter are published with the permission of APICS—The Educational Society for Resource Management, 10th edition, 2002.

material. Other names used to describe the JIT process are *short-cycle manufacturing, stockless production,* and *zero-inventory manufacturing.*

JIT is much more than a material-ordering plan that schedules deliveries at the time of need. JIT supports the new approach to value-added manufacturing. The JIT concept, developed in Japan following World War II, focuses on the elimination of all waste. Robert Hall's 1987 classic, *Attaining Manufacturing Excellence,* presents a summary of the "seven wastes" that become the target of elimination in a JIT process (see Figure 9–1). A JIT approach leads to operations having only what is needed, nothing more. JIT takes the "problem of manufacturing" to the extreme level of having *only* the right materials, parts, and products in the right place at the right time. It is a relentless approach where waste at any point in the operation (even at the management level) is not tolerated.

The definition of JIT outlines the two fundamental objectives: waste elimination and *kaizen,* an attitude of continuous improvement. Supporting these objectives are three JIT elements that help management keep the JIT focus and foster an environment conducive to successful implementation: technology management, people management, and system management (Figure 9–2). The graphic in the figure was developed by David W. Buker, a management consultant on manufacturing resource planning (MRP II) and JIT implementation. The following description of the JIT elements was adapted from a manual published in 1991 by David W. Buker, Inc.: *7 Steps to JIT.*

Waste of overproduction: Make only what is needed now—reduce set-up time, synchronizing quantities and timing between steps, compacting layout.

Waste of waiting: Synchronizing work flow as much as possible, balance uneven loads by flexible workers and equipment.

Waste of transportation: Establish layout and locations to make transport and handling unnecessary. If possible reduce what cannot be eliminated.

Waste of processing itself: Question why this part should be made at all—why is this process necessary?

Waste of stocks: Reduce stocks by reducing set-up times and lead times, reducing other wastes reduces stocks.

Waste of motion: Study motion for economy and consistency. Economy improves productivity. Consistency improves quality. Be careful not to just automate a wasteful operation.

Waste of making defective products: Develop process to prevent defects from being made. Accept no defects and make no defects. Make the process "fail safe."

Figure 9–1   The "Seven Wastes."
(Source: Robert W. Hall, Attaining Manufacturing Excellence, *The McGraw Hill Companies, 1987.)*

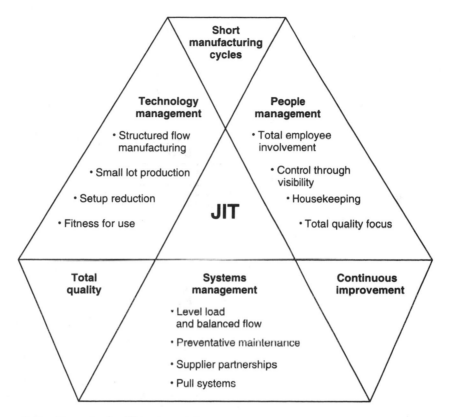

Figure 9–2 Elements of a JIT Implementation.
(Source: Courtesy of David W. Buker, Inc.)

JIT Elements

Technology Management. The first JIT element, *technology management,* calls attention to the production environment and emphasizes the need for a responsive manufacturing system. A responsive production system results when the following four areas are addressed.

Structured-Flow Manufacturing. In flow manufacturing, the machines and work cells are organized and grouped to maximize the velocity of parts through production and to minimize the transportation and queue time for parts. The three types of production layouts shown in Figure 9–3 illustrate this concept. In the initial layout (a), the ratio of value-added work to part movement is very low because machines are grouped by function. The structured-flow layout (b) provides increased throughput because the machines are organized by product process and assembly requirements. In the last layout (c), the structured flow has additional refinements, with consideration of group technology principles.

Figure 9–3 Structured Production Flow: (a) Initial Layout; (b) JIT Layout; (c) JIT Layout with Group Technology.

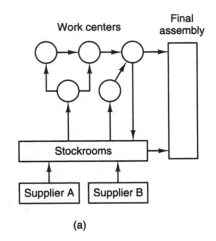

(a)

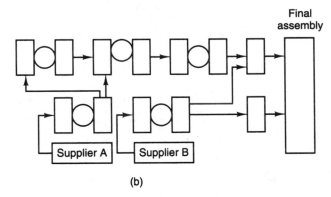

(b)

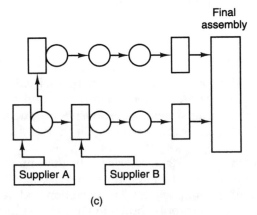

(c)

Small Lot Production. Reducing lot sizes to the smallest quantity possible is supported by structured production and short setup times. The goal is a lot size of 1, or the smallest customer order.

Setup Reduction. Setup time is the total time from the completion of the last piece of the previous production run to the first good part on the new production job. Reduction in setup time increases capacity and production capability while reducing inventory.

Fitness for Use. *Fitness for use* means that the product satisfies the customer's requirements perfectly. Customers are the external users of the finished products or the next workstation in the structured-flow production line. Every operation in the enterprise is a customer for someone and a vendor for someone else. In each case, the needs of the customer must be met precisely.

People Management. The second area, *people management,* is critical for the continuous-improvement objective in JIT. This element creates an environment in which all employees, from the president to the hourly workers, have the responsibility and authority to suggest and implement improvements to the production system. Creation of this type of environment requires the following items.

Total Employee Involvement. The company that has every employee working on solutions to performance problems will outperform the competition. The Japanese term for continuous improvement, *kaizen,* implies that all employees are valuable resources for the solutions to problems. JIT is built on the premise that everyone works on continuous improvement of the process through functional and cross-functional corrective-action teams.

Control Through Visibility. *Control through visibility* means that simple visible means are used to communicate goals and identify problems: for example, progress charts for work-center goals, control charts for tracking critical process variables, and flashing lights to indicate a machine problem that needs immediate action.

Housekeeping. Housekeeping focuses on the work center or workstation, with an emphasis on cleanliness, simplification, discipline, and organization to eliminate wasted time, motion, and resources.

Total Quality Focus. Total quality focus addresses broad quality issues from suppliers to customers at every element of the production chain. The emphasis is on the quality of the process at every function and work center in the enterprise because the quality of the product is determined by the quality of the process.

Systems Management. The third element, *systems management,* addresses the effective distribution and application of the limited enterprise resources. This integration element is supported by the following items.

Level Load and Balanced Flow.   Both of these elements focus on effective utilization of manufacturing resources. The first, *level load,* deals with scheduling products in roughly equal quantities for a given period, such as a week or month. The second, *balanced flow,* works toward a continuous flow of products through manufacturing.

Preventive Maintenance.   Preventive maintenance works to eliminate equipment failure as a source of process defects by maintaining machines at the highest level of operational performance. Preventive maintenance, a requisite for structured-flow manufacturing and quality products, ensures that machines operate on demand and that operational performance meets specification levels.

Supplier Partnerships.   Supplier partnerships are key elements because a healthy supplier-user relationship is critical to a JIT operation. Long-term vendor partnerships cut cost for all members of the partnership through shared quality goals and design cooperation, frequent product deliveries, and a total cost perspective.

Pull Systems.   *Pull system* is synonymous with *JIT manufacturing* because it describes in one word, *pull,* how JIT manufacturing works. Parts are produced only when the next workstation in the structured-flow production system indicates that parts are required. This requirement implies that parts are produced on demand with very short lead times. As a result, some of the technology management elements, such as rapid setup and small lot production, are important.

These twelve JIT elements in the three management groups represent the environment that must be present for a JIT system to function effectively. A JIT implementation requires that each element be addressed through a systematic plan.

Implementing JIT

JIT concepts are implemented for two reasons. First, they are implemented to improve current manufacturing efficiencies with no intention of implementing JIT across all operations. For example, a company may have an effective MRP II system but may need to reduce lot sizes and to increase the flexibility of the production system. Working on the JIT elements would improve an already-operational MRP II system. The second reason is to install an operational JIT manufacturing system. The installation may be one or just a few work cells from a large manufacturing operation. These work cells may be judged critical for rapid response to customer needs. Another JIT installation may focus on a product line or cover the entire factory. In each case, the work cell, product line, or factory is changed from a *push*-type manufacturing system that produces inventory per some schedule for future use to a *pull*-type system that produces parts only when they are needed in the next level of the bill of materials. A seven-step implementation process developed by David W. Buker, Inc., illustrates the critical success factors for the installation:

1. A four-stage education plan covers every employee. Education ranges from a working overview of JIT for top management to focused group education on how to implement a system.

2. Assessment of the twelve JIT elements in the technology, people, and systems management areas is a major step in the implementation process. The assessment process determines the readiness level for JIT. In addition to the twelve elements, the level of management commitment and effectiveness of education and training must be evaluated and the results of the entire assessment published.

3. When the assessment is complete, an implementation plan is developed that uses the results of the self-study to target areas requiring work. The plan includes implementation activity, status of activity, person responsible, start and completion dates, and resources required to complete the activity.

4. A recommended practice in JIT installations is the use of a small pilot project as the initial implementation. The smaller project permits analysis of the impact of JIT at a departmental or unit level. In addition, the pilot permits unexpected problems that arise to be solved before JIT is applied across the enterprise.

5. Continuous-improvement activities critical to the long-term health of the JIT implementation are begun. One technique, called *small group improvement activities (SGIA)*, is frequently used to organize and provide structure to the continuous-improvement process. In this step, it is critical to provide the leadership and education needed to work in the group setting, to limit the scope of projects so that the group can achieve success, and to develop a method for evaluating the success of the process.

6. The success of the implementation must be measured through a performance evaluation process. Baseline measurements are usually taken in the following areas: customer service, elimination of cost-added operations, product and process cycle time, inventory levels and turns, quality, number of employee suggestions, manufacturing output, and employee productivity.

7. In the final step, the process is moved to other parts of the enterprise. This internalization and companywide transition is a signal to the organization that a journey has been started to bring the organization in line with world-class standards—and that it will never end.

Many U.S. and offshore companies have implemented JIT successfully for the production of various products. One of the most advanced JIT production systems was developed by Toyota. The Toyota JIT model is illustrated in Figure 9–4. The system has two objectives: *cost reduction* and *increase in capital turnover ratio*. Elimination of waste, called *unnecessaries* in the Toyota model, is the mechanism to be used to reach these objectives. The production method utilizes a continuous-flow process with two elements: JIT and *self-stop automation*. If a quality problem is detected, any employee has the power and authority to stop the production line if necessary. The JIT production system has two components: *production methods* and *information system*. The items listed under production methods are consistent with the definition of JIT provided in this section of the book. The information system used for JIT is called *kanban*.

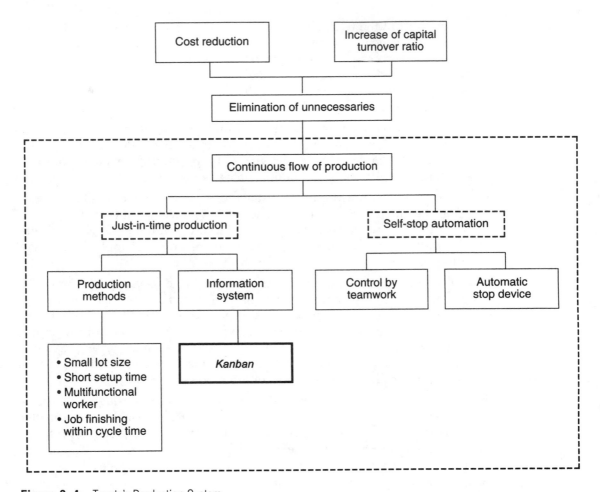

Figure 9–4 Toyota's Production System.
(Source: From the European Working Group for Production Planning and Inventory Control, Lausanne, Switzerland, July 1982.)

Kanban

The information system block in the Toyota model (Figure 9–4) represents all of the manufacturing production and control (MPC) functions necessary to run the JIT system. Within the MPC system, *kanban* controls the flow of production material. *Kanban* is a Japanese word that means "card." One- and two-card *kanban* systems are in common use. A two-card *kanban* system may be used to move material through production. The first card is a *transport*, or *conveyance*, card, and the second is a *production* card. Sample *kanban* cards are illustrated in Figure 9–5. The *process* refers to the production work centers, and the *issue number* represents the number of containers released. The JIT pull system using *kanban* is best described by an example.

A *kanban* system is illustrated in Figure 9–6; study the figure until work centers and container labels are familiar. Each work center has some raw material inventory on the left and finished-products inventory on the right in the figure. The work cells

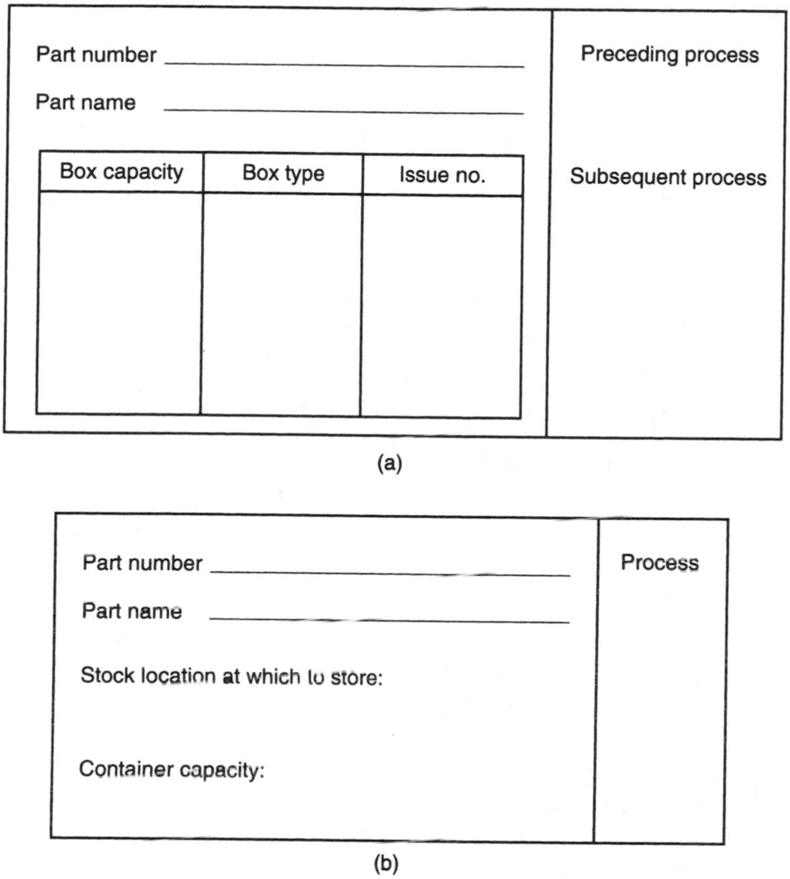

Part number _____			Preceding process
Part name _____			
Box capacity	Box type	Issue no.	Subsequent process

(a)

	Process
Part number _____	
Part name _____	
Stock location at which to store:	
Container capacity:	

(b)

Figure 9–5   A Two-Card *Kanban* System: (a) Withdrawal *Kanban* Card; (b) Production *Kanban* Card.

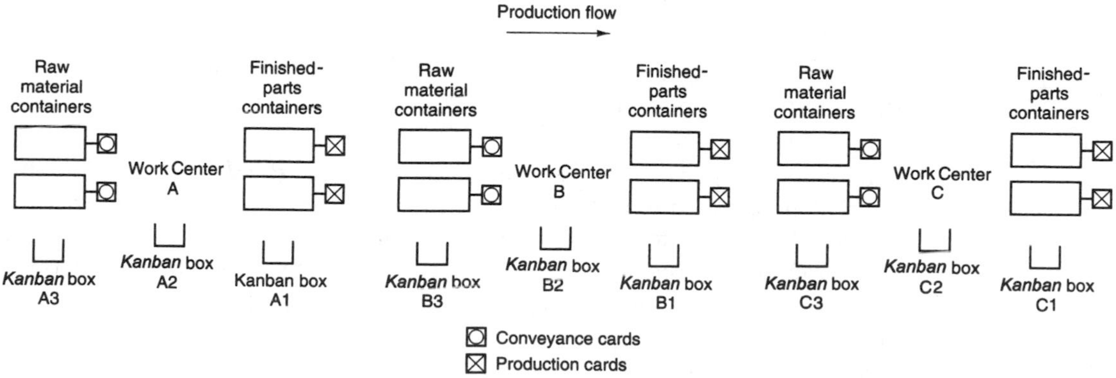

Production flow
→

Raw material containers — Work Center A — Finished-parts containers — Raw material containers — Work Center B — Finished-parts containers — Raw material containers — Work Center C — Finished-parts containers

Kanban box A3 — *Kanban* box A2 — Kanban box A1 — *Kanban* box B3 — *Kanban* box B2 — *Kanban* box B1 — *Kanban* box C3 — *Kanban* box C2 — *Kanban* box C1

◻ Conveyance cards
⊠ Production cards

Figure 9–6   Two-Card *Kanban* System.

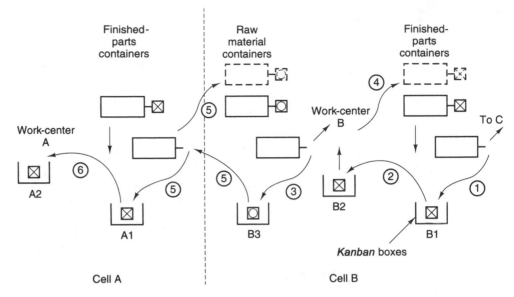

Figure 9–7 Movement of Cards and Inventory in a *Kanban* System.

are labeled A, B, and C; and *kanban* boxes are located in the raw material inventory, work-center, and finished-product inventory areas. Each raw material container has a *conveyance* card, and a finished-goods container has a *production* card. The work centers shown in Figure 9–6 are idle at the start of this example. The process starts with movement of a finished-parts container from work-center B to work-center C, where the parts are used for an assembly operation. Changes in the work cells as a result of this movement are illustrated in Figure 9–7. Understanding the movement of the cards is critical for insight into the *kanban* production process. The sequence of card movement is as follows (the circled numbers in Figure 9–7 correspond to the *kanban* sequence described next):

1. The production card is removed from the production container before the container with finished parts is moved from cell B to cell C. The card is placed in *kanban* box B1.

2. An inventory specialist moves the production card from *kanban* box B1 to the work-cell *kanban* box B2. The arrival of a production card authorizes the work-cell operator to produce another container of finished parts to replace the container just removed. The production card is retained by the operator.

3. Production is started by moving a container of raw materials into cell B from the input side. The transport, or conveyance, card in the raw material container is removed and put into *kanban* box B3.

4. The operator in cell B finishes production on the container of parts and moves the finished-goods container to the output side of the cell. The production card held by the operator is placed on the finished container.

5. The inventory specialist retrieves the conveyance card from *kanban* box B3. The card authorizes the movement of a finished container from the output side of cell A to the input side of cell B. The production card in the container is placed in *kanban* box A1, and the conveyance card is placed in the container on the input side of cell B.
6. A similar process is followed in cell A.

This system is a *pull* type of production system because a work center is authorized to produce parts only when the operator has a production card. A card is present only when the next work center in the production sequence pulls finished parts away. Work is paced by the flow of the *kanban* system, and no work center is allowed to work just to keep operators busy. In the preceding example, the input and output queue had two containers; however, depending on the demand rate, multiple containers could be used for raw material and finished parts. In that case, additional cards for each container would be necessary.

The cards replace all work orders and inventory move tickets. Note that production cards just circulate around the output side of the work cell from a finished-goods container to the production cell operator and then back to the finished-goods container. The conveyance cards circulate at the input side of the work cell in a similar fashion. The system is visual, is manual in operation, and operates with less inventory, and the problem of sequencing jobs is greatly reduced. Experience in the Toyota system indicates that movement of cards can also extend to external suppliers.

JIT is used most often in a high-volume *repetitive* manufacturing operation; however, low-volume *nonrepetitive* production systems can apply JIT with some modifications. The most difficult problem in low-volume applications with many product models is leveling the workload at the cells.

Claims in favor of JIT manufacturing include reduced inventory levels, reduced work-in-progress inventory, shorter manufacturing lead times, and increased responsiveness to customers. Shorter lead times help compress schedules and lead to less work-in-progress material. Elimination of any non-value-added activity leads to spending less time and money. The streamlined process works better and faster because the waste has been removed.

JIT will not work if management only announces that the operation is now JIT. Successful JIT manufacturing is the result of dull, boring, incremental, methodical, but highly effective work in the plant and with the material suppliers that support production.

MRP II and JIT

Elements in the operational philosophy for MRP II and JIT are frequently opposed. For example, inventory is minimized in JIT and is planned in MRP II. MRP II originated as a push system producing parts inventories for future use. JIT is a pull system. However, the two are not mutually exclusive because the manufacturing system improvements necessary for JIT, such as reduced setup time, would also make MRP II more efficient. In addition, some parts of an MRP II production system could have cells that operate with a JIT philosophy.

9–2 SYNCHRONIZED PRODUCTION

Synchronized production is defined in the *APICS Dictionary* as follows:

> **synchronized production**: *A manufacturing management philosophy that includes a consistent set of principles, procedures, and techniques where every action is evaluated in terms of the global goal of the system. Both kanban, which is a part of the JIT philosophy, and drum-buffer-rope, which is a part of the theory of constraints philosophy, represent synchronized production control approaches.*

(Source: © *APICS Dictionary*, 10th edition, 2002)

Kanban, based on the JIT philosophy, and drum-buffer-rope (DBR), based on the theory-of-constraints philosophy, are two synchronized production control approaches. However, in many manufacturing systems, the JIT philosophy cannot be adopted because disturbances in the system do not permit effective implementation. Synchronous manufacturing using DBR techniques can be used in these situations because the production is synchronized by a philosophy that is slightly different from that of JIT.

History of the Problem

In the past, product cost was the dominant metric used by manufacturing managers to determine the best manufacturing process; as a result, efficiency took precedence over product flow. To keep product cost down, managers targeted the product cost of individual end items. To achieve this end, managers mandated large batches with few setups. However, this approach is in direct conflict with the demands for fast, smooth material flow that requires smaller batches and more setups. In addition, the premise that reducing individual product cost will reduce total enterprise cost is not valid because of the complex interaction present in manufacturing operations. Therefore, the procedures used to evaluate and manage synchronized manufacturing in operations had to be changed. A broader view of manufacturing beyond single products is necessary when the global goals of the system are considered.

One approach to achieving this result is to change the measurements from individual product cost to *throughput, inventory,* and *operational expense.* In most situations, production managers have little control over the throughput requirements received from the business planning area. Therefore, the goal for managing the synchronized production flow system is restated as *meeting the throughput requirements while efficiently managing inventory and production expenses.*

Managing Inventory and Operational Expenses

Inventory and operational expenses generally have an inverse relationship. For example, operational expenses in production can be reduced if inventory is allowed to rise. Certainly, larger batch sizes and fewer setups reduce operational expenses, but they result in a larger work-in-process inventory. In contrast, inventory costs are reduced to a minimum level when all parts are produced in a manufacturing system that supports a lot size of 1. However, the lot-size-of-1 system often has the highest operational expense. The goal is to increase throughput while decreasing inventory and operational expenses. A manufacturing control system supporting synchronous manufacturing is key to reaching this goal.

Drum-Buffer-Rope System

The *drum-buffer-rope (DBR) system* is a synchronous manufacturing process in which a planned production flow is executed during a given period of time and compensates for the disturbances commonly found in most production flow systems. The disturbances usually result from three *critical constraints: market demand, capacity,* and *material limitations.* The production plan, developed around these critical constraints, must (1) not exceed projected market demand, (2) ensure a sufficient supply of materials, and (3) ensure that the planned production flow does not overload the processing capabilities of the resources.

The DBR process starts with a preliminary production plan and identification of *capacity constraint resources (CCRs).* The production flow is analyzed by using input from operators to determine what operations in the flow for a family of products have serious capacity constraints. Operations with constraint problems are designated CCRs, and the remaining operations are labeled *non-CCRs.* The preliminary production plan is modified according to the schedule for the CCR, and the resulting document, called the *master production schedule (MPS),* is used to schedule actual production on the shop floor and establish customer order delivery dates. The MPS sets the beat for production flow in the synchronous manufacturing system; as a result, the process used to establish the MPS is referred to as the *drum.*

Every manufacturing system has disruptions to the process; as a result, the actual flow is often different from the planned flow. To ensure that customer delivery dates are honored, managers build in a protective cushion, or *buffer,* for the operations most likely to cause a disturbance in the flow. The planned lead time is then the sum of the setup and process times plus the time buffers required.

The product structure, process sheets, buffers, and production procedures are used to set the planned production at each resource. Therefore, the planned production schedule at each production resource is tied to the tempo set by the MPS, or the *drum beat.* All non-CCRs are synchronized to support fully the product rate set by the MPS. Since the MPS was established from the CCR schedules, a link between the non-CCRs and the CCRs is built. Linking the resources together in this fashion is analogous to the practice of tying mountain climbers together with ropes so that they climb at the same rate and support each other. This process of linking CCRs and non-CCRs is the *rope* in DBR.

DBR Process

Techniques and processes are well established to identify the CCRs in a flow production system and to determine the production plan, MPS, and buffers required for a functioning DBR synchronized manufacturing system. A detailed analysis of the process used to develop a fully functional DBR system is available in several texts covering the topic. In short, the process is as follows:

1. Identify the CCRs.
2. Set the *drum beat* by developing an MPS that effectively schedules the CCRs. The MPS plans sequence and lot size that do not overload operations while maximizing throughput and minimizing work-in-process inventory and operational expenses.
3. Set the *buffers* by identifying the amount of inventory at the front of each CCR to avoid disruptions in the material flow.
4. Identify schedule release points at both the CCRs and the non-CCRs.
5. Set the *rope* by developing schedules for release points that are back scheduled from the MPS, using planned lot sizes for the CCRs. Schedule non-CCRs by using simple flow control to determine sequence and transfer lot size.

Synchronous Manufacturing: DBR Versus JIT/*Kanban*

JIT requires a change in the management philosophy for successful implementation because in time, every work center is affected. This type of wholesale change in management is difficult for many top managers to accept. However, DBR's focus on the bottlenecks in the production flow is closer to the emphasis found in many U.S. companies. In addition, U.S. managers take a project management approach to improvements in production, and DBR fits this type of thinking better than JIT does.

DBR overcomes many of the following limitations present in JIT/*kanban*:

- JIT/*kanban* is better suited for a highly repetitious process environment with a limited number of processes.
- Disruptions in the process flow are disastrous for a JIT/*kanban* implementation.
- JIT/*kanban* implementation is often lengthy and difficult.
- The JIT continuous-improvement process has a systemwide focus, so that CCRs with the greatest possibility for productivity are not singled out.

9–3 THE EMERGENCE OF LEAN PRODUCTION

Competition between manufacturing companies in Japan and those in the United States reached a startling new level in the late 1980s. U.S. companies found themselves losing market share, even whole product segments, to foreign competition that did not exist just a few years before. Hundreds of executives and managers made the trip to Japan to see firsthand what was taking place. What they saw was the result of a production system that was significantly different from the systems used in the United States. Companies scrambled to figure out what was happening and to try to replicate the Japanese system.

The Japanese system has become known as the *Toyota production system.* It was created at Toyota by Taiichi Ohno. When pressed by U.S. executives about the source of this revolutionary system, Ohno is reported to have laughed and said that he learned it all from Henry Ford's book *(Today and Tomorrow,* first published in 1926). Ford's book was republished in 1988. Ford's factories implemented a new system of production and produced quality automobiles at prices that nearly every worker could afford. The Ford system led to fantastic productivity improvements that allowed car prices to be cut in half and worker wages to double. Norman Bodek, former president of Productivity Press, commented as follows in the preface to the new printing of Ford's book:

> *[Ford] insisted that work environments be spotlessly clean; that business leaders think in terms of serving their communities and society at large; that production techniques not be taken for granted but continuously change and improve. He said that primary industries should help their suppliers and service industries to produce cheaper and better products in less time; and that managers should not remain in their offices but should walk around, know their workers, and be capable of doing the work themselves. He emphasized that workers should be trained and have the opportunity to better themselves and make product improvements. (p. viii)*

Taiichi Ohno explained the motivation for the Toyota production system as an attempt to catch up with the automobile industries of the advanced Western nations following World War II. Remember that in the postwar period, the Japanese economy was trying to recover from extensive bombing. Resources in Japan were limited at best. Ohno knew that to become a true competitor to the West, his operations had to become more productive and produce quality goods at low cost. His system focused on the ferocious elimination of waste in all manufacturing operations and on the effects demonstrated through actual practice on the shop floor. There was no recipe showing how to be an effective producer. Toyota learned how to be effective by practicing, making lots of changes, and learning along the way.

The production processes developed and used in the Toyota factories have been documented by Yasuhiro Monden, a professor at the University of Tsukuba, Japan. Monden's book *The Toyota Production System* was first released in 1983, and the second edition was published in 1993. It describes in detail the procedures, techniques, and tools that have been developed at Toyota. The editions of Monden's book are important sources of information on the Toyota system and the tools that the Japanese developed. This system appears to be too simple, just an extension of common sense. There is much more to the Toyota system, however, than simple tools and techniques. It is important to look deeper at the work of Ohno and Monden.

James Womack and a team of educators from the Massachusetts Institute of Technology (MIT) completed a major study of the automobile industry and published their findings in a landmark book, *The Machine That Changed the World,* in 1990. This outstanding book provides a look at the history of the auto industry and the tremendous changes that have taken place since 1975. The authors pulled many concepts that have been mentioned in this book together under the overarching title of *lean* production. Lean production is defined as follows:

> **lean production:** *A philosophy of production that emphasizes the minimization of the amount of all the resources (including time) used in the various activities of the enterprise. It involves identifying and eliminating non-value-adding activities in design, production, supply chain management, and dealing with the customers. Lean producers employ teams of multiskilled workers at all levels of the organization and use highly flexible, increasingly automated machines to produce volumes of products in potentially enormous variety. It contains a set of principles and practices to reduce cost through the relentless removal of waste and through the simplification of all manufacturing and support processes. Syn: lean, lean manufacturing.*
>
> (Source: © *APICS Dictionary*, 10th edition, 2002)

The heart of the concept is the aggressive elimination of waste throughout all parts of the operation and the organization. It is a great testimonial to the power of the Toyota approach and tools.

James Womack and Daniel Jones teamed up again in 1996 to write a follow-up book titled *Lean Thinking*. Womack and Jones helped to package the important concepts invented by Ohno and documented by Monden. Their second book provides case study examples and lays out the five principles that summarize the lean production concept:

1. Specify precisely the value of a certain product.
2. Identify how the value is realized by the customer.
3. Make value flow without interruptions.
4. Allow the customer to pull value from the producer.
5. Pursue perfection and improve continuously.

Lean thinking addresses all the areas of waste presented earlier in the discussion of JIT; however, the lean approach is more focused. An expert in waste elimination known as a *sensei* is a critical part of the lean approach. The experience of the *sensei* leads to the identification of problems and the design of solutions in a short time. The elimination of waste and the redesign of a work cell, including the relocation of machines, is often done in one day. The benefits of the efficient production cells are immediately visible and often lead to additional changes in other areas. The five steps sound fairly simple and a lot like the commonsense ideas that we should all embrace. Taken together and used as the basis for a plan of action with the help of a skilled *sensei,* the preceding five steps have helped companies remove amazing amounts of waste from their operations in a short time. The cases presented in *Lean Thinking* demonstrate how the five principles have been applied in small, medium-size, and large companies in the United States, Germany, and Japan. The examples build a credible argument for a move toward lean production, and the concept of lean production should remain popular in manufacturing for years to come.

Jim Womack is the founder and leader of the Lean Enterprise Institute. Through the institute, Womack and his followers lead the crusade for lean production. Visit the Web site at *http://www.lean.org* for additional news and information.

Americans are notorious for searching for the quick fix and not taking the actions to ensure problem elimination. Too often people skip the critical work of describing or mapping the value stream and really understanding it. The Lean Enterprise Institute has started to develop training materials to help people do lean the right way. The first how-to book in a series planned by the institute is *Learning to See* (1998), by Mike Rother and John Shook. *Learning to See* provides detailed instructions on how to map the value stream to add value and eliminate waste. Womack has seen too many companies start on a lean production project that never really comes together the way it should. A crisis arises and a champion leads the charge. Too often a company does a lot of work related to implementing the lean techniques and tools but does not realize the benefits they were hoping for. Womack encourages the use of an experienced waste fighter called a *sensei*. The *sensei* provides a jump start to the improvement. The people involved in the improvement process replace a long learning curve with the insight and knowledge of the *sensei*. The improvement team picks something they think is important and starts removing waste immediately.

American companies tend to be impatient with process changes. The critical steps of understanding a product's value and mapping the value stream for all product families is often omitted or done poorly. As Rother and Shook stated in their book's title, it all begins with "learning to see." Companies have seen some short-term savings by applying the tools of lean production. An important key to long-term success is truly understanding the product's value and the related value stream from the customer back to the producing company and the company's suppliers. Weaving the beliefs and concepts of lean thinking into the fabric of the organization leads to ongoing improvements and savings that are impossible any other way. You can expect to hear more about the lean concepts and the Lean Enterprise Institute in the future.

9–4 MODERN MANUFACTURING SYSTEMS IN A LEAN ENVIRONMENT

In lean production, the elimination of waste in all parts of the operation is stressed. Clerical time and waste from excess transaction processing is a big problem in many companies. Traditional MPC systems can be very labor intensive, forcing people into a lot of work to feed data to the system. Rick Anderson, President of TTW, Inc. (which markets and supports the WinMan enterprise resources planning [ERP] software), has demonstrated the link of ERP with lean production. The following discussion was developed from a presentation made by Anderson at an APICS chapter meeting. Think of ways a computer-based system can help people work smarter.

In a lean production environment, all attention is on providing product value to the consumer. Value-added activities are identified and preserved. Non-value-added activities are exposed and eliminated. There is a drive toward perfection through nonstop continuous improvement. Production is based on customer demand that is pulled through the system, not pushed. Events are sequenced to optimize the flow of material and information. Departmentalized boundaries are removed. The corporate culture supports constructive change and a can-do attitude.

Manufacturing software should support the lean production process and the lean environment. The computer-based systems must be tuned to the customer's requirements. The system should not be used to keep all machines busy or to justify large batch sizes. Emphasis should be on reducing inventory, not on system capabilities that support the management of larger-than-required inventories. The software that drives production should support the lean culture and not become a barrier to change.

The goal of manufacturing, of course, is to make money. *Profit* is not a dirty word; in fact, it is essential to the survival of our lifestyle. Modern manufacturing systems should provide an integrated management tool set that allows its user to achieve profitability for the company. TTW clients have reported the following significant benefits from their systems:

- Inventory turns exceeding twenty times per year
- Fifty percent less manufacturing floor space required
- Sixty percent reduction in delivery lead times
- Huge reductions in manual data processing through the use of automated transaction processing
- Significant reductions in indirect labor and overhead costs
- Greatly improved quality due to "making it right the first time"
- Increased customer satisfaction
- Improved ability to compete in the marketplace
- Significantly improved corporate profitability

Parts of the traditional system methodologies, such as reorder points, optimum order quantities, material requirements planning (MRP), and so forth, still have a place in company systems. However, a new set of innovative methodologies such as *kanban* and *backflushing* must be a part of new planning and control systems. *Backflushing* is a method of inventory bookkeeping in which component levels are automatically reduced by the computer after completion of activity on upper-level items. The stock reductions are based on the items and quantities specified in the bill of materials and the production quantities reported. Systems need to allow users to select the tools and methods that best fit their business situation. Management at TTW, Inc., believes new ERP software needs to provide a "bridge" for the users from the old methods to the new—in other words, a "transitional tool set."

The following tool sets are representative of the techniques provided in the WinMan software. They provide examples of how modern manufacturing system software can complement lean production operations. System transactions are linked to specific events or signals. Once a signal is received, the system is configured to place orders, receive material, update material, finish goods inventories, and track the associated costs of each transaction. The tools help enable the software to better fit the lean production methods and eliminate system waste.

Tool Set #1: External *Kanban*

WinMan provides the ability to link a purchased component to a specific supplier. Total quantities and lot sizes are prenegotiated with the supplier, and the system is configured to allow employees on the production floor to place orders when they are needed. The use of a *kanban* (signal) card on each container eliminates the need for traditional order paperwork. Please note that this approach does require a thorough understanding of the production process and careful planning and design of the ordering system and the signal cards.

Figure 9–8 shows a sample *kanban* card prepared by the WinMan system. When a container is empty, the *kanban* card identifier is scanned into the computer system. The computer then sends a fax *kanban* to the designated supplier, automatically placing the order for the prenegotiated quantity and delivery details.

Figure 9–8 *Kanban* Signal Card Prepared by the WinMan System.

(Source: From WinMan ERP, developed by TTW, Inc. 2003.)

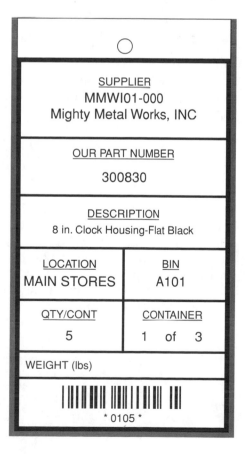

SUPPLIER
MMWI01-000
Mighty Metal Works, INC

OUR PART NUMBER
300830

DESCRIPTION
8 in. Clock Housing-Flat Black

LOCATION | BIN
MAIN STORES | A101

QTY/CONT | CONTAINER
5 | 1 of 3

WEIGHT (lbs)

* 0105 *

This is often a part of the system's overnight routines, but it can be set up for processing during the day if necessary. The *kanban* card is placed on a display board designated for incoming orders and becomes a signal that material is expected to come in. After receipt of the product, the *kanban* card identifier, a bar code, is scanned into the system. The card is taken off the display board and attached to the incoming container of new material. Inventory records update automatically upon entry of the receipt into the system by another scan of the bar code, and the contract quantity decreases accordingly. Careful planning and design of this type of system eliminates transaction and time waste.

Note the key elements of the lean (no-waste) system. Computer automation reduces the daily purchasing grind of non-value-added activities by taking on routine activities. The functional boundaries between procurement and the production floor have been reduced, which allows production to coordinate its own inventory flow. The pull system for production reduces the amount of inventory in material stores. The planned *kanban* system brings material inventory to the point of use on the shop floor, so there is no need for additional warehousing and handling. *Kanban* cards with bar coding minimize data entry to two bar code scans: one signal to place an order and one for the receipt of new material.

Tool Set #2: "Orderless" Manufacturing with Internal *Kanbans*

The ability to utilize internal *kanban* mechanisms to signal the completion of production allows the use of backflushing techniques for the control of the components used within the production process. Material consumption by manufacturing is based on the reported completions of parts and assemblies and the bill of material details maintained in the system. Orderless manufacturing is dependent upon a 100 percent accurate bill of materials, and that is part of an effective lean environment.

Once again, note the key elements of the lean system. Production control personnel are spared from the daily grind of non-value-added activities through the automation of system transactions. The functional boundaries between the production control function and the production floor have been reduced, which allows production to coordinate its own inventory flow. The pull system reduces the amount of work in the process inventory. Transactions throughout the operation trigger specific system events. Communications throughout the enterprise are improved. Figure 9–9 maps the key transactions in the manufacturing process. Automation of these transactions and well-executed manufacturing operations will reduce the need for inventory adjustment transactions and reduce waste.

Tool Set #3: Lean Accounting Practices

The automation of system transactions for accounts payable provides the ability to automatically process invoices and make payment to a certified supplier without additional clerical work. System action again begins with the processing of a *kanban* signal. The system's self-billing invoicing routine generates payments

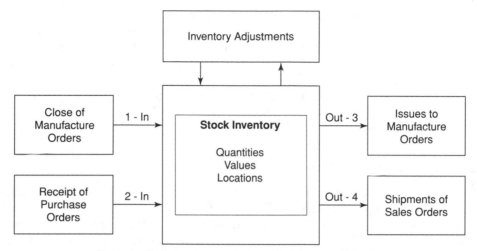

• Control point 3 pertains to only those that are not utilizing lean
manufacturing concepts (whereby the components are
backflushed at the time of closing out the manufacture order).

Figure 9–9   Key Transactions in a Lean Manufacturing Process.
(Source: Adapted from WinMan system documentation TTW, Inc., 2003.)

based on the prenegotiated pricing. A mass payments routine can be used to take advantage of supplier discounts for the timely payment of bills.

The key elements of the lean system are demonstrated once again. The accounting clerical function is spared from non-value-added labor. Goodwill is generated with suppliers that improves the relationship and gives the purchasing agent greater ability to negotiate favorable pricing and terms. The tremendous potential for significant cost savings by timely processing of the accounts payable loop can now be realized.

Similar savings from the routine are found in the application of modern system techniques to accounts receivables. The system can provide the ability to invoice shipments automatically. The integration of the system with third-party shipping software such as that of United Parcel Service (UPS) or FedEx can improve the scheduling of freight pickups and order tracking. The non-value-added labor of the accounting clerks is reduced. Once again, the barriers between functional departments are broken down as the invoicing function is linked to shipping transactions. The customer invoice can be automatically generated at the time of shipment and electronically issued to the customer by fax, electronic data interchange (EDI), or XML e-mail through the Internet. The use of the automated system enhances customer relationships with timely notification of shipments. Cash flow is improved by the timely processing of accounts receivable transactions and customer invoicing.

The Lantech Corporation is a recognized leader in the application of lean concepts to production and business functions. Jean Cunningham, CFO for Lantech, has teamed with Orest Fiume to write a book titled *Real Numbers: Management Accounting in a Lean Organization.* This book describes the important changes that need to be made to accurately measure a lean operation. The ERP system that Lantech uses is WinMan, by TTW, Inc. The WinMan system has many advanced features that apply the fundamental concepts of lean operations and help system users eliminate wastes that are common in traditional push systems. Visit the WinMan site on the World Wide Web at *http://www.winman.com* for the latest information.

Additional System Benefits

Automated functions to reduce non-value-added transactions take advantage of system functionality. The use of a common operating system such as Windows helps support enterprise-wide sharing of data and functional integration. The one-time keying (entry) of data into a single, integrated database and the sharing of system information provides significant savings and elimination of non-value-added work. Multiple entries of data during the sequence of events—from the preparation of a quote to the receipt of a sales order through the production processes and on to shipping and invoicing—are eliminated. Communications using e-mail can be integrated into customer relationship management functions. Integration with computer-aided design (CAD) or product data management (PDM) systems can eliminate database duplications that cause confusion and lead to errors.

9–5 SUMMARY

The environment of manufacturing is continually changing. Just-in-time (JIT) manufacturing is a new approach to planning and operations that is yielding companies outstanding results. JIT and the various systems based on the system of production used by Toyota Motor Manufacturing (often called *lean production*) are challenging the thinking behind traditional computer-based planning systems.

JIT is a production philosophy based on the elimination of all waste and the continuous improvement of productivity. JIT requires that three management areas (technology, systems, and people) be addressed together to create an effective manufacturing system. A collection of tools and techniques for organizing, controlling, and communicating the needs of production throughout the organization is available. Effective implementation of JIT or any of the lean production approaches begins with education on the proper application of the new methods and tools.

Signaling systems first developed using printed cards are commonly called *kanban* systems. Visual signal systems control the replenishment of materials, the work done, and the movement of materials through the production processes. The *kanban* signals are a critical part of a system that pulls materials and assemblies through the production process. Products are made and material is replenished on the basis of what is actually used to satisfy the customer requirements.

The concept of waste elimination throughout the operation extends beyond the production and purchasing departments. Modern computer-based systems can help eliminate waste in the manufacturing system operation. Transactions normally entered into the system manually can be automated, and much of the routine paperwork and information flow can be improved. The detailed planning benefits that result from a formal manufacturing production and control (MPC) system can, with some conceptual and program changes, be beneficial in a lean production system. The planning push that comes from manufacturing resource planning (MRP II) and the controlled execution of a system driven by *kanban* signals can come together to complement each other. These systems will develop and improve through ongoing efforts that simplify the operating processes of manufacturing and take advantage of expanding software and system capabilities.

BIBLIOGRAPHY

BUKER, D. W. *7 Steps to JIT.* 2nd ed. Antioch, IL: David W. Buker, Inc., & Associates, 1991.

COX, J. F., and J. H. BLACKSTONE, eds. *APICS Dictionary.* 10th ed. Alexandria, VA: American Production and Inventory Control Society (APICS)—The Educational Society for Resource Management, 2002.

CUNNINGHAM, J. E., and O. J. FIUME. *Real Numbers: Management Accounting in a Lean Organization.* Durham, NC: Managing Times Press, 2003.

EVANS, J. R., and W. M. LINDSAY. *The Management and Control of Quality.* 5th ed. Cincinnati, OH: South-Western College Publishing, 2002.

FORD, H. *Today and Tomorrow.* Portland, OR: Productivity Press, 1988.

HALL, R. W. *Attaining Manufacturing Excellence.* Chicago: Dow Jones–Irwin, 1987.

MONDEN, Y. *The Toyota Production System.* 2nd ed. Norcross, GA: Industrial Engineering and Management Press, 1993.

ROTHER, M., and J. SHOOK. *Learning to See.* Brookline, MA: The Lean Enterprise Institute, Inc., 1998.

SRIKANTH, M. L., and M. M. UMBLE. *Synchronous Manufacturing Principles for World Class Excellence.* Cincinnati, OH: Southwestern Publishing Co., 1990.

WOMACK, J. P., and D. T. JONES. *Lean Thinking.* New York: Simon & Schuster, 1996.

WOMACK, J. P., D. T. JONES, and D. ROOS. *The Machine That Changed the World.* New York: HarperPerennial, 1990.

QUESTIONS

1. Define *just-in-time (JIT) manufacturing*.
2. Describe the two fundamental objectives associated with JIT.

3. Describe the three JIT elements that must be addressed for successful JIT implementation.

4. Describe the four areas in technology management that must be considered for successful JIT implementation.

5. Describe the four areas in people management that must be considered for successful JIT implementation.

6. Describe the four areas in systems management that must be considered for successful JIT implementation.

7. Describe the seven-step process used to implement a JIT solution.

8. Compare the production systems illustrated in Figure 9–3 and describe the advantages of continuous flow.

9. Describe a two-card *kanban* system.

10. Compare and contrast a pull production system and a push production system.

11. Define *synchronous manufacturing* in your own words.

12. Compare and contrast the operation and limitations of the *kanban* and drum-buffer-rope (DBR) production systems.

13. Describe a DBR system in your own words.

PROJECTS

1. Using the list of companies developed in project 1 in chapter 1, determine which companies use JIT or lean production systems.

2. Select one of the companies in project 1 that uses JIT or lean production and compare the world-class standards achieved by that company with the standards presented in chapter 1.

3. Select one of the companies in project 1 that uses JIT or lean production and describe the process used to achieve continuous improvement.

4. A few years ago, isolated functional areas of excellence called *silos* were visible in many organizations. The functional silo provided a perfect environment for a group to work together to become the world's best at its specialty. However, the silo helps keep people isolated from the outside world and the other functions in the organization. Identify the "silos" in your organization, company, or school and report on how the tools of JIT, lean production, and CIM can be used to eliminate the silo effect.

CASE STUDY: PRODUCTION SYSTEM AT NEW UNITED MOTOR MANUFACTURING, PART 1

On February 17, 1982, General Motors (GM) Corporation and Toyota Motor Corporation reached agreement on the formation of a joint company, later named New United Motor Manufacturing, Inc., to produce a subcompact car. The location chosen for the operation was Fremont, California, at the site of a GM plant that was closed because of poor production and labor troubles.

Background

Each company wanted to achieve specific objectives from the joint venture. GM wanted firsthand experience with the cost-effective and efficient *Toyota production system* and wanted a high-quality vehicle for the Chevrolet Division. Toyota needed experience working with American unionized labor and wanted immediate access to and information about U.S. suppliers. The venture was an experiment in the production of an automobile that blended the Toyota production philosophy with U.S.-supplied parts and a unionized workforce.

After 2 years and an estimated $350 million for renovations and improvements to the existing facility, the first Nova cars were produced in December 1984. The Nova was produced from 1984 to 1988, then production was shifted to the Toyota Corolla, and eventually to the Chevrolet Geo Prizm. In 1989, the workforce included 2,300 hourly and 400 salaried team members organized into approximately 340 teams with 100 group leaders. Production totaled 485 Geo Prizms and 335 Corolla sedans per day, with approximately 1,100 vehicles in the system at one time. At the present time, the total time to produce a vehicle is about 19 hours, so the facility has the capacity to produce 220,000 cars annually.

The Production System

The production system at New United Motor covers 60 acres with over 3 million square feet of covered production space. The final assembly line is 1.3 miles long and includes the installation of over 3,000 automotive parts. Parts are supplied from U.S. and Japanese vendors, with the engine and transmission coming from Toyota's Japanese production facility. The joint venture uses 74 U.S. parts vendors, with 17 located in California. JIT production requires forty-five truck shipments daily and four ships from Japan weekly.

Most of the body parts are produced in the stamping plant at the site, where twenty-six stamping presses produce eighty-one different body parts from approximately 300,000 pounds of steel per day. The body shop assembles the body-in-white with 95 percent of the welds on each vehicle performed using automation. Flexible automation utilizing 210 robots produces 70 percent of the 3,800 welds on each vehicle.

The function of the system is to manufacture cars with quality as high as that anywhere in the world while ensuring that product costs are the most competitive of any manufacturer. The philosophy of the production operation is that quality should be ensured in the production process itself. The system is built on the concept of *work teams,* with an average team composed of six members. Team meetings, led by the *team leader,* are held in forty team rooms placed across the production facility. The team leader is directly responsible for the performance of the individual team. The team leader is a member of the bargaining unit and an integral part of the team-building process. Three to five teams are organized into a group with a *group leader*. This person is the first line of salaried supervision at New United Motor. The key responsibility of the group leader is to ensure open two-way communications between team members and managers.

In addition to production responsibilities, team members also follow the four *S*s: *seri* (clearing), *seiton* (arrangement), *seiketsu* (cleanliness), and *seiso* (sweeping and washing). These four Japanese words help team members focus on the practice of work-cell order and control. If the workplace is clean, well organized, and neat, production will be more efficient, easier, and safer. To achieve this efficiency in production, leaders emphasize three concepts: *just-in-time* production; *jidoka*, a Japanese term meaning "the quality principle"; and full utilization of the worker's abilities. A description of these principles and other aspects of the Toyota production system are included in the continuation of this case study, which follows.

CASE STUDY: PRODUCTION SYSTEM AT NEW UNITED MOTOR MANUFACTURING, PART 2

The basics of the Toyota production system used at the joint venture between General Motors Corporation and Toyota Motor Corporation are described in part 1 of this case study. The joint venture company, called *New United Motor Manufacturing,* relies on the following three concepts for effective and efficient operation: *just-in-time* (JIT) production; *jidoka,* a Japanese term referring to the quality principle; and full utilization of the *worker's ability*.

The Three Concepts for Effective and Efficient Operation

JIT Production.   The philosophy of JIT at New United Motor is not to sell products produced but to produce products to replace those that are sold. The first process, selling products produced, implies that cars are made for a finished-goods inventory and represents a *push* type of production system. The latter approach, producing cars to replace those that are sold, suggests that cars are produced only when a demand for the vehicle is present and represents a *pull* type of production system.

In the production environment, JIT is a concept designed to supply the right parts at the right time in exactly the correct amount during each step in the production process. The primary tool used to control the production system at New United Motor is called *kanban*. *Kanban* is an information system that controls production and manages the JIT pull manufacturing system. As a result of the JIT implementation, *muda*, a Japanese term meaning "waste," is significantly reduced and the product is delivered at the lowest possible cost.

The non-value-added operations in production usually fall into one of seven categories: *corrections, overproduction, processing, conveyance, inventory, motion*, and *waiting*. When *muda* is implemented effectively, all slack is removed from the production system; as a result, problems are exposed that may otherwise be hidden by excess inventories. Production teams use another Toyota method, *kaizen*, in which they search on a continuous basis for production improvement. To be successful, *kaizen* must include the workers in the process, using their ideas and suggestions.

Jidoka: **The Quality Principle.**   A basic principle at New United Motor is that quality should be ensured by the production process itself. When equipment and operators function under normal working conditions, they focus on approaching a zero quality-problem operation. The principle applied to the production system is called *jidoka* and means "the quality principle." *Jidoka* refers to the ability of production machines or the production line to shut down automatically when abnormal conditions are present. For example, if a machine starts to produce parts that fall outside the allowed tolerances, the system is shut down. In this system, no defective parts move to the next work center because the system stops production. A current phrase used to describe this technique is "quality at the source." In addition to the production machines' halting the production system by using automated quality-checking sensors, every worker has the responsibility of checking quality problems and stopping the line to prevent poor-quality products from leaving the work center. The objectives of *jidoka* are (1) 100 percent quality at all times, (2) prevention of equipment breakdowns, and (3) efficient use of every worker.

Full Utilization of the Worker's Ability.   Team members have the authority to make decisions in their work area; are expected to be multifunctional and to solve problems; and are treated with consideration, with respect, and as professionals. Job rotation is a regular part of the learning and training process. In many older production systems, the machines run the workers; however, at New United Motor, the team member operates the machine. Wherever possible, automation has been introduced to eliminate the use of workers in monotonous, difficult, and dangerous operations.

Production Techniques and Methods

The goals of high quality and low cost are addressed with various production techniques and methods. The major techniques and methods include *kanban, production leveling, standardized work, kaizen, baka-yoke, visual control,* and *the team concept.*

Kanban.   *Kanban* is the Japanese word for "card." The card is designed to prevent overproduction and to ensure that finished parts are pulled through the production system as needed. With the card system, one process produces only enough parts to replace those drawn from the following process. The *kanban* (1) gives work instructions, (2) provides for visual control of the production volume, (3) prevents over-production, and (4) identifies problems for correction.

Production Leveling.   Production leveling attempts to average the highest and lowest variations in orders so that the production resources remain relatively constant and at a low cost level. The production-volume changes normally associated with automotive production cause waste at the work site. At New United Motor, the leveling process focuses on three causes of volume changes: the total volume, the models of cars produced, and the options added. Without this leveling process, the *kanban* implementation of JIT would not function because of disruptions caused by volume variations. The leveling goes beyond the Fremont plant and includes the many vendors supporting the New United Motor operation.

Standardized Work.   *Standardized work* is defined as follows:

> *Standardized work is work done at highest efficiency, with a minimum of waste, as a result of all tasks at the work site being organized into perfect sequences.*

The three goals for standardized work are (1) high productivity, (2) line balancing for all processes from a production timing standpoint, and (3) elimination of excessive work-in-process inventory. Each team member is trained in standardized work processes and principles; as a result, teams are responsible for the efficient layout of work assignments. Therefore, there is no need for industrial engineers in that function at the facility.

Kaizen.   Visitors from the United States who toured manufacturing plants in Japan in the 1980s found companies that were driven to improve quality. This effort was initially thought to be focused on product quality. The philosophy of *kaizen* is much richer. *Kaizen* drives improvement in all functional areas of the business. It lifts quality above the manufacturing floor and involves all areas in an effort to enhance the overall quality of the company. The drive to continually improve all aspects of the business has been woven into the fabric of the entire company. Employees no longer have to stop and think about quality decisions; quality improvement is ingrained in their thinking.

The successes driven by *kaizen* are often the result of the implementation of many small suggestions. The sum of this accumulated effort, with time, has produced outstanding results. *Kaizen* programs are built on the principles of documented operating practices, the total involvement of every employee, and a commitment from the company to provide training in the philosophy and tools of the *kaizen* process.

Kaizen continues to gain popularity in the United States and it has proven to be a successful approach for many companies. Visit the following Web site for additional information: *http://www.kaizen-institute.com.*

Baka-yoke.   *Baka-yoke* is a Japanese word that means "machine sensors" and is used to identify malfunctions in production machinery. These devices improve in-process quality and serve as a backup in the event of human operator error. When problems are detected, machines are stopped automatically.

Visual Control.   *Visual control* means that the status of production can be determined by only visual inspection of the manufacturing operation. The concept applies to the work of both team members and production machines. The goal of visual control is to spot problems as quickly as possible and correct them immediately. The principal device used in plants for visual control is the *andon signboard,* an electrical board that shows the current state of production operations by means of lighted indicators. The indicators on the board identify the type and location of the problem, and chimes or a musical melody is often played to alert team members to the presence of a problem. The andon signboard is triggered by either sensors on production machines or by team members who spot a problem. Additional visual control occurs in the form of graphs, charts, data, and status displays indicating production effectiveness and efficiency.

The Team Concept.   Successful problem solving in manufacturing today often requires the inputs of people from different functional areas. Individual effort has traditionally been the focus in many organizations. The dynamic environment of manufacturing operations today makes team-based solutions more important. A single person in an organization will seldom have all the answers. Thinking of the best solutions often involves people from many parts of the organization. Members of problem-solving teams must work together effectively, and successful teamwork is more difficult to achieve than one might think. Companies have found that training in and development of team dynamics and teamwork skills is often required.

Teams may take on many formats in an organization. Quality experts James Evans and William Lindsay have identified several of the most common: quality circles, problem-solving teams, management teams, work teams, project teams, and now even "virtual" teams. Teams should be formed to meet or solve a particular task or problem. The goal should be clear, and team members should support each other. The team needs a level of empowerment that allows it to get the job done.

The subject of teams is too large for detailed discussion in this text. However, the significance and power of teams cannot be overlooked. The days of the lone problem solver are gone in most cases. Decisions require the inputs and consideration of others in the organization. Readers with an interest in a career in manufacturing should make a point to study more about teams and teamwork skills. The Internet offers access to various Web sites related to teamwork, team-building, and team-skills development. Try researching these key words using a search engine such as AltaVista: *http://www.altavista.com.*

Summary

The Toyota production system used at New United Motor effectively integrates the Japanese production system into the U.S. labor and supplier market. The experiment has yielded many more successes than failures, and the results indicate that many of the production strategies used by Asian manufacturing companies can be adopted by Western management.

Enabling Processes and Systems for Modern Manufacturing

PART GOALS

The goals for this part are to introduce the reader to the manufacturing processes, machines, and systems used to implement a successful automation project in computer-integrated manufacturing (CIM). In addition, the reader is presented with the need for an enterprise information network, quality, and the development of the most important enterprise resource: its people.

After you complete part 4, the following will be clear to you:

- Properly designed manufacturing processes, machines, and systems are the CIM elements found on the shop floor.
- Industrial robots, automated material-handling devices, machine control computers, and programmable controllers are the automation foundation of the factory floor.
- The technology required for automated control of manufacturing cells and production areas is available and within reach of manufacturing operations implementing CIM.
- Systems for effectively controlling machines and systems are elements of the CIM system.
- An enterprise-wide data and information network is a necessity for successful implementation of CIM.
- Quality is a major element in the CIM system.
- Successful CIM depends on the people of the organization.

CAREER INSIGHTS

The careers described in part 2 focused primarily on the development of the product by using technologies and software to create the best design. The skills described in part 2 were primarily associated with software applications. In part 4, the career focus is on the technologies and hardware necessary to manufacture the product. A wide variety of job skills are associated with the technologies described in this part

CAREER INSIGHTS (contd.)

of the text. These skills can initially be grouped into categories such as programming, system design, system installation, and system troubleshooting. In addition, there are careers with the vendors that supply the automation complements. These careers would have groupings similar to those just listed, but the work would be in just one complement of automation, such as industrial robotics. As in part 2, the educational requirements for a career in this area vary from a technical trade school diploma to a graduate degree in engineering. For example, technicians usually perform system troubleshooting and some parts of system installation. Technicians are usually employees with a diploma from a trade school or an associate's degree from a technical college. However, a bachelor's degree in engineering or engineering technology is usually required for the individuals performing design and programming. A career working with the hardware and software associated with manufacturing is not restricted to any particular geographic region, because manufacturing and vendors producing manufacturing hardware and software are located across the United States and in every developed country. The amount of travel required in a career of this type depends on the particular job function. If you work for a vendor producing automation hardware and software, it is possible that some travel would be required to support the installation of the systems. However, if you work at a manufacturing site in this career area, the travel may be limited to trips for training on the systems you are working on. If the topics and concepts covered in this part of the text are interesting to you, then a career in this area could be interesting, challenging, and rewarding.

Graduates trained in the design process and in the application software used to create and analyze product designs have an opportunity for an interesting and rewarding career in the product development segment of the CIM wheel. If you find the material presented in part 4 interesting, then this is a career area that you should explore.

Production Process Machines and Systems

OBJECTIVES

After completing this chapter, you should be able to do the following:

- Describe the value-added and cost-added activities present during process machine operation
- List and describe the primary processes for manufacturing
- List and describe the secondary processes for manufacturing
- Define *flexible manufacturing system* and describe the operation of a typical system
- Define *flexible manufacturing cell* and describe the operation of a cell
- Describe a fixed automation system and compare the operation of an inline system with that of a rotary system
- Describe the selection process for fixed automation systems

Production machines and systems are the fundamental building blocks on the shop floor. The concepts presented in this chapter provide a broad overview of the types of machines used for the production of finish goods. The many types of production systems and machine configurations are too numerous to cover in detail in a computer-integrated manufacturing (CIM) introductory text; however, the production processes and machines commonly used are identified and described in detail. In this chapter, we focus on the materials-processing part of factory automation; the remaining components are addressed in chapter 11.

Material and parts are actively processed only a small percentage of the time that they are in production. For example, the spindle shown in Figure 5–1 could be machined from bar stock in approximately three minutes of value-added time. In many manufacturing operations, however, the bar stock could spend three or more hours of non-value-added time at the machine for setup, which includes setting up the production tooling, verifying production programs, and checking for the first good part. During this period, metal is not being removed from production parts. If additional work at other work centers is required after this operation, the ratio of value-added to non-value-added time is repeated and the wasted time

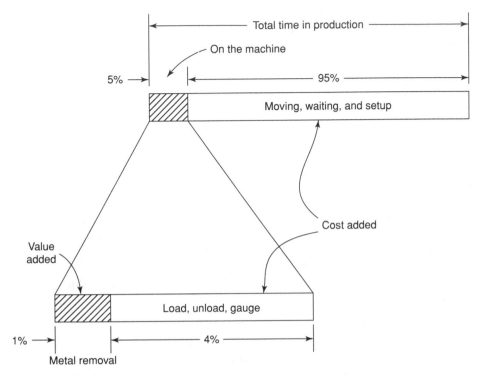

Figure 10–1   Production Efficiency.

increases. This ratio is illustrated graphically in Figure 10–1. The figure indicates that total production time is divided between 5 percent on the machine and 95 percent moving, waiting, and setting up. All of the 95 percent is cost added. The 5 percent is made up of 1 percent metal cutting and 4 percent in additional cost-added activities. Therefore, only 1 percent of the production time is value added. Reducing the move time, shortening the travel distance, reducing the setup time, and decreasing the need for inspection eliminates cost-added operations, which is the goal of world-class manufacturing.

The types of production machine or system typically used in the five manufacturing systems categories are illustrated in Figure 10–2. Comparison of the manufacturing systems shown in Figure 2–3 with the graph shown in Figure 10–2 indicates that the project has the lowest-volume output and continuous processes have the highest. The graph shows that *manual* and *programmable machines* are adequate for low-volume production. However, *fixed automation* systems are required for the large volumes associated with continuous production systems, as in the petroleum and chemical industries. A special type of fixed automation system, called a *transfer line*, is often used for high-volume production of a complex machined part or family of parts. The production of components such as air-conditioning compressors for the automotive industry typically uses a repetitive manufacturing system. Therefore, the manufacturing floor has different production systems, ranging from *programmable machines* to *flexible automation* systems. Let's start the

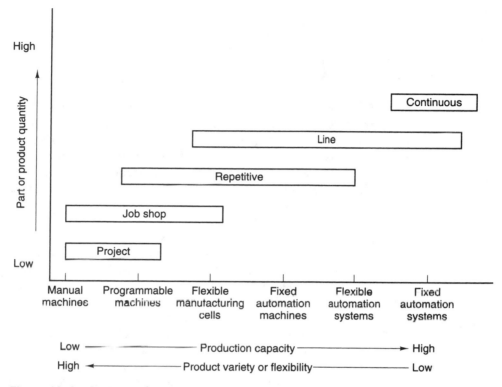

Figure 10–2   Production Systems.

description of production machines and systems with the basic processes used to change raw material into finished parts and products.

10–1 MATERIAL AND MACHINE PROCESSES

A review of the production process chart shown in Figure 3–18 is a logical starting point for this study of machine processes. Study the production processes in the figure until the names and basic operations are again familiar.

Process Operations

The process operations transform raw material into a finished product without adding other materials or components. The shape, physical properties, and surface of the material are altered in the manufacturing process by the application of energy. For example, the mechanical energy expended by a cutoff saw when it cuts the end from a 2-inch-diameter steel bar changes a steel bar into a cylinder 2 inches across and 5 inches in height. The process operations performed in manufacturing are classified into one of four categories:

1. Primary operations
2. Secondary operations

3. Physical properties operations
4. Finishing operations

A review of these four operation areas in manufacturing is provided in the following sections.

Primary Operations.   A *primary operation* converts raw material into the basic geometry required for the finished product. Metal casting is a good example of a primary operation. The raw material is metal in a molten state, and a void in the mold cavity is the rough geometry of the product. When the molten metal is poured into the mold and cools, the primary operation is complete, and the raw casting has the general shape of the finished product. Typical primary operations include *casting, forming, oxyfuel and arc cutting,* and *sawing.*

Casting.   Casting is a manufacturing process that uses molten metal and molds to produce parts with a shape close to that of the finished product. The molten metal is poured or forced into a mold with a cavity shaped like the part desired. After the metal fills the cavity and returns to a solid state, the mold is removed and the casting is ready for secondary processing. The *green-sand molding process* is used for 75 percent of the 23 million tons of castings produced in the United States annually. A cross-sectional view of a typical green-sand mold is provided in Figure 10–3.

The process starts with a pattern of the desired part, usually constructed from wood. The lower half of the mold, called the *drag,* is filled with green sand, a sand-like material that sets up or hardens. The pattern is placed on the surface of the

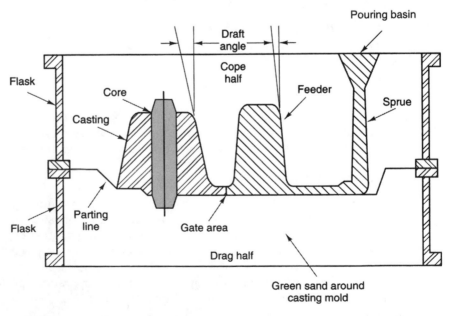

Figure 10–3   Cored-Casting Cross Section.

lower mold half (Figure 10–3), and the upper half of the mold box, called the *cope*, is placed on top and filled with additional green sand. After the green sand sets, the upper and lower halves of the mold are separated at the *parting line*, and the pattern of the casting is removed. The mold halves are put back together and the molten metal is poured into the pouring basin and enters the cavity through the *sprue* and the *runners* (the runners are not visible in the view shown in Figure 10–3). The molten metal fills the cavity and forms the part.

The green-sand method just described is one of several traditional casting processes. Other traditional casting techniques include *sand mold, dry sand, shell molding, full mold, cement mold,* and *vacuum mold.* Casting operations vary from totally automated systems that produce over 300 castings per hour to manual productions. The types of metal used in casting include most of the ferrous metals, along with aluminum, brass, bronze, and other nonferrous materials.

Several casting processes that do not rely on sand to produce the mold have been developed in the last thirty years. These contemporary casting and molding processes are generally used for smaller-geometry parts and often do not require additional processing or finishing on the surfaces with noncritical dimensions. The most frequently used process, called *high-pressure die casting,* uses metal molds. The basic die-casting machine has four components: (1) a two-part die, (2) a die mounting and closing mechanism, (3) a molten-metal-injection device, and (4) a source for the molten metal. The sequence of operation for the process is illustrated in Figure 10–4. Note that in (a) the two-part die with two movable cores is closed and metal is being poured into the cylinder. In (b) the molten metal is forced into the mold by the ram, and the metal cools. The die separates in (c), and when the die is open and the cores are withdrawn, the part is ready for removal (d).

Die-cast machines can utilize either a *hot-chamber* or a *cold-chamber* process. The older, hot-chamber process incorporates a furnace in the machine to keep a reservoir of molten metal ready for the injection device. Machines of this type can typically produce 100 castings an hour weighing up to 50 pounds (23 kilograms) or 1,000 parts in the 1-ounce range. The hot-chamber process uses zinc alloys primarily, because aluminum and copper alloys chemically attack the submerged metal-injection device.

Cold-chamber machines have the molten metal introduced from an external source, as in the operation shown in Figure 10–4. A complete cold-chamber machine has a *traveling plate* and a *front plate*. Half of the die is attached to the traveling plate and the other half to the front plate. A closing cylinder moves the traveling plate with one-half of the die attached so that the die is closed. Molten metal is introduced by the shot cylinder, and a part is produced. Other contemporary casting processes include *permanent mold, centrifugal, plaster molding, investment,* and *solid ceramic.*

Forming.   In the *forming* process, the shape of raw material is changed by the application of force. For example, in the *extrusion* process shown in Figure 10–5, pressure from the pressure stem forces a hot, round billet of metal into a die, and the extruded part that results takes the shape of the hole in the die. A cookie press

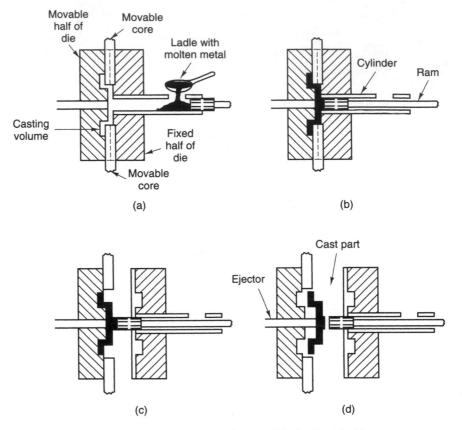

Figure 10–4 Sequence of Operation for a Cold-Chamber Die-Casting Machine.

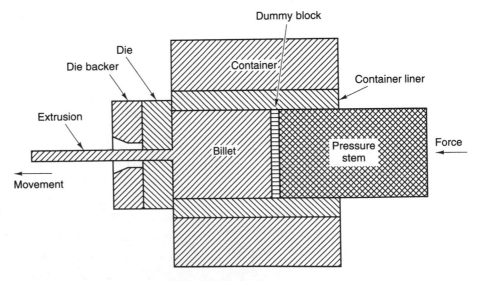

Figure 10–5 Cross Section of a Forward Extrusion Press.

operates in a similar fashion to produce cookies with different shapes. Other forming processes frequently used include forging (hot and cold), roll and *spin forming, shearing, punching, drawing, upsetting,* and *swaging.*

Oxyfuel and Arc Cutting.   *Oxyfuel gas cutting* constitutes a group of four cutting processes (*oxyacetylene, oxyhydrogen, air-acetylene,* and *pressure gas*) in which the base metal is heated to an elevated temperature and then cut by the chemical reaction of oxygen with the base metal. In some cases, a chemical flux or metal powder is used to cut oxidation-resistant metals.

    Plasma arc cutting is an arc-cutting process that uses the difference in the voltage between the cutting electrode and the material to cut through the metal (Figure 10–6). The area where the arc penetrates the metal causes localized melting, and the molten metal is removed by the high-velocity ionized gas directed by the nozzle. The process is used to cut any material that conducts electricity. Other arc-cutting methods include *oxygen arc, gas metal arc, gas tungsten arc, shielded metal arc,* and *carbon arc.*

Sawing.   Every manufacturing operation uses raw material in the form of wire, bars, tubes, pipes, extrusions, sheets, plates, castings, or forgings. In many cases, the material must be cut to the required length to prepare for secondary operations.

Figure 10–6   Gas-Shielded
Plasma Arc Cutting.

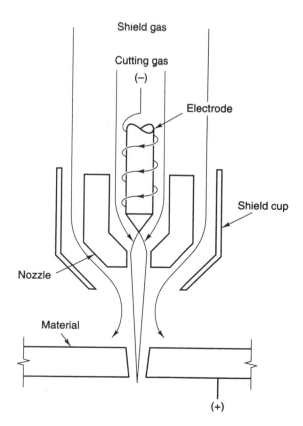

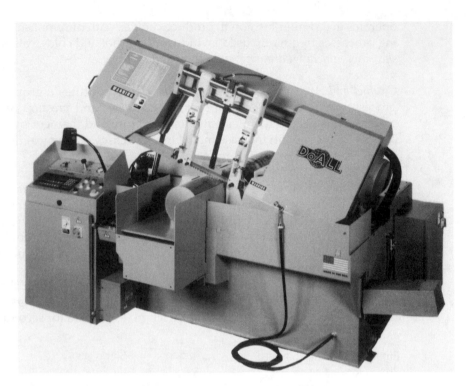

Figure 10–7   High-Production Power Saw with 10-hp Drive Motor and NC Control.
(Source: Courtesy DoALL Company, Des Plaines, Illinois.)

Sawing is a primary machining process using straight, band, or circular blades to cut the raw material to the rough shape required for the secondary machining operations. The cutoff operation, sometimes called *slugging,* is performed most often on power hacksaws, band saws, and circular saws. Hacksaw machines are available in both manual and numerical control (NC) models. Band saws come in the same range of control, from manual to fully automated, like the saws shown in Figures 10–7 and 10–8. Cutoff is also performed on lathe-type machines using single-point and parting tools.

Secondary Operations.   *Secondary operations* are performed to give the raw material the final shape required for the finished part or product. For example, after iron is cast into the general shape of an automotive engine, the cylinder walls are machined so that the pistons fit smoothly inside. As in this example, secondary operations often follow primary manufacturing operations. The main difference between the two types of operations is that secondary operations take the material to the final geometry required for use. Also, many secondary operations, using different types of machine tools, are required to achieve the proper shape and dimensional accuracy in the finished part. Typical secondary operations include *turning, boring, milling, drilling, reaming, grinding,* and *nontraditional machining processes.*

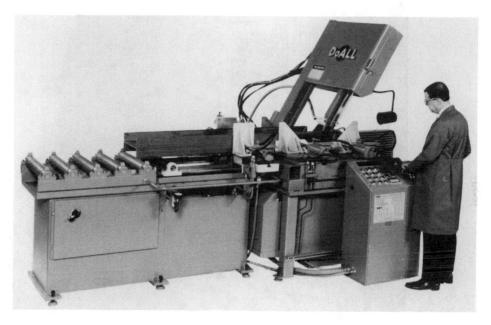

Figure 10–8 Tilt-Frame Band Saw with NC Controls.
(Source: Courtesy DoALL Company, Des Plaines, Illinois.)

Turning.   Turning is performed on a machine tool called a *lathe* and can be manual or fully automated with sophisticated NC or computer numerical control (CNC). In *turning* operations, the material or workpiece is held in a spindle and rotated about the longitudinal axis while a cutting tool is fed into the rotating workpiece to remove material and produce the desired shape. The workpiece is held in the lathe spindle by a chuck, collet, fixture, or faceplate, or between centers. In a turning operation, the material can be removed from either the external surface (outside diameter, or OD) of the workpiece or from the internal part (internal diameter, or ID) of the material in the spindle. The OD operations include facing, chamfering, grooving, knurling, skiving, threading, and cutoff (parting). The ID operations are called *recessing, drilling, reaming, boring,* and *threading.*

The NC and CNC machines generally fall into three categories: (1) center-type machines, where primarily OD cutting is performed with the workpiece held between centers; (2) chucking-type machines, with larger and lower-speed spindles, with cutting performed on both the ID and the OD; and (3) universal machines, which have characteristics of both preceding types. The workpiece and cutting tool turret are visible on the CNC Cincinnati Milacron machine shown in Figure 10–9.

Boring.   *Boring* is a machining process for producing precision internal cylindrical surfaces by using either manual and NC or CNC lathes or special-purpose boring machines. In the lathes, the workpiece rotates and the single-point or multiedge cutting tool is stationary. However, in special-purpose boring machines

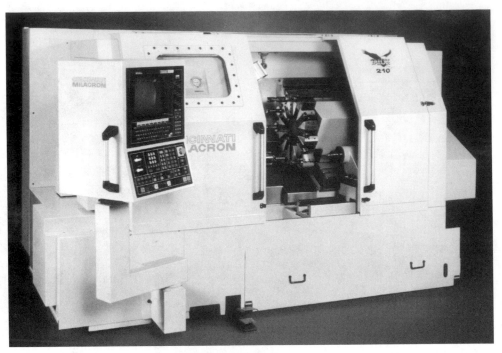

(a)

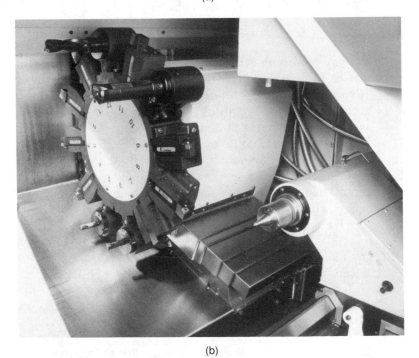

(b)

Figure 10–9 (a) CNC Turning Center; (b) Indexing Tool Turret.
(Source: Courtesy Cincinnati Milacron, Inc.)

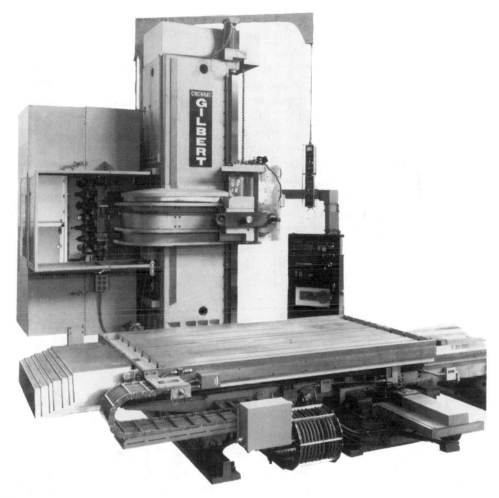

Figure 10–10   CNC Boring Machine.
(Source: Courtesy Cincinnati Gilbert, Inc.)

(Figure 10–10), the tool is rotated and fed into the stationary workpiece. Common applications include the precision finishing of cored, pierced, or drilled holes and contoured internal surfaces. In addition to the precision operation, the massive rigid construction of boring machines is ideal for precision heavy cutting operations.

Milling.   *Milling* is the process of removing material by using a rotating cutter with multiple cutting edges. The primary characteristic of milling is that each cutter tooth takes a small individual chip of material as the cutter and workpiece move relative to each other. Cutting is achieved by feeding the workpiece into a stationary rotating cutter, feeding a rotating cutter into a stationary workpiece, or a combination of the two. In most applications, the workpiece is mounted on the machine table, and the table is advanced at a relatively slow rate past a cutter turning at a relatively high speed.

The number of machine *axes* refers to the different number of directions in which the workpiece can be moved simultaneously past the cutter. A simple single-axis machine can move the table only in the x direction, while a six-axis machine could move the table in the x-y-z directions and a three-axis move of the workpiece. Multiaxis machines of this type are used to mill the complex shapes required in wing spars and ribs required by the aircraft industry. Milling is one of the most widely used secondary processes and includes a large number of different types of machines. Most of the machines are grouped into two categories: horizontal and vertical. The orientation, horizontal or vertical, identifies the direction of the center axis of the rotating tool. Figures 10–11 and 10–12 show examples of vertical and horizontal CNC mills. Milling methods are generally grouped in two

Figure 10–11   Vertical CNC Mill.

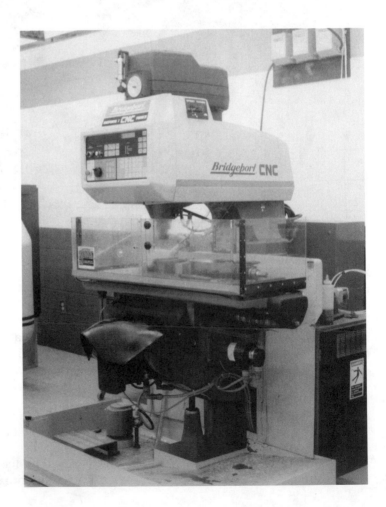

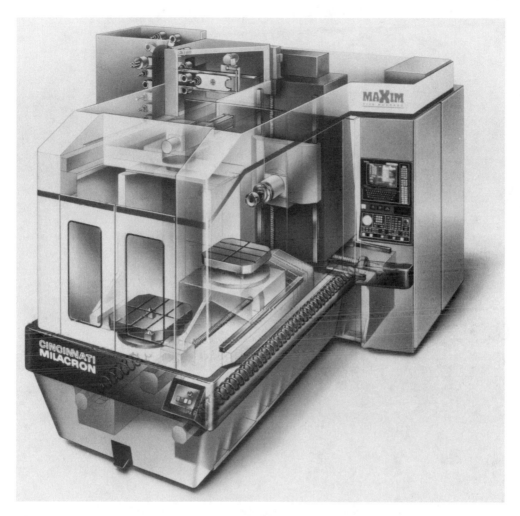

Figure 10–12 Horizontal CNC Mill.
(Source: Courtesy Cincinnati Milacron, Inc.)

major categories: *peripheral* and *face*. While milling is one of the most widely used secondary processes, it is also one of the most complex machining methods.

Drilling.   *Drilling* is a process used to produce or enlarge a hole in a workpiece by using one of the following drilling methods: conventional, deep hole, or small hole. The tool used in the drilling process, called a *twist drill*, is a rotary end-cutting tool with one or more cutting lips and flutes. Twist drills are classified according to the following characteristics: drill material, shank type, number of flutes, direction of rotation for cutting, length, diameter, and point configuration. When a drilling operation is performed on a vertical CNC mill, the workpiece is fixed, and the

Figure 10–13   Drill and Carbide Milling Cutter Tools in Tool Turret of Turning Center.
(Source: Courtesy Cincinnati Milacron, Inc.)

rotating drill is fed into the part. Figure 10–13 shows a drill mounted on the tool turret of a turning center. In this type of machine, the part is rotated and the drill is stationary.

Reaming.   The drilling operation just described is not considered a precision process; rather, a drilling operation is selected because it is fast and economical. If hole precision is necessary, the drilling operation is followed by a *reaming* process. Reaming is used for enlarging, smoothing, and accurate sizing of existing holes.

Grinding.   Grinding ranks first according to the number of machine tools in use, exceeding processes such as turning and drilling. *Grinding* is an abrasive process that removes material by means of an abrasive grain. Thousands of types of grinding wheels and surfaces are available to the production engineer determining the optimum production method. In addition to being available in many varied shapes, grinding wheels differ on the basis of five elements: (1) type of abrasive agent used; (2) size of the grain particle used; (3) type of bonding material used to hold the grains in operational shape; (4) grade, which determines the hardness or strength of the wheel; and (5) structure, which sets the proportion and configuration of the grains and bond. A wheel's porosity is determined by a combination of structure and grade. A manual surface grinder is illustrated in Figure 10–14; the wheel, the magnetic chuck surface to hold the part, and a small block ready for grinding are visible.

Grinding machines are available in a wide range of sizes with manual and CNC controls. In addition, machines are designed to grind cylindrical stock or

Figure 10–14   Manual Surface Grinder.

rectangular material. Grinding of cylindrical surfaces is performed with the part held between centers, as in a lathe operation, or by using a centerless grinding process. Grinding of rectangular parts is performed on a surface grinder (Figure 10–14) or jig grinder. The grinding process produces smoother surface finishes than those achieved through either milling or turning. As a result, grinding is frequently the last operation in a surface-finishing production sequence.

A standard part of the grinding operation includes *trueing* and *dressing* of grinding wheels. The trueing process adjusts the surfaces, OD, and sides of the grinding wheel so that they are at the correct angle and distance relative to the drive shaft. The dressing process removes metal particles produced from previously ground parts that collected on the wheel and restores the wheel to the original geometry.

Metal-cutting machines such as lathes, mills, and grinders use cutting fluids to lubricate and cool the cutting tools and grinding wheels during the production process. However, grinding exhibits characteristics that separate it from other types of material removal. As a result, grinding different material may require a specific type of cutting fluid to ensure minimum friction, long wheel life, minimum heat buildup, and optimum finish on the part.

Nontraditional Machining Processes.   The category of *nontraditional machining* includes a large number of processes developed after 1940 that use mechanical, electrical, thermal, or chemical energy to remove material. Many of the processes remain experimental; however, several have become the mainstay of specialty machining areas.

The first, and most frequently used, mechanical process in the nontraditional area is *hydrodynamic machining*. Hydrodynamic machining is used primarily for slitting and contour cutting of nonmetallic materials such as wood, paper, asbestos, plastic, gypsum, leather, felt, rubber, nylon, and fiberglass. A narrow cutting width, or *kerf*, is produced as workpiece material is removed by a high-pressure (usually 60-psi), high-velocity stream of water or a water-based cutting fluid. Frequently, the motion of the cutting heads is computer numerically controlled so that complex contours are cut with a high degree of repeatability. A second mechanical process in the nontraditional machining area is *ultrasonic machining*. In this process, the tool is vibrated at ultrasonic frequency to enhance the conventional cutting action in the workpiece.

The nontraditional machining electrical process, called *electrochemical* or *electrolytic processing*, uses chemical electrolytes in combination with electrical energy to create a cutting action. As Figure 10–15 illustrates, the cutting action is the result of a reverse plating action between the tool and the workpiece. The chemical electrolyte is forced down through the tool into a gap between the tool and the workpiece. The high-current dc power source and the electrolyte force electrons from the workpiece material, and the metallic bonds of the workpiece atoms are broken at the surface. The workpiece atoms enter the electrolyte solution as metal ions and are washed away by the constant flow of electrolyte. Hard metals with the ability to conduct an electric current are frequently processed with this

Figure 10–15   Electrochemical
Machining.

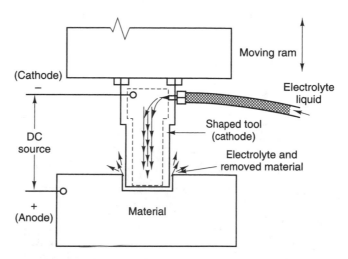

method. In a similar process, called *electrochemical discharge grinding,* an ac or pulsating dc source of electrical energy is used. Again, most of the metal removal results from the electrochemical process; however, sparking between the graphite grinding wheel and the workpiece removes the oxide developed by the electrochemical process. This grinding process is used routinely to grind and sharpen carbide tools.

The nontraditional machining thermal processes use thermal energy to remove metal from the workpiece. The four methods used to generate the necessary thermal energy for machining include *electrical discharge, electron beam, laser beam,* and *plasma beam.* The first method is widely used to machine intricate parts by using two methods: *electrical discharge machining (EDM)* and *electrical discharge wire cutting (EDWC).* The latter process, EDWC, is often called *wire EDM.* The components of an EDM machine are illustrated in Figure 10–16. Metal is removed from the work-piece as a result of rapid electrical discharges (arcing) between the tool electrode and the workpiece immersed in a liquid dielectric. Small, hollow metal chips are produced as the workpiece material is removed by melting and vaporization. The resulting shape of the workpiece matches the contour of the tool electrode.

The second popular electrical discharge method, wire EDM, uses a fine metal wire as the cutting electrode (Figure 10–17) to cut metal sheets and plates into intricate shapes. Note the fine vertical wire visible in Figure 10–17. The parts shown in Figure 10–18 are examples of wire EDM work performed on the CNC wire EDM pictured in Figure 10–19. This machining method is frequently used in the production of stamping and extrusion dies.

In the chemical nontraditional machining process, chemical energy is used as the primary source for metal removal. Although this process was used for many years, it was after World War II that a process was developed by the North American aviation industry to use chemical milling in volume production. The two major chemical processes used in industry are *chemical milling* and *photochemical*

Figure 10–16 Components of an Electrical Discharge Machine.

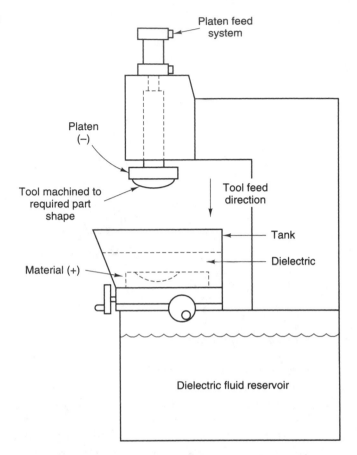

Platen feed system

Platen (−)

Tool machined to required part shape

Tool feed direction

Material (+)

Tank

Dielectric

Dielectric fluid reservoir

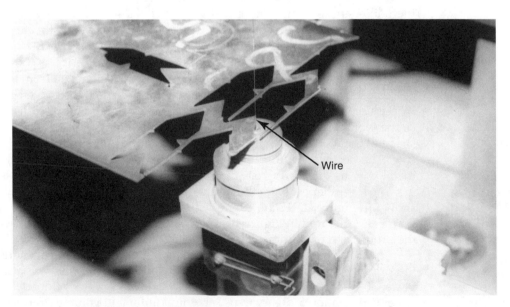

Wire

Figure 10–17 Wire EDM Cutting Wire.

Figure 10–18   Parts Produced with Wire EDM.

machining. Chemical milling is an etching process where metal is removed by a chemical. The amount of metal removed or the depth of the etch is related to the chemical used and the immersion time in the solution. Photochemical machining uses a photographic process to place a chemical-resistant image on the surface of a metal sheet. The metal not protected by the resistant coating is then removed through a chemical milling process. Both of these processes are used in a wide variety of applications where removal of a relatively small amount of material is required.

The primary and secondary operations produce the finished geometry for the production part. In many applications, the processing phase of manufacturing ends with the last secondary operation; however, it is often necessary to change the properties of the material or to add a finish to the exterior of the product.

Physical Properties Operations.   The process that changes the physical properties of the part does not change the part geometry significantly. The most common physical property change performed in manufacturing is *heat treating*. Heat treating of ferrous and nonferrous metals is an operation or combination of operations

Figure 10–19   Wire EDM Machine with Fanuc Controller.

involving the heating and cooling of solid metals and alloys to change the microstructure of the material. The changes in the material are generally grouped into one of two categories:

1. Processes that increase the strength, hardness, and toughness of metals through either hardening of the entire part or hardening of the surface metal on the part
2. Processes that decrease the hardness of metals through annealing and normalizing of the part to improve homogeneity, machinability, and formability or to relieve stresses

Setting up heat-treating processes requires a mixture of science and practice. The science of heat treating includes a complex analysis of the chemical and metallurgical properties of metals and metal alloys. Although the science of metal structure and behavior is an important part of the process, much of the success in heat-treating operations results from experience gained in heat treating parts for many years.

Finishing Operations.   Like physical properties, finishing operations do not significantly change the geometry of the finished parts. In most cases, finishing operations add a thin layer to the surface of the part to improve the operational life and serviceability. The finishing operation frequently prevents surface oxidation of the metal used to produce the part, especially in parts produced from steel. The most frequently used finishing operation to reduce oxidation of parts is painting. Other finishing operations commonly used include (1) plating of the parts by using either hot dips or electrolysis, and (2) acid etching or pickling of the surface.

The four basic material process operations use general-purpose production machines to support manufacturing requirements. In many manufacturing situations, however, specialized machines are required for high-volume production. A discussion of these machines and manufacturing systems follows.

10–2 FLEXIBLE MANUFACTURING

Rapid response and production flexibility in manufacturing were identified in chapter 1 as world-class measurement standards for productivity and important *order-winning criteria*. Take a few minutes to review Figures 1–4 and 1–6. In Figure 1–6, *flexibility* refers to the number of different parts that a workstation can produce under normal production conditions. Flexibility in manufacturing is often described as (1) the ability to adapt easily to engineering changes in the part, (2) an increase in the number of similar parts produced on the system, (3) the ability to accommodate routing changes that allow a part to be produced on different types of machines, and (4) the ability to change the system setup rapidly from one type of production to another. In each case, the critical point is that properly designed production areas make production planning and scheduling easier and result in a quicker response to customer needs. The production systems illustration in Figure 10–2 shows how selection of the production system affects manufacturing flexibility.

The *order-winning criteria* listed in Figure 1–4 include flexibility and lead time. Both of these criteria, necessary to maintain or increase market share, are affected by the design of the production area. Group technology, covered in chapter 5, focused on the design of production cells to handle a family of parts with common production characteristics. Further evidence that integrated production cells are critical elements of CIM is provided by the movement toward self-directed work teams. This important human resource development activity, described in chapter 13, emphasizes the grouping of machines into production cells that are managed by a team of workers. Realizing the benefits of CIM and the integrated enterprise in many product areas requires implementation of *flexible manufacturing cells (FMCs)* or *flexible manufacturing systems (FMSs)* on the shop floor.

Flexible Manufacturing Systems

Global market pressures in the early 1980s demanded higher production efficiency, lower cost, and a faster response; as a result, manufacturers installed FMSs for mid- and low-volume production products. An FMS is defined in the *Automation Encyclopedia* (Graham, 1988) as follows:

> *A flexible manufacturing system is one manufacturing machine, or multiple machines that are integrated by an automated material handling system, whose operation is*

managed by a computerized control system. An FMS can be reconfigured by computer control to manufacture various products.

The first FMS, designed in the mid-1960s by a British firm, was called *System 24.* Because of insufficient control and computer technology, the system was never completely installed. The most notable early installation of an FMS in the United States was at Caterpillar Inc. by Kearney & Trecker Corporation. The goals set for the FMS were met or exceeded because the system goals were specific and addressed specialized applications. The FMS did not exhibit the flexibility described by the preceding definition; however, it did satisfy the Kearney & Trecker definition, which follows:

An FMS is a group of NC machine tools that can randomly process a group of parts, having automatic material handling and central control to balance resource utilization dynamically so that the system can adapt automatically to changes in parts production, mixes, and levels of output.

A review of the two definitions indicates some common elements, such as *NC* or smart production machine tools, *automatic material handling, central computer control, production of parts in random order,* and *data integration.* As the definitions indicate, an FMS is a collection of hardware linked together by computer software. The process hardware frequently includes NC and CNC machines like many of those described in section 10–1. In addition, an FMS has production tooling and setup systems, part-cleaning and -deburring stations, raw material and finished-parts automatic storage and retrieval systems (AS/RSs), and coordinate measuring machines (CMMs). The systems are linked with automated material handling, which ranges from less sophisticated belt conveyors to highly sophisticated robots and automatic guided vehicles. A typical FMS layout is illustrated in Figure 10–20. Study the figure until you locate all the elements in the FMS.

The FMS shown in Figure 10–20 is designed to produce a family of machined metal parts that can be manufactured with three-axis vertical machining centers. Five machining centers are required to meet the production demands; locate them in the figure. Raw material for the parts is delivered to the automatic work changers (number 5 in the figure) and is loaded onto pallets or fixtures that will hold the material for one or more milling operations. Frequently, a partially finished part will be removed from one fixture and placed on another in a different orientation for additional machining. The pallets or fixtures are delivered to the correct machining center cell by the computer-controlled cart or automatic guided vehicle (AGV). The vehicles have no person on board for navigation and use a current-carrying wire embedded into the floor and electronics on the cart for path, direction, and speed control. The material-handling AGVs are also used to deliver tools from the setup and calibration area (9 in the figure) to the tool interchange stations (2 in the figure). From the tool interchange stations, tools are changed automatically on the machining centers to match the requirements of the parts to be produced. Finished parts still mounted to the pallets are delivered by the AGV system to the part-washing station (8 in the figure) prior to inspection and shipping. To track the quality of

1. Five Milacron 5-axis T-30 CNC Machining Centers, 90 tools each

2. Five tool interchange stations, one per machine, accepting tool delivery via cart

3. Three computer-controlled carts, with wire-guided path

4. Cart maintenance station

5. Two automatic work changers, 10 pallets each, with dual load/unload positions with 90° tilt, 360° rotation

6. Two material review stands, for on demand part inspection

7. Inspection module, with LK Tool Co. Metre Four Microvector horizontal arm coordinate measuring machine

8. Automatic part-washing station

9. Tool chain load/unload, tool gauge and calibration gauge stands

10. Elevated computer room, with DEC VAX 8200 central computer

11. Centralized chip/coolant collection/recovery system, with dual flume

------- Flume Path

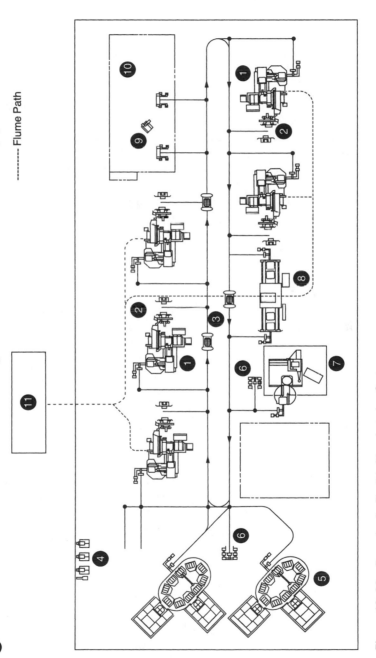

Figure 10–20   Example of a Flexible Manufacturing System.

(Source: Courtesy Cincinnati Milacron, Inc.)

407

finished parts, the AGV delivers machined parts to the CMM (7 in the figure) for automatic inspection or to the manual inspection stations (6 in the figure). A centralized chip and coolant recovery system collects metal removed in machining and filters the cutting fluid used in the cells. The cells in the FMS are under area computer control from a central computer (10 in the figure).

The hardware in the FMS is interfaced to computer controllers at several levels: (1) the sensors, gauges, switches, and controls supporting the operation of a machine are linked directly to the computer in the machine through the discrete signal input-output interface (Figure 10–21a) or the devices are connected to a cell control computer that interfaces with the machine computer (Figure 10–21b), (2) computer-controlled machines located in a manufacturing cell are linked together by a cell computer controller (Figure 10–21b), (3) the automated cells form an FMS when the cells are linked to an area controller (Figure 10–22); and (4) the FMS is linked to the system controller or enterprise computer and the intranet/Internet (Figure 10–22). Sophisticated computer software is required in this complex FMS

Figure 10–21   FMS Computer Control Options.

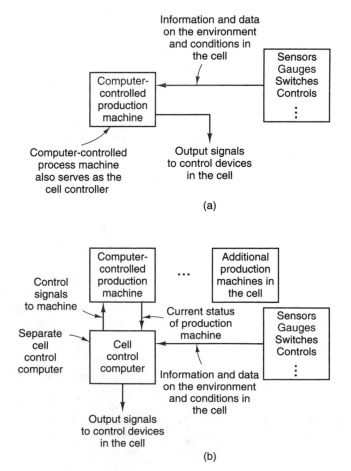

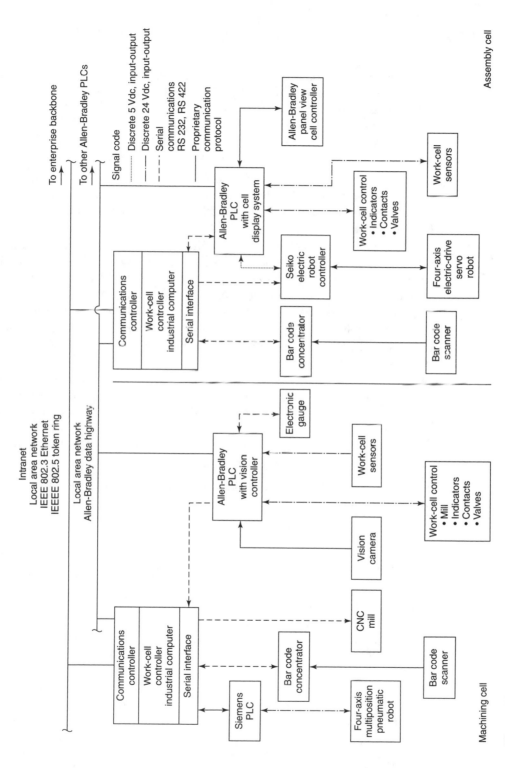

Figure 10–22 Multicell Control Through an Area Computer Controller.

with multiple control levels to handle the high degree of variability associated with the production of many different parts. The degree of complexity is evident when the levels of hardware and software present in the FMS are identified. Typically, a minimum of five technology levels are present in an FMS. For example, the five levels of technology present in the FMS shown in Figure 10–20 are listed next (the numbers refer to the equipment in the figure):

1. *Enterprise level:* Scheduling production requirements for the FMS, preparing computer programs and code for the production system and machines, generating purchase orders for raw material, and creating shipping documents for finished goods

2. *System level:* Centralized coolant and chip collection system (11); control and scheduling of computer-controlled carts (10); downloading of computer code for production machines (10); synchronization of all cell operations (10); central calibration and setup of tools and tooling for the machines (9); and tracking of tooling, raw materials, and finished-goods inventory (10)

3. *Cell level:* Machining cells (1), tool gauge and calibration station (10), material load and unload stations (5), testing and quality control cell (7), and part-washing cell (8)

4. *Machine level:* CNC machining centers (1), manual operations (6), automatic wire-guided carts (3), work holders and changers (5), quality testing machines (7), automatic part-washing machine (8), and tool interchange stations (2)

5. *Device level:* Sensors, ac and dc motors, pneumatic and hydraulic components, tools, fixtures, electrical components, connectors, wire, and fiber optics

The level of complexity is magnified by the number of different hardware and software vendors required to assemble an FMS. Some of the most successful FMSs have been built and integrated by machine tool manufacturers. A review of the operation of the typical system shown in Figure 10–20 and the technology levels just identified indicates that the complexity associated with an FMS is the major disadvantage and a serious obstacle to implementation. However, successfully implemented FMSs offer several tangible advantages:

- *Inventory reduction.* In some implementations, the work-in-process inventory and finished-goods inventory are reduced by over 80 percent. If the FMS permits production of parts in lot sizes of 1, a just-in-time operation is possible, and the finished-goods inventory for that operation is nearly zero.

- *Direct labor cost reduction.* The automation associated with FMS implementations permits a large production system to operate during three shifts with fewer workers than would be necessary when manual moving and loading of raw materials into production machines is practiced. The manual direct labor component for the cost of goods sold (COGS) is typically only about 10 percent, so the reduction in direct labor due to the automation is not as significant an advantage.

- *Machine utilization increase.* Another benefit of automation is the ability to operate expensive production machinery during three shifts and seven days a week. This continuous operation causes improved capital utilization and increased capacity without a significant increase in labor cost. As a result of this increase in capacity, fewer machines are required and the need for production floor space decreases.

- *Support of world-class standards.* The standards used by world-class enterprises to measure manufacturing effectiveness were described in chapter 1 and illustrated in Figure 1–6. A properly designed FMS improves each of these standards: low setup time, high quality, manufacturing space a large percentage of total factory space, low in process and finished-goods inventory, flexibility, small travel distance for material through production, and good machine uptime.

- *Support of order-winning criteria.* When properly designed and implemented, an FMS directly supports the *order-winning criteria* discussed in chapter 1 and listed in Figure 1–4: low price, high quality, short lead times, high percentage of on-time delivery, and good flexibility in manufacturing a product family.

The advantages associated with a successful FMS are impressive, and several systems have been installed around the world that prove FMS technology can work. However, the cost, complexity, and level of technology required to implement an FMS limits FMS solutions to the very large manufacturers. As a result, a greater emphasis in automation design is currently placed on *flexible manufacturing cells* (FMCs).

FMC Versus FMS

An *FMC* is defined in the *Automation Encyclopedia* (Graham, 1988) as follows:

> *An FMC is a group of related machines that perform a particular process or step in a larger manufacturing process.*

According to this definition, the production building blocks used to assemble an FMS are just FMCs. The manufacturing root of the term *cell* is difficult to trace; however, in most cases, the term referred to a production area that had one or more NC or CNC machines. Today, production hardware grouped in the formation of cells is segregated for reasons that include raw material needs, operator requirements, manufacturing cycle times, and group technology. In general, cells can be grouped into two classifications:

1. The traditional stand-alone production cell
2. The automated and integrated production cell

In both cases, the cells can have more than one production machine and operation. In addition, the production machines may be a combination of manual and computer controlled. For example, a traditional stand-alone system may have two CNC machines, a manual punch and brake, and an assembly station so that a complete subassembly is produced from several parts made in the cell. The machines shown in Figures 10–9, 10–10, and 10–12 are good examples of CNC machines used in traditional production cells. Since the automatic tool change capability is not always present, it is important to have an operator available to change tools as required by the machining sequence. The traditional cell usually relies on human operators to load raw material and run the machines. If the number of operators is equal to the number of machines, it is a *one-to-one operation*. Frequently, one operator runs two CNC machines, and the process is called a *two-to-one operation*. The two-to-one operation works especially well on production jobs with long machine cycle times.

As the name implies, the automated and integrated production cell usually has automated material-handling hardware to load and unload one or more computer-controlled machines in the production cell. In cases where a milling operation is placed into a fully automated cell, the machine tools have tool carousels that can hold over 100 tools. As a result, a large variety of milling cutters and drill sizes are available to the programmer as the part cutting geometry and tooling are specified. However, human operators are not excluded; for example, in some automated cells, operators work with industrial robots in the machine load and unload cycle. The welding operation shown in Figure 10–23 is a good example of this type of operation. Study the picture until you can identify the robot and the circular welding table. The robot shown in Figure 10–23 is performing metal inert gas (MIG) welding. The welding table that holds the parts is on the left of the picture and has a plastic screen that separates the robot side of the table from the operator side. The circular table has two fixtures to hold the parts, one on each side of the screen. The operator loads the parts that need to be welded into the fixture and then presses a switch that rotates the circular table to put parts in front of the robot. The rotation of the table delivers to the operator the parts just welded by the robot. While the robot welds the new parts, the operator removes the finished parts and loads up another set to be welded. When the robot finishes the welding process, the table is again rotated by the operator and the process is repeated. The screen allows the worker to see the welding process to verify that everything is correct but protects the operator from the sparks, flash, and vision-damaging ultraviolet light from the welding process. The operator and robot are performing as a team.

In other cases where workers are in an automated cell, operators perform intricate secondary operations, such as deburring or inspection, on production parts after a robot unloads the part from the machine. The automated cell is also characterized by multiple levels of computer control and a complex control algorithm. Integrating the automated cells into the CIM computer network and database is usually easier than the task of bringing an entire FMS up to speed. The cells that form the FMS in Figure 10–20 are good examples of automated and integrated production cells.

It is difficult to determine when a large FMC becomes an FMS. One distinguishing factor is the level of control above the cell control hierarchy for the synchronization of the cells. If area controller software is used to control the sequence of operation at the cell level, the system is crossing from an FMC to an FMS. Other

Figure 10–23   Welding Robot Work Cell.

factors that differentiate FMCs from FMSs are tool capacity and flexibility in manufacturing. As the software used to control the production environment improves, the number of automated production systems that are true FMSs will increase. The control strategies used for FMCs and FMSs are described in chapter 11.

10–3 FIXED HIGH-VOLUME AUTOMATION

A manufacturing system capable of producing large volumes of discrete parts is necessary for the economic production of some products. A good example of this type of product is the disposable razor or lighter. Although the production rate is much less than that of disposable razors, discrete-parts production in the automotive industry for items such as engine crankshafts also falls into this category of manufacturing. Manufacturing systems capable of satisfying this type of production are called *transfer machines* or *transfer lines*. The many different systems used for large-volume discrete-parts production are collectively called *Detroit-type automation*.

In-Line Fixed Automation

In-line fixed automation systems, similar to the system illustrated in Figure 10–24, have the following characteristics:

- A series of closely spaced production stations are linked by material-handling devices to move the parts from one machine to the next.
- There is a sequential production process, with each station performing one of the process steps and the cycle time at each machine usually about equal.
- Raw material enters at one end and finished parts exit at the other.
- The number of stations in the system is dictated by the complexity of the production process implemented.
- The production stations used in the system could include process machines such as mills, drills, and lathes; automated assembly stations; and automated inspection equipment.

Although most machines of this type are fully automated, it is possible to include manual stations along the line to handle operations that are either not economical or are difficult to automate. Figure 10–25 shows an in-line fixed automation production system that performs a series of manufacturing and assembly operations on small products. Frequently, buffer storage is placed in the in-line production system to smooth out work-flow irregularities and to continue to supply parts when maintenance is necessary on an upstream station.

Rotary Fixed Automation

Rotary fixed automation locates the production stations around a circular table or dial. As a result, this type of production system is called an *indexing machine* or a *dial index machine*. The system pictured in Figure 10–26 has seven production stations around the machine. In both this system and the in-line system shown in Figure 10–25, automatic parts feeders supply the pneumatic robots with components for the final assembly. One type of parts feeder, a *vibrating bowl feeder,* is pictured in Figure 10–27. A rotary fixed automation system is limited to smaller workpieces and fewer production stations than those in the in-line type.

Selection Criteria for Fixed Automation Systems

Automation of this type generally requires a product with a high degree of stability and a long life. For example, the production of a specific type of automotive engine crankshaft could have a five-year production life because the same

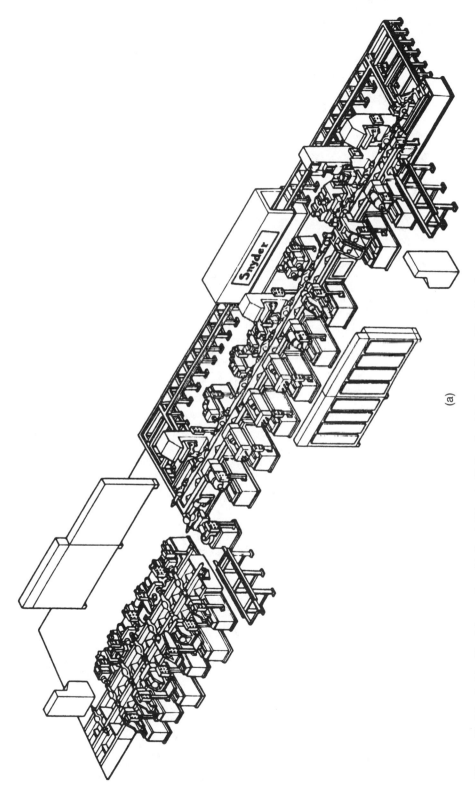

Figure 10–24 In-Line Fixed Automation: (a) Special Twenty-Station Combination Pallet ar d Free Transfer Machining System for Truck Rear-Axle Housing; (b) Truck Rear Axle Produced on the System.
(Source: Courtesy Snyder Corp.)

(a)

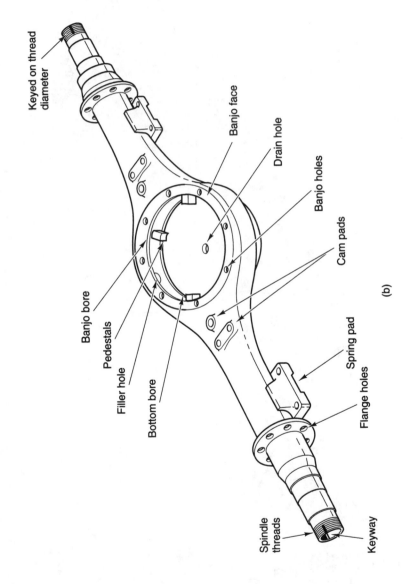

Keyed on thread diameter

Banjo face

Drain hole

Banjo holes

Cam pads

Banjo bore

Pedestals

Filler hole

Bottom bore

Spring pad

Flange holes

Spindle threads

Keyway

(b)

Figure 10–24 Continued.

Figure 10–25   In-Line Fixed Automation System for Small Parts.
(Source: Courtesy of Metro-fer, Inc., Pittsburgh, PA.)

engine is used in several different car models. Even different engine models could use the same crankshaft. In addition, products made on these machines have high production rates that are driven by high product demand. Also, this type of automation is selected when the labor content would be excessive if the part were produced manually. This type of automation improves on the following world-class metrics from Figure 1–6: short setup time, improved quality, small floor-space requirement, reduced in-process inventory, and minimum distance traveled in production. In addition, *order-winning criteria* from Figure 1–4 that are supported include lower product cost, better quality, less lead time, and improved delivery performance.

10–4 SUMMARY

The processes in manufacturing are grouped into four operation categories: primary, secondary, physical properties, and finishing operations. Typical primary operations include casting, forming, oxyfuel and arc cutting, and sawing. Typical secondary operations include turning, boring, milling, drilling, reaming, grinding,

Figure 10–26   Indexing, or Dial Index, Production System.
(Source: Courtesy of Metro-fer, Inc., Pittsburgh, PA.)

and nontraditional machining. The physical properties of the part are often changed with a heat-treating process where the material is hardened in some cases and in other cases the hardness is reduced. The final processing group is finishing, where painting, plating, and pickling are used to change the surface of the part.

To achieve the flexibility and rapid response required by customers, many manufacturers are turning to flexible manufacturing systems (FMSs). An FMS is one or more machines integrated with automated material handling and managed by a computer control system. Typically five levels of technology are present in an FMS: enterprise, system, cell, machine, and device. Successful implementation of FMSs provides the following benefits: inventory reduction, direct labor cost reduction, machine utilization increase, and support for world-class standards and order-winning criteria.

A building block of the FMS is the flexible manufacturing cell (FMC). The FMC is a group of related machines that perform a particular process or step in a larger manufacturing process. The FMC operation falls into one of the following two categories: (1) traditional stand-alone production cell, or (2) automated and integrated production cell.

Figure 10–27   Vibrating Bowl
Feeder.
*(Source: Courtesy of Metro-fer, Inc.,
Pittsburgh, PA.)*

Fixed automation is a manufacturing system focused on high-volume production of very specific parts or part mixes. Machines capable of this type of production are often called *transfer lines* or *Detroit-type automation*. Fixed automation systems are generally configured by using either in-line or rotary geometries.

BIBLIOGRAPHY

GOETSCH, D. L. *Modern Manufacturing Processes*. New York: Delmar Publishers, 1991.

GRAHAM, G. A. *Automation Encyclopedia*. Dearborn, MI: Society of Manufacturing Engineers, 1988.

GROOVER, M. P. *Automation, Production Systems, and Computer-Integrated Manufacturing.* 2nd ed. Upper Saddle River, NJ: Prentice Hall, 2001.

HAGGEN, G. L. "History of Computer Numerical Control." *Industrial Education,* August/September 1990, 14–16.

LUGGEN, W. W. *Flexible Manufacturing Cells and Systems.* Upper Saddle River, NJ: Prentice Hall, 1991.

OWEN, J. V. "Flexible Justification for Flexible Cells." *Manufacturing Engineering,* September 1990.

PALFRAMAN, D. "FMS: Too Much, Too Soon." *Manufacturing Engineering,* March 1987.

REHG, J. A. *Introduction to Robotics in CIM Systems.* 5th ed. Upper Saddle River, NJ: Prentice Hall, 2003.

SOBCZAK, T. V. *A Glossary of Terms for Computer-Integrated Manufacturing.* Dearborn, MI: CASA of SME, 1984.

QUESTIONS

1. What are *process operations*?
2. Name the four categories of process operations.
3. What is the function of the primary operations?
4. Describe two primary operations.
5. What is a *secondary operation*?
6. Describe two secondary operations.
7. What is included in the physical properties operations?
8. Describe the two categories of physical property changes.
9. What are *finishing operations*?
10. What are the attributes of flexible manufacturing?
11. Compare and contrast flexible manufacturing cells and flexible manufacturing systems.
12. Define an FMS in your own words.
13. Describe the five technology levels that may be present in an FMS.
14. What advantages do FMSs offer?
15. Define *FMC*.
16. How are cells typically grouped in an FMC?
17. What distinguishes a large FMC from an FMS?
18. What is *Detroit-type automation*?
19. What are the characteristics of in-line fixed automation?
20. Describe the selection criteria for fixed automation systems.

PROJECTS

1. Using the list of companies developed in project 1 in chapter 1, develop a matrix to show the process operations used at the companies. Organize the operations in the primary, secondary, physical properties, and finishing groups.

2. Using the list of companies developed in project 1 in chapter 1, determine which companies are using flexible manufacturing cells or flexible manufacturing systems.

3. Select one of the companies in project 2 that uses either an FMC or an FMS and develop a case study on the implementation.

4. Select one of the companies in project 1 that uses either an in-line fixed automation system or rotary fixed automation and develop a case study based on the implementation.

The following projects are a continuation of projects 15 through 18 at the end of chapter 5.

5. A company that manufactures kitchen products wants to add a set of plastic storage bowls or containers to its product line. The company wants to distribute the product through Wal-Mart, Target, or a chain of stores in your region. The production cell will use plastic-injection-molding machines to form the bowls. Use the Internet to locate vendors of this type of machine, select machines that could produce the bowl set, and design a work-cell layout for the product.

6. A company that manufactures kitchen products wants to add a high-end manual can opener to its product line. The company wants to distribute the product through Wal-Mart, Target, or a chain of stores in your region. Do the following:

   a. Determine the manufacturing process suitable for the production of each part of the product.

   b. Design a rotary fixed automation system that would assemble the product. Draw a top view of the system, illustrating the number of stations.

7. A company that manufactures kitchen products wants to add a manual hand-held eggbeater to its product line. The company wants to distribute the product through Wal-Mart, Target, or a chain of stores in your region. Do the following:

   a. Determine the manufacturing process suitable for the production of each part of the product.

   b. Design a rotary fixed automation system that would assemble the product. Draw a top view of the system, illustrating the number of stations.

8. A company that manufactures kitchen products wants to add a manual handheld garlic press to its product line. The company wants to distribute the product through Wal-Mart, Target, or a chain of stores in your region. Do the following:

   a. Determine the manufacturing process suitable for the production of each part of the product.

b. Design a rotary fixed automation system that would assemble the product. Draw a top view of the system, illustrating the number of stations.

APPENDIX 10–1: HISTORY OF COMPUTER-CONTROLLED MACHINES

CIM is a process, adopted by a manufacturer of goods and/or a provider of services, to increase productivity and serve customers more effectively. The definition of CIM in chapter 1 emphasizes the integration of systems and the use of data communications to achieve these increased productivity and customer service goals. Although the term *computer* is absent from the definition, the implementation of any CIM system requires the use of computer-driven devices and machines. Therefore, an overview of the development of computer-controlled machines provides a perspective on the development of CIM technologies and some insight into the direction of future technology initiatives.

Computers were included in production machines to speed the processing of data, perform arithmetic calculations, and make decisions on the basis of production conditions. One of the first examples of a counting machine was the *abacus,* used over 1,500 years ago. Many centuries later, in 1620, William Aughtred and others invented the *slide rule* to handle calculations more rapidly. In that same century, Blaise Pascal, a French philosopher and mathematician, and Gottfried Wilhelm developed wheel-and-cog devices that performed the four basic mathematical operations.

It was not until the eighteenth century that production machines were interfaced with crude control devices. In 1804, Joseph M. Jacquard built an automated loom capable of complex weaving designs, using hole patterns in wooden boards as a program. Later, in 1833, an Englishman, Charles Babbage, designed two devices: the *difference engine* and the *analytical engine.* Although not built successfully by Babbage, these two devices were the forerunners of mechanical adding machines and cash registers. In 1816, another British mathematician, George Boole, developed what was to be the universal language for computers. His *Boolean algebra* used statements that were either true or false and involved three basic operations: *AND, OR,* and *NOT.* The counting of the U.S. census data was automated in 1890 by placing the data on punched cards and reading the data from the cards with an automated counter. The previous census processing time was reduced from 7½ years to 6 weeks for the tabulation and 2½ years for the statistical analysis. As a result of the success of this project, the developer of this punched-card process, Herman Hollerith, formed the Tabulating Machine Company, which later became the International Business Machine Corporation (IBM).

The first application of Boole's universal language to the solution of computational problems occurred in 1936, when Claude Shannon, a 21-year-old graduate student, published a paper entitled "A Mathematical Theory of Communications." During the next four years, George Stibitz made three significant contributions to computing: (1) the first electromechanical circuit using Boolean algebra, (2) the first working information network in the United States, and (3) remote computing through the use of a terminal. Following this early work, a series of fully functional computers were built for the first time. The *Mark I*, an electromechanical computer built by IBM in 1944, was 51 feet long and had 750,000 parts connected with 500 miles of wire. These early machines were designed for computational solutions; applications in machine control for manufacturing was only a dream.

The development of the transistor in the early 1950s and the integrated circuit in the early 1960s began the first revolution in machine control. For the first time, computer technology could be applied practically to the problem of machine control in manufacturing. Some of the first machines to benefit from computer technology were the machine tool and the robot. In 1952, a project at the Massachusetts Institute of Technology (MIT) sponsored by the Air Force produced the first numerically controlled (NC) milling machine. Each machine movement was assigned a code. The code, along with the distance to travel, was fed into the machine on 1-inch-wide paper tape on which the location of punched holes provided the program information. In the late 1950s, the first NC programming language, called *automated programmed tools* (*APT*), was developed by MIT. The computer simplified the manufacturing automation problem by converting the English language–like statements in APT into the elaborate codes needed to cut complex parts on a machine.

In the early 1950s, George Duval, the inventor of the industrial robot, developed a drum programming unit that controlled the motion of his early robot designs. The programming drum had adjustable protrusions on the surface that triggered motion in the robot arm as the drum rotated. Borrowing technology from the developing computer industry, this mechanical programming device was later replaced by electronic programming units. As in machine tools, mechanical programming gave way to high-level computer languages used to program robots, in which the computer translated the language statements into robot motion.

The development of the microprocessor in the early 1970s started a second revolution in machine control for manufacturing automation. *Direct numerical control* (*DNC*), which connected machine tools directly to host computers, was replaced by *computer numerical control* (*CNC*). A CNC machine tool has a dedicated computer integrated into the machine controller and is capable of storing programs to cut complex parts inside the machine tool. DNC operation is still used, but now the host is only a storage repository for part programs that are downloaded to the CNC machine's computer for execution. The development of the microprocessor and the microcomputer throughout the 1970s and 1980s permitted computer control to be added to almost every manufacturing device and system on the shop floor. The capability of current machine controllers exceeds the demands of most current applications. For example, controllers such as the BostoMatic and GMFanuc have the following features:

- Spindle speeds as fast as 40,000 revolutions per minute
- Simultaneous 200-inch-per-minute, simultaneous five-axis contouring
- Analog and discrete cell control
- Cell control of twenty axes and over 200 discrete inputs and outputs

- Full-color three-dimensional operator interfaces with real-time display of cutting and control sequences
- Storage of up to 999 part programs in the machine
- Interface capability to the enterprise's computer networks
- Software to support the manufacturing database and management information system
- A full 32-bit computer architecture for speed and accuracy

Prior to 1980, applications for computer control in manufacturing were limited by the ability of the technology to support the application. Since that date, the only limiting factor has been the ability of manufacturers to apply technology tools present in the marketplace.

Production Support Machines and Systems

OBJECTIVES

After completing this chapter, you should be able to do the following:

- Define *industrial robotics* and describe a basic robot system
- Name and describe the four robot geometries
- Describe the operation of the two major types of robot tooling—standard grippers and vacuum devices—and give some examples of process tools
- Describe the operation of a servo robot controller and a robot teach station
- List and describe the three categories of robot applications and give an example of each
- Describe the selection process used in selecting a robot for an application
- Write a robot program to perform a basic manufacturing application
- Name the two goals of a transfer function
- List and describe the three categories of transfer systems and give an example of each
- Describe the function and operation of the six types of automatic guided vehicles
- Describe the function and operation of automatic storage and retrieval

The production machinery and systems described in chapter 10 add value to the product when the process is acting on the raw material. However, in many manufacturing operations, partially finished parts and raw material spend days or even months in the shop, being moved, waiting for an available production machine, being loaded and unloaded from production machines, and being inspected. Reducing or eliminating these non-value-added operations is the goal for world-class companies; however, completely eliminating these cost-added operations is often not possible. As a result, production hardware such as robots, material-handling systems, and storage systems is necessary. Therefore, analysis of automation systems for ways to minimize the impact of these cost-added operations is important and starts with a description of industrial robots.

11–1 INDUSTRIAL ROBOTS

Industrial robots are given a separate section in this chapter because they are unique production machines. Although several different production machines may appear to have the capacity to work like a robot, on closer inspection the robot is distinctly different. Without a definition, it would be difficult to differentiate industrial robots from the millions of automated machines used by industry. The Information Standards Organization (ISO) defines an *industrial robot* in standard ISO/TR/8373-2.3 as follows:

> *A robot is an automatically controlled, reprogrammable, multipurpose, manipulating machine with several reprogrammable axes, which may be either fixed in place or mobile for use in industrial automation applications.*

The key words are *reprogrammable* and *multipurpose* because most single-purpose machines do not meet these two requirements. *Reprogrammable* implies two characteristics: (1) the robot's motion is controlled by a written program, and (2) the program can be modified to change the motion of the robot arm significantly. Programming flexibility is demonstrated, for example, when the pickup point for a part is located by a vision system, and the location is sent to the robot while the robot is moving to the part. *Multipurpose* emphasizes the fact that a robot can perform many different functions, depending on the program and tooling currently in use. For example, a robot could be tooled and programmed in one company to do welding, and in a second company the same type of robot could be used to stack boxes on a palletizer.

The Basic Robot System

A *robot system* is defined by the Robotic Institute Association as follows:

> *An industrial robot system includes the robot(s) (hardware and software), consisting of the manipulator, power supply, and controller; the end-effector(s); any equipment, devices, and sensors with which the robot is directly interfacing; any equipment, devices, and sensors required for the robot to perform its task; and any communications interface that is operating and monitoring the robot, equipment, and sensors.*

A basic robot system is illustrated in Figure 11–1. The system includes a *mechanical arm* to which the *end-of-arm tooling* is mounted, a *computer-based controller* with attached *teach station, work-cell interface,* and *program storage device*. In addition, a source for pneumatic and/or hydraulic power is part of the basic system. The work-cell controller, a part of the manufacturing system, connects to the robot through the robot work-cell interface.

Mechanical Arm.   The arm is a mechanical device driven by *electric drive motors, pneumatic devices,* or *hydraulic actuators*. The basic drive elements will be either

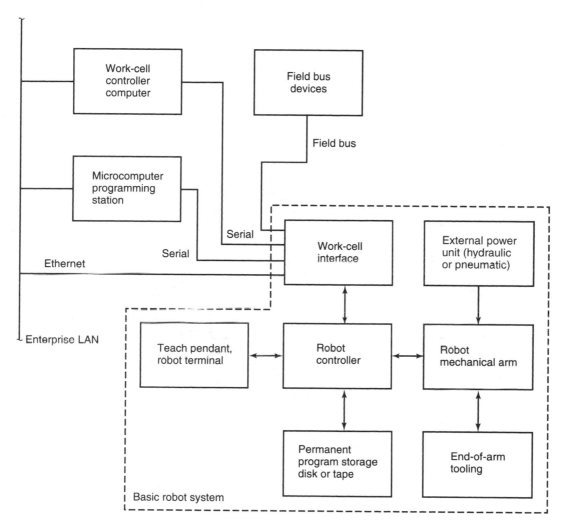

Figure 11–1 Basic Robot System.

(Source: James A. Rehg, Introduction to Robotics in CIM Systems, *Fifth Edition, © 2003, p. 26. Reprinted by permission of Prentice Hall, Upper Saddle River, New Jersey.)*

linear or *rotary* actuators. The combination of motions included in the arm will determine the type of arm geometry that is present. The basic geometries include *rectangular, cylindrical, spherical,* and *jointed spherical.* Figure 11–2 shows a jointed-spherical mechanical arm with all the motions indicated.

The six motions are divided into two groups. The first group, called the *position motions,* includes the *arm sweep, shoulder swivel,* and *elbow extension.* With a combination of these three motions, the arm can move to any position required within the work area. The second group of motions, associated with the wrist at

Figure 11–2 Jointed-Spherical Mechanical Arm.
(Source: Courtesy of ABB Robotics, Inc.)

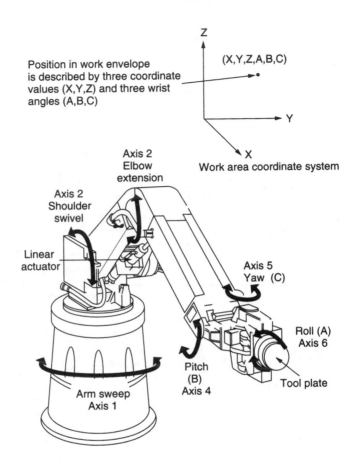

Position in work envelope is described by three coordinate values (X,Y,Z) and three wrist angles (A,B,C)

(X,Y,Z,A,B,C)

Work area coordinate system

Axis 2 Elbow extension

Axis 2 Shoulder swivel

Linear actuator

Axis 5 Yaw (C)

Roll (A) Axis 6

Tool plate

Pitch (B) Axis 4

Arm sweep Axis 1

the end of the arm, includes the *pitch, yaw,* and *roll.* A combination of these three motions, called *orientation,* permits the wrist to orient the tool plate and tool with respect to the work.

Robot Arm Geometry Classification. In general, the basic mechanical configurations of the robot manipulator are categorized as *Cartesian, cylindrical, spherical,* and *articulated.* The Cartesian is divided into two groups, *traverse axes* and *gantry,* and the articulated is divided into the *horizontal* and *vertical* groups. A description of these configurations follows.

Cartesian geometry. A robot with a *Cartesian geometry* can move its gripper to any position within the cube or rectangle defined as its working volume. Two configurations form this geometry, *traverse* and *gantry* machines. Figure 11–3 shows an example of a gantry machine that can move in the *x, y,* and *z* directions as indicated. The tooling is attached to the tool plate, so work is performed from above. The rectangular work envelope of this type of robot is often used to move parts from conveyor systems into production machines. In the gantry illustration, the

Figure 11–3   Gantry-Type Robot.

three degrees of freedom for positioning are indicated by arrows to show movement in the x, y, and z directions. Three degrees of freedom, A, B, and C, are provided on the wrist to orient the tool mounted on the tool plate. The second type of Cartesian geometry, traverse, is pictured in Figure 11–4.

Cylindrical geometry.   *Cylindrical* coordinate robot systems, like the one pictured in Figure 11–5, can move the gripper within a volume that is described by a cylinder. The cylindrical coordinate robot gripper is positioned in the work area by two linear movements from the vertical z-axis and horizontal extension, and finally one angular rotation (base rotation) about the z-axis. The axes on cylindrical coordinate robots are driven pneumatically, hydraulically, or electrically. The robot shown in Figure 11–5 is a small cylindrical coordinate robot.

Spherical geometry.   The *spherical geometry* arm, sometimes called *polar,* is illustrated in Figure 11–6. Spherical-arm-geometry robots position the wrist through two rotations and one linear actuation. As in the previous cases, the orientation of the tool plate is achieved through three rotations in the wrist (A, roll; B, pitch; C, yaw). In

Figure 11–4   Traverse
Cartesian Geometry Robot.
*(Source: Courtesy of Metro-fer, Inc.,
Pittsburgh, PA.)*

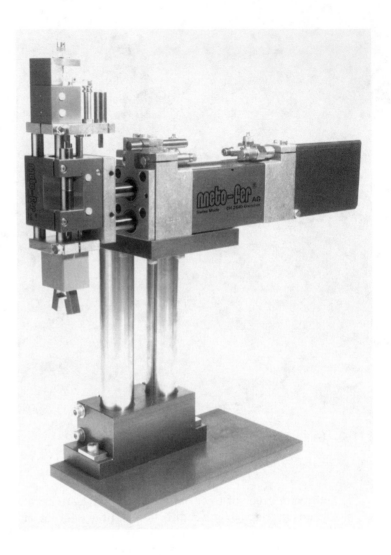

theory, the rotation about the *y*-axis could be 180 degrees or greater, and the wrist rotation about the *z*-axis could be 360 degrees. Then if *R*, robot reach, went from the retracted to the fully extended position, the volume of operating space defined would be two concentric half-spheres. The actual working volume is much less than the theoretical volume of the machine because of mechanical design constraints. Spherical geometry machines use either *hydraulic* or *electric* drives as the prime movers on the six axes, with pneumatic actuation used to open and close the gripper. Few machines of this type are sold currently.

Articulated geometry.   *Articulated* industrial robots, often called *jointed-arm, revolute,* or *anthropomorphic* machines, have an irregular work envelope. This type of robot has two main variants, *vertically articulated* and *horizontally articulated*. The

Figure 11–5 Small Cylindrical Coordinate Robot from Seiko Corp.

(Source: James A. Rehg, Introduction to Robotics in CIM Systems, Fifth Edition, © 2003, p. 61. Reprinted by permission of Prentice Hall, Upper Saddle River, New Jersey.)

vertically articulated robot (Figure 11–7) has three major angular movements, consisting of a base rotation (axis 1), shoulder joint (axis 2), and forearm joint (axis 3). The irregular work envelope is illustrated in Figure 11–8. As in the previous arm designs, the orientation of the tool plate is provided by the three rotations in the wrist. Electric drives with feedback control systems are used on most machines. An example of a jointed-spherical configuration is illustrated in Figure 11–9. Note

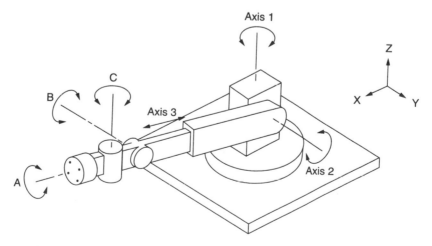

Figure 11–6 Spherical Coordinate Robot Geometry.

(Source: James A. Rehg, Introduction to Robotics in CIM Systems, Fifth Edition, © 2003, p. 62. Reprinted by permission of Prentice Hall, Upper Saddle River, New Jersey.)

Figure 11–7 Vertically
Articulated Jointed-Arm Robot.

*(Source: James A. Rehg, Introduction to
Robotics in CIM Systems, Fifth Edition,
© 2003, p. 64. Reprinted by permission
of Prentice Hall, Upper Saddle River, New
Jersey.)*

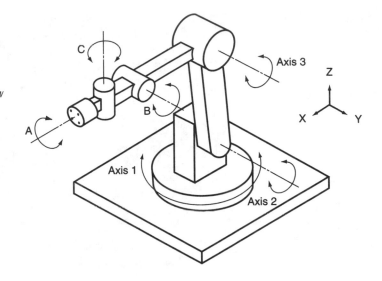

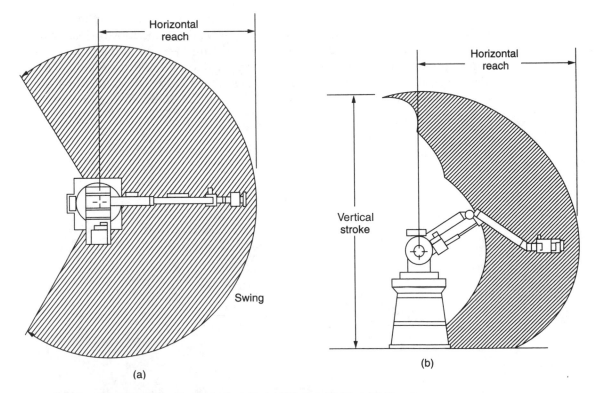

(a)

(b)

Figure 11–8 Work Envelope of a Vertically Articulated Robot: (a) Plan; (b) Elevation.

(Source: James A. Rehg, Introduction to Robotics in CIM Systems, Fifth Edition, © 2003, p. 64. Reprinted by permission of Prentice Hall, Upper Saddle River, New Jersey.)

Figure 11–9 Spot Welding Body Assemblies with a Jointed-Arm Robot.

the ball-shaped *three-roll wrist* that generates the orientation motion in all three wrist axes for the spot-welding tooling attached to the tool plate.

The *horizontally articulated* robot arm has two angular movements for positioning the tooling (arm and forearm rotation), and one-position linear movement that is a vertical motion. Horizontally articulated arms are implemented with two mechanical configurations: (1) the *SCARA (selective compliance articulated robot arm)* illustrated in Figure 11–10, and (2) the *horizontally based jointed* arm.

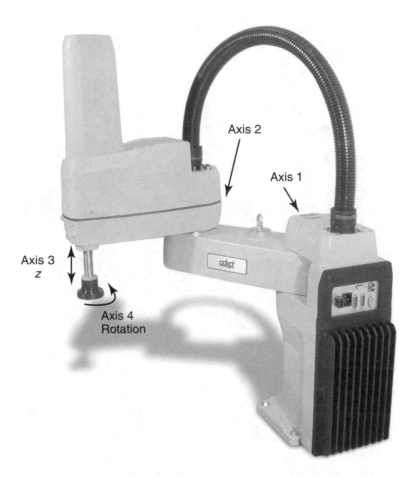

Figure 11–10   Selective Compliance Articulated Robot Arm (SCARA).
(Source: Courtesy of Adept Technology, Inc.)

The SCARA machine (Figure 11–10) has two horizontally jointed arm segments fixed to a rigid vertical member. Positions within the cylindrical work envelope are achieved through changes in axis numbers 1 and 2. Vertical movement of the gripper plate results from the z-axis (axis 3) located at the end of the arm. SCARA machines usually have only one wrist axis (axis 4) rotation. This arm geometry is frequently used in electronic circuit board assembly applications because this geometry is particularly good at vertical part insertion.

The horizontally based jointed arm uses the same construction as the SCARA with one exception; the vertical z-axis is located between the rigid vertical base and the shoulder joint of the upper part of the arm. With this configuration, the wrist and the gripper mounting plate are located at the end of the forearm. Robots produced by Reis Robotics made this cylindrical geometry popular.

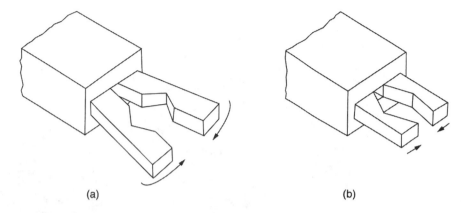

(a)   (b)

Figure 11–11   (a) Angular Gripper; (b) Parallel Gripper.
(Source: James A. Rehg, Introduction to Robotics in CIM Systems, *Fifth Edition, © 2003, p. 163. Reprinted by permission of Prentice Hall, Upper Saddle River, New Jersey.)*

Production Tooling.   The robot arm alone has no production capability, but the robot arm interfaced to production tooling becomes an effective production system. The tooling to perform the work task is attached to the tool plate at the end of the arm.

The tooling is frequently identified by several names. The term used to describe tooling in general is *end-of-arm tooling* or *end effector*. Tooling with an open-and-close motion to grasp parts is most often called a *gripper*. Figure 11–11 illustrates two standard gripper configurations: angular and parallel. The end-of-arm tooling used on current robots can be classified in the following three ways:

1. According to the method used to hold the part in the gripper
2. On the basis of the special-purpose tools incorporated in the final gripper design
3. On the basis of the multifunction capability of the gripper

The first classification of grippers includes *standard grippers,* which have either an angular or a parallel mechanism, as illustrated in Figure 11–11. A set of dual parallel grippers, shown in Figure 11–12, is used to pick up cylinder blanks for loading into a turning center. When two grippers are mounted on the tooling for machine-loading applications, motion efficiency occurs in the robot program. The robot picks up a blank part in one gripper and moves to the machine. The other gripper removes the finished part from the machine, then the robot loads the blank before moving away with the finished part. The load and the unload operations are performed with one move to the machine.

Standard grippers are the most frequently used tooling, followed by *vacuum tooling* in the form of vacuum cups and vacuum surfaces. In Figure 11–13, a robot with vacuum tooling is placing a front window into a car on an assembly line. In this

Figure 11–12 Two Parallel Grippers Loading a Turning Center with Steel Cylinders.

application, four vacuum cups are used to lift the glass and hold it securely while installation occurs. There are also a number of less frequently used types of tooling, including *magnetic grippers*, which use electromagnets for lifting, and *air-pressure grippers*, which use an air-filled bladder or bellows to squeeze parts for lifting.

Figure 11–13   Vacuum Grippers Used to Install the Front Windshield in Cars.

The second classification of tooling includes *drills, spot-welding tooling* (Figure 11–9), *weld torches* (Figure 11–14), *paint sprayers* (Figure 11–15), and *grinders*. Figure 11–9 shows a robot performing spot welding on a car body as the car moves past the robot work cell. The robot program moves the robot at the same speed as that of the moving car, so the weld is placed correctly and the assembly line is not required to stop. In Figure 11–14, two robot arms weld a tubing product on a weld-positioning device. A robot with a third type of tool, a paint-spraying gun, is pictured in Figure 11–15 with a second robot arm that is used to open car trunk lids and doors. In this application, robots that travel horizontally at the same speed as that of the car paint the moving car. One robot opens trunk lids and doors when the paint-spraying robot needs access to the interior to paint surfaces inside the vehicle.

The third type of gripper tooling includes *special-purpose grippers* that are designed for a specific task. For example, a robot with special tooling could lift a folded box and open it for filling.

When any manufacturing system is being set up, some criteria must be used for selection of the equipment. When robots are included in the automation, an

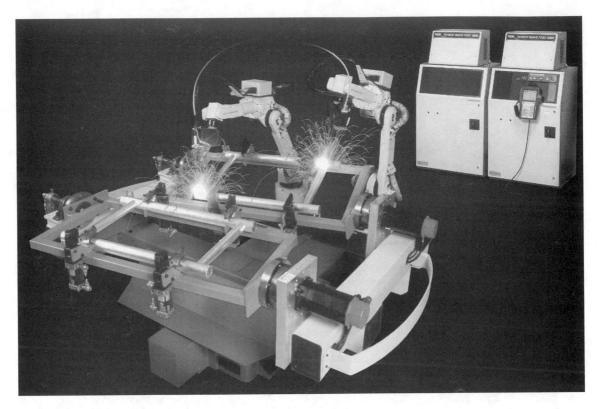

Figure 11–14   Dual-Arm Welding Robot Work with a Part on a Fixture.
(Source: Courtesy of FANUC Robotics, America, Inc.)

analysis of each sequence will establish the requirements of the gripper for each step of the manufacturing process. Every sequence in the manufacturing process should be examined and the relative difficulty established. The robot system capability, including the end-of-arm tooling, must be equal to or greater than the most demanding sequence in the process.

Robot Controller.   Of all the hardware building blocks of a robot system, the controller is the most complex. It is also the unit with the greatest degree of variation from one manufacturer to the next. Figure 11–16 is a block diagram of a typical controller used on electric robots.

The controller, basically a special-purpose computer, has all the elements commonly found in computers, such as a central processing unit (CPU), memory, and input and output devices. Most controllers have a network of CPUs, usually standard microprocessors, each having a different responsibility within the system. The distributed microcomputer network in the controller has the primary responsibility for controlling the robot arm and controlling and/or communicating with the work cell in which it is operating.

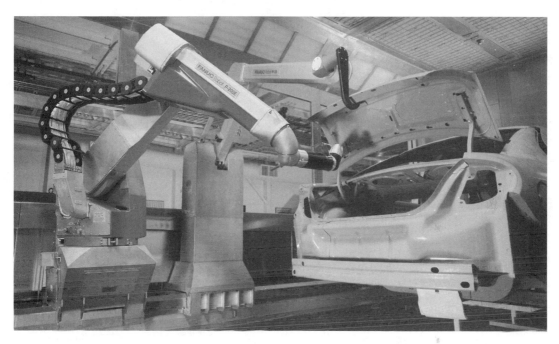

Figure 11–15   Paint-Spraying Robot.
(Source: Courtesy of FANUC Robotics, America, Inc.)

Robot controllers fall into two categories: *closed loop,* with feedback from every axis to the controller to indicate the current status of the arm, and *open loop,* where no feedback is used. Closed-loop machines are often called *servo* robots, and open-loop machines, *nonservo.* Servo robots are most often powered by alternating current (ac) or direct current (dc) servo drive motors, while nonservo robots are pneumatic drive–type machines. In servo robots, the controller receives velocity and position feedback from sensors on the robot arm. On the basis of a comparison between the current position and velocity of the arm and the program describing the desired path, the controller changes the voltage to the servo drive motors to correct for movement and velocity errors. Servo robot controllers are usually special-purpose computers developed by the robot arm manufacturers to control the arm motion. The controller in servo robots also communicates with external devices through the input-output interface shown in Figure 11–16.

The controllers for nonservo, or open-loop, robots are frequently programmable logic controllers (PLCs) supplied by PLC vendors like Allen-Bradley, Siemens, or AutomationDirect. The PLC is a special-purpose industrial computer that can be programmed to control a sequence of discrete operations by turning on and off output devices on the basis of discrete inputs from sensors. The most frequently used programming language is call *ladder-logic programming.* With this programming language, pneumatic valves are cycled open and closed, which causes pneumatic cylinders to extend and retract. These cylinders or axes are stacked together to form

Figure 11–16 Robot Controller Block Diagram for Electric Robots.

(Source: James A. Rehg, Introduction to Robotics in CIM Systems, Fifth Edition, © 2003, p. 29. Reprinted by permission of Prentice Hall, Upper Saddle River, New Jersey.)

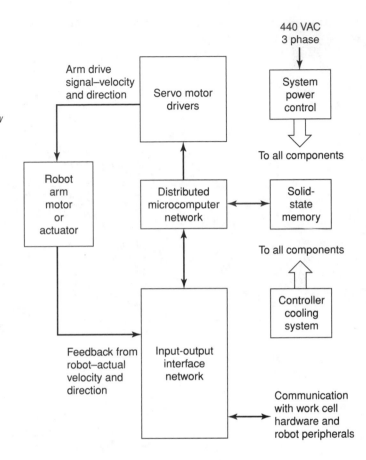

an arm with multiple axes. Figure 11–4 shows a Cartesian geometry robot with a horizontal cylinder, a vertical cylinder, and a gripper attached to the end of the vertical axis. Three pneumatic valves would be used to move the three pneumatic devices, and the valves would be controlled by a PLC program.

Teach Stations.   Teach stations on robots may consist of teach pendants, programming units, controller front panels, and DOS- or Windows-based software running on microcomputers. Some robots permit a combination of programming devices to be used in programming the robot system, whereas others provide only one system programming unit. In addition, some robots have a UNIX or Microsoft operating system in a microcontroller that is used for program creation. Some robot vendors have programming units that are teach pendants with enhanced program creation and editing capability. All of the vendor teach stations support three activities: (1) robot *power-up* and preparation for programming, (2) *entry* and *editing* of programs, and (3) *execution* of programs in the work cell. The development of a program includes keying in or selecting menu commands in the robot controller programming language, moving the robot arm to the desired position in the work cell, and recording the position and command in memory.

Robot Applications

The robot is a very special type of production tool; as a result, the applications for which robots are used are quite broad. These applications can be grouped into the following categories:

- Material processing
- Material handling
- Assembly and fabrication

In the first category, robots use tooling to *process* the raw material. For example, the robot tooling could include a drill and the robot would be performing drilling operations on raw material. The robots shown in Figures 11–14 and 11–15 are performing the process operations of welding and paint spraying. These are two of the most frequently used end-of-arm tools for processes with robots.

Material handling, the second application group, is the most common application for robots. In many cells, robots replace human operators for loading and unloading production machines. The robot in the foreground of Figure 11–17 is loading and unloading two turning centers in a flexible manufacturing system making printing press rollers.

Figure 11–17   Robot Serving as a Material Handler in a Flexible Manufacturing Cell.

Figure 11–18   Robot Used to Assemble Candy Boxes.
(Source: Courtesy Adept Technology, Inc.)

Assembly is another large application area for using robotics. Robots are used to assemble a wide range of products, from cars on assembly lines (Figure 11–13) to the boxes of candy shown in Figure 11–18.

Table 11–1 indicates the wide range of applications for which robots are used. The applications are listed across the top, and the left column lists the models of ABB robots (ABB Robotics, Inc.) available. The bullets in the center indicate which robot models are best suited to the application. Note that one robot works for only one application—poke welding—but some models can be used for most of the applications. This chart indicates the versatility of the robot across a host of manufacturing applications.

Selecting and Justifying Robot Applications—New Design.   A major step in the design process for a *new* automation system or work cell that includes robotics is the selection of the robot system. It is important to understand that the process is different for a new automation cell than it would be to convert a manual cell over to robotic automation. In this type of bottom-up design process, the designer of the work cell(s) has the liberty to set or adjust most of the manufacturing and production variables as part of the design process. As a result, the work cell can be configured (1) to maximize the impact of automation components on the quality of finished products and the payback for the project, (2) to permit the lowest-cost but most-suitable robot model to be used in the work cell(s), and

Table 11–1 Robot Applications

Model of ABB Robot	Application											
	Arc welding	Glueing/ sealing	Coating/ painting	Cleaning/ spraying	Cutting/ deburring	Grinding/ polishing	Assembly	Spot welding	Poke welding	Material handling	Machine tending	Palletizing/ packaging
IRB 1400	●						●				●	
IRB 1400H	●	●			●		●			●	●	●
IRB 2400L	●	●									●	
IRB 2400/10	●	●		●	●		●				●	●
IRB 2400/16				●	●	●				●	●	●
IRB 4400/45		●	●	●	●	●		●		●	●	●
IRB 4400/60			●			●				●	●	
IRB 510		●	●		●	●	●			●		
IRB 6400/ 3.0-75					●						●	
IRB 6400/ 2.4-120						●	●			●	●	
IRB 6400/ 2.4-150							●	●		●	●	●
IRB 6400PE									●	●		
IRB 6400S								●			●	
IRB 6400C								●				
IRB 8000L										●	●	
IRB 8000A										●	●	

(3) to ensure that the likelihood of a successful automation project is close to 100 percent. In the bottom-up design of a new system, the robot is selected on the basis of specifications developed for the overall integration and production requirements of the work cell.

General Work-Cell Design.   The design of an automated work cell requires a detailed study of the attributes of the cell that influence the quality of the automation solution. The checklist in Table 11–2 shows manufacturing variables that should be addressed during the design process. Depending on the function of the cell and the type of automation proposed, the variables most critical to a successful design will change.

The list is divided into six major areas: *performance requirements, layout requirements, product characteristics, equipment modifications, process modifications,* and

Table 11–2   Work-Cell Design Checklist

Performance requirements

Cycle times	Tolerance of parts
Part-handling specifications	Dwell time of tools
Feed rate of tools	Pressure on tools
Product mix	Maximum repair time
Equipment requirements	Malfunction routines
Human backup requirements	Allowable downtime
Future production requirements	

Layout requirements

Geometry of the facility	Service availability
Environmental considerations	Floor loading
Accessibility for maintenance	Safety for machines and people
Equipment relocation requirements	

Product characteristics

Part orientation requirements	Gripper specifications
Surface characteristics	Part size, weight, and shape
Unique handling requirements	Inspection requirements

Equipment modifications

Requirements for unattended operation	Maximum and minimum machine speeds
Requirements for increased throughput	Requirements for automatic operation

Process modifications

Lot-size changes	Routing variations
Process variable evaluation	Process data transfer

System integration

Data interfaces and networks	Interface requirements
Hardware integration requirements	Software integration requirements
Data integration requirements	

system integration. All parameters present in the cell should be addressed by the items in these six categories. To use the list, place the sentence fragment "What is/are the" in front of the checklist element, and place the fragment "for the work cell?" at the end. For example, the questions "What are the *cycle times* for the work cell?" and "What is the *allowable downtime* for the work cell?" illustrate how the checklist elements are used to cover design variables. The data gathered from checklist element questions are comprehensive in scope but contain sufficient detail for you to make design decisions. Study all six sections, and be sure that you understand what work-cell parameters are covered by each element.

Robot Selection Criteria. Development of this more general work-cell design process leads to the robot selection criteria listed next:

- *Positioning resolution, repeatability, accuracy.* These parameters are established by the product's parts, production requirements, and assembly demands.
- *Work envelope size.* This parameter is set by the size and layout of all the production equipment in the work cell.
- *Arm geometry.* The geometry required is dictated by the cell layout and the type of dexterity required by the production and assembly process.
- *Degrees of freedom.* The number of joints is also dictated by the machine locations and the types of moves the tooling must make during production.
- *Positioning flexibility.* This parameter, set by the production requirements, determines the type of wrist movement required and the number of axes necessary for the wrist.
- *Maximum and rated payload.* The *payload,* or lifting limit, for the robot is set by the maximum weight for the parts and gripper.
- *Maximum and payload-dependent velocity.* The robot velocity limits with the gripper loaded and empty are determined from the product cycle time requirements for the work cell.
- *Downward force.* The downward force capability of the robot is set by an analysis of the insertion requirements present in the production process.
- *Compliance requirements.* Robot compliance demands are usually determined from the work-cell parts-assembly analysis; however, in some machine-loading applications compliance between the raw material and the fixtures needs to be considered.
- *Tool change requirements.* This specification is determined from a study of all the tooling needed to complete the production and assembly operations in the work cell.
- *Force or torque sensing requirements.* These specifications are usually determined from the work-cell parts-assembly analysis.
- *Programming (on- and off-line).* The flexibility and versatility needed in the robot programming language and operating system are dictated by an

analysis of the data integration present in the production system into which the robot work cell will be integrated.

- *Cost.* The maximum cost is an important parameter that is determined in the justification analysis for the system and set by the allowed payback period.
- *Special options.* These requirements are dictated by special needs in the production and assembly process.
- *Vision integration.* This requirement is also dictated by special needs in the production and assembly process and by inspection and quality requirements.

The robot selection process starts with an analysis of the entire production cell. On the basis of that analysis, some of the robot selection criteria can be ignored and others become critically important. For example, in a machine-tending application, repeatability, work envelope size, arm geometry, degrees of freedom, positioning flexibility, cost, and maximum and rated payload are most likely the critical parameters. However, if the parts need to be palletized, then special programming options will also be critical. If cycle times are critical, then the velocity will also have to be checked.

Selecting and Justifying Robot Applications—Existing Production System. Integrating an industrial robot into an existing production station requires a detailed design process that is different from the process used for a new work-cell design. The cost of a robot automation project depends on the complexity of the cell, the quantity and quality of existing production equipment, and the type of robot selected. The three basic steps in the robot cell development process are as follows: (1) pick the best manufacturing situation for the implementation, (2) pick the best robot for the specific job identified in step 1, and (3) build the best work cell possible around the robot selected. Although overly simplified, the three-step process emphasizes an important point: *the first step is to identify the best manufacturing application for the robot project.* Identification of the best application area involves a two-step process: (1) implement a process to identify all the production areas where the possibility of success is between 90 and 100 percent, and (2) study in detail the production cells selected and choose the one with the highest return on investment.

Robot Survey.   The designer of automation projects on existing production systems is often not permitted to change manufacturing and production variables. The current production system could be a manual operation that must be automated or an automated cell(s) that requires more current technology for improved productivity or a lower burden cost. Selection of robot automation for this type of automation project requires an additional step in the robot selection process. Before the design can start, a determination must be made about the feasibility of using a robot in the production cell. One quick way to determine if robot automation is an option for a work cell(s) is to use the robot survey form shown in Figure 11–19. This survey is

---

Robot applications—initial plant survey

Answer the following questions for each workstation in the plant survey:

1. Can inspection by operators be eliminated from this workstation?   Yes   No
    It is difficult and expensive to include parts inspection in a
    robot work cell.
2. Is the shortest machine cycle 3 seconds or longer?   Yes   No
    Robot speed is limited. Human operators can work faster than
    robots when demanded by the process.
3. Can the robot displace one or two people for three shifts?   Yes   No
    If the average robot project costs $100,000, then it will
    be necessary to save the cost of one or two operators for three
    shifts to get a one- or two-year payback.
4. Can the parts be delivered in an oriented manner?   Yes   No
    Picking parts from a tote bin is easy for humans but difficult
    for robots. If the parts can come oriented for easy robot pickup,
    then robot automation is possible.
5. Can a maximum of six degrees of freedom do the job?   Yes   No
    Robots have one arm that moves through a restricted work space
    compared with the two-armed human. A single-armed robot must
    be able to do the job.
6. Can a standard gripper be used or modified to lift the part or parts?   Yes   No
    The tooling is a major part of the work-cell expense. The simpler
    the tooling, the greater the likelihood of a successful project.
    Also the weight of the part plus the gripper must be consistent
    with the robot's capability.

If all the answers are yes, then you have a prime candidate for a robot application.

Figure 11–19   Robot Survey Form.

(Source: James A. Rehg, Introduction to Robotics in CIM Systems, Fifth Edition, © 2003, p. 111. Reprinted by permission of Prentice Hall, Upper Saddle River, New Jersey.)

especially useful when robot automation is considered for a manually operated cell. The process for using the form is outlined as follows:

1. Form an automation team (three to five people) to perform the initial plant survey. It is important that the team members have some training in robotics and automation. In addition, all shop-floor employees should know why the survey is being performed and know that the written company policy for automation-related job loss will be followed. To start the process, team members make independent visits to every production area and complete a survey form (Figure 11–19) for each. After all production cells are surveyed, the production areas that received six *yes* answers from every team member are identified as the areas for additional analysis. All the cells selected that meet the following criteria are identified:

    ■ Production areas where maintaining safe working conditions is difficult or costly

- Production areas where excessive protective equipment and clothing must be used by the operators
- Production areas with a history of worker opposition and worker discontent
- Production operations where material savings are possible with automation

2. All the production areas identified in step 1 have a high implementation success factor. A work area is selected for robot automation from this group of work cells, and a detailed design study is performed. The study must include a detailed job analysis that gathers data in three areas: (1) technical considerations, (2) economic considerations, and (3) human factors. The study includes the selection of a robot and all the necessary work-cell hardware to support the automated production process.

After the work cell is selected, the design process is similar to that outlined for a new work-cell design.

11–2 PROGRAM STATEMENTS FOR SERVO ROBOTS*

Robot program development follows an eight-step process. Study each step before reading the detailed description.

1. Establish a basic program structure.
2. Analyze the manufacturing process in which the robot will work.
3. Divide the robot action into tasks and subtasks.
4. Draw a *task point graph* that describes the desired motion and identify the translation points on the task point graph.
5. Assign values for all system variables to control the motion.
6. Write and enter the command statements.
7. Create or teach the translation points.
8. Test and debug the program.

These steps require clarification.

Basic Program Structure—Step 1

The first step in developing a robot program is the establishment of the basic program structure (Figure 11–20). All programs start at the robot's *HOME* position and move out to a start point in the cycle called *CYCLE START*. The translation point of CYCLE START should be located as close as possible to the main motion path of the tooling because the robot returns to this point after every manufacturing cycle.

The final point in the program, called *END OF CYCLE*, provides a branching command to direct the program execution back to the CYCLE START point.

---

*This robot programming procedure was adapted from Rehg (2003).

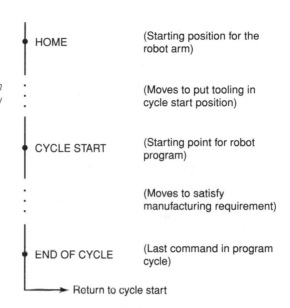

Figure 11–20 Basic Robot Program Structure.
(Source: James A. Rehg, Introduction to Robotics in CIM Systems, *Fifth Edition, © 2003, p. 330. Reprinted by permission of Prentice Hall, Upper Saddle River, New Jersey.)*

HOME — (Starting position for the robot arm)

— (Moves to put tooling in cycle start position)

CYCLE START — (Starting point for robot program)

— (Moves to satisfy manufacturing requirement)

END OF CYCLE — (Last command in program cycle)

Return to cycle start

The program points and commands that solve the manufacturing problem are located between the CYCLE START and END OF CYCLE points. The operations are divided into tasks and subtasks, with the tasks embedded into the main body of the program and the subtasks included as subroutines called from the main body.

Process Analysis—Step 2

The development of the program starts with an analysis of the production problem that will be solved by the robot. The programmer must have complete knowledge of the manufacturing process in which the robot functions. From this knowledge of the operation of the process, the programmer can identify the required motion and commands, divide the motion into tasks and subtasks, and establish the values of the system variables, such as tool velocity, for each translation point.

Tasks and Subtasks—Step 3

On the basis of a complete understanding of the production process, the programmer divides the required robot motion into tasks and subtasks. The following application of an injection-molding machine illustrates this concept. A robot unloads a part from the injection-molding machine, passes the part under a vision inspection system, and places good parts on an exit conveyer and bad parts in the recycling bin. Consider the following task breakdown. The major tasks include unloading the machine and submitting the part to the vision inspection system. Placing good parts on the conveyer and bad parts in the rework area would be subtasks. In this example, the major tasks are in the main program; the subtasks are included as subroutines called from the main program. The following rules of thumb are used to separate tasks and subtasks:

- If alternative production actions are selected according to input from external sources, then make each alternative path a subtask in a separate subroutine. In the previous example, the finished part is moved either to the exit conveyor or to the rework area, so both of these would be subtasks.
- If a particular robot motion or command sequence (e.g., turning on an alarm to call an operator) is required in two or more places in the program, then make the action a subtask, and put it in a subroutine.
- If the program requires two or more sequences of major tasks, then put each major task sequence in separate subroutines called from the main program. Assign subtasks for each major task to subroutines called from the major task subroutine.
- If part of the robot's action is likely to require frequent modification or updating, then make that portion of the program a subtask, and place it in a subroutine.

After the robot action is divided into tasks and subtasks, the structure of the total robot program is established.

Task Point Graph—Step 4

A *task point graph (TPG)*, step 4 in the programming steps, is a visual tool to illustrate the program flow and arm motion required for a manufacturing problem. The TPG includes all translation point data (programmed locations), the motion variables (e.g., gripper conditions, delays, and tool point velocity) used at every point, and the logic used to make decisions during the program execution. A TPG for the robot unloading an injection-molding machine is shown in Figure 11–21. Each program point on the TPG has several pieces of information: the name of the programmed point and a brief description, where the tooling will move next, tool speed, and other commands as required. Note that the subroutines are not shown.

System Variables—Step 5

The motion variables associated with the translation points shown in Figure 11–21 include *velocity, tool center dimensions, Cartesian coordinate values, language functions,* and *commands.* These variables are usually included on the TPG as part of the program development.

Write and Enter the Program—Step 6

With the translation points identified and named and the TPG developed, the final step is to write the robot program code using the command structure and

Figure 11–21 Task Point Graph.

(Source: James A. Rehg, Introduction to Robotics in CIM Systems, *Fifth Edition, © 2003, p. 332. Reprinted by permission of Prentice Hall, Upper Saddle River, New Jersey.)*

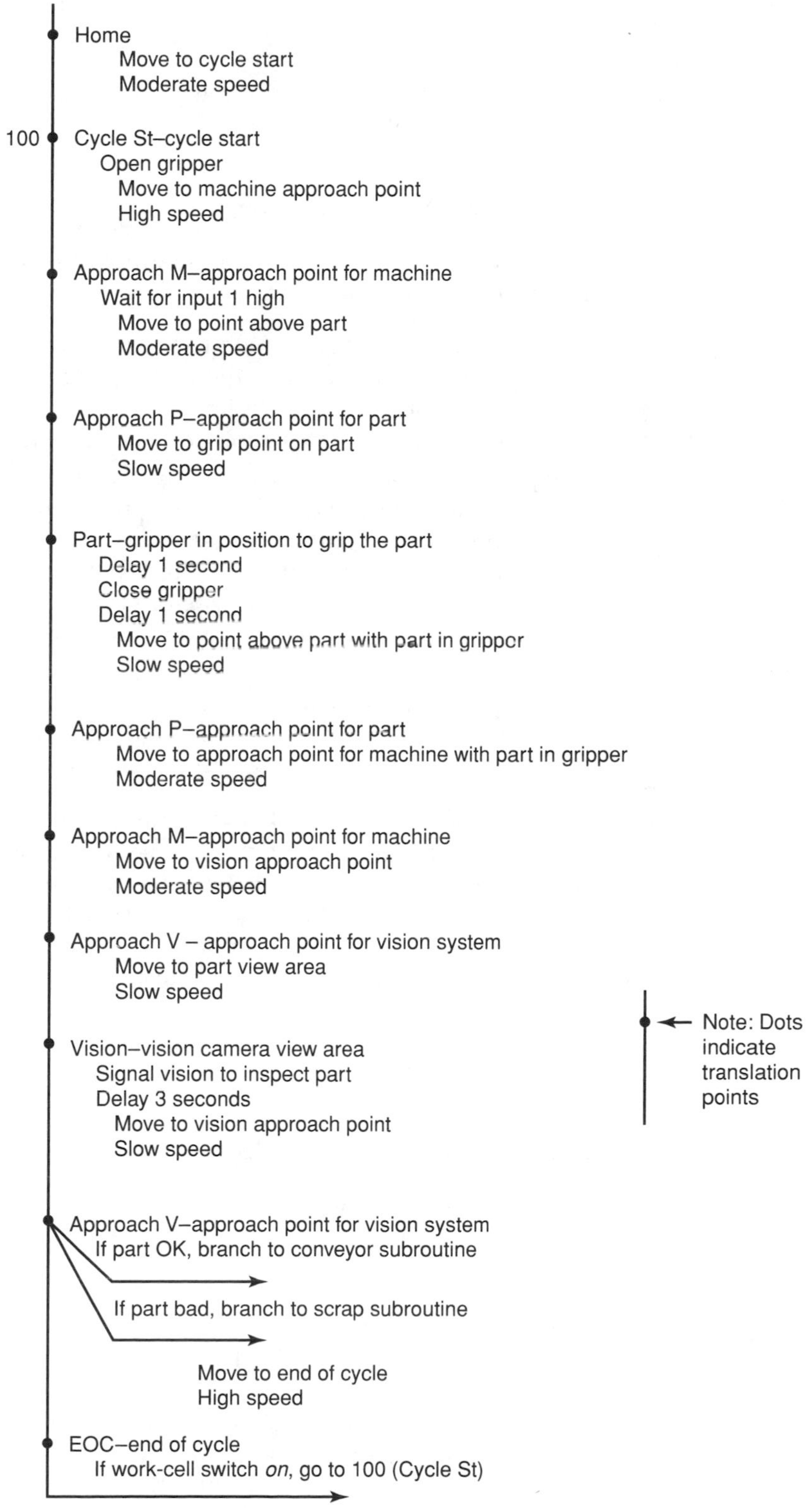

Home
   Move to cycle start
   Moderate speed

100   Cycle St–cycle start
   Open gripper
    Move to machine approach point
    High speed

Approach M–approach point for machine
   Wait for input 1 high
    Move to point above part
    Moderate speed

Approach P–approach point for part
    Move to grip point on part
    Slow speed

Part–gripper in position to grip the part
   Delay 1 second
   Close gripper
   Delay 1 second
    Move to point above part with part in gripper
    Slow speed

Approach P–approach point for part
    Move to approach point for machine with part in gripper
    Moderate speed

Approach M–approach point for machine
    Move to vision approach point
    Moderate speed

Approach V – approach point for vision system
    Move to part view area
    Slow speed

Vision–vision camera view area
   Signal vision to inspect part
   Delay 3 seconds
    Move to vision approach point
    Slow speed

Approach V–approach point for vision system
   If part OK, branch to conveyor subroutine

    If part bad, branch to scrap subroutine

      Move to end of cycle
      High speed

EOC–end of cycle
   If work-cell switch *on*, go to 100 (Cycle St)

← Note: Dots indicate translation points

syntax for the controller. An example of this process for the TPG shown in Figure 11–21 is provided later in this chapter.

Teach the Translation Points—Step 7

The translation points, step 4 in the programming steps, are taught or created by using one of the following methods:

- In an online programming technique, visual judgment establishes the fit and alignment between the tooling and the parts. Although production must be stopped, this technique is the most frequently used method for establishing the translation points in a cell.
- Either an off-line programming technique allows translation points to be established, or the locations of translation points are calculated inside a program. In either case, a translation point is established by using coordinate data relative to the robot reference frame. Therefore, these techniques require some type of system to accurately measure the position of work-cell hardware and calibrate the robot arm for positional accuracy.
- An external signal can be used to interrupt the current MOVE command and trigger the creation of a new translation point. The new point is the location of the TCP at the time of the interrupt.

The creation of the translation points is the most costly part of the programming if production must be stopped; therefore, as few points as possible should be taught for a given program. Four techniques can be used to reduce the number of points that need to be programmed with the teach pendant: (1) programming a point by entering the coordinate point values for the desired point from the keyboard, (2) using algebraic expressions to combine previously programmed points to create new points, (3) referencing new positions by using offset values for each coordinate from a previously programmed point, and (4) downloading points through the RS-232C port.

Test and Debug the Program—Step 8

The final step in the development of a robot program is to test the program and correct any problems. Several suggestions make this last step safer for the robot and the other work-cell hardware:

- Keep the speed between translation points at a low value during the initial run so that any unexpected moves can be aborted by using the emergency stop function.
- Use a single-step function. If the controller supports that mode of operation, it permits the program to be executed one move at a time with a pause at the end of each move.

- Do not have any parts in feeders or in the machines.
- Keep one hand on the emergency stop button for all moves.

Online and Off-line Programming

The terms *online programming* and *off-line programming* define the location where the robot program is developed. For online programming, the production operation is stopped and the programmer puts the robot into the programming mode. Then the programmer teaches the robot the required position, motion, and control sequences. The positions, or translation points, are taught by visually moving the production tooling to the exact work-cell location and entering the position into the program with a teach button on the teach pendant. In this method, the exact location of the work-cell components and the native accuracy of the robot system are not critical for good operation. The automation will work as long as the robot's repeatability is good and the location of work-cell machines and parts does not change. The major disadvantage of online programming is the lost production time.

Some robot languages use variable names for the translation points and permit the control structure, moves, and program logic to be developed on a word processor on a microcomputer or engineering workstation. The program, complete except for the assigning of work-cell locations to translation point variables, is downloaded to the robot controller over a serial communication channel, and production is stopped only to teach translation points for all the location variables named in the program. A significant reduction in lost production time results from this modified online programming technique.

The term *off-line programming* means that *all* of the programming is performed away from the robot and the production area. All translation points are calculated by the robot controller from translation point coordinate values entered into the program in the off-line mode. For this technique to work, several conditions must exist:

- The accuracy of the robot and the controller must be excellent and the same for all the robots used in the production area.
- The exact distance from the robot reference frame to all the production equipment and machine fixtures must be known.
- Work-cell simulation software or a robust robot programming language must be available to program the cell off-line.

The first condition rarely occurs in standard industrial robots. For example, when a robot is programmed off-line to go to a translation point in the work envelope, it misses the point because of the mechanical tolerances in the arm linkages and feedback mechanisms. Effective off-line programming is possible only with highly accurate placement of work-cell hardware and calibration of the robot arm. This type of robot and work-cell calibration is possible with integrated systems such as WORKSPACE and ROBOTRAK from Robotic Workspace Technologies, Inc. ROBOTRAK provides the calibration of the robot arm and work-cell hardware used in the WORKSPACE program. WORKSPACE is simulation software that builds a manufacturing automation simulation and integrates the work-cell equipment location data and robot arm signature captured with ROBOTRAK. The robot signature data include the zero position of each joint, the length of each link, the distance offset at each link, and compliance at each joint. These types of systems measure static position and motion paths to an accuracy of 0.2 millimeter (0.008 inch) in three dimensions. As a result, a functional off-line program is developed with WORKSPACE by integrating the location data for all work-cell devices and compensating for the variations present in the mechanical linkages in the robot.

11–3 PROGRAMMING A SERVO ROBOT*

The material provided in this chapter offers some guidelines for programming servo robots. The robot languages from all of the major suppliers are different, so it is not possible to learn one standard that would apply to all models and types. However, if you know how to develop a program using the structure and syntax from one vendor, then learning to program a different robot is not so difficult. With that in mind, some of the most often used commands and syntax for the Yaskawa Performer MK3 robot are provided next. As the commands are introduced, the syntax used to program the TPG shown in Figure 11–21 is also discussed.

Command Modes

The software programming language, called *ACL*, has two command modes, *DIRECT* and *EDIT*. In the DIRECT mode, commands are executed immediately after they are entered from the keyboard, whereas in the EDIT mode, the commands are not executed until the entire program is run.

Two modes of program operation are supported, *RUN* and *STEP*. In the RUN mode, programs are executed starting at the first program line and continuing until the program is terminated. In the STEP mode, the user can execute the program one line at a time.

Coordinate Systems

The robot controller supports three coordinate systems, *JOINT* (resolver), *XYZ* (Cartesian), and *TOOL*. Figure 11–22a illustrates the joint coordinate system, Figure 11–22b shows the *XYZ* coordinates system, and Figure 11–22c displays the tool coordinate system. These three modes are used to move the robot in the work envelope and are selected by using a key on the robot teach pendant.

---

*This section was adapted from Rehg (2003).

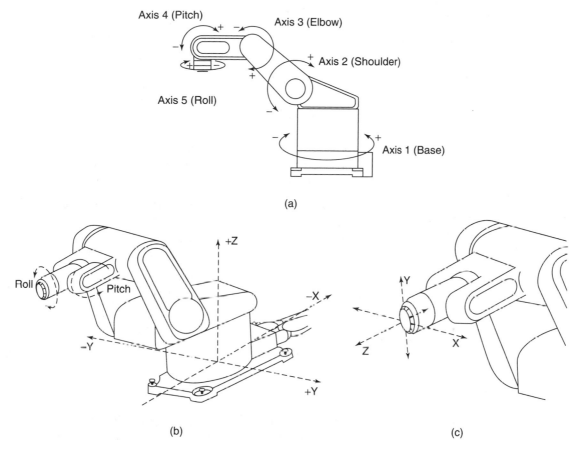

Figure 11–22   (a) Joint, (b) *XYZ*, and (c) Tool Coordinate Systems.

(Source: James A. Rehg, Introduction to Robotics in CIM Systems, *Fifth Edition, © 2003, p. 335. Reprinted by permission of Prentice Hall, Upper Saddle River, New Jersey.)*

Data Types

The ACL language has two types of variables: *global,* which can be referenced in all programs, and *private,* which can be used only in the program edited at the time the variable is defined. The following examples illustrate how these data types are defined:

DEFINE *X*	Defines the private variable *X*.
GLOBAL *Top*	Defines the global variable *Top*.
DIM A[20]	Defines an array named *A* with 20 private variables.
DIMG App[5]	Defines an array named *App* with 5 global variables.

Axis Control Commands

A number of commands provide for the control of the robot axes, joints, and gripper. Some of the commands are used together, such as the MOVE and SPEED commands; therefore, that is how they are described.

MOVE W   Moves the axes to the taught point W, using the current joint speed.

SPEED *var*   Sets the axes speed to the value of the variable *var*, where *var* is a value (percentage of full speed) from 1 to 100 with a default at 50. After the speed is set, it remains at that value until changed in a later program step.

The following example illustrates how these commands would be used for the TPG shown in Figure 11–21.

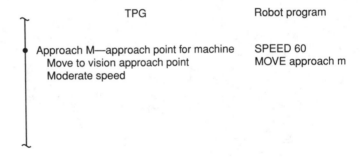

TPG	Robot program
Approach M—approach point for machine Move to vision approach point Moderate speed	SPEED 60 MOVE approach m

MOVEL Y   Moves the robot's tool center point (TCP) to the taught point Y along a linear path, using the current linear speed.

SPEEDL *var*   Sets the linear speed to the value of *var*, where *var* is a value expressed in millimeters per second up to 500 mm/sec.

The following example illustrates how these commands would be used for the TPG shown in Figure 8–21.

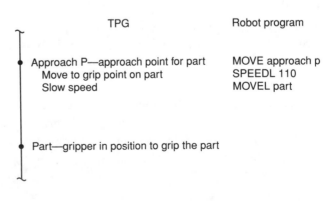

TPG	Robot program
Approach P—approach point for part Move to grip point on part Slow speed	MOVE approach p SPEEDL 110 MOVEL part
Part—gripper in position to grip the part	

Five other commands are frequently used as well:

OPEN	Opens the gripper.
CLOSE	Closes the gripper.
DELAY *var*	Suspends program execution for the time specified by the *var* value, with *var* defined in hundredths of a second.
HOME	Searches for microswitch home positions for all robot axes.
SET *var1* = *var2*	Assigns the value of *var2* to *var1*.

The following example illustrates how these commands would be used for the TPG shown in Figure 11–21.

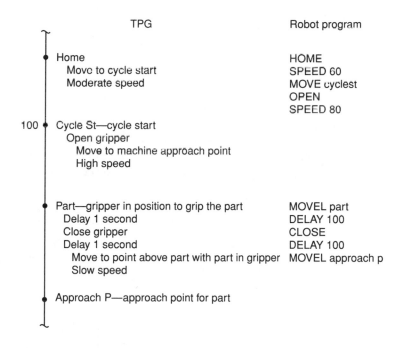

	TPG	Robot program
	Home	HOME
	Move to cycle start	SPEED 60
	Moderate speed	MOVE cyclest
		OPEN
		SPEED 80
100	Cycle St—cycle start	
	Open gripper	
	Move to machine approach point	
	High speed	
	Part—gripper in position to grip the part	MOVEL part
	Delay 1 second	DELAY 100
	Close gripper	CLOSE
	Delay 1 second	DELAY 100
	Move to point above part with part in gripper	MOVEL approach p
	Slow speed	
	Approach P—approach point for part	

Input-Output Control and Program Flow Commands

The following commands have to do with the sixteen discrete input and output ports on the robot controller.

SET OUT [*n*] = {0/1}	Sets the state of output *n* to either on {1} or off {0}, where *n* is an output between 1 and 16.
IF IN [*n*] = {0/1}	A conditional IF command that checks to see if the variable IN [*n*] is equal to the value in the brackets {0/1}. Any of the standard operation signs can be used in place of the = sign. Operators include greater than (>), less than (<), equals (=), less than

or equal to (<=), greater than or equal to (>=), and not equal (<>). If the condition is met and the result true, then the next sequential program line is executed. If it is false, then the execution jumps to after the ELSE command if it is used or to after the ENDIF command if no ELSE is present. The ANDIF and ORIF commands allow for additional AND and OR conditions to be considered.

The following example illustrates how these commands would be used in general and for the TPG shown in Figure 11–21.

```
If A = B
    AND IF C > 1
MOVE top
ELSE
MOVE bottom
ENDIF
```

The preceding code demonstrates how the IF command is used. If the values A and B are equal AND if C is greater than 1, then the MOVE to location *top* is executed. If any of those equalities is false, then the MOVE to location *bottom* occurs. ORIF would be used in a similar fashion. Combinations of ANDIF and ORIF can be used to get any logical combination required.

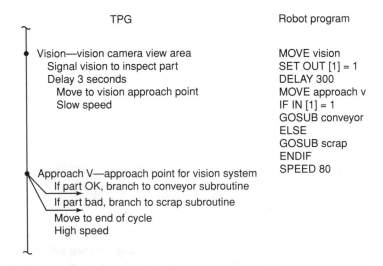

TPG	Robot program
Vision—vision camera view area	MOVE vision
Signal vision to inspect part	SET OUT [1] = 1
Delay 3 seconds	DELAY 300
Move to vision approach point	MOVE approach v
Slow speed	IF IN [1] = 1
	GOSUB conveyor
	ELSE
	GOSUB scrap
	ENDIF
Approach V—approach point for vision system	SPEED 80
If part OK, branch to conveyor subroutine	
If part bad, branch to scrap subroutine	
Move to end of cycle	
High speed	

The SET OUT [1] = 1 command tells the vision system to check the part. The vision system puts a signal on robot input number one, and the IF command uses the IN [1] = condition to know if the part is good (IN [1] = 1) or bad (IN [1] = 0) so that the correct subroutine is run. The following command is used with the

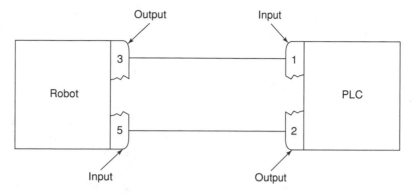

Figure 11–23   Robot and PLC Interface.

(Source: James A. Rehg, Introduction to Robotics in CIM Systems, *Fifth Edition, © 2003, p. 339. Reprinted by permission of Prentice Hall, Upper Saddle River, New Jersey.)*

SET OUT command to create a handshaking exchange with a programmable logic controller (PLC).

WAIT *var1* oper *var2* — Suspends program execution until the condition *oper* is satisfied. The two variables *var1* and *var2* can be any allowed variable or system parameter, and the *oper* can be any of the standard operators defined earlier. The value of *var1* in the handshaking routine will be the IN [*n*] command. The interface between the robot and the PLC is shown in Figure 11–23.

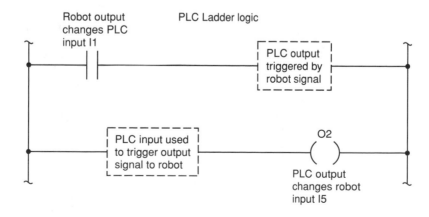

The robot output bit 3 is wired to the PLC input bit 1, and the PLC output bit 2 is wired to the robot input bit 5. The following commands create FOR loops, permit program jumps, and branch to subprograms.

FOR *var1* = *var2* TO *var3* Executes all of the program lines between the FOR command and ENDFOR command as many times as the *var2* and *var3* values indicate. The *var1* must always be a variable, but *var2* and *var3* can be either a variable or a constant.

Robot program

```
FOR L = M to N              Assume M = 1 and N = 5
    MOVE POS [L]
ENDFOR
```

The code between the FOR and ENDFOR commands will execute five times. During each loop, the value of L will change from 1 to 2 then 3, 4, and 5. The POS [L] term is an array of five taught points (POS [1], POS [2], POS [3], POS [4], and POS [5]). This FOR loop will cause the robot to move to all five array positions.

LABEL *n* Marks a line in the program to be executed after a GOTO *n* command is executed, where *n* is any value between 0 and 9,999.

GOSUB *prog* Transfers control of the robot to another program with the name *prog*. The execution of the main program is suspended until the subroutine is completed.

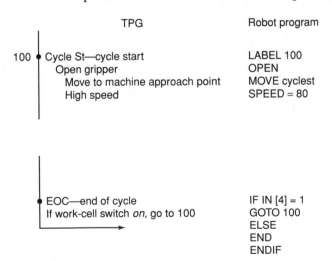

TPG	Robot program
100 ● Cycle St—cycle start	LABEL 100
Open gripper	OPEN
Move to machine approach point	MOVE cyclest
High speed	SPEED = 80
● EOC—end of cycle	IF IN [4] = 1
If work-cell switch *on*, go to 100	GOTO 100
	ELSE
	END
	ENDIF

The work-cell switch is connected to robot input 4. While there are more language commands available to the programmer, the commands covered in this section are the ones most frequently used.

11–4 AUTOMATED MATERIAL HANDLING

In most production systems, the part or raw material is either in transit, waiting in a queue, being processed, or undergoing inspection. Except for the processing time, all of the other operations add cost, not value, to the product. As a result, the material-handling process for parts and raw material should be automated only after every unnecessary inch of material transport distance has been removed from the production process.

The work simplification and analysis process that precedes the design and selection of material-handling automation starts with a diagram of the production flow. Process flow analysis uses symbols, such as those illustrated in Figure 11–24, to diagram the production system. Starting with the symbolic representation of the production flow (Figure 11–24), the analysis of part movement and queues leads to the elimination of any unnecessary elements. With the distance traveled by parts and raw material reduced to the minimum possible, the selection of effective automation to transport and handle parts is possible.

The transfer mechanism used to move parts between work cells and stations has two basic functions: (1) to move the part in the most appropriate manner between production machines, and (2) to orient and position the part with sufficient accuracy at the machine to maximize productivity and maintain quality standards.

Automated Transfer Systems

The many mechanisms used to achieve the two goals just mentioned are grouped into three categories: *continuous transfer, intermittent transfer,* and *asynchronous transfer.*

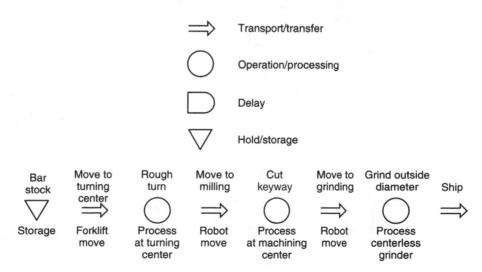

Figure 11–24   Symbols Used to Diagram a Production Flow.

Continuous Transfer.   In a *continuous-transfer operation*, the parts or material moves through the production sequence at a constant speed. The work does not stop at a workstation for a production operation to be performed. A good example of this type of transfer is in an automated car wash. The car is pulled through the car wash at a constant speed, and every washing mechanism, such as the brushes that clean the wheels, must perform its operation synchronized with the moving vehicle. In the manufacturing area, a prominent example is the assembly of cars on an automotive assembly line. The cars move at a constant rate through 95 percent of the assembly process, with the automated assembly machines and manual operations performed on a moving vehicle. The picture in Figure 11–25 shows a spot-welding operation performed on a moving car by a robot. In this case, the robot has the capability to move the spot-welding tooling connected at the end of its arm at a rate synchronized with that of the material-handling system carrying the car. The

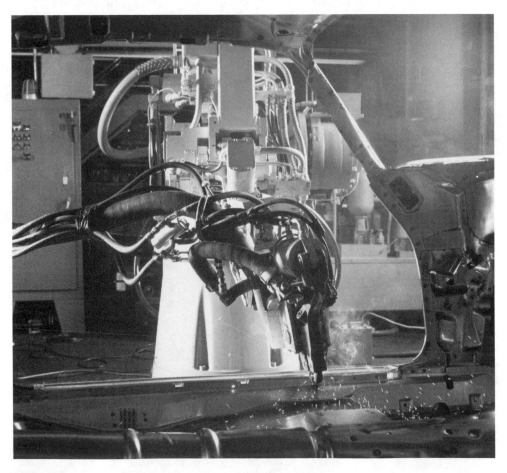

Figure 11–25   Spot-Welding Operation on a Moving Car Body.

material-handling system moves each car through the same line past the robot. The robot senses the speed of the car and controls the motion of the welding gun with sufficient accuracy that the welds are placed with a repeatability of one-quarter inch. For the continuous-transfer system to function, the workstations must support part or material movement during the work process.

The types of mechanisms used to achieve continuous motion in manufacturing include the following:

- *Overhead monorail.* This mechanism is a continuous series of interconnected hooks attached to free-turning rollers that ride in overhead tracks. Parts are hung from the hooks to transport material between workstations or through a production operation. This type of system is frequently used to bring parts past human or robot paint sprayers so that they can complete a painting or coating operation.
- *Monorail tow systems.* This system is the same as the overhead monorail system, except that wheeled carts are attached to the hooks, and the overhead system pulls the carts from one destination to another.
- *In-floor tow systems.* This operation is similar to the monorail tow except that the towing mechanism is a continuous chain riding in a channel placed below the surface of the floor. The chain has hooks placed at regular intervals that pull a wheeled carrier through production. This method is the most popular for towing car carriers through the main automotive assembly process.

Intermittent Transfer.   The *intermittent,* or *synchronized, transfer system* has the following characteristics:

- Workstations are fixed in place.
- The motion of the transfer device is intermittent or discontinuous so that parts cycle between being in motion and being stationary.
- The motion of the parts is synchronized so that all parts move at the same time and then are motionless during the same time frame.

This type of material and parts transfer system is usually found in machining applications, especially progressive die operations.

The most common type of mechanisms used to achieve synchronized intermittent motion is the *walking beam transfer system,* illustrated in Figure 11–26. Study the figure until you are familiar with the names of the various parts. The transfer system has a *fixed rail* that has part carriers resting in notches. The *transfer rail* has a similar set of notches and is free to move. The synchronized motion occurs when the transfer rail picks up all the parts and moves them forward by one set of notches. Notice that in Figure 11–26a, the six parts are in the first six sets of notches on the left side of the fixed rail. After one cycle of the transfer rail, all parts are located one notch to the right of their previous position (Figure 11–26d).

Figure 11–26   Walking Beam Transfer System.

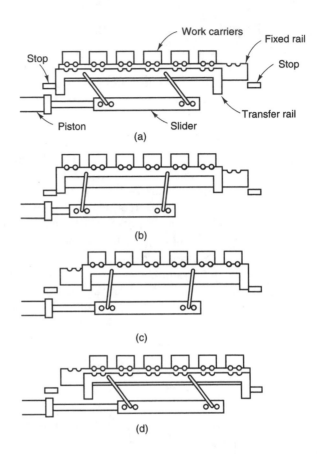

Asynchronous Transfer.   The term *asynchronous* means "not synchronized"; therefore, an *asynchronous transfer system* is one where each part moves independently of other parts. The asynchronous transfer is often called a *power-and-free system* to indicate that parts can be either free from the transfer mechanism or powered by the transfer device. Several advantages of this type of transfer are evident in the pallet shuttle type of asynchronous system pictured in Figure 11–27. Two pallets that hold parts for machining are held at a stop while the twin-belt conveyor continues to move under the stationary pallets. A third pallet is carried by the conveyor to a part loading station. In-process work storage is provided by asynchronous transfer, with some pallets in production and others ready to move into production as soon as the previous part is machined. The process permits finished parts to move to the next production operation while other parts are still in production.

Other advantages of this type of automated transfer include the ability to aid in line balancing by using parallel stations for operations with long cycle times and single production stations where the cycle time is shorter. This type of transfer is also used where manual production stations are integrated with fully automated operations. The singular disadvantage is that the cycle rates for asynchronous transfer systems are usually longer than those for the two previous types of material transfer.

Figure 11–27 Pallet Shuttle Conveyor System.

The most common type of mechanism used to achieve asynchronized transfer motion is the *power conveyor*. The driven surface on the conveyor is frequently a belt made from steel, impregnated fabric, slats, or interlocking plates. In many applications, pallets, pulled by the frictional force between the pallet and the conveyor belt (Figure 11–27), move with the conveyor and hold a single workpiece or a family of parts.

The transfer system just described performs over 60 percent of all material-handling functions in automated production systems.

11–5 AUTOMATIC GUIDED VEHICLES

Automatic guided vehicles (AGVs) are defined by the Material Handling Institute as follows:

> *An AGV is a vehicle equipped with automatic guidance equipment, either electromagnetic or optical. Such a vehicle is capable of following prescribed guide paths and may be equipped for vehicle programming and stop selection, blocking, and any other special functions required by the system.*

As the definition indicates, an AGV is a driverless vehicle capable of performing all the operations formerly available only with forklift trucks and other types of human-operated delivery vehicles. Like the vehicle pictured in Figure 11–28,

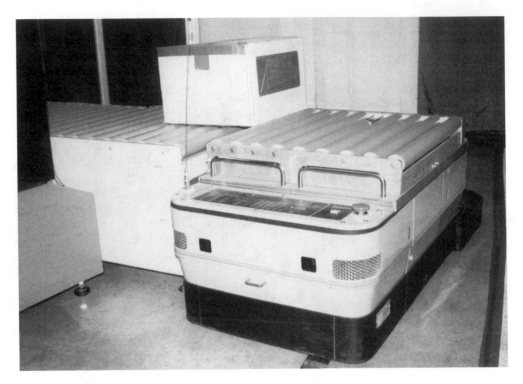

Figure 11–28   Automatic Guided Vehicle.

AGVs are usually powered by electric motors that receive energy from electric batteries on the vehicle. AGV technology dates to the early 1950s.

History of AGVs

The AGV concept was developed in the United States in the 1950s by Barrett Electronics. The early technology used towing vehicles to pull a series of trailers, predominantly in the warehousing environment, to predefined locations. The negative climate for productivity automation prevented wide acceptance and installation of the system in the United States. However, the European manufacturing community aggressively adopted the concept and agreed on one standard pallet size (800 by 1,200 mm), called the *Euro pallet*. With most of Europe using this common skid-type pallet without bottom boards, the development of driverless material-handling devices was enhanced in the European community.

In the 1970s, several factors caused widespread use of AGV technology in the United States. Competition from foreign markets forced a reduction in direct labor cost and a new interest in promoting productivity through automation. In addition, advances in the computer, especially the microprocessor, permitted the development of sophisticated AGV systems. Using a "land-based" AGV computer and microprocessor computers onboard, AGVs could provide bidirectional operation between points in manufacturing, operate on open- and closed-loop paths using FM radio signals for communication, handle traffic control and queuing in multiple-vehicle systems, and perform material tracking on all parts moved throughout the plant. Today, AGVs perform a valuable material-handling service on part transfers that are necessary for efficient production.

Types of AGVs

There are six basic types of AGVs: *towing, unit load, pallet truck, fork truck, light load,* and *assembly line.* An outline of each type is provided in Figure 11–29; review each outline until you are familiar with the design. The design incorporated into each type makes the AGV ideal for a specific manufacturing application. Four of these types of AGVs are pictured in Figure 11–30. The key features of each of the six types are described next.

Towing Vehicles.   This type was the first introduced and continues to be used in many applications. The primary function is to tow trailers (Figure 11–30) with capacities from 5,000 to 50,000 pounds at speeds up to 3 miles per hour. The number and variety of trailers that are pulled in these applications is widely varied, but ten is usually the limit. The primary application for this type of AGV is bulk movement of product into and out of warehouse areas.

Unit Load Vehicles.   Unit load vehicles are equipped with a deck that holds a pallet of material and generally have some type of automatic transfer device to deliver and pick up material. The primary application is in warehouse and distribution systems, with the AGVs transferring a high volume of material and parts to conveyors and work cells. Figure 11–31 shows a unit load type of AGV.

Pallet Truck Vehicles.   The application for pallet truck vehicles is primarily in distribution and movement of palletized loads at floor level. The AGV is frequently loaded by an operator who boards the vehicle and backs it into a loaded pallet. With the pallet onboard, the operator drives the AGV back to the guide path, inputs a destination code into the AGV, gets off the vehicle, and then uses a start command to send the AGV to the programmed destination.

Fork Truck Vehicles.   Fork truck AGVs (Figure 11–32) emulate the operation of a manual fork truck, with the flexibility to pick up and deliver loads at different heights at each stop. This flexibility makes this type of AGV the most expensive to install and requires greater attention to the path location and accurate positioning of loads. Unlike the pallet truck AGV, the guided fork trucks operate without human intervention.

Light-Load Vehicles.   As the name indicates, the light-load AGV has only a several-hundred-pound-capacity. This type of AGV is ideal for moving small parts trays or bins from small parts storage areas to individual workstations

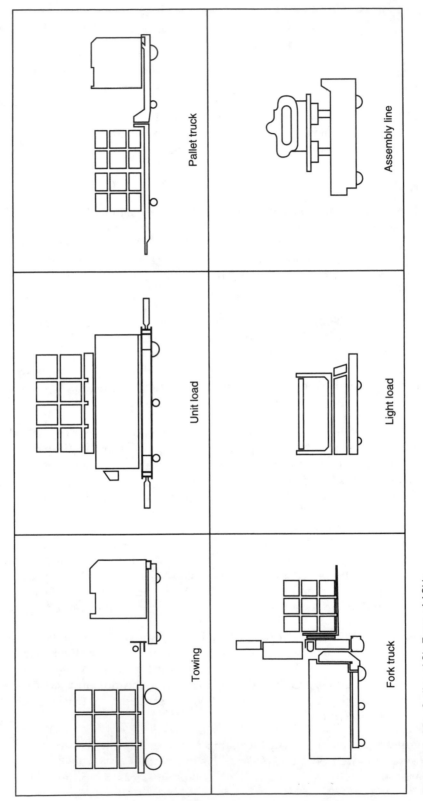

Figure 11–29 Outline of Six Types of AGVs.

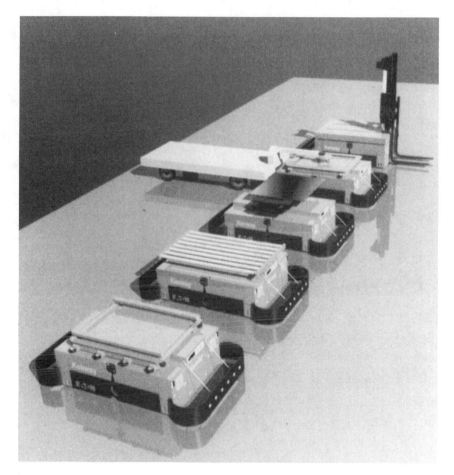

Figure 11–30   Picture of AGV Types (Front to Back): Unit Load Vehicle, Unit Load Vehicle with T-Slot Top, Assembly Line Vehicle, Vehicle Towing a Trailer, Fork Lift AGV.
(Source: Courtesy Eaton-Kenway, Inc., Salt Lake City, UT.)

where operators perform light assembly. Applications that justify this type of AGV include electronic assembly and small product assembly.

Assembly Line Vehicles.   The assembly line AGV (Figure 11–33) carries subassemblies, such as automotive engines, through a serial assembly line process until the assembly is complete. In some applications, the AGV passes through a parts staging area where trays of components required for assembly are loaded onto the AGV. This process is repeated until the subassembly is complete and ready for delivery to the main assembly line.

The six types of AGVs are designed to satisfy different manufacturing requirements. However, each of the different types shares the same operational characteristics.

Figure 11–31   Unit Load AGV.
(Source: Courtesy Eaton-Kenway, Inc., Salt Lake City, UT.)

AGV Systems

AGV systems must perform five functions: *guidance, routing, traffic management, load transfer,* and *system management.* The significance of each of these functions on the effective operation of an AGV system is described in the following subsections.

Guidance.   The vehicle must follow a predetermined path that is optimized for a given material flow pattern. The path is most often fixed by embedding a current-carrying wire just below the surface of the floor. Wire-guided AGV systems use onboard sensors to detect the magnetic field surrounding the current-carrying wire to keep the vehicle on the path. The wires are placed in the floor by cutting a narrow slot about one-half-inch deep, laying the wire into the slot, and then sealing the cut with epoxy for a smooth finish. Figure 11–34 illustrates the relationship between the buried wire in the floor and the magnetic sensors onboard the AGV.

In other applications, optical sensors on the AGV track a line painted on the floor to mark the path. The optical technique, called a *chemical guide path,* uses an invisible fluorescent dye approximately one-inch wide. In clean-room environments, the chemical guide path can be used; however, in heavy industrial manufacturing, the wire-guided systems are preferred.

Another classification of guidance techniques is called *off-wire technology.* Off-wire systems use various techniques to give the AGV path location information along the route. The chemical guide path is an example of this type of path control. Other techniques used in specialized applications include *optical triangulation, dead*

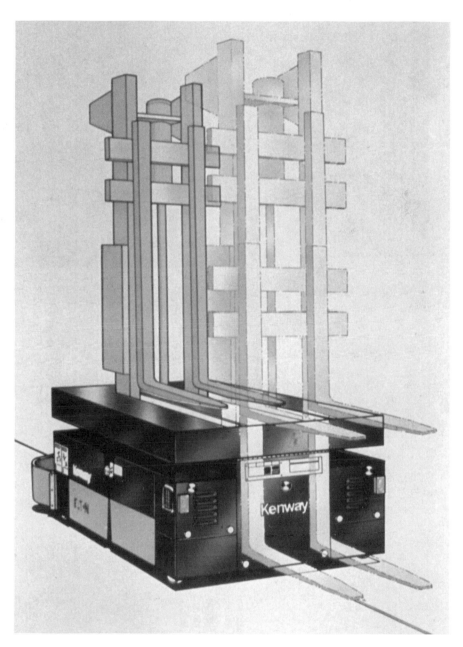

Figure 11–32 Fork Truck AGV.

(Source: Courtesy Eaton-Kenway, Inc., Salt Lake City, UT.)

Figure 11–33   Assembly Line AGV.
(Source: Courtesy Eaton-Kenway, Inc., Salt Lake City, UT.)

reckoning, laser guidance, inertial guidance, position reference beacons, and *ultrasonic imaging.*

Routing.   The AGV must be capable of changing the route to a destination on the basis of current conditions and needs in manufacturing. Study the simple AGV guide-path layout in Figure 11–35. If an AGV is at *stop 1* and must proceed to *stop 4*

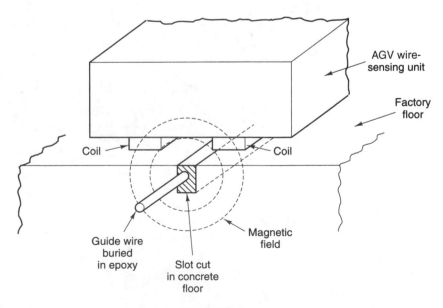

Figure 11–34   Operation of a Buried-Wire AGV System.

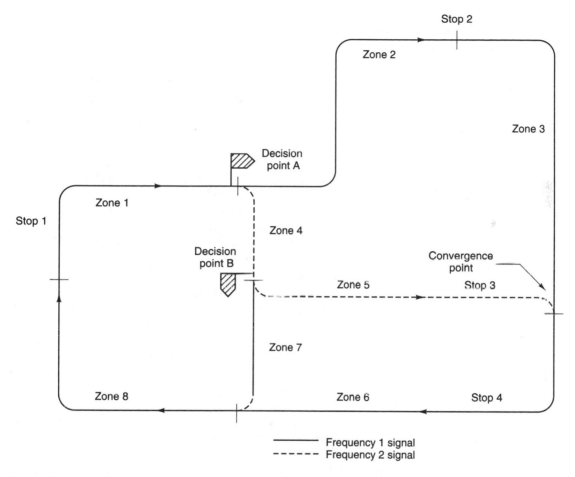

Figure 11–35   Typical AGV Path.

across the shortest distance, the vehicle must pass two *decision points* at A and B. In addition, the vehicle must pass through one location, called a *convergence point*, where two guide paths merge into a single path. AGV routing uses two techniques, *frequency select* and *path switch select*, to direct the vehicle to the correct path at a decision point.

An AGV is alerted that a decision point has been reached by a passive marker in the floor. The marker is usually an embedded permanent magnet, a metal plate, or some other coding device. In the *frequency select method*, the current in each guide wire leading away from a decision point is changing at a different ac frequency. The AGV is programmed to follow the guide wire with the frequency that will cause the vehicle to reach the next stop over the shortest path. Study Figure 11–35 and note that in going from stop 1 to stop 4, the AGV would choose frequency 2 at *decision* point A and continue to use that frequency at the second decision point. At the *convergence* point, the AGV would again track the guide wire with frequency 1.

The second routing method, *path switch select,* indicates the correct path for the AGV by switching the current *off* for the incorrect direction at the decision point. The AGV always follows the path at a decision point where the guide-wire current is *on.* For example, as the AGV approaches decision point A in Figure 11–35, the ac guide-wire current for the straight route would be switched off and the current in the guide wire heading toward decision point B would remain on. At the decision point, the AGV would sense the path with current and turn right toward decision point B.

Traffic Management.   The system must maximize material flow through the production system while minimizing interference and collisions with other AGV vehicles. Three traffic management techniques are generally used: *zone control, forward sensing,* and *combination control. Zone control* is the most frequently used traffic management technique. Study the path shown in Figure 11–35 until you have identified all eight zones. When a vehicle occupies a zone, a trailing AGV cannot enter the zone; therefore, AGVs are separated by a minimum of one zone to prevent collisions. Three methods are used to implement zone control: *distributed zone control, central zone control*, and *onboard zone control.* In the first method, distributed zone control, *zone control boxes* sense the presence of an AGV in a zone and activate a *zone hold beacon* to stop entry into the occupied zone. The second technique, central zone control, achieves the same results but performs the control from one central control box. In the last method, onboard zone control, each AGV has the intelligence to recognize the zone it currently occupies and transmit that information to all the other AGVs on the guide wire. The first method is cost effective on small systems and the second is used on larger systems. Onboard zone control is becoming popular for any size system because it eliminates the interface wiring between the zones.

In addition to using zone control, AGVs use *forward sensing* to perform traffic management. In this technique, each AGV has onboard sensors and electronics to detect that another AGV is on the guide wire in front of an approaching AGV. Three types of sensors are used to detect the vehicle in front: (1) *sonic,* which works like radar; (2) *optical,* which uses reflected infrared light; and (3) *bumper,* which uses contact. Forward sensing works well when the AGV system has several straight-line paths.

As the name indicates, the *combination control* is a combination of the other two types of control systems. Forward sensing is used for long straight sections of the path, and zone control is used for the curved sections.

Load Transfer.   The delivery and removal of material and parts from a terminal point must be performed without disturbance to the other production systems. The transfer of the load to and from the AGV is performed by using one of five methods:

1. Manual labor
2. Automatic coupling and uncoupling
3. Power rollers and belts

4. Horizontal power lifts
5. External power push and pull

The type of method used is a function of the type of AGV in use.

System Management.   The operation of the AGVs throughout the site must be managed by the AGV system in a cost-effective manner. The system management function has two components: vehicle dispatch techniques and system monitoring methods. The *dispatch techniques* include *onboard dispatch, off-board call system, remote terminal, central computer,* and *combination.* The onboard dispatch system uses an operator interface panel on the AGV to input destination information. The off-board call system uses an operator interface external to the AGV to direct the vehicle to path locations. The remote terminal is an extension of the off-board control with the addition of a monitor screen or an AGV locator panel to display current vehicle locations. Computer control, the highest level of AGV control, uses a central computer to control the movement of all the AGVs in the system. The last type, combination, uses a combination of control techniques in the same system environment. *System monitoring* focuses on visual displays of current AGV positions and the manufacturing environment surrounding the AGV movement.

The design of production systems that use an AGV requires consideration of each of these five functions during the design of the AGV material-handling system.

Justification of AGV Systems

The primary advantages of using AGV systems include a net reduction in direct labor, floor-space requirements, maintenance, peripheral material-handling equipment, and product damage resulting from material handling. In addition, AGVs also provide better control of material flow and inventory, higher throughput, and improvements in safety records.

The disadvantages, similar for any new technology, include the complexity of the hardware and software, which places new demands on training and increased skill levels for operators and maintenance personnel. Despite the problems associated with the adoption of high-technology equipment, the implementation of AGV systems is easily justified when specific conditions exist in manufacturing.

The justification of an AGV investment is based on current and future material-handling requirements. The presence of five or more of the following conditions, adapted from Miller's (1987) text on AGVs, warrants consideration of AGV technology:

- There is lost or late delivery of material and parts to work centers more than 5 percent of the time with manual material delivery systems.
- Ten or more pickup and delivery locations exist in the facility.
- Total material movement activity at all work centers falls between 35 and 200 loads per hour.

- Three or four forklifts are currently required on a minimum of two shifts.
- Over 300 feet of roller or power-and-free conveyor is currently in use.
- The production activity control or production control software requires online real-time material tracking.
- The level of in-process inventory on the shop floor makes increased productivity and throughput impossible.
- Automation introduced in the work centers or in parts and material storage requires an automated material movement system.
- Damage during material handling in manual systems creates significant quality problems.

The use of current AGV technology in computer-integrated manufacturing (CIM) implementations will continue to increase. However, as computer technology expands and intelligent AGV systems permit autonomous navigation without guide wires, applications for driverless material-handling systems will expand in all aspects of information, material, and parts distribution.

11–6 AUTOMATED STORAGE AND RETRIEVAL

In a manufacturing facility, the raw material is in one of two places: storage or production. When the raw material is in production, four substates are possible: in transit, waiting in a queue, being processed, or being inspected. Unsold finished products are categorized as finished-goods inventory. The storage of raw material, the holding of production parts in a queue, and the warehousing of finished goods all add cost to the end product. In addition to these three major inventory items, several other types of inventory are normally carried in a manufacturing facility: (1) purchased parts and subassemblies used in the assembly of products, (2) rework and scrap that result from production operations, (3) tooling used in production, (4) spare parts for repair of production machines and the facility, and (5) general office supplies. The elimination of all inventory is neither possible nor in many cases desirable; however, the application of modern storage technology minimizes the added costs associated with this necessary inventory.

The term used to describe modern mass storage technology, *automatic storage and retrieval systems (AS/RSs)*, is defined by the Material Handling Institute (1977) as follows:

> *AS/RS is a combination of equipment and controls that handles, stores, and retrieves materials with precision, accuracy, and speed under a defined degree of automation.*

Systems that satisfy this definition are frequently used in CIM systems to store and retrieve raw materials, production tools and fixtures, purchased parts and subassemblies, in-process inventory, and finished products. The systems are often unique because every manufacturer has distinctive production system requirements and products. The AS/RS has four basic components: a *storage structure*, a *storage and retrieval machine, unit modules,* and *transfer (pickup and deposit) stations.* Each of these components is described in the following sections.

AS/RS Components

The primary function of the *AS/RS structure* is to support the material placed in the storage system. In many cases a secondary function, roof and building support, is present when the AS/RS structure is integrated into the structural system for the storage or manufacturing building. An AS/RS furnishing structural support for the building roof and providing a large storage capability is pictured in Figure 11–36. An AS/RS consists of a series of storage aisles running from the floor to the ceiling. Each aisle separates large storage walls with numerous compartments or bins to hold the inventory items listed in the preceding section. The AS/RS is often a substantial structure because the full weight of a fully loaded system plus roof loading must be supported safely.

Material is stored and retrieved from the AS/RS compartments by a *storage/ retrieval (S/R) machine* or crane that moves horizontally and vertically on guide rails in the aisles between the storage walls. The S/R machine performs two operations: (1) the vertical and horizontal movement permits the carriage on the S/R machine to reach any compartment for delivery or pickup of material, and (2) mechanisms inside the carriage pull material from the storage compartment into the carriage or push material from the carriage into the storage bin. Also, the S/R machine must maintain a high degree of position accuracy between the carriage and the storage compartment over the entire range of travel from one end of the AS/RS to the other. The carriage in some systems travels at horizontal and vertical speeds of 500 and 100 feet per second, respectively.

Figure 11–36   Integrated AGV and AS/RS System.
(Source: Courtesy Eaton-Kenway, Inc., Salt Lake City, UT.)

Figure 11–37   AS/RS System.
(Source: Courtesy Eaton-Kenway, Inc., Salt Lake City, UT.)

The material moved into and out of the storage compartments is placed on *unit modules*. The unit modules have a standard base size so that storage in any compartment is possible. A unit load system is shown in Figure 11–37. In addition, the modules are usually held in the compartments by interlocking tracks or rails. The unit modules are usually configured with one of the following: pallets, wire baskets, tote pans, or special drawers. The carriage moves the unit modules between a *single delivery point*, called the *pickup and deposit (P&D) station*, and the storage compartment in the AS/RS.

The P&D stations are located at one or both ends of the AS/RS and aligned with the S/R machine traveling in the aisle (Figure 11–37). The material-handling capability present in the P&D station must be compatible with that of the S/R machine and the material-handling system used to bring storage modules to the AS/RS. Load transfer to the P&D stations from manufacturing is performed with manual load/unload, forklift vehicles, gravity and power conveyors, and AGVs.

Types of AS/RSs

AS/RSs are designed in various configurations. For example, some are totally automated and operate on a unit load transfer basis; others, called *human-on-board AS/RSs*, have a human riding in the carriage to facilitate the picking of specific parts from the compartments. Other systems, called *miniload* and *automatic item retrieval*

AS/RSs, are designed to permit single-item retrieval from the AS/RS. In the miniload type, for example, a bin is retrieved from the storage location in the AS/RS and delivered to the P&D area, an operator takes out the necessary parts, and the bin is returned to storage. The second type, automatic item retrieval, permits individual items to be retrieved from a storage location without the need for a human operator.

Regardless of the type of configuration, the performance of any AS/RS is evaluated according to four criteria: *storage capacity, material throughput, percentage of time used,* and *reliability.*

Carousel Storage Systems

Carousel storage systems offer a lower-cost alternative in applications requiring a miniload type of AS/RS. Operation of the carousel system is exactly opposite that of the AS/RS. In the carousel system, the S/R machine is fixed in place at the P&D station, and the storage compartments rotate on a track that passes by the P&D station. A typical carousel storage system is illustrated in Figure 11–38. Carousel systems are frequently used in the following types of applications:

■ *Small parts assembly.* Storage of a large variety of small parts used in mechanical, electrical, and electronic assembly is a typical application for a carousel system. A parts kit containing all the components required for an assembly is built by selecting parts from different bins in the carousel system.

■ *Assembly transport.* In this application, the manual and automated assembly stations are located around the periphery of the carousel. At each assembly station, the partially completed assembly and parts for the product are removed from the carousel. The parts are added to the assembly and the partially finished product is placed back into the carousel. At the

Figure 11–38 Carousel Storage System Supporting an Assembly Operation.

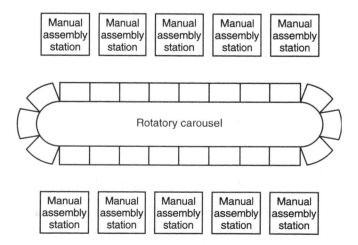

next assembly station, the partially finished assembly is retrieved along with the additional components that must be added. At the end of the process, the assembly is complete.

In world-class enterprises, the level of inventory in storage is reduced to as low a level as possible. However, when large levels of inventory are necessary to support make-to-stock production operations, it is important to minimize the added cost to the product. AS/RS is one technique used to minimize the cost-added nature of inventory.

11–7 SUMMARY

Industrial robots are part of most automated production systems, where they perform specialized material handling or some production process. The basic robot system includes a robot arm in one of four geometries, production end-of-arm tooling, controller electronics, and a teach mechanism. The numerous robot applications are grouped into three categories: material processing, material handling, and assembly and fabrication.

Robots are grouped into two categories according to the type of controller used. Servo, or closed-loop, controllers, designed and manufactured by the robot vendor, have feedback from the arm to indicate how close the arm movement is tracking the programmed path. Error detected in the controller is used to correct the arm movement. Nonservo, or open-loop, systems use off-the-shelf programmable logic controllers (PLCs) to control the discrete pneumatic valves that move the pneumatic cylinders at each robot axis. These pneumatic robots are moved through a series of positions by the PLC's sequential control program, called *ladder logic*. Programming is performed either off line (away from the robot controller without stopping the production process) or online (at the robot controller with production halted). Off-line programming is preferred since lost production is avoided.

The robot programming process for servo robots follows an eight-step development process. When the process is completed, a list of robot commands using the robot language syntax is created that moves the robot arm through a desired path. Nonservo robots are programmed by creating a sequential control program in ladder logic on a PLC.

Material-handling automation is a necessity for an automated computer-integrated manufacturing (CIM) facility. The many mechanisms used to automate material movement are grouped into three categories: continuous transfer, intermittent transfer, and asynchronous transfer. The automatic guided vehicle (AGV) is a special-purpose material movement system. The types of AGVs in common use include towing, unit load, pallet truck, fork truck, light load, and assembly line. Some systems require no human operator intervention; others use operators to maximize production efficiency. AGV systems perform five functions: guidance, routing, traffic management, load transfer, and system management.

Automatic storage and retrieval systems (AS/RSs) automate the placement and removal of the necessary inventory in a manufacturing facility. The AS/RS has four basic components: a storage structure, a storage and retrieval (S/R) machine, unit modules, and transfer (pickup and deposit, or P&D) stations. The types of AS/RSs include human on board, miniload, and automatic item retrieval. Another type of AS/RS, called *carousel storage,* provides a lower-cost alternative to the storage and retrieval of parts, especially for small parts assembly and applications where a partially assembled product must be transported through assembly.

BIBLIOGRAPHY

GOETSCH, D. L. *Modern Manufacturing Processes.* New York: Delmar Publishers, 1991.

GRAHAM, G. A. *Automation Encyclopedia.* Dearborn, MI: Society of Manufacturing Engineers, 1988.

GROOVER, M. P. *Automation, Production Systems, and Computer-Integrated Manufacturing.* 2nd ed. Upper Saddle River, NJ: Prentice Hall, 2001.

LUGGEN, W. W. *Flexible Manufacturing Cells and Systems.* Upper Saddle River, NJ: Prentice Hall, 1991.

MATERIAL HANDLING INSTITUTE. *Consideration for Planning and Installing an Automated Storage/Retrieval System.* Pittsburgh, PA: Material Handling Institute, 1977.

MILLER, R. K. *Automated Guided Vehicles and Automated Manufacturing.* Dearborn, MI: Society of Manufacturing Engineers, 1987.

REHG, J. A. *Introduction to Robotics in CIM Systems.* 5th ed. Upper Saddle River, NJ: Prentice Hall, 2003.

SOBCZAK, T. V. *A Glossary of Terms for Computer-Integrated Manufacturing.* Dearborn, MI: CASA of SME, 1984.

WIERSMA, C. H. *Material Handling and Storage Systems: Planning to Implementation.* Mansfield, OH: Self-published, 1984.

QUESTIONS

1. What distinguishes a robot from other types of automation?
2. Draw a basic robot system diagram.
3. Describe the four basic arm geometries used for robots.
4. Compare and contrast position movement and orientation movement.
5. Describe the two types of Cartesian geometries.
6. Describe the two types of articulated geometries.
7. What are the terms used to describe the production tooling used on robots?
8. Describe the three classifications used for robot tooling.
9. Describe the basic components in a robot controller.

10. Describe the three functions served by robot teach stations.
11. Describe the three categories of robot applications and give industrial examples of each.
12. How do servo robots and nonservo robots differ?
13. How many programming points are available on an axis of the servo robot? How many on a nonservo robot?
14. What type of controller is used for nonservo robots? What is the most commonly used programming language?
15. Briefly describe the eight-step servo robot programming process.
16. Describe the difference between online and off-line robot programming.
17. How is the robot survey form used to plan a robot installation in an existing work cell?
18. Describe the selection and justification process used to select a possible robot implementation.
19. How does fixed high-volume automation differ from flexible automation?
20. What are the characteristics of fixed in-line automation?
21. Compare rotatory automation to in-line automation.
22. Describe the selection process used for fixed automation systems.
23. Why is automated material handling a critical technology for a company that wants to become a world-class manufacturer?
24. What are the two basic functions of material and part transfer mechanisms?
25. Describe the three types of transfers.
26. What are the three types of mechanisms used to achieve continuous-motion transfer?
27. Describe a walking beam transfer system.
28. Define *automatic guided vehicle*.
29. Describe the six basic types of AGVs.
30. Describe the five functions that AGV systems must perform.
31. What is off-wire AGV technology?
32. Compare frequency select and path switch select routing.
33. Describe how wire-guided AGVs operate.
34. Describe the three traffic management techniques used for AGVs.
35. What are the five methods used to transfer the load between the AGV and a fixed site?
36. Why is system management of AGVs more difficult than the system management of robots?
37. What are the basic elements present in most AS/RSs?
38. Why is the study of AS/RSs important if inventory is an unwanted quantity?
39. What are the four metrics used to evaluate the effectiveness of an AS/RS?
40. Describe the different types of AS/RSs.

PROJECTS

1. Using the list of companies developed in project 1 in chapter 1, determine which companies are using robotics, fixed automation systems, automatic guided vehicles, and automatic storage and retrieval systems.
2. Select one of the companies in project 1 that uses robots and classify the systems by arm geometry, control techniques, and type of tooling used.
3. Select one of the companies in project 1 that uses AGVs and describe the types, guidance, route control, traffic control, and management used in the system.

The following projects are a continuation of projects 5 through 8 at the end of chapter 10.

4. A company that manufactures kitchen products wants to add a set of plastic storage bowls or containers to its product line. The company wants to distribute the product through Wal-Mart, Target, or a chain of stores in your region. Design a cell layout that would include a robot to unload finished bowls from a plastic-injection-molding machine, trim excess mold material from the bowls with a trim press, and place the finished bowls on an exit conveyor. A single mold die is used for all four bowls. Use vendor Internet sites to select production hardware. Document all assumptions made, indicate how robot selection criteria were used in the selection process, and list vendor names, models, and descriptions for all machines. Draw a top view of the cell, illustrating the footprint of all machines and the work envelope of the robot.
5. A company that manufactures kitchen products wants to add a high-end manual can opener to its product line. The company wants to distribute the product through Wal-Mart, Target, or a chain of stores in your region. Do the following:

   a. Design a servo robot assembly cell for the product. Use vendor Internet sites to select production hardware. Document all assumptions made; indicate how robot selection criteria were used in the selection process; and list vendor names, models, and descriptions for all machines. Draw a top view of the cell, illustrating the cell layout and the work envelope of the robot.

   b. Develop a robot program for the assembly process showing the task point graph and the robot program commands.

6. A company that manufactures kitchen products wants to add a manual hand-held eggbeater to its product line. The company wants to distribute the product through Wal-Mart, Target, or a chain of stores in your region. Do the following:

   a. Design a servo robot assembly cell for the product. Use vendor Internet sites to select production hardware. Document all assumptions made; indicate how robot selection criteria were used in the selection process; and list vendor names, models, and descriptions for all machines. Draw a top view of the cell, illustrating the cell layout and the work envelope of the robot.

   b. Develop a robot program for the assembly process showing the task point graph and the robot program commands.

7. A company that manufactures kitchen products wants to add a manual hand-held garlic press to its product line. The company wants to distribute the product through Wal-Mart, Target, or a chain of stores in your region. Do the following:

   a. Design a servo robot assembly cell for the product. Use vendor Internet sites to select production hardware. Document all assumptions made; indicate how robot selection criteria were used in the selection process; and list vendor names, models, and descriptions for all machines. Draw a top view of the cell, illustrating the cell layout and the work envelope of the robot.

   b. Develop a robot program for the assembly process showing the task point graph and the robot program commands.

CASE STUDY: AGV APPLICATIONS AT GENERAL MOTORS

The number of automatic guided vehicles (AGVs) used worldwide exceeds 13,000 in a wide variety of machine types and applications. AGV installations span every conceivable production industry and include several service industries. The worldwide automotive manufacturing industry is a major user of this technology. The following case information is drawn from experiences in AGV applications at General Motors and is adapted from case studies in *Automated Guided Vehicles and Automated Manufacturing,* by R. K. Miller (1987).

Oldsmobile Division

The General Motors Oldsmobile Division in Lansing, Michigan, installed the first AGV system for automotive assembly. The 185-vehicle AGV system had over 10,000 feet of guide wire and used Eaton-Kenway unit load carriers in several areas. The engine dress system area used 95 AGVs on 7,500 feet of guide wire, the chassis system used 65 AGVs in two areas, and the engine stuff system used 25 AGVs in two areas. The AGVs supported the production of front-wheel-drive Oldsmobiles, Buicks, and Pontiacs by carrying engines through assembly until the engines were merged with the car bodies.

The AGVs offered the following advantages over previous assembly methods:

■ The AGV carriers permitted each engine to remain stationary at a workstation until the assembler was sure that the assembly operation was performed correctly. Therefore, engines released to final assembly had fewer quality problems.

■ Engines held in the stationary position during assembly supported the application of automated assembly technology, like robots.

■ Assembly efficiency was enhanced because the AGV permitted the height of the engine from the floor to be adjusted to the needs of each assembler and assembly operation.

Assembly Plant.   The General Motors assembly plant at Orion, Michigan, used a 22-vehicle AGV system to move 70 percent of the incoming stock on a just-in-time basis to 69 drop zones in two departments. The system used 24,000 feet of guide path and Conco-Tellus vehicles with various containers from 30 to 54 inches wide. The sophisticated vehicles had automatic alignment capability for pickup and drop-off of material using photocells in the ends of the lift forks. In addition, the AGVs could go off wire under some conditions, using an FM radio communications system of data exchange with a central computer controller. All the vehicles were "smart devices" because they carried onboard microcomputers. Recharging of AGV batteries was automatic. When the battery life dropped below 20 percent, the AGV took itself out of service and reported to a recharge station for a full charge that would provide another 16 hours of continuous service.

Machine and System Control

OBJECTIVES

After completing this chapter, you should be able to do the following:

- Describe the network and subnetwork structure used to link machines and devices on the automated shop floor
- Describe the primary function of DeviceNet, ControlNet, FOUNDATION Fieldbus, and Ethernet/IP
- Describe the function of cell controllers and cell control software
- Describe the difference between proprietary and open system interconnect (OSI) software
- Define *programmable logic controller* (*PLC*) and name all the parts of a typical PLC system
- Develop a PLC ladder-logic program for the simple sequential machine
- Describe the PLC languages listed in the 61131-3 international standard
- Develop a numerical control (NC) program for a simple cutting requirement
- Describe the function and operation of two types of bar codes and bar code readers
- Describe the operation of radio-frequency (rf) tag technology

The design and analysis systems and software introduced in this book, and the production control and process technology discussed earlier must operate in harmony, using a shared database in the computer-integrated manufacturing (CIM) enterprise. The control mechanisms for this diverse set of technologies are spread across the entire enterprise and are organized in a hierarchical structure. On one extreme, this hierarchical structure links the "on" condition of a mechanical limit switch that senses the presence of parts on a conveyor to the machine control that must respond. At another point in the hierarchy, an enterprise computer is interfaced through an external communication network, like the Internet, to pass production orders electronically to computers at vendor locations around the world. Developing and implementing this complex mix of hardware and software technologies is an evolutionary process.

Review the automation time line illustrated in Figure 4–2. Note on the time line that computer control in each of the three major functions described by the Society of Manufacturing Engineers (SME) enterprise wheel started many years ago. However, only in the last twenty years has the control of the individual elements matured sufficiently to allow integration of the many diverse enterprise functions. The technologies used to integrate the CIM enterprise and to establish the hierarchical control are described in this chapter.

12–1 SYSTEM OVERVIEW

A good starting point for an understanding of the control hierarchy used by the CIM enterprise at the shop-floor level is a comparison of the machine and device interfaces and network structures found in earlier automated cells and the techniques used in systems built today. The cell structure illustrated in Figure 10–22 is an example of earlier interface techniques. Review the description of the figure in chapter 10 so that you can compare the earlier integration approach with the techniques available today.

Older Work-Cell Interface Techniques

The two work cells shown in Figure 10–22 use a number of signal types in the exchange of data between devices and controllers. The cell includes proprietary communication protocols and discrete signals from 0- to 5-volt and 0- to 24-volt levels, which are RS-232 and RS-422 standards. The discrete voltage levels are used to exchange data between discrete devices, like sensors and actuators, and the programmable logic controller (PLC). For many of the other devices, like bar code readers and electronic gauges, the data are transferred with a serial data interface. A serial interface is also used to link various machine controllers, like the robot controllers, to the work-cell computer. Proprietary protocols are used between devices and machines made by the same vendor.

The interface techniques used in Figure 10–22 were quite common in the past. While they performed well, they had the following limitations:

- The serial interface was slow—usually less than 100 kilobits per second for data transfer.
- The discrete interfaces required large wire bundles since every sensor added two wires to the interfaced cable.
- Duplication of similar networks occurred as a result of the use of the proprietary network.
- All data had to pass through the cell control computers, which limited the speed of information reaching the enterprise network and reduced the rate at which programs could be downloaded to production machines.
- The numerous interfacing techniques added complexity to the solution, which may have been more difficult to maintain.

The network illustrated in Figure 10–22 continues to be used in existing work cells in some automated systems in industry. While the solution illustrated is an acceptable solution, some of the new networking technologies eliminate many of the shortcomings just mentioned.

Control and Information Protocol

DeviceNet, *ControlNet*, and *Ethernet/IP* are members of the control and information protocol (CIP) family. These newer network standards are used to streamline the interface between machines and devices in automation systems. Before describing how the new network technologies are used in the work cells shown in Figure 10–22, we present a description of each.

DeviceNet.   *DeviceNet* is a low-cost communications link that connects *smart* input devices—such as sensors, limit switches, and blocks of input terminals—and *smart* output devices—such as blocks of valves, light bars, and blocks of output terminals—through a network to a PLC or computer. A *smart device* is one that has an embedded microprocessor so that it can communicate across a network, which eliminates time-consuming and costly hardwiring. In addition, this linking of computer-controlled sensors and actuators on a single network permits data exchange between the devices as well as with the PLC processor or computer. Also, technicians can troubleshoot device problems and use device-level diagnostics from a single remote computer.

DeviceNet is an open network standard developed by Rockwell Automation in 1993, based on a proven network technology called *controller area network* (*CAN*). Currently, the Open DeviceNet Vendor Association, Inc., governs this open network standard, which means that any vendor can develop and supply DeviceNet-compatible devices without paying any licensing fees. As a result, products from different vendors can be used to build a DeviceNet network. This permits the designers to select the best combination of devices from multiple vendors to solve the control problem. As a result, DeviceNet is the fastest-growing device network in the world, with over 500,000 devices installed.

ControlNet.   *ControlNet* is another open network standard. It lies one level above DeviceNet in the control hierarchy. While the primary function of DeviceNet is the networking of input and output (I/O) devices, ControlNet uses the *producer-consumer* network model to efficiently exchange time-critical application information for both processes and manufacturing automation. This model permits all nodes on the network (consumers) to simultaneously access the same data from a single source (producer). In addition, ControlNet's *Media Access Method* uses the producer-consumer model to allow multiple controllers to control I/O on the same wire. This provides a significant advantage over other networks, which allow only one master controller on the wire. ControlNet also allows *multicast* (simultaneous broadcast of data to multiple devices on the network) of both inputs and peer-to-peer data, which reduces traffic on the wire and increases system performance.

ControlNet is highly *deterministic* and *repeatable*. *Determinism* is the ability to reliably predict when data will be delivered by using token-passing technology, and *repeatability* ensures that transmit times are constant and unaffected by devices connecting to or leaving the network. These two critical characteristics ensure dependable, synchronized, and coordinated real-time performance. As a result, ControlNet permits data transfers, such as program uploads and downloads and monitoring of real-time data, in flexible but predictable time segments. Management and configuration of the entire system can be performed from a single location on ControlNet or from one location on an information-level network, such as Ethernet. ControlNet can link a variety of devices, including *motor drives, motion controllers, remote input-output modules, PLCs,* and *operator interfaces.* In addition, ControlNet can provide a link to other networks, such as DeviceNet and FOUNDATION Fieldbus.

ControlNet, developed in 1995, is an open network standard like DeviceNet, but is managed by ControlNet International, an independent body setting the standards for vendors supplying ControlNet hardware. As a result, control devices from multiple vendors can be interconnected on the same ControlNet network.

Ethernet/IP. Ethernet is not new because it was developed by Xerox Palo Alto Research Center in the 1970s for use as a local area network (LAN) technology for office environments. Digital Equipment Corporation and Intel joined Xerox in 1979 to promote the new network and publish the first Ethernet specification. The Ethernet specification was passed to the Institute for Electrical and Electronic Engineers (IEEE), which released IEEE standard 802.3 in 1983. In 1985, Ethernet was released as an International Standards Organization (ISO) standard. Today, Ethernet is used for almost all network data communications. In addition, the introduction of the gigabit Ethernet and the use of more reliable Ethernet switches opened up shop-floor applications for a new version called *Ethernet/IP.* The *IP* stands for *"industrial protocol,"* and it is being used on the plant floor for control and even device-layer applications. Figure 12–1 illustrates how the new system is used on the shop floor.

Ethernet/IP is fast, moving data at 100 megabits per second, which is better than ControlNet. However, it lacks the interchangeability between competing vendors' devices and the level of determinism that is currently available in the ControlNet standard. However, the use of fast Ethernet switches and the evolution of Ethernet/IP toward ControlNet-type interchangeability give this network standard a place in shop-floor control. However, it doesn't mean that ControlNet and DeviceNet will disappear, and future networks will most likely be a mix of Ethernet/IP, ControlNet, and DeviceNet.

Study the three levels of network control illustrated in Figure 12–1. You can see that plant-floor architecture has flattened into three layers: Ethernet, ControlNet, and DeviceNet/FOUNDATION Fieldbus. The highest level, Ethernet or Ethernet/IP, provides the *information layer* for plantwide data collection and program maintenance. The middle level, ControlNet, supplies the *automation and control layer* for real-time I/O control, interlocking (coordinating update times

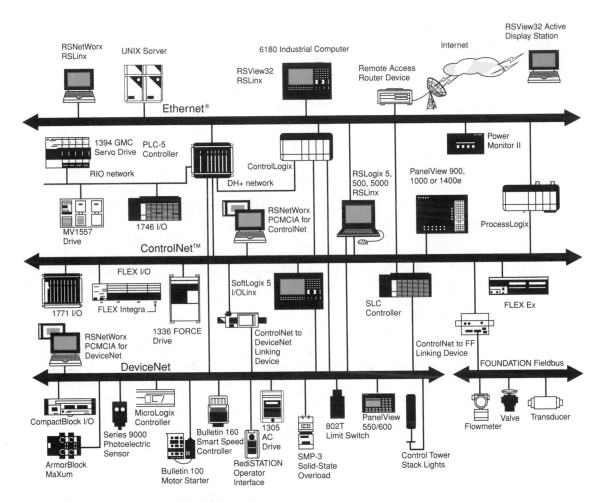

Figure 12–1 Three Levels of Network Control.
(Source: Courtesy of Rockwell Automation, Inc.)

between applications), and messaging. The lowest level supports a *device layer* for cost-effective integration of individual devices. In the lowest level, DeviceNet provides the primary discrete interface and FOUNDATION Fieldbus is most often associated with the analog control segment in process control. Networking and distributed control technology are starting to allow the best solutions from various vendors to be mixed in some industrial solutions.

Cell Control with the New Standards

Figure 5–52 shows the enterprise network with typical enterprise functions linked by LANs. Remember, a LAN is a nonpublic communications system that allows devices connected to the network to exchange data and information electronically

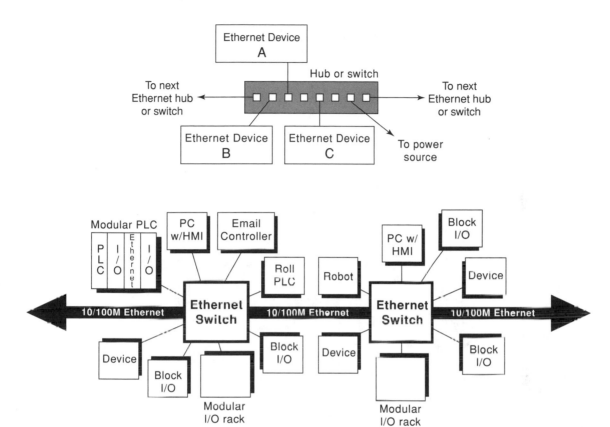

Figure 12–2   Ethernet Switches.

across large distances. Each of the area LANs is connected to the backbone through fast Ethernet switches (Figure 12–2) so that everyone has fast access to the shared data and links to external resources and systems. The enterprise network links to the shop floor to allow departmental access to shop-floor data and bidirectional data flow.

The cell control system illustrated in Figure 10–22 is redrawn with the new network standards in Figure 12–3. Compare Figures 10–22 and 12–3 to see how the new network strategy changed the layout of the cells. Some devices are linked by DeviceNet, others are linked by ControlNet, and some are interconnected directly by the Ethernet/IP. The result is a much cleaner network interface with fewer technologies used to exchange data and information.

Some key characteristics differentiate the cell control from the higher levels in this typical hierarchy. Table 12–1 illustrates these differences in five different categories: *controller hardening*, *response time*, *control versus information processing*, *number of users*, and *communications interface*. The hardening column indicates the percentage of applications where a system must be able to operate in a harsh environment

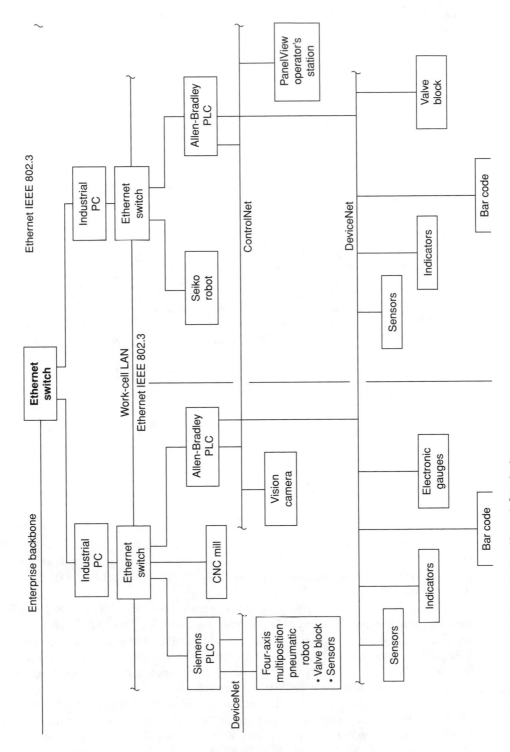

Figure 12–3 Cell Control with New Network Standards.

Table 12–1   System Characteristics at Various Enterprise Levels.

Enterprise level	Hardening applications (%)	Response time	Control versus information processing	Number of users	Communicaton system
Plant host	0–3	Days to seconds	$\dfrac{0\text{–}5\%}{95\text{–}100\%}$	Hundreds	LAN
Area controllers	8–12	Minutes to seconds	$\dfrac{5\text{–}10\%}{90\text{–}95\%}$	Tens	LAN
Cell controllers	80–90	Seconds to milliseconds	$\dfrac{10\text{–}20\%}{80\text{–}90\%}$	1–12	LAN and ControlNet
Device controllers	90–95	Milliseconds to microseconds	$\dfrac{90\text{–}95\%}{5\text{–}10\%}$	1–2	DeviceNet

in which extreme changes in temperature, shock and vibration, power fluctuations, airborne contaminants, and moisture exist. The response time indicates how fast the system must be capable of generating an output on the basis of input conditions. Control versus information processing addresses the percentage of control problems versus the percentage of information-processing problems. The last two parameters indicate the typical number of users and the type of communication system. The system needs at the cell and device levels stress very fast response times with a heavy emphasis on the control of a small number of devices. At this low level, point-to-point communication is still used, but DeviceNet is used frequently and will become the device network for manufacturing.

The control hierarchy illustrated in Figure 12–3 establishes the interface between production machines, production machine controllers, and the enterprise computer network. The cell controller links computer-controlled machines in the cell with the other enterprise departments. A good starting point for a description of CIM hierarchical control is at the lowest level, the production cell.

12–2 CELL CONTROL

The hardware in the work cells falls into two general categories: intelligent and nonintelligent. Most of the devices, like sensors and valves, in Figure 10–22 are nonintelligent, while all the devices used in Figure 12–3 are intelligent. Intelligent equipment has some type of internal computer control, which permits programming and network operation. In some applications, intelligent machines are called *smart devices*. Like the sensors and bar code readers illustrated in Figure 10–22, nonintelligent hardware often uses solid-state electronics for signal conditioning and conversion, but it is not programmable. Automated cells utilize an infinite variety of intelligent and nonintelligent hardware, control software, and every conceivable process machine. While production cells are rarely the same, the concepts learned from a study of the two cells shown in Figures 10–22 and 12–3 will transfer to the operation of other production cells. Therefore, an analysis of the operation of these two cells is a convenient starting point for the study of cell control techniques.

Cell Controllers

The cell controllers in the machining and assembly cells shown in Figure 12–3 are usually industrial computers using Intel microprocessor chips. For example, the Siemens computer shown in Figure 12–4 is an industrial computer serving as the programming station for a computer numerical control (CNC) mill and is designed for operation in the harsh production environment. These boxes usually run the Microsoft NT/2000/XP operating system. The system configuration for cell controllers includes everything found in a standard PC.

Cell Control Software Structure

The operating system software used for cell controllers in most applications is usually Windows NT/2000/XP. The Windows environment permits multiple programs to run at the same time. Software in the cell controller may be interacting with numerous machines in the cell through the Ethernet network links. For example, a machine operator must regularly use the cell controller's direct numerical control (DNC) software to download a new part program to the CNC milling machine. As this DNC function is occurring, the following programs continue to execute in the background: production data from the PLC data registers are uploaded to production control, and bar code data are supplied to automated product tracking software. Work-cell controller applications that demand this level of flexibility require multitasking operating systems.

The cell controller interfaces with other intelligent machines and devices using standard DeviceNet, ControlNet, and Ethernet protocols. The cell controllers shown in Figure 12–3 have network connectivity to the work-cell LAN and the enterprise backbone. The computer is the link between these two LANs. A typical software configuration in the cell controller to manage these interfaces is illustrated in Figure 12–5. Study the figure until you are familiar with the descriptions in each box.

The operating system (OS) is usually one of the three listed at the bottom of the figure. Application program interface (API) and system enabler (SE) software reside between the OS and the applications. The API and SE handle some of the operational overhead common to all applications; in addition, an SE helps manage the resources, like hard-drive data storage. For example, text and graphics dumps to the monitor are handled by an API program. Therefore, developers spend less time programming for the OS and more time on the specific application. The applications can include a wide variety of software, like those listed in Figure 12–5.

The cell controller frequently has various interface cards to support network standards such as Ethernet and token ring. Review the interfaces listed in

Figure 12–4 Siemens
Industrial PC on a CNC Machine.

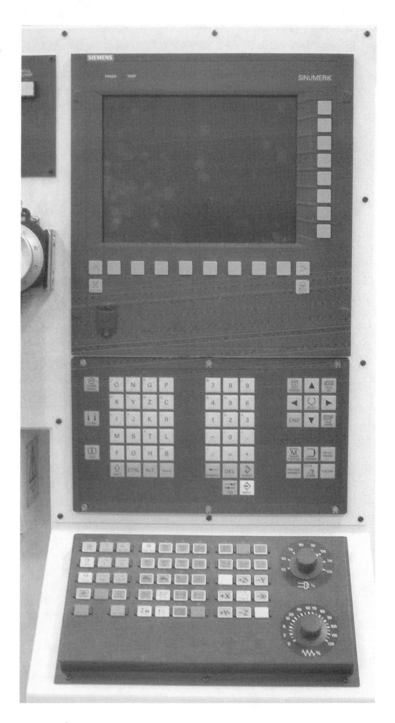

Figure 12–5   Work-Cell
Controller Software Layers.

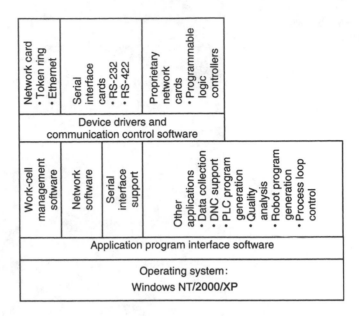

Figure 12–5. Device drivers and communications control software reside between
the interface cards and the application software.

Work-Cell Management Software

The major application software present in cell controllers is the software used to
manage all of the cell activity. The primary role of the cell controller is communications and information processing. For upstream LAN communications, the cell
controller is a concentrator of information from the cell and a link to area controllers and other applications, such as manufacturing resource planning (MRP II)
and computer-aided design (CAD) product design. On the downstream side, the
cell controller is a distributor of data and information files to programmable control devices and production equipment from the product data management system (PDM). This communications workload, transferring data files ranging from
single-bit to megabyte files, often uses over 50 percent of the resources of the
processor in the cell controller. Data management from real-time memory resident
file storage to support for the CIM shared databases is another cell control function. In addition, management of program libraries for devices in the cell, support
for engineering change control, and production tracking represent other cell
control responsibilities. The information and communication tasks frequently
consume 80 to 90 percent of the resources of the cell controller.

Applications typically loaded into the cell controller include the following:

- Production monitoring
- Process monitoring
- Equipment monitoring

- Program distribution
- Alert and alarm management
- Statistical quality and statistical process control
- Data and event logging
- Work dispatching and scheduling
- Tool tracking and control
- Inventory tracking and management
- Report generation on cell activity
- Problem determination
- Operator support
- Off-line programming and system checkout

12–3 PROPRIETARY VERSUS OPEN SYSTEM INTERCONNECT SOFTWARE

Software to manage the information flow through the cell controller falls into one of three categories: (1) in-house-developed systems, (2) application enablers, and (3) adoption of an open system interconnect (OSI) such as the *manufacturing message specification* (*MMS*) ISO standard 9506. The first two are proprietary software systems because they are either written for a specific application or developed around a third-party software management shell, called an *application enabler*. The last technique uses an OSI standard such as MMS to develop solutions. A more detailed description of each of these approaches in the following sections illustrates the advantages and disadvantages of each approach.

In-House-Developed Software

Before 1980, most cell control and management software was written by the user using the C programming language. In-house development was necessary because satisfactory third-party software options were not available. The cost of these programming projects was substantial. In addition, the complex and cryptic nature of the code drove up the costs of modifications and updates to existing programs.

The in-house-developed programs were written to address the specific control needs of the cell, and the software interfaced with other in-house-developed software in other enterprise areas. In addition, the lack of available interface software or drivers for equipment in the cell was not a problem because all the code was generated as part of the initial program development. However, these advantages were offset by a major disadvantage in cost and the inability to change the software easily when the cell hardware or configuration changed.

Enabler Software

A common impediment to the development of a fully implemented CIM work cell is the exorbitant cost, time, complexity, and inflexibility of custom-programmed CIM solutions. However, with the introduction of *enabler software,* application development has shifted from software engineers to manufacturing engineers. This shift was possible because enabler software provided a set of software productivity tools to develop control software for the CIM cells. Products such as *Plantworks* from IBM, *Cell Control* from Hewlett-Packard, *FIX DMACS* from Intellution, *FactoryLink* from USDATA Corporation, and *InTouch* from Wonderware are available to reduce the difficulty in developing cell control and management applications.

Enablers such as Wonderware and FIX DMACS have a library of driver programs for many commonly used production machines and machine controllers. In addition, they offer LAN and serial data communications support, mathematics and logic functions for internal computations, links to commonly used mainframe and microcomputer relational databases, real-time data logging, real-time generation of graphics and animation of cell processes, alarm and event supervision, statistical process control, batch recipe functions, timed events and intervals, counting functions, and the ability to write custom applications. The screen shown in Figure 12–6 shows how an enabler dynamically displays a process. The graphic of the process is built by using graphic drawing tools supplied with the enabler or from a third-party vendor.

The advantage of enablers is an estimated tenfold improvement in cell control and ease of program development. The primary disadvantage is that the cell

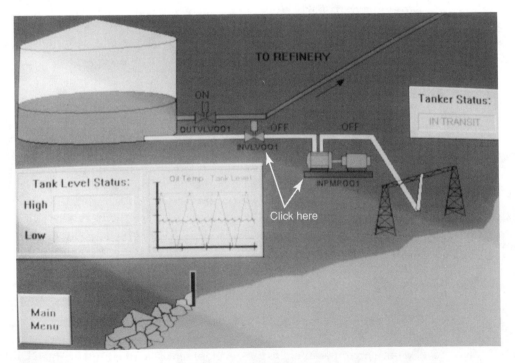

Figure 12–6   Screen from FIX DMACS Showing Cell Enabler Software.

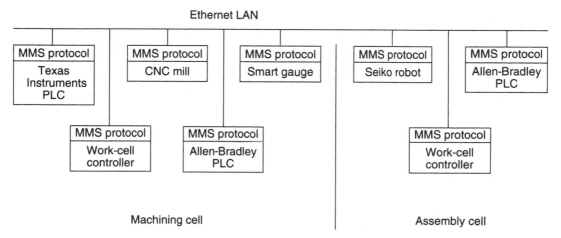

Figure 12–7 MMS Implementation.

control is tied to a third-party software solution, so selecting the best solution is critical. In addition, an enabler is written to satisfy the needs of the average company; therefore, if a specific requirement is not covered, additional Visual Basic or C code programming is necessary to obtain a solution.

OSI Solution

The most frequently used OSI solution for cell control is the MMS. MMS is a standard (ISO 9506) for communicating information between intelligent devices in a production environment across networks based on the OSI model.

Implementing an MMS solution requires that all the devices support the MMS protocol and that they are linked over a manufacturing automation protocol (MAP) Ethernet or broadband manufacturing network. The work-cell system shown in Figure 12–3 is reconfigured as an MMS solution in Figure 12–7. Take a few minutes to compare the two implementations. Implementation of the new network standards—DeviceNet, ControlNet, and Ethernet/IP—have all but eliminated work by hard vendors on MMS protocol. The new Ethernet/IP evolution will come very close to the MMS goal.

12–4 DEVICE CONTROL

Device controllers fall into two categories: *proprietary* and *generic. Proprietary controllers* are generally special-purpose microcomputers built around a common microprocessor chip and programmed to control the target device. For example, the CNC machine tools described in chapter 10 all have special-purpose computers built into the machine tool for control. In another example, the robot system outlined in Figure 11–16 has a special-purpose controller to move the arm under program control. However, the robot computer controller is located in a remote enclosure and connected to the robot through an electrical interface. The *generic controllers* are *general-purpose* devices designed to interface to different work-cell hardware to provide control.

Both proprietary and generic device controllers support *discrete* and *analog* control requirements. Discrete control is used, for example, to turn a motor on or off. Therefore, *discrete* control implies just two operational states for the device under control.

Analog controllers allow devices to operate over the range from on to off. Analog controllers used in production machines and cells generally control the motion or position of an object or the value of a process parameter such as pressure, temperature, level, or flow. For example, an analog temperature controller could hold the temperature of liquid in a tank at any temperature between 20° and 90°C. Achieving this type of control requires that the energy supplied to the heating element by the controller be set at a level that produces the correct temperature. In contrast, the discrete controller would either turn the heating element on (heat transferred to the liquid) or off (no heat transferred).

Describing all the special-purpose controllers used in manufacturing is difficult in an overview. However, two types of device controllers—*programmable logic controllers* (*PLCs*) and *computer numerical controllers* (*CNCs*)—are used frequently in production automation, so understanding their operation is important.

12–5 PROGRAMMABLE LOGIC CONTROLLERS

The best way to begin a study of PLCs is with several definitions. The term *PLCs* is defined as follows:

> *PLCs are special-purpose industrial computers designed for use in the control of a wide variety of manufacturing machines and systems, using a proprietary programming language.*

The definition states that PLCs are *industrial computers*, which means that they are designed to operate in the harsh physical and electrical noise environments present in many production plants. They use a proprietary programming language, called *ladder logic*, which is unique to the PLC brand. PLC applications vary from the on-off control of a single pump motor with a liquid-level switch to the control of an entire package conveyor system used by United Parcel Service to sort packages to multiple loading areas by using destination zip codes.

In 1978, the *National Electrical Manufacturers Association* (*NEMA*) published a standard for PLCs, which included the following definition:

> *A PLC is a digitally operating electronic apparatus, which uses a programmable memory for the internal storage of instructions for implementing specific functions such as logic, sequencing, timing, counting, and arithmetic to control, through digital or analog input/output modules, various types of machines or processes. A digital*

computer, which is used to perform the functions of a programmable controller, is considered to be within this scope. Excluded are drum and similar mechanical type sequencing controllers.

This early U.S. standard, NEMA ICS3-1978, defined PLCs as any computer-controlled device used for sequential control but excluded mechanical sequencers, like *drum controllers*, from this group. With the rapid growth in PLCs, a broader standard was necessary that cut across national boundaries.

In 1979, the *International Electrotechnical Commission (IEC)* established a working group to look at the complete standardization of PLCs. The new initial standard, called *IEC 1131* (later changed to *IEC 61131*), had five parts:

1. General information
2. Equipment requirements and tests
3. Programming languages
4. User guidelines
5. Messaging service specification

Part 3, *Programming Languages,* the language standard for *Programmable Controllers,* was released in 1993 and updated in 2003. It specifies the standards for PLC software, covering PLC configuration, programming, and data storage. IEC standard 61131-3 provides a very specific and detailed definition of PLC programming. However, for our study of PLCs, the simpler earlier definition will suffice.

PLC History

Dick Morley introduced the first programmable controller in the United States on New Year's Day of 1968. Both the new device and the company supplying the system were named *Modicon*, which is short for *MOdular DIgital CONtroller*, the term used for the new technology. The first installed Modicon in industry was the model 084. While the first installation at the Oldsmobile Division of General Motors Corporation and the Landis Company in Landis, Pennsylvania, occurred in 1970, the fledgling company's growth was slowed because of industry concern about replacing relays with computer-controlled logic. However, Modicon's growth increased as a result of engineer Michael Greenberg's development of the model 184, a much more sophisticated version of the original model. Gradual industry acceptance and the success of this new technology created the global PLC industry that we know today.

The PLC Industry Today

While the PLC business is more than a $6.5 billion industry that is growing at a rate of 20% per year, few people beyond manufacturing automation experts know it exists. The PLC has been a strong, silent partner in promoting manufacturing automation around the globe. A search of the Thomas Register online revealed 110 listings under the heading "Programmable Controller Vendors," and there are over

1,000 vendors offering PLC add-in boards and other peripherals. In addition, more than twice this number of vendors develop PLC solutions. This industrial-strength microcomputer is responsible for controlling a wide variety of industrial processes, from automobile production to stamping out Oreo cookies. Reliability is critical in all industrial applications because for every minute of downtime, a company loses money. For example, downtime on an automotive production line can be as costly as $20,000 per minute. The PLC's record for reliability is unmatched when compared with that of other microprocessor-controlled devices. For example, the mean time between failures is close to five years for some of the PLCs sold by major PLC vendors.

Major PLC vendors have promised compliance with the language standard of IEC 61131 and are delivering systems that meet the standard. Network standards permit manufacturing automation system designers to mix vendor systems and system components to a much greater degree. Customer acceptance of the technology, vendor work on standards, and increases in the technology delivered per dollar spent have allowed PLC applications to be implemented in areas never imagined.

PC Versus PLC

The original design for the PLC was just called a *programmable controller*, or *PC*. The use of this PC abbreviation for a programmable controller caused no confusion until the personal computer became widely used and was frequently referred to as a *PC* in the literature. To avoid confusion, many programmable controller vendors added the word *logic* in the title, producing the new term *programmable logic controller*, or *PLC*. You will still find references in the literature to PCs with the reference implying a programmable controller, not a personal computer.

In addition to the name abbreviation, the PC and the PLC have some things in common, as well as many that make them different. The architecture of the PC system and that of the PLC system are similar. Both feature a motherboard, a processor, memory, and expansion slots. The processor in the PLC includes a central processing unit (CPU) composed of a microprocessor and the computer-type architecture. PCs are generally designed to do many jobs, but the PLC performs only one chore: sequential and process control.

12–6 RELAY LADDER LOGIC

The early PLCs were special-purpose industrial computers designed to eliminate relay logic from sequential control applications. To understand how PLCs accomplished this task, you must understand the operation of relays and relay ladder logic. Study the side-view illustration of a relay in Figure 12–8 before continuing with the description of this electromechanical device.

Figure 12–8   Electromagnetic
Relay.

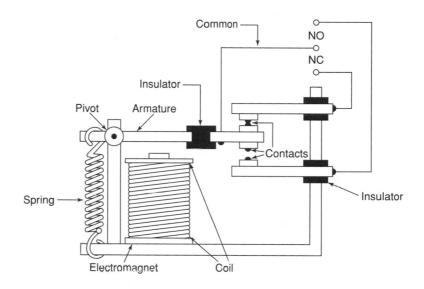

Electromagnetic Relay

The relay has three main components: (1) the *electromagnet* produced by the coil wound around the steel core, (2) the *armature*, called a *clapper*, that is pulled down when the coil is energized, and (3) the *contacts* that create one electrical path, called *normally closed* (NC), when the coil is off (armature up) and a second path, called *normally open* (NO), when the coil is energized (armature down). This configuration is called *single pole, double throw* because one common contact (single pole or armature) is present with two positions (NC and NO) called *throws*. Note that the spring holds the armature in the *up* position when the coil is *not* energized, which creates a low-resistance connection between the common armature and the NC contact. When the coil is energized, the armature pivots down so that the contact with the NC contact is broken, and a low-resistance connection between the common armature and the NO contact is established. *Insulators* are used to isolate the electrical switching contacts of the relay from the rest of the relay components.

Figure 12–9 illustrates two schematic representations for the relay shown in Figure 12–8. The two schematics for a relay indicate that the symbols used for electronic circuits are often different from those used in control-type schematics. Compare the electronic and control schematics shown in Figure 12–9 with the relay shown in Figure 12–8 until you see the relationship between the actual device and the schematic representation of the device. Note the symbols used for NO and NC contacts in the control schematic. The NO has two parallel lines indicating an open circuit, and the NC has two parallel lines with a line across them indicating closed contacts.

Sequential Control

To understand how relays are used in sequential control, consider the following simple control problem. A tank, illustrated in Figure 12–10a, is filled through an

Figure 12–9 (a) Electronic Symbol and (b) Control Symbol for a Single-Pole, Double-Throw Relay.

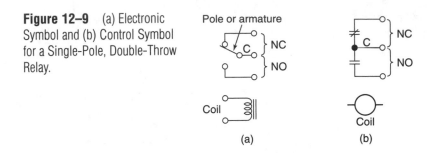

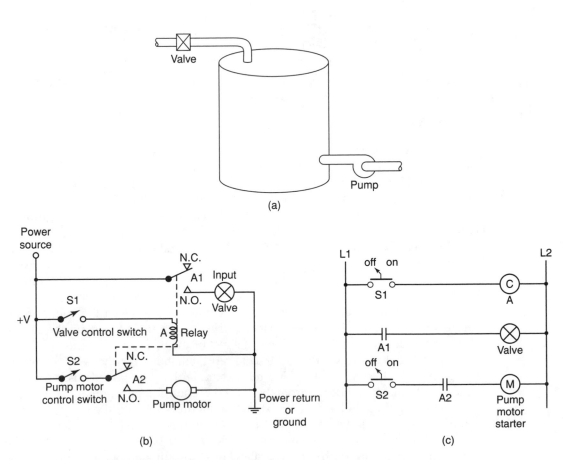

Figure 12–10 (a) Sequential Control of a Storage Tank; (b) Electronic and (c) Control Schematics for Valve-Pump Control.

electrically operated valve and emptied by a motor-driven pump. Control of the valve and pump must satisfy the following logic:

1. The pump can operate only when the input valve to the tank is open.
2. The input valve can be opened when the pump is either operating or not operating.

The electronic schematic shown in Figure 12–10b illustrates a solution to the control problem. The same solution is represented in Figure 12–10c with control schematic symbols in a drawing style called a *two-wire control diagram*. Study both illustrations and verify that both schematics satisfy the logic required for this control problem.

The circuits work as follows:

1. Switch S1 is closed manually and causes the magnetic relay A to be energized.
2. When the relay is energized, the poles on contacts A1 and A2 (dashed lines indicate that both contacts are linked to the status of the coil) move from the NC positions to the NO positions.
3. The change in contact A1 energizes the input valve and allows liquid to flow into the tank.
4. The change in contact A2 causes no immediate action.
5. Switch S2 is closed manually and causes the pump to operate.

Note that in step 4 the NO relay contact A2 has to close for the pump to operate when the pump switch is activated. If the valve switch S1 is opened manually while the pump is operating, the pump motor stops. It stops because the relay is not energized, so poles move from the NO contacts back to the NC contacts. The change in contact A2 opens up the circuit that is energizing the pump, and the pump stops. Review this section and study Figure 12–10 until operation of this control circuit is clear.

The control diagram shown in Figure 12–10c uses control-type symbols for the components in a configuration called a *two-wire diagram*. The two-wire diagrams have a vertical line at the left (marked L1) and right (marked L2) sides of the paper. The left vertical line usually represents the positive source for power, and the right vertical line represents the power return or ground. All the circuits containing the switches, sensors, and control devices used to operate a machine are drawn between the two vertical lines. The circuit shown in Figure 12–10c is the two-wire-diagram equivalent of the electronic schematic for the pump control circuit illustrated in Figure 12–10b. The schematic shown in Figure 12–10c is often called a *relay-ladder-logic* circuit because it uses *relays*, it looks like a *ladder,* and it satisfies the *logic* control requirements for control of the output device. Standard control-drawing symbols are used to represent the different input and output devices, such as mechanical switches, sensors, magnetic contactors and relays, and electrical contacts. In Figure 12–10b, the motor is connected directly to the switch and relay contacts, but this is rarely done in industrial control applications. The control schematic, shown in Figure 12–10c, indicates the preferred solution, which is to use a special-purpose motor starter relay (identified with an *M*), called a *contactor,* to switch the power for the pump motor.

12–7 PLC SYSTEM AND COMPONENTS

Before the way PLCs replace relay ladder logic is discussed, it is important to understand the architecture of a PLC system. Study the block diagram for the PLC system shown in Figure 12–11. Notice that the PLC system is the sum of a number of different components and modules. For the larger systems, all of these modules are mounted in a *rack,* as illustrated in Figure 12–12. The rack provides mechanical support, the electrical interconnections, and the data interface between all of the PLC modules and parts.

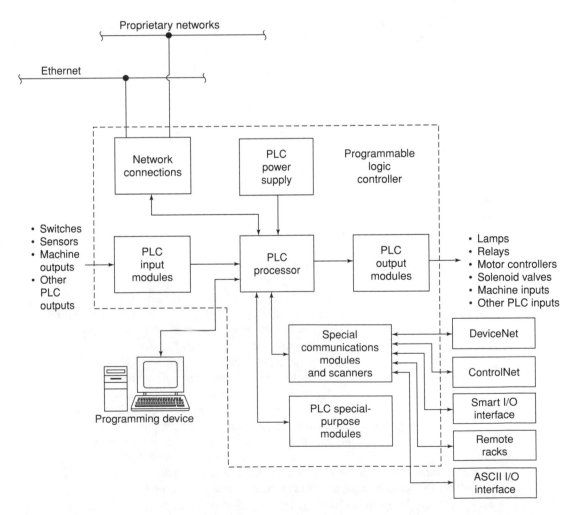

Figure 12–11 PLC Component Block Diagram.

Figure 12–12 PLC Rack and Module System.

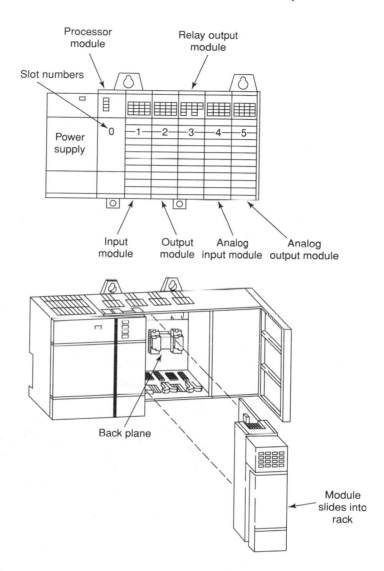

Back Plane

The power and data interface for the modules is provided by the *back plane* in the rack (Figure 12–12, bottom). The back plane is a printed circuit board with copper wires, called *lands*, that delivers power to the modules and that provides a data bus to exchange data between the modules and the CPU. Modules slide into the rack and engage connectors on the back plane to access the back plane's power and data buses. The number of slots in the rack is determined by the number and type of modules required for the control application. Now, let's study each component in the block diagram for the PLC system illustrated in Figure 12–11.

Processor and Power Supply

The center box in Figure 12–11, called the *PLC processor*, is the CPU, or the computer, that handles all logical operations and performs all the mathematical computations. The processor occupies the zero slot position in the rack illustration in Figure 12–12. The PLC processor has a microprocessor in the CPU, which creates the computer-type architecture. A picture of an Allen-Bradley PLC 5 processor is shown in Figure 12–13 in the far left slot.

The *PLC power supply* is pictured above the processor in Figure 12–11 and is the left-most box in the rack illustration in Figure 12–12. The power supply provides power to the processor and to the modules plugged into the rack. In addition to the processor and the power supply in Figure 12–11, there is a symbol for a programming device and five blocks that interface with external devices. The five modules are *input, output, special communications, network connections,* and *special purpose.* These PLC elements are described in the following paragraphs.

Programming Device. The programming device connected to the CPU (PLC processor) shown in Figure 12–11 is used to load ladder-logic control programs

Processor
module

Figure 12–13   Allen-Bradley PLC Rack and Modules.

into the PLC in the language supported by the vendor. A computer is indicated as the programming device in the figure; however, a special-purpose teach pendant or terminal provided by the PLC vendor is also used for programming.

Input and Output Interface. The input provides the interface between the PLC processor and the external input devices, such as switches, sensors, machine outputs, and other PLC outputs. Input devices are often called *field devices*, which indicates that they are not part of the PLC hardware. The input is both a *physical interface* for the connection of wires and an *electrical/data interface* to determine the on-off state or level of the signal from the attached field device. A variety of inputs are available in both discrete (0 or 24 volts) and analog (0–5 volts or 4–20 milliamps) signal types. In addition, modules have different numbers of input terminals, have direct current (dc) and alternating current (ac) voltage preferences, and can be either *current sourcing* (current flows out from the input when active) or *current sinking* (current flows into the input when active). Several input modules are shown in the rack illustration in Figure 12–12.

The output provides the interface between the PLC processor and the external devices or actuators, such as lamps, relays, motor controllers, solenoid valves, machine inputs, and other PLC inputs. The term *field device* is also used to address the wide range of output devices attached to a PLC system. Again, the output is a location both for *termination of wiring* and for *providing the proper level and type of output drive power* to activate the output field device. The selection of output (fixed and modular) types and models is similar to that offered on the input side. Several output modules are shown in the rack illustration in Figure 12–12.

Special Communications Modules and Network Connections. The *special communications modules* included in Figure 12–11 provide a link for the PLC processor to other computer-controlled machines and devices that must share data and control requirements with the PLC. Five examples are listed: *DeviceNet, ControlNet, Smart I/O interfaces, remote racks, and ASCII I/O interfaces.* In addition, the *network connections* box indicates that the PLC is linked to *information-level* networks like the *Ethernet*, and vendor-specific networks, called *proprietary networks*. The use of these communications modules permits the PLC to act as a data hub or data concentrator for these five subnetworks and a gateway or link to the enterprise Ethernet. In some applications, the PLC is linked to other controllers from the same vendor by using a vendor-specific network called a *proprietary network* (Figure 12–11, top). The Allen-Bradley *data highway* is an example of this type of vendor-specific network. A description of other network technology follows.

PLCs located in this network architecture often use a *smart I/O interface, remote racks,* and an *ASCII I/O interface* to further distribute data communications to remote control locations. A description of the last three special communications modules follows.

Smart I/O Interface. PLC vendors have proprietary network protocols to allow devices from the same vendor to communicate by using that vendor's specific

Figure 12–14   (a) Traditional I/O Interface; (b) Smart I/O Interface.

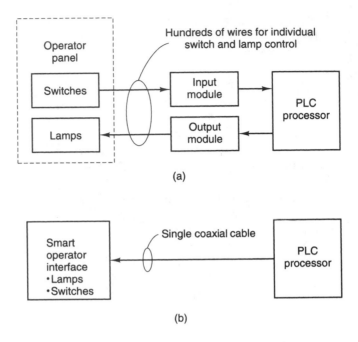

(a)

(b)

network interface. One common application for a proprietary network is the use of *smart I/O devices*. The term *smart* again implies that the device or interface includes a microprocessor so that it can be programmed for network data exchange.

Figure 12–14 illustrates the concept of the smart I/O interface using an operator panel. Operator panels have switches for control of process machines and devices, and the panels have lights to indicate the condition of process equipment. The operator panels are frequently located in a control room away from the process itself. The traditional approach to building an operator panel (Figure 12–14a) requires a minimum of one wire per switch and lamp plus several return wires between the operator panel and the PLC input and output modules. As a result, the wire bundle between these two devices often has hundreds of wires that must be enclosed in conduit across distances of hundreds of feet.

In contrast, the smart operator interface (Figure 12–13b) uses the same number of switches and lamps but controls them with a small microcomputer located in the operator panel. The interface between the operator panel and the PLC is just a single coaxial network cable that permits the operator panel's microcomputer to communicate with the PLC processor. Many smart external devices are available, including motor drives, process controllers, text readout devices, programmable cathode ray tube (CRT) displays supporting full color and graphics, voice input and output devices, and discrete input and output devices.

Remote Racks.   In an effort to distribute the control capability across a large automation system, the PLC vendors provide *remote rack* capability. The rack, similar to the rack illustrated in Figure 12–12, uses the standard I/O modules for control of machines and processes; however, the processor module is replaced with a remote rack communications module. The processor in the main PLC rack sends control instructions across the single network cable to the communication module in the remote rack and then to the I/O modules included in the remote rack. Use of this technology permits the I/O modules to be located close to the point of control, which eliminates the long wire runs required if the sensors are connected to I/O modules in the main PLC rack.

ASCII I/O Interface.   The last special communications module, listed in Figure 12–11, is the ASCII I/O interface. This communication resource is either built into the processor module or comes as a separate module. In both cases, the ASCII interface permits serial data communication using several standard interfaces, such as RS-232 and RS-422. This interface is used to connect devices such as the smart gauges and bar code readers that must transfer data between the device and the PLC.

PLC Special-Purpose Modules.   The final element in the PLC system block diagram (Figure 12–11) is labeled *PLC special-purpose modules*. This term represents a broad collection of modules developed by PLC vendors for PLC control of a variety of automation devices. Examples of modules include analog I/O, temperature measurement and control (thermocouple and resistance temperature device), multiple proportional-integral-derivative (PID) loop control, servo motor control, stepper motor control, high-speed counter, and hydraulic ram control. Most vendors offer some special-purpose modules, like the analog I/O, while only a single vendor supports other modules.

An Internet search using the words *programmable logic controller* returns hundreds of sites with hardware or software supporting sequential control applications. Included in this number are many vendors offering PLC hardware and software solutions ranging from general control applications to controls for specific types of manufacturing systems. The PLC systems described have proprietary hardware and software and offer minimal interchangeability between different vendor systems. While each vendor has a proprietary system and software, the block diagram in this section successfully describes the operation of each vendor's system.

12–8 PLC TYPES

There are three operational classifications into which all PLC-like devices can be grouped—namely, *rack-* or *address-based systems, tag-based systems,* and *soft PLCs,* or *PC-based control.* The first type, rack- or address-based control, was implemented in the initial PLC systems and is still the most frequently used type.

Rack- or Address-Based System

The PLC system illustrated in Figure 12–12 is a rack- or address-based system because the location of the I/O card in the rack establishes the PLC address for the input or output signal attached to the card. Let's explore this concept in greater detail. The cards placed in the rack to the right of the processor are most often some type of input or output card. The type of device that is connected to the PLC and the signal type—such as ac, dc, discrete, or analog—associated with that device dictate the type of I/O card chosen. While numerous types of I/O cards are used, each has two functions: (1) it provides the interface terminals to which device wires are attached, and (2) it provides an electronic circuit that interfaces the PLC with the type of signal presented. The signal level presented at each input is represented inside the PLC by a variable, which is usually the letter *I* followed by an address reference number. While each PLC vendor of rack- or address-based systems has a different addressing scheme, the address is determined by the type of card present (input or output), the *rack slot number occupied* by the card, and the *terminal number used* for the connection. For example, the following syntax is used to address discrete inputs on an Allen-Bradley SLC 500 system:

$$I:(\text{rack slot number})/(\text{terminal number})$$

The *terminal number* on the input module to which the input is attached, and the *slot number* in the PLC rack in which the input card is placed determine the input address for the SLC system. The letter *I* indicates that it is an input, the colon (:) is a delimiter separating the card-type letter and the address reference, and the slash (/) is a delimiter between the rack number and the terminal number. The following example illustrates this addressing concept.

Example 12–1
An Allen-Bradley SLC system uses a rack like the one illustrated in Figure 12–12. Determine the address for a discrete input signal attached to terminal 5 of the dc input module.

Solution
The address for the discrete dc input would be as follows:

$$I:1/5$$

The letter *I* indicates that it is an input, the 2 indicates that the dc input module is in slot 1, and 5 indicates that the discrete (on or off) input signal is connected to terminal 5.

The addressing for a discrete output module would be similar to that for the input module, except that the letter *I* would be replaced by the letter *O*. The following example illustrates the rack or address scheme for outputs.

Example 12–2
An Allen-Bradley SLC system uses a rack like the one illustrated in Figure 12–12. Determine the address for a discrete output signal attached to terminal 12 of the analog output module.

Solution

The address for the discrete ac output would be as follows:

$$O:5/12$$

The letter *O* indicates that it is an output, the *5* indicates that the analog output module is in slot 5, and *12* indicates that the discrete output signal is present on terminal 12.

Tag-Based System

Tag-based systems are used in all of the new models of PLCs offered today. In a tag-type system, inputs and outputs are represented by variables in the design of the control program. Later, the variables are referenced to the appropriate input and output terminals.

Soft PLCs, or PC-Based Control

A PC-based control system, called *soft PLC*, is usually described as the emulation of a PLC using software on a PC. This implementation uses an industrial PC, some type of remote or distributed I/O interface to the system, and application software that makes the PC operate like a PLC.

A second implementation uses a standard PLC with an industrial PC module placed in one of the PLC rack slots. This version puts the PC on the PLC back plane and gives the soft PLC application running in the PC access to all of the I/O modules in the PLC rack. Soft PLC solutions use one of these two implementations in most applications.

When is a soft PLC the optimum solution? In general, discrete control with a small number of inputs and outputs would be an application for a rack-type PLC solution. However, if large data storage and manipulation is required along with sequential or process control, then a soft PLC implementation may be the better choice. If the power of the PC can be utilized, such as in data storage and processing or in displaying the graphics of the process, then the higher cost associated with a soft PLC solution is justified.

12–9 RELAY LOGIC VERSUS LADDER LOGIC

Ladder logic is a PLC graphical programming technique that was introduced with the first PLCs more than thirty-five years ago. There are many similarities between relay logic, as illustrated in Figure 12–10c, and the ladder logic used to program PLCs. However, the subtle differences between the two systems often create confusion for the new user of PLCs. To understand the differences, start with a review of the sequential control problem that generated the relay-logic diagram illustrated in Figure 12–10. It is important to note that the contact symbols in the diagram represent the contacts of mechanical switches present in the system. As a result, an

NO contact in the system would be represented by an NO contact symbol in the diagram. In summary, all the symbols in the diagram represent actual components present in the control system.

PLC Solution

When a PLC is introduced into the tank control problem (Figure 12–10), the PLC replaces the mechanical relay but not the input switches or output actuators. The switches and actuators are interfaced to the PLC as illustrated in Figure 12–15a. Note that the input switches (illustrated using the standard symbol for a selector switch) are connected to the PLC input module, and the actuators are wired to the output module. The terminations at the input and output modules are identified by terminal numbers. For example, the switches are connected to terminals 1 and 2 on the input module and the valve and pump are connected to the similar terminal numbers on the output module. The illustration indicates that the PLC processor and the ladder-logic program are located between the input and output modules.

The PLC ladder-logic program that would provide the same logical control as that of the circuits shown in Figures 12–10b and 12–10c is illustrated in Figure 12–15b. Compare Figures 12–10c and 12–15b to identify similarities and differences. They include the following:

- The control circuit in Figure 12–10c exists in the form of physical comments and wire, but the ladder logic in Figure 12–15b exists only as a set of logical statements in the PLC memory.
- The mechanical relay in relay logic has been replaced by a software control relay (CR) in the ladder logic. The software, or virtual, relay exists only in the PLC memory.
- Each rung of the ladder logic represents a logical statement with inputs on the left and outputs on the right. If the inputs are true, then the output is true. For example, if input contact I1 is true, then output O1 will be active. Input contact I1 will be true if switch S1 is closed and a voltage is present at terminal 1 of the input module. If output O1 is active, then terminal 1 of the output module applies zero volts at the output terminal and the solenoid valve is turned on.
- The number of contacts for the control relay in ladder logic is limited only by the size of the PLC memory, while the number of contacts for the mechanical relay is limited to the size of the relay used.
- The input contacts and output symbols in the ladder logic do not represent the switches and actuators directly. The input contacts are symbols associated with the input signal presented to the input module by the switches. The output symbol is associated with the signal that will be presented to the actuator at the output module.
- The input and output devices shown in Figure 12–15 have separate power sources that are isolated from the power for the PLC processor.

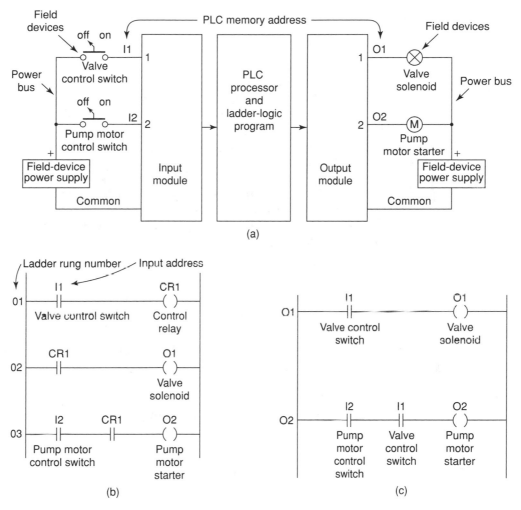

Figure 12–15   (a) PLC I/O Interface; (b) PLC Ladder-Logic Program; (c) Alternate Ladder-Logic Program.

PLC Operation

The PLC processor *scans* all the inputs and writes a binary *1* or *0* (input terminal with voltage present is a *1* and input with zero voltage is a *0*) into a memory location that corresponds to that input. Every input contact on the ladder-logic rungs that has a corresponding 1 in its memory location is put in the *true* state. If the input logic on the ladder rung is true, then the output ladder logic symbol on that rung is active, and a 1 is written into the corresponding output memory location. Then the processor scans the output memory locations and activates each output that has a 1 in memory. Let's apply this operational concept to the ladder-logic solution for the tank problem in Figure 12–15.

The solution in Figure 12–15b has three rungs, with input contacts on the left and outputs on the right. The input contacts in rungs 1 and 3 have data addresses of I1 and I2, so the voltage at input terminals 1 and 2 determines if these contact states are true or false. The input terminal voltages are set by the position of external switches S1 and S2 in Figure 12–15a. The input contact in rung 2 is a contact from the control relay, CR1.

The outputs in rungs 2 and 3 have data addresses of O1 and O2, so their state determines the state of the outputs connected to output terminals 1 and 2. If the logical outputs are active, so are the external devices associated with them. The output in rung 1 is a control relay (virtual relay created in the PLC software), CR1, with two contacts in rungs 2 and 3. If the output CR1 is active, then all the contacts associated with that control relay are true. The operation of the ladder logic is summarized as follows:

Rung 1: Output CR1 is active because contact I1 is true (input terminal 1 has a voltage present because switch S1 is closed). If CR1 is active, then both of the contacts associated with CR1 in rungs 2 and 3 are true.

Rung 2: Output O1 is active because input contact CR1 is true. If O1 is active, then the valve solenoid is on and the inflow valve is open.

Rung 3: Output O2 is active only if input contacts CR1 and I2 are both true. If output O2 is active, then the pump motor starter connected to output terminal 2 is on and the pump is running. Thus, the pump operation requires CR1 to be true (switch S1 to be closed and the input valve to be open) and switch S2 to be closed (pump control switch to be on).

Spend time studying the operation of the program and PLC interface in Figure 12–15 until you are familiar with the notation and operation.

In summary, the PLC solution requires the following:

- Input switches and sensors are wired to terminals on the input module.
- Output devices are connected to terminals on the output module.
- Contacts with addresses for each input device are placed in the ladder logic along with output symbols addressed for each device wired to the output module.
- Control relays and combinations of input contacts are placed on the ladder rungs to provide the desired control of the outputs.

An Alternate Solution

Another solution to the ladder logic shown in Figure 12–15b is illustrated in Figure 12–15c. Note that the control relay, CR1, has been removed, and multiple input contacts with the address I1 have been used to achieve the same logical control. The valve control switch had only one set of contacts, but had to control two devices. As a result, a relay was necessary in the relay ladder logic to provide the extra contact. The I1 contact in the PLC ladder logic is not a physical contact but one created in memory. The physical valve control switch establishes the condition at terminal 1 (voltage present or no voltage present), and a corresponding 1 or 0 is saved in memory. A nearly unlimited number of I1 contacts can be used in the PLC ladder logic, and each will be true if there is a 1 in memory or false if there is a 0 in the terminal 1 memory location. This alternate solution illustrates that engineering problems often have multiple solutions, and one is optimum. Note that the alternate solution has only two rungs, compared with three in the original solution. Fewer rungs use less memory and run faster. Study the solution in Figure 12–15c, and verify that it works equally well.

In the PLC solution, the only physical wires in the system are the interfaces between the external inputs and actuators and the PLC I/O modules. All the elements on the ladder program rungs—namely, the control relays, control relay contacts, addressed input contacts on the left, and addressed outputs on the right—exist only in software in the PLC memory. As a result of these operational features, the PLC has many advantages when it is compared with conventional relay logic. Understanding these advantages is important and is addressed in the next section.

PLC Advantages

The best illustration of the advantages offered by PLCs over relay logic is provided by an example.

Example 12–3
The tank control system illustrated in Figures 12–10 and 12–15 was modified to include a second pump that operates only when the valve is open, when pump 1 is on, and when a new, pump 2, control switch is closed. Determine the changes required in the relay-logic solution shown in Figure 12–10 and in the PLC solution illustrated in Figure 12–15.

Solution
The second pump and pump control switch must be added in both the PLC and the relay-logic solution. No other changes to the physical system are necessary for the PLC solution. However, the relay-logic solution requires a new relay with three contacts, replacement of the original single-pole pump control switch with a double-throw type, and extensive modifications to the relay control wiring. The new relay-logic solution is illustrated in Figure 12–16, and the PLC solution is shown in Figure 12–17. Study both solutions until you understand the operation and verify that all solutions presented meet the control requirements.

While both systems satisfy the control requirements, the relay logic requires extensive changes to the physical system, compared with the PLC implementation. Both diagrams are notated with an *R* to indicate wiring that must be reworked and an *N* for new wiring. Compare the work required for each solution. The PLC solution has just the new wiring for the additional input switch

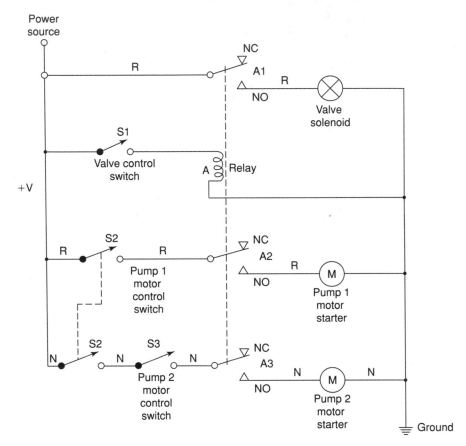

Figure 12–16 Relay-Logic Solution for the Two-Pump Configuration.

and pump motor starter, while the relay-logic solution has the wiring for the new components plus extensive wiring changes because the original relay and pump switch had to be replaced. There are extensive changes in the ladder-logic program for the PLC solution; however, that is performed in software so it is both fast and at much lower cost. Study the two solutions illustrated in Figures 12–17b and 12–17c, and note that Figure 12–17c uses only contacts referenced to input or memory values set by the external field devices.

It is clear from this example why PLCs are the choice for sequential control over relay logic. In addition, the similarity between the PLC ladder-logic program and the two-wire control diagram (Figure 12–10c) used for relay logic provided an easy transition to PLCs for electricians, technicians, and system designers.

Other advantages of PLCs include the following:

■ *Reliability.* Relays are electromechanical devices, and physical wear in relay-logic controls occurs every time the devices are turned on. PLCs

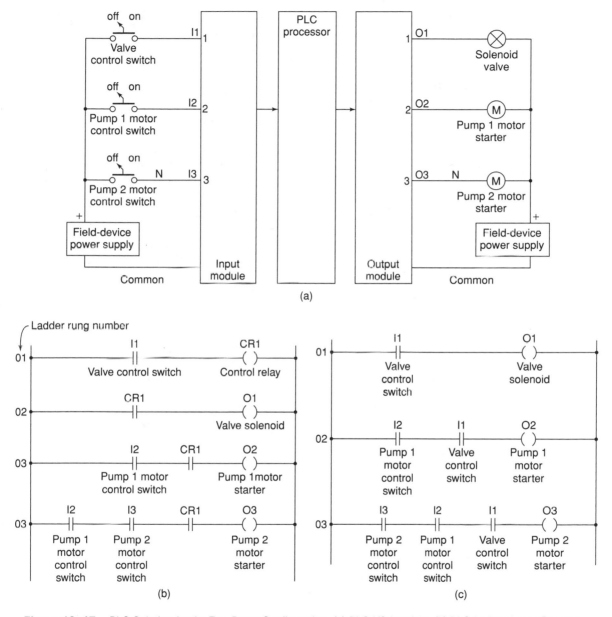

Figure 12–17 PLC Solution for the Two-Pump Configuration: (a) PLC I/O Interface; (b) PLC Ladder-Logic Program; (c) Alternate PLC Ladder-Logic Program.

have the reliability inherent in all electronic devices. In addition, reliability and long mean time between failures are designed into PLCs.

■ *Improved maintenance and troubleshooting.* If a problem does occur in any PLC module or in the processor, the module or processor can be changed in a matter of seconds without any changes in wiring. The

PLC also makes troubleshooting the entire control system less difficult because a technician can check the status of each input and force a change in each output to identify the input or output device causing the problem.

■ *Off-line programming.* In the past, PLCs could be programmed only with special programming terminals supplied by the vendor; however, the present systems use microcomputers running special PLC programming software as programming terminals. The microcomputer either is attached directly to the PLC processor or is a node on the PLC network. In both situations, most of the application programming is performed by using microcomputer hardware and software resources and does not require the PLC to be in the *program mode.* This process, called *off-line programming*, allows new program development and current program modifications without the need to take the PLCs out of the production process. Some program modifications, such as changing variable values or set points, are performed while the processor is in the *run mode,* and no production time is lost. In contrast, installing and modifying relay-logic circuits often takes days or weeks, with considerable lost production as the control circuits are interfaced to the production system.

■ *Broad application base.* PLC software supports a broad range of discrete and analog applications in various industries. With just program and interface module changes, a PLC can be moved from sequential control of discrete actions in an assembly cell application to control of the liquid level and temperature in a process control system.

■ *Low cost and small footprint.* The cost and the size of PLCs have dropped significantly in the last ten years. For example, a microPLC, which would fit in the palm of your hand, offers powerful machine control for less than $300.

■ *High-end control grows exponentially.* While cost and size have been dropping on the low end, the capability of the large PLC system has been expanding as well. The capability of large PLC controllers and the ability to network and distribute the control by using numerous proprietary and international network standards permit PLCs to take control of entire manufacturing systems and production plants.

The IEC 61131 Programming Standard

As mentioned previously, during the early 1990s, the IEC published the IEC 61131 standard, which covers PLCs. Part 3 of that standard covers programming. Until the IEC 61131-3 standard was published in March 1993, there was no suitable standard that defined the way control systems, such as PLCs, could be programmed. The ladder programming had numerous problems that included the following:

■ The ladder programming varied among PLC products.
■ Limited structured or hierarchical programming options were available.

- Software could not be recycled or reused.
- Little or no support for data manipulation and data structures existed.
- Use of arithmetic operations was awkward.

The IEC 61131-3 PLC languages eliminate these weaknesses. The languages in the standard include the following:

Structured Text (ST). ST is a high-level textual language that resembles Pascal so that it encourages structured programming. In addition, it supports a wide range of standard functions and operators.

Function Block Diagram (FBD). FBD is a graphical language used to show the signal and data flows through the use of reusable software elements. FBD program structure is excellent for expressing the algorithms and logic used to describe control systems.

Ladder Diagram (LD). LD is a graphical language much like the ladder logic currently used. However, the IEC LD language allows the programmer to integrate user-defined function blocks and functions into the ladder diagram and to include hierarchical design.

Instruction List (IL). IL is a low-level, more cryptic language that is similar to the instruction list languages found in current PLCs.

Sequential Function Chart (SFC). SFC is a graphical language similar to GRAFCET, which was developed in France. It is good for programming control systems that have a sequential behavior when sequences are time and event driven. It can use other 61131 languages to define output control structures.

12–10 COMPUTER NUMERICAL CONTROL

The definition of *computer numerical control* (*CNC*) was restricted initially to numerical control (NC) machines that incorporated internal computers to handle machine control and program execution. Currently, CNC represents all production machines that use onboard computers to control the movement of tooling with production programs. The basic robot system described in Figure 11–1 is an example of a machine that would fit the definition of a CNC. A generic block diagram for CNC machines is provided in Figure 12–18. Compare the robot controller in Figure 11–1 with the CNC block diagram and verify that the functions are similar.

The input to the CNC system comes either from an operator or from another machine. Frequently, the input for CNC machines is another computer; for example, many of the CNC machines described in chapter 10 received the program they used to cut parts from a computer that translated the part drawing from a CAD vector file to a machine code program file used by the CNC machine to cut the part. The processing that converts part geometry to an intermediate machine code is performed either by an NC software module inside the CAD software or by a separate computer-aided manufacturing (CAM) software program such as SmartCAM or MasterCAM. The translation from the intermediate machine code to a program for

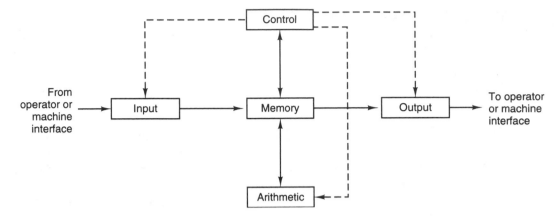

Figure 12–18 Five Major Functional Units of a CNC.

a specific brand and model of CNC machine is performed by a program called a *postprocessor*. The postprocessor produces a program with the tooling and move commands required for a specific CNC machine. When this CNC machine code program is loaded into the memory unit shown in Figure 12–18, the control and arithmetic elements use the program to drive the tooling through the desired motions. A large number of production machines use CNC. Included in the group are robots, coordinate measuring machines (CMMs), and most material-processing machines. One of the first, and still a common, CNC programming application includes NC of lathes and mills. An overview of the programming techniques used on these types of process machines follows.

CNC Programming

CNC programming of mills and lathes is like any other machine programming process; it requires a complete understanding of the programming language. The language used for NC of mills and lathes is often referred to as *G codes*. The process used for milling machines and machining centers provides a good example of G-code use because it encompasses about 75 percent of all NC operations. The following five categories of programming commands and techniques are used for mill NC programming.

Basic Commands. The following command codes and their functions are used to write an NC program for a noncomplex part:

- Motion commands (G00, G01, G02, G03)
- Plane selection (G17, G18, G19)
- Positioning system selection (G90, G91)
- Unit selection (G70 or G20, G71 or G21)
- Work coordinate setting (G92)

- Reference point return (G28, G29, G30)
- Tool selection and change (Txx M06)
- Feed selection and input (Fxxx.xx, G94, G95)
- Spindle speed selection and control (Sxxxx, M03, M04, M05)
- Miscellaneous functions (M00, M01, M02, M07, M08, M09, M30)

Compensation and Offset.   The following commands are used to define work coordinate systems, perform tool diameter compensations, and compensate for tool length differences:

- Work coordinate compensation (G54–G59)
- Tool diameter (radius) compensation (G40, G41, G42)
- Tool length offset (G43, G44, G49)

Fixed Cycles.   These commands in the following three categories provide a method of executing a series of repetitive machining operations with a single code block:

- Standard fixed cycles (G80–G89)
- Special fixed cycles
- User-defined fixed cycles

Macro and Subroutine Programming.   Modern NC controllers support basic computer programming syntax that includes defined variables, arithmetic operations, and execution of logical decisions to define part geometry mathematically.

Advanced Programming Features.   These are generally considered to be user-defined controls.

A simple example of G-code programming illustrates how NC programming codes are used.

Example 12–4

Write a G-code program to cut the path illustrated in Figure 12–19 with a cutter speed of 9 inches per minute (IPM).

Solution

N05	G90	Absolute coordinate system
N10	G00 X1.0 Y1.0	Rapid mode move to A (1.0, 1.0)
N15	G01 Y6.0 F9.0	Linear interpolation, cut from A (1.0, 1.0) to B (1.0, 5.0) at a rate of 9 IPM
N20	X7.0	Cut to C (7.0, 5.0)
N25	X1.0 Y1.0	Cut to A (1.0, 1.0)

Figure 12–19 G-Code Programming.

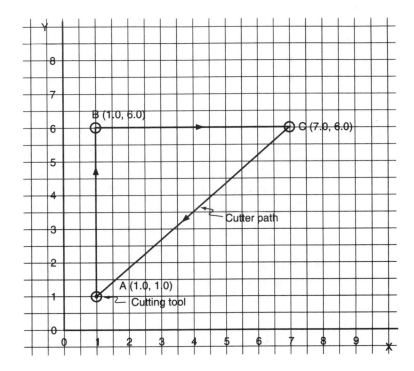

Note that some commands in the previous example remain in effect after they are used. For example, X1.0 did not have to be repeated in line N15 if the X distance did not change, and the Y distance was not needed in line N20 for the same reason. Also, the feed rate was not repeated for lines N20 and N25 because no change in the cutting rate was desired.

As for all programming languages, a full programming course is necessary for you to be able to use all of the NC G-code commands for mills and lathes. Complete coverage of G-code programming is beyond the scope of this book. However, a list of G codes for a Fanuc lathe controller is provided in appendix 12–1.

Manual Versus Automated Programming

Programming of NC machines takes two forms: manual programming and code generation with the support of CAM software. Example 12–4 is an example of manual programming. Starting with a drawing of the milled part, the programmer creates a G-code sequence that drives the machine's cutting tool over the desired path. CAM-generated NC code uses a postprocessor for the target machine tool to convert the drawing of the part directly to the G-code program that will run on the machine selected. The CAM software and postprocessors, described in Chapter 5 (Figures 5–42 and 5–48), fell into two categories. One type, MasterCAM and SmartCAM, is stand alone and works with drawing files from all the major CAM vendors. The second type, developed by the CAD vendor, is integrated with the CAD program and runs as part of the integrated CAD/CAM design software.

12–11 AUTOMATIC TRACKING

Tracking the movement of parts and tools through production is a major task. The primary technique used to follow movement of parts is *bar codes*. The concept of giving products a unique label started in the grocery industry in the early 1900s. The first bar code, a circular design, was patented in 1949; however, over fifty unique bar code symbols have followed that first circular design.

Bar Code Symbols

Bar code scanning is rapidly replacing the keyboard and other data-entry devices for recording manufacturing information and data. Sophisticated automation on the shop floor uses bar code technology to provide the information for material resource planning, statistical process control, production program selection, and other CIM system applications.

A bar code is a symbol composed of parallel bars and spaces with varying widths. The symbol is used as a graphical code to represent a sequence of alphanumeric characters that includes punctuation. For example, the three bar codes pictured in Figure 12–20 represent the numbers written below the codes. The codes can also represent combinations of letters and numbers. More than fifty codes were developed and each offers one or more of the following code characteristics: robust

143800-001

24105277P165

MONITOR PRODUCT OF KOREA

H000

CONFIGURATION:

Figure 12–20   Sample Bar Codes.

character set, structural simplicity, generous tolerances for printing and reading, and a high density-to-size ratio. The two most popular codes used in industry are the *interleaved two-of-five code* and *code 39*.

Interleaved Two-of-Five Code.   The *interleaved two-of-five code* is a numeric code, which means that only numbers can be represented by the code. A sample interleaved two-of-five code is illustrated in Figure 12–21. The code was designed with two key features: (1) the code has a high density, so that a large amount of information can be encoded in a short space; and (2) the code has a wide tolerance for printing and scanning. This code is a continuous code, which means that both the bars and the spaces are coded. The odd-number digits are represented by the bars, and the even-number digits are represented by the spaces. The specification requires that the code represent an even number of digits. Therefore, a zero is placed in front of all numbers with an odd number of digits before conversion to the interleave code.

Figure 12–21   Interleave Two-of-Five Bar Code.

Code table	
Character	Code
0	00110
1	10001
2	01001
3	11000
4	00101
5	10100
6	01100
7	00011
8	10010
9	01010

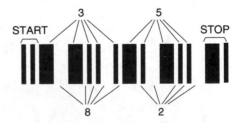

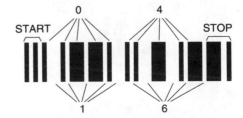

Figure 12–22   Code 39 Bar Code.

The interleaved two-of-five code is effective for coding information on corrugated boxes, where bars must be wide to permit automatic scanning while the boxes are moving on conveyors; however, space on the boxes is usually limited. The code is used widely in the automotive industry and in warehousing and heavy industrial applications.

Code 39.   *Code 39*, the most widely used and accepted industrial bar code symbol, was developed in 1975 by Dr. David Allais and Ray Stevens for what is now Intermec, Inc. The code, illustrated in Figure 12–22, has become the de facto industrial standard for item identification. Code 39, also called a *three-of-nine code*, is a complete code system. The code can represent the twenty-six letters of the alphabet, the ten digits, and seven additional characters. The code has unique start and stop bits and can vary in length. All characters are self checking, and as a discrete code, the intercharacter space is not part of the code.

Major advantages of this code are the high level of data security provided by the self-checking feature and the wide tolerance for printing and scanning. For example, with good dot matrix bar code printers and properly selected bar code scanners, the error rate is less than one character substitution in 3 million characters scanned. If high-quality printers and scanners are used, the value drops to one substitution error in 70 million characters scanned. For even greater data security, an additional check character is added to the code. As a result of this high security, code 39 was added to standards developed by health care and automotive industries and by the Department of Defense.

Reading Bar Codes.   The scanners used to read bar codes have three major components: (1) a light source to illuminate the code, (2) a photodetector to sample the light reflected from the code, and (3) a microcomputer to convert the photodetector output into the series of letters and numbers represented by the code. The operation is straightforward. A source of light from a laser, a light-emitting diode

Figure 12–23   Bar Code Scanner.

(LED), or an incandescent lamp is aimed at the bar code symbol. The dark bars absorb the light and the spaces reflect the light back to the photodetector. From these data, the width of the bars and the spacing are determined by a microcomputer. In addition, the computer decodes the bar code and provides a serial data output of the code in ASCII format. This output is passed to another computer or a PLC to use in a host of production applications.

Scanners fall into three categories: *handheld contact type, handheld noncontact type,* and *fixed-station systems.* Figure 12–23 shows a handheld-type scanner used in industrial applications. Handheld scanners are connected to portable data storage devices that are connected to a computer, or transmit the data over radio frequency links to a host computer. The simplest form of scanners used in a large number of applications are the handheld type with a *light wand.* In the contact type of scanner, the wand has a light source (LED or incandescent lamp) and photodetector in the pencil-like wand and two fiber-optic cables connecting these devices to the tip of the wand. As the wand is dragged over the bar code, the width of bars and spaces is read from the reflections of the light through these fiber links. Handheld scanners using laser light sources are used to make noncontact readings. The laser's pinpoint light beam permits scans with the reader some distance from the bar code. The principle of operation is the same as that of the contact-type scanner. Light reflected back to the photodetector provides information on bar and space width and orientation.

Two other types of scanners are the *moving-* and *fixed-beam* devices. The moving beam typically uses a helium-neon laser as a light source and sweeps the laser beam across the object at a rate of 1,440 passes per second in search of the bar code. The advantage of this type of scanner is that bar code symbol placement is not critical as long as the code falls into the field of view of the scanner. The fixed-beam type of scanner using a laser, an LED, or an incandescent lamp illuminates a large area of the bar code without moving the beam. However, the fixed beam requires that the bar code be moved past the scanner to generate signals required by the photodetector. Again, reflected light is used to decode the information in the code.

Other Tracking Mechanisms

In many applications where parts or tools need to be tracked, the environment does not permit bar codes to be applied or read. For example, the cutting tools used on CNC machines are costly and their location in the production area must be tracked. The environment does not support bar codes, however, because of the cutting fluids and the general handling the tools receive. A common tracking technique for cutting tools and other devices that cannot use bar codes is the use of *radio-frequency tags*. The tags are passive electronic circuits that transmit a code when subjected to radio-frequency (rf) energy. The passive electronic circuit, shaped like a small computer chip, is held in a slot in the tool base by epoxy. Each circuit transmits a unique code when the tracking device focuses rf energy at the tool. The tracking device receives the tool code and records the number. In addition to being used for cutting tools, the rf tag is useful for many other harsh application environments.

Another tracking technique, frequently used on production pallets, utilizes a binary code to track objects. The pallets have a sequence of holes placed somewhere on the surface. At preselected points along the production process, the holes are aligned with proximity sensors or limit switches with a sensor or switch over each hole. A pallet is given a unique code by filling some combination of holes with pins. When the pallet passes under the sensors or switches, the holes with pins trip the sensor or switch. With four holes, sixteen unique codes or pallet identification numbers are available.

Tracking of products using one of the standard bar codes or other types of tracking techniques is a common practice in automated work cells. In addition to providing data on the location of products, parts, and tools, the unique part code can also identify the part to the machine. In an application at Allen-Bradley, for example, a bar code placed on the base of a blank motor contactor tells the production system what type of contact to build. At every step in the totally automated process, machines add the correct parts to the contactor on the basis of the information read from the bar code. Material, tools, and product tracking play a major role in most world-class production facilities.

12–12   NETWORK COMMUNICATIONS

A key requisite for any CIM installation is access to a product data management (PDM) system that archives all the product data for equal access by all departments. The machine, cell, and area controllers described in the last sections must be linked to departments across the enterprise for successful implementation of a shared database system. Developing the database and network to achieve this goal is not a trivial task; as a result, most organizations view the process as evolutionary and spread the work over three or more years.

Identifying the requirements for the enterprise network is a prerequisite for the design of the physical configuration and selection of network hardware and software. Important considerations are identified and discussed next.

Enterprise Data in the PDM

The type of data files communicated across the network and stored in the central data repository dictates network characteristics and configuration. The term *data file* is often used to describe the information stored in a computer, and an appreciation for the difference between data files is requisite for understanding network characteristics.

The storage of manufacturing information or data in a computer is analogous to storing the same data in a file cabinet. In the file cabinet, the information or data printed on paper is placed in folders, which are then placed into drawers in a predetermined order. The type of manufacturing information placed in the folders can be words or numbers, pictures, and line drawings. The CIM central database has similar data storage requirements. In the computer, the storage device is divided into sections similar to the drawers of the file cabinet, and the data are stored electronically in a predetermined order. The major difference between the paper and the computer storage process is the conversion of the information and data from paper to the electronic format.

The computer data files generated across the enterprise fall into three classifications: *text* (words and numbers), *vector* (line drawings), and *image* (pictures). The data type produced and the size of the data file required for storage are functions of the process; for example, two-dimensional (2-D) product drawings developed with CAD software are stored as vector files. Text files use the least amount of computer memory or storage space, and image files use the greatest. A comparison of the size for each file type illustrates the complex problem surrounding the development of a central CIM data repository for all enterprise data. A standard page of text would take no more than 2,500 storage locations in a central data repository. A moderately complex 2-D CAD drawing that would fit on the same-size page could require 50,000 storage locations, and an image file for the same-size page could require over a million storage locations. These numbers will fall with technology innovations; however, the relative position of each file type with respect to size will not change.

In general, the product and process element where product design and documentation occur generates all three types of data files in the design and documentation of products. However, the majority of the files are vector type with large data files. The manufacturing planning and control (MPC) area generates text files the majority of the time; however, the MPC function works with drawing files (vector type) from the design area. The shop-floor area generates text files in most applications but uses files in all three formats.

The required data transfer rate for the network is higher if large data files must be exchanged between users in the enterprise. Also, the network design is affected by the requirement for a large number of concurrent users requesting large data files.

12–13 SUMMARY

In this chapter, we described the hierarchy used by the computer-integrated manufacturing (CIM) enterprise for control of devices, cells, areas, and corporate-wide systems. The cell control hardware is classified as either intelligent or nonintelligent, with intelligent systems including an onboard computer. Most cells are organized with a cell controller at the top of the control hierarchy, with the remaining cell hardware interfaced to the controller directly or through other intelligent devices. The cell controller is usually a microcomputer running one of the following operating systems: Windows NT/2000/XP. The major application present in most cells is the cell control management software that manages the bidirectional data flow between production equipment and the remainder of the enterprise. The software falls into two categories: proprietary and open system interconnect (OSI). The proprietary group includes in-house-developed software and application enablers provided by third-party software developers. Internal system compatibility is a major benefit of the software written specifically for the cell control application. However, the cost of internally developed packages is forcing companies to switch to enabler software. Enabler software, developed by third-party vendors, is a control shell that is tailored for each company's specific control application. The OSI cell control solution most frequently selected is manufacturing message specification (MMS). MMS is a standard for communicating information between intelligent devices in a production environment over networks based on the OSI model. The new network standards—DeviceNet, ControlNet, and Ethernet/IP—provide all features present in MMS. In addition, they have wide acceptance in the industry.

Device control uses a combination of proprietary and generic control devices. The proprietary controllers are built specifically to control a single type of machine, such as a computer numerical control (CNC) machine tool. Generic controllers are general-purpose devices that are applied to the control of various automation machines. One frequently used generic type of controller is a programmable logic controller (PLC). PLCs were designed initially to replace mechanical relays in sequential control logic circuits in industrial applications. PLC technology has advanced significantly in the last twenty-five years, and PLCs are now used for a broad range of discrete and analog control applications. The applications range from a single PLC for machine control to networks of PLCs that run entire production systems. Although the representation of PLC programs in the United States continues to use a ladder-logic format, the European community has moved to a more structured format called *GRAFCET*. In addition, the method of programming has changed from special-purpose programming units to microcomputers connected to the PLC or on the PLC network. The advantages of PLCs include easy interface modifications, improved maintenance and troubleshooting, off-line programming, a broad application base, low cost, and a broad range of PLC types and peripheral devices.

Another type of special-purpose controller frequently found in industry is the computer numerical controller. This type of controller, most often associated with metal-cutting machine tools, is used for several production machines that require high-level control of operation or motion. The robot, for example, uses a type of computer numerical controller to drive the arm through the programmed motions required for a production operation.

Automatic tracking is a requirement present in every manufacturing operation. Bar codes, the most frequently used tracking technology, are symbols composed of parallel bars and spaces with varying widths. Over fifty different bar codes are used to represent letters, numbers, and punctuation. The interleaved two-of-five code and code 39 are the two most commonly used codes. Bar codes are read with a wide variety of scanners that fall into three categories: the handheld contact type, the handheld noncontact type, and fixed-station systems. In addition to bar codes, radio-frequency tags and mechanically generated binary-coded tracking systems are used.

A communications network for data and information is necessary for successful operation of a central database in a CIM implementation. All the data stored in the CIM database fall into one of the following three categories: text (words and numbers), vector (line drawings), and image (pictures).

BIBLIOGRAPHY

GRAHAM, G. A. *Automation Encyclopedia.* Dearborn, MI: Society of Manufacturing Engineers, 1988.

GROOVER, M. P. *Automation, Production Systems, and Computer-Integrated Manufacturing.* 2nd ed. Upper Saddle River, NJ: Prentice Hall, 2001.

INTERNATIONAL ELECTROTECHNICAL COMMISSION (IEC). *Programmable Controllers.* Part 3, *Programming Languages* (IEC 61131-3). 2nd ed. Geneva, Switzerland: IEC, 2003.

LIM, S. C. J. *Computer Numerical Control.* Albany, NY: Delmar Publisher, 1994.

LLOYD, M. "Graphical Function Chart Programming for Programmable Controllers," *Control Engineering,* October 1985, 37–41.

LUGGEN, W. W. *Flexible Manufacturing Cells and Systems.* Upper Saddle River, NJ: Prentice Hall, 1991.

REHG, J. A. *Introduction to Robotics in CIM Systems.* 5th ed. Upper Saddle River, NJ: Prentice Hall, 2003.

SOBCZAK, T. V. *A Glossary of Terms for Computer-Integrated Manufacturing.* Dearborn, MI: CASA of SME, 1984.

WEBB, J. W. *Programmable Controllers: Principles and Applications.* Columbus, OH: Charles E. Merrill, 1988.

QUESTIONS

1. Describe the interface between the enterprise network and the factory LAN shown in Figure 12–3.
2. Why are point-to-point communication systems and fast response necessary for the data interfaces at the cell control level?
3. Describe the difference between intelligent and nonintelligent cell control devices in Figure 12–3.
4. Describe how cell controllers link the manufacturing floor with the enterprise data base.
5. Describe the different types of interfaces used in cell controllers.
6. What is the major type of application software found in most cell controllers, and what is the major function served by the software?
7. Describe the difference between proprietary and open system interconnect software.
8. What are the advantages and disadvantages of in-house-developed cell control software?
9. Describe application enabler software and list the advantages and disadvantages.
10. Describe the most frequently used OSI software for cell control applications.
11. What are the advantages of using MMS for cell control applications?
12. Describe how DeviceNet, ControlNet, and Ethernet/IP fit into the control hierarchy for automated cells.
13. Describe the difference between proprietary and generic device controllers.
14. What is the difference between discrete and analog control requirements?
15. What are the characteristics of a basic sequential logic control application in industry?
16. List and describe the system components for a programmable logic controller.
17. Describe the basic PLC operation.
18. Describe the data communication options available in PLC applications.
19. Describe how smart I/O interfaces operate and how they can significantly affect automation applications.
20. Describe the process for programming a PLC for an industrial application.
21. Why is the GRAFCET programming technique better than ladder logic?
22. What advantages do programmable controllers offer in cell and area control applications?
23. Describe the differences between an address-based PLC and a tag-based system.
24. What does standard 61131 define?
25. Describe the five languages defined in IEC 61131-3.
26. Describe the characteristics of CNC.

27. Describe three types of automatic tracking techniques used in industry.
28. Compare and contrast the interleaved two-of-five and code 39 bar codes.
29. What are the major components in a bar code scanner?
30. Describe the different types of bar code scanners and the accuracy possible with different configurations.
31. Describe the three types of data files stored in the CIM central database.

PROJECTS

1. Using the list of companies developed in project 1 in chapter 1, determine which companies are using cell controllers, PLCs, automatic tracking, and factory automation networks.
2. Select one of the companies in project 1 using cell controllers and describe in detail the hardware and software used in the system.
3. Select one of the companies in project 1 that uses automatic tracking and describe in detail the type of system implemented and the operational characteristics present in the system.
4. Select one of the companies in project 1 that has a factory network and describe in detail the type of system implemented and the operational characteristics present in the system.
5. Analyze the control circuit shown in Figures 12–17a and 12–17b, then answer the following questions:

   a. To what input terminal is the field device wired that controls input contact I1?
   b. What field-device condition(s) causes the CR1 contacts to be true?
   c. What input conditions are necessary for the output O2 to be active?
   d. The valve and pump 1 field switches are in the on position (voltage is present at the input terminals), the pump 1 motor starter is on, but no fluid is flowing into the tank. What device failures could produce this symptom?
   e. All three field-device switches are closed. There is fluid flowing into the tank and pump 1 is operating; however, the pump 2 motor starter is not active. What device failures could produce this symptom?

6. Draw the input and output PLC interface and the ladder-logic program for the relay ladder logic shown in Figure 12–24. Interface the devices to the following terminals on their respective modules: selector switch to 1, float switch to 2, flow switch to 3, thermo switch to 4, and pilot light to 1.
7. Modifications are necessary for the control system shown in Figure 12–17. A flow switch is installed in the fluid pipe to verify that the valve is open and fluid is flowing. The flow switch (see problem 6 for symbol) is connected to

Figure 12–24 Relay Ladder
Logic for Problem 6.

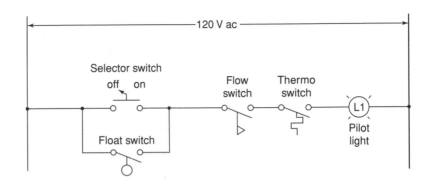

PLC input terminal 4. Draw the new I/O interface and change the ladder logic shown in Figure 12–17c on the basis of the following control requirements:

a. Valve solenoid is on when valve control switch is closed.

b. Pump 1 is on if the pump 1 motor control switch is closed and if the valve control switch is closed and the flow switch is closed.

c. Pump 2 is on if pump 1 is on and if the pump 2 motor control switch is closed.

8. Additional modifications are required for the system solution from problem 6. A float switch (see problem 6 for symbol) is installed in the tank at the 80 percent full level and is wired to PLC input terminal 5. Draw the new I/O interface and change the ladder logic developed in problem 6 on the basis of the following control requirements. What function does the float switch have in the control? Change the ladder logic as follows:

a. Valve solenoid is on when valve control switch is closed and float switch is open.

b. No change in the pump 1 control logic.

c. Pump 2 is on if pump 1 is on and if the pump 2 motor control switch is closed or if pump 1 is on and if the float switch is closed.

9. Redraw the pump 2 control rung from problem 8 using an input addressed to the pump 1 ladder-logic output. (*Hint:* Ladder-logic input contacts can have the address of a ladder-logic output. When the output is active, the referenced input contact is true.)

APPENDIX 12–1: TURNING G CODES

G codes are sometimes called *preparatory functions* and are specified by the G address followed by a two-digit identification number. Two types of G codes are commonly used: modal codes and nonmodal codes. A modal code is effective until it is overridden by another G code in the same group; a nonmodal code is effective only for the block in which it is used. The code and function of some G codes used in turning are the same for milling; however, others have different assigned functions. The following list shows the Fanuc G codes used in the United States for turning with controllers:

G00   Rapid traverse (positioning)
G01   Linear interpolation
G02*   Circular interpolation (clockwise)
G03*   Circular interpolation (counterclockwise)
G04   Dwell (temporary stop)
G20*   Inch data input (G70 for some systems)
G21   Metric data input (G71 for some systems)
G27   Reference point return check
G28   Reference point return
G30   Second, third, fourth reference return
G33   Thread cutting
G34   Variable-lead thread cutting
G40   Tool nose radius compensation cancel
G41   Tool nose radius compensation right
G42   Tool nose radius compensation left
G65   Macro call command
G70   Finishing cycle
G71   Stock removal in turning
G72   Stock removal in facing
G73   Pattern repeating
G74   Peck drilling in z-axis
G75   Grooving on x-axis
G76   Multiple threading cycle
G77   Diameter cutting cycle
G78   Thread cutting cycle
G79   End face turning cycle
G92   Coordinate system setting, or maximum spindle speed setting
G94   Feed rate per minute
G95   Feed rate per revolution
G96   Constant surface speed control
G97*   Constant surface speed control cancel

---

*Modal operation.

Chapter

Quality and Human Resource Issues in Manufacturing

13

OBJECTIVES

After completing this chapter, you should be able to do the following:

- Describe the foundations and philosophies that support quality in a manufacturing operation
- Define the key terms and acronyms that make up the technical language of the quality movement
- Select the appropriate quality tools and techniques to be used to identify and solve problems in manufacturing operations
- Sketch a normal distribution and relate the properties of this fundamental statistical concept to process capability and process control in manufacturing
- Describe ways the management of manufacturing operations is changing in response to increased competition
- Describe the growing use of teams in manufacturing operations and outline the characteristics of effective work teams
- Present a plan for the establishment and installation of self-directed work teams

Many manufacturers in Europe and the United States experienced a loss in market share in the mid-1970s; even producers who were able to maintain their market share felt the pressure of foreign competition. At the same time, manufacturing technology started to mature with the introduction of sophisticated and reliable hardware such as robots, computer numerical control machine tools, programmable logic controllers, material-handling systems, coordinate measuring machines, and automatic storage and retrieval systems. Computers became smaller and less expensive, and distributed computing became a reality. In addition, the number of off-the-shelf software solutions for manufacturing and production control increased dramatically. Many manufacturers enthusiastically invested in manufacturing resource planning (MRP II) technology solutions, assuming that greater data capture-and-control power would automatically produce the efficiencies necessary to restore the enterprise to world-class performance. Thus, it appeared that technology had matured just in time to save the day.

Although the introduction of technology caused some improvement in performance, many potentially first-rate U.S. manufacturing companies failed to soar as a result of advanced manufacturing systems. In some cases, manufacturing efficiencies dropped when technology was introduced. The lesson learned by manufacturing in the mid-1970s is that technology alone cannot improve business performance, and a new emphasis on *quality* and the development of the *workforce* was required. The delivery of higher performance is achieved only through people operating in a quality-rich environment on good production processes, applications that use technology effectively. In this chapter, we provide an overview of the quality concept and the development of human resources.

13–1 QUALITY FOUNDATIONS

The move to *quality* means different things to different people. In some industries, the term *quality* refers to the processes and tools used to measure and record the degrees of quality achieved in production. A technique such as *statistical process control* (*SPC*), which records the degree of variation in a product on charts with statistically determined limits, is an example of a process. The emphasis on SPC in the early 1980s was a reaction to the superior quality of products received from offshore competition. U.S. industries recognized the need for quality products in the early 1970s, and isolated fixes such as SPC were adopted in an attempt to match the quality of offshore competition.

The quality revolution began in the late 1940s with the rebuilding of manufacturing in Japan and Europe. Many experts in quality and manufacturing were involved in the effort to rebuild the economies of countries that were devastated during World War II. Several individuals deserve particular attention.

W. Edwards Deming

Perhaps the most well known of the quality experts who led the transformation in Japan was Dr. W. Edwards Deming. Deming described a "chain reaction" that would take place if quality improved. Improving quality would reduce rework and mistakes, and this would lead to improved production effectiveness and higher productivity. Customers respond to better quality and lower prices, which results in market share increases that help make companies profitable and provide more jobs. Companies in Japan that followed Deming's instructions have demonstrated the power of this sequence of events.

Deming taught companies to develop an appreciation for their operating and production systems and to understand any variations found in the system. He demonstrated the need for consideration of cause-and-effect relationships that can lead to predictions of future outcomes. Perhaps his most surprising teaching addressed the need to understand the psychology of the people working in the system. Management is responsible for the company's systems. Management must develop a sincere trust and belief in its people and understand how people work in its systems.

Variation is the enemy of effective systems. Natural variations (common causes that are inherent to the process) are a part of all processes, and these levels need to be understood. Sources of special (assignable) causes of variation must be identified and eliminated. The techniques of SPC help identify special causes of process variation. Deming stressed the need to use statistical methods to find the facts. It is important to understand the causes of variation and their impact on the system. Elimination of the cause of the variation leads to real process quality improvements.

Dr. Deming's philosophy and teachings about quality have had an impact on generations of people in Japan. In recognition of his contributions, the Union of Japanese Scientists and Engineers (JUSE) instituted the Deming Prize in 1951. This is the highest quality award given in Japan.

Deming's work was largely ignored in the United States until quality problems hit the automobile industry in the 1980s. Deming's book *Out of the Crisis*, published in 1986, brought his quality philosophy to a new audience that was ready to learn and apply it in the United States. One of the most famous parts of Deming's philosophy is his "Fourteen Points for Management." The abridged view of Dr. Deming's Fourteen Points condenses years of work and thought into just a few words. This view provides an overview that can jog your memory, but it does not do a good job of providing the rich detail Deming provides to expand on each key point. Deming's books, the Deming Institute (**http://www. deming.org**), and writings in other quality textbooks provide sources of more detailed explanations of each of the fourteen points that follow.

Deming's Fourteen Points for Management

1. Create constancy of purpose toward improvement of the product and service to all customers.
2. Adopt the new philosophy that comes from the new economic age.
3. Cease dependence on mass inspection for quality products by building quality into the product and process.
4. End the practice of awarding business on the basis of price and change to a meaningful measure of quality.
5. Find problems and continuously improve the system of production and service.
6. Institute modern methods of training on the job.
7. Institute leadership at all levels of management.
8. Drive out the fear of suggesting changes to company policy and operations.
9. Break down the barriers between departments.
10. Eliminate numerical goals, posters, and slogans for the workforce that work for higher levels of productivity without improving the production methods.

11. Eliminate work standards that prescribe numerical quotas.
12. Remove barriers that reduce pride in workmanship for hourly workers, engineers, and managers.
13. Institute a vigorous program of education and self-improvement.
14. Make everyone in the company responsible for achieving the transformation.

Joseph Juran

Joseph Juran, a contemporary of Deming's, is recognized as one of the true scholars of quality. He wrote extensively and served as the lead author of *Juran's Quality Handbook* and many articles on quality and management. Visit the Juran Institute Web site (**http://www.juran.com**) for more information on these publications.

A clear understanding of the costs that are associated with quality is a fundamental part of Juran's process for quality management. He defined four categories of cost related to quality that companies can use to better understand where they spend money on quality. The four cost categories are prevention, appraisal, internal failure, and external failure costs. The categories can be used to highlight problem areas and begin the process of quality-cost reduction. The work of Juran on these quality-related costs was expanded and systematized by Philip Crosby in the late 1970s.

Armand Feigenbaum and Kaoru Ishikawa

The concept of *total quality management* (*TQM*) became popular with American companies in the 1980s, but Armand Feigenbaum first described the idea in the 1950s. Feigenbaum recognized (as did Juran and others) that quality leadership with a strong focus on planning was essential to a successful quality system. In the postwar economy, the technology of quality was going to change to involve the entire workforce. Quality was no longer a problem that was addressed only on the shop floor. The commitment to quality was going to be supported by skilled consultants and the continuous training and motivation of employees.

Kaoru Ishikawa was one of the instrumental leaders in developing the Japanese quality strategy and approach. Like Feigenbaum, he advocated the concepts of TQM, but with some important additions. He was influential in the development of participative approaches to quality problems involving plant workers. He replaced the outside consultants with trained employee experts. He advocated the use of simple visual tools and statistical techniques. His approach produced an empowered workforce that was highly skilled in the problem-identification and problem-solving methods.

Philip Crosby

Philip Crosby is the relative newcomer to this list of quality experts. He did most of his work to consolidate and package the philosophy and tools for causing quality in the 1970s while he was an executive at ITT Corporation. In the 1980s, Crosby formed a consulting and teaching company that helped bring the quality revolution into many companies. Crosby stressed that quality is more than statistics and SPC methods. His experiences at ITT had shown him that before companies would be ready to effectively use the statistical tools, they would have to create a receptive corporative culture. The new culture begins with a clear definition of quality and a plan for causing quality that everyone in the company can understand.

The Absolutes of Quality.   Crosby was convinced that getting everyone to agree and embrace quality the same way would not be easy without a framework for the effort. He defined his "absolutes of quality" in the book *Quality Is Free* in 1979. Four statements define his quality framework:

1. Quality must be defined as conformance to requirements, not goodness.
2. The system for causing quality is prevention (of defects).
3. The performance standard for quality is zero defects.
4. The measurement of quality is the price of nonconformance, not indexes.

These four statements have a far-reaching impact. Clearly defined requirements become the driver of quality. Systems to cause quality concentrate on the prevention of defects—as Crosby would say, "Doing it right the first time." The performance standard of zero defects tells everyone that we expect it to be done right (no defects) every time. (*Note:* This is not the same as the motivational approach to zero defects used in some industries in the 1960s. This approach is driven by the prevention system.) Quality levels and percent defectives as measures of quality are replaced by accounting for the money spent on reworks, repairs, scrap, and defect-related waste. Crosby's contention from the beginning has been that quality is not the cost driver; it is the cost of *nonquality* (the price of nonconformance) that drives up costs. Get these costs out in the open and get people working on eliminating their root cause and the money once spent on nonquality turns into profits.

Crosby is a master at pulling the important concepts together and packaging them so that companies can learn them and put them to use effectively. Similar to Juran, Crosby directs companies to identify costs related to causing quality and fixing nonquality. He calls these simply the *price of conformance* (*POC*) and the *price of nonconformance* (*PONC*). Companies can choose to spend their resources to cause quality, prevent defects, improve processes, and improve designs and such. They may also choose to spend money on finding, fixing, or repairing products that do not meet the specified requirements. The Crosby consultants found companies spending 10 to 20 percent of gross sales to fix the nonquality problems. At the same time, spending to cause quality and prevent defects was less than 5 percent of gross sales. Identifying costs and implementing plans to eliminate the causes led to significant cost savings in just a few months.

Companies often search for directions or a plan to guide them on the journey to improved quality. Crosby recognized this need in the marketplace and developed his fourteen-point plan for a quality improvement process (QIP). This plan is also presented in the book *Quality Is Free*. It has been proven to work by companies around the world and in all types of industries. The Crosby consultants support these fourteen steps and other tools and systems developed by the Philip Crosby Associates. The following steps describe a QIP process. All steps must be addressed to ensure success.

Fourteen Steps for a QIP

1. *Management commitment.* Review corporate quality commitment and solidify and communicate management's position on quality and its commitment to the QIP.

2. *Quality improvement team (QIT).* Establish a team of the executive management leaders from all areas who will cause the necessary actions to happen. The QIT is established to manage the QIP and make sure the attitude of the Four Absolutes becomes part of the company culture.

3. *Quality education.* Formal training is designed to provide a common understanding of the Four Absolutes at all levels of the organization and to teach employees their roles in the QIP.

4. *Quality measurement.* Establish procedures and a standard of quality measurement for use throughout the company that displays identified nonconformance problems in a manner that permits objective evaluation and corrective action.

5. *Cost of quality.* Quantify the ingredients of the cost of quality and explain its use as a management tool. Establish the PONC as the one true measure of quality.

6. *Quality awareness.* Develop a new quality-awareness attitude in the mind of each employee and new communications between employees and their supervisors: Our quality reputation depends on us all.

7. *Corrective actions.* Establish a formal company system for resolving forever the problems identified internally and with the suppliers. Ensure that all employees understand and can effectively use the system.

8. *Committee for planning Zero Defects Day.* A special group is selected to investigate the zero defects concept and plan ways to implement or communicate the concept to the entire company on Zero Defects Day.

9. *Zero Defects Day.* A "big deal" celebration is held to emphasize the company's new commitment to zero defects—a "new attitude day" for the employees.

10. *Goal setting.* Turn pledges and commitments into action. Ask each employee to participate in establishing the goals (personal and departmental) that they would like to strive for—specific, measurable, and time specific—and build on the foundation of step 5.

11. *Error cause removal.* Develop a procedure for capturing employee descriptions of problems that prevent them from performing error-free work. Provide acknowledgment of the input to the employee within 24 hours. Error cause removal should evolve into the "entrance" to the corrective action procedure. Error cause removal is a statement of a problem, not a suggestion or a solution. Solutions are the responsibility of an assigned individual, supervisor, or

functional group (such as maintenance, engineering, quality circle, data processing, etc.).

12. *Recognition.* Award programs are implemented to recognize and appreciate those who participate, meet their goals, or perform outstanding acts. Emphasis is placed on employee action and participation. Awards are primarily for recognition but must consider any company precedents.

13. *Quality council.* Regular meetings (bimonthly or quarterly) of quality professionals and selected employees are held to keep in touch with the people and ensure that the QIP is properly perceived at all levels of the organization.

14. *Do it over again.* This is an ongoing process, not one that is finished in twelve to eighteen months. The ongoing program of quality never ends.

It is clear from the inputs of all the quality leaders that people play a critical role in any company's quality effort. Quality and the people that cause it are vital parts of a computer-integrated manufacturing (CIM) approach to production.

The quality focus in the enterprise is not just the adoption of technology such as SPC, smart gauges, coordinate measuring machines, and vision systems. The quality focus is a broad view of all operations that blends the effort of all employees and the necessary technology into a comprehensive plan to approach a fault-free production environment. This quality process includes four major elements: (1) appropriate technology at all levels in the enterprise to control and measure critical quality parameters, (2) good design that permits manufacturing and assembly with near-zero defects, (3) management understanding of all quality-related issues and the commitment to strive for a defect-free operation within the enterprise, and (4) broad workforce training and involvement of every employee in the effort to reach the quality goals. This broad view of quality is addressed by the TQM process, and development of these elements is the focus for the next sections.

13–2 TOTAL QUALITY MANAGEMENT

Total quality management (TQM) has two components: the principles of the TQM process and tools to get the job done. The principles or philosophy associated with TQM allows an organization to overcome the traditional barriers and impediments that prevent the management group from utilizing the potential, skills, and knowledge of every employee. Following the TQM philosophy, the institution sets higher goals, recognizes and eliminates barriers to change, and solicits the opinions and ideas of every employee. TQM helps everyone see more of the big organizational picture by focusing on the common quality goal to "get it right the first time."

In addition to a guiding philosophy or principle, TQM is also driven by tools. These qualitative and quantitative quality tools are not new; however, the method of application is new when used with the TQM philosophy. The tools help an organization to understand better the way it does business and to measure the

Figure 13–1   Production and TQM System Cycle.

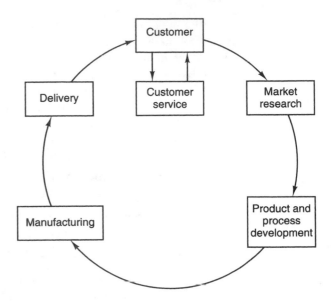

degree of improvement along the TQM journey. The tools help individuals, departments, and the total enterprise recognize when the goals for improved *productivity, performance, efficiency, customer satisfaction, work life,* and *quality* are reached.

The domain for TQM is defined in the illustration in Figure 13–1. The cycle illustrated in the figure shows the product moving from customer demand through the many phases of manufacturing until a finished product is shipped. This illustration is also a TQM cycle because a concern for the quality at every step in the sequence is a focus of the TQM process.

Definition of TQM

According to the *APICS Dictionary*, TQM is defined* as follows:

> **total quality management (TQM):** *A term coined to describe Japanese-style management approaches to quality improvement. Since then, total quality management (TQM) has taken on many meanings. Simply put, TQM is a management approach to long-term success through customer satisfaction. TQM is based on the participation of all members of an organization in improving processes, goods, services, and the culture in which they work. The methods for implementing this approach are found in teachings of such quality leaders as Philip B. Crosby, W. Edwards Deming, Armand V. Feigenbaum, Kaoru Ishikawa, J. M. Juran, and Genichi Taguchi.*
> (Source: © *APICS Dictionary*, 10th edition, 2002)

Study the definition and identify the three dominant themes present.

---

*The *APICS Dictionary* terms and definitions in this chapter are published with the permission of APICS—The Educational Society for Resource Management, 10th edition, 2002.

The three components critical for the TQM process to flourish are (1) a management group willing to allow everyone in the organization to participate in the decision process, (2) a process of continuous improvement for *all* elements in the enterprise, and (3) the use of multifunctional teams. The first element, participative management, is an evolutionary process that requires trust from both management and labor. Management must look for examples of labor's ability to participate positively in the management process and must recognize the capabilities and contributions that employees make to improve the production process and enhance the enterprise business. Management must take the first step to lower the barriers between itself and labor by giving employees the opportunity to participate in decision making. It is an act of faith that management must take to launch the TQM process in the slow journey toward full participative management.

The continuous-improvement process is an admission that every activity or process in the enterprise could be improved. This commitment never changes, regardless of the length of time that continuous improvement is practiced. No process will ever be perfect because some improvement will always be possible.

The TQM teams are a cross section of members from inside and outside the enterprise who represent some part of the process under study. The team frequently includes enterprise employees associated with the process; vendors who supply raw materials, services, or production equipment; and the customers who use the products. Two types of team processes occur in the CIM enterprise. The first, just described, is cross-functional teams with representation from a broad cross section of the enterprise and external sources. These teams are primarily process-troubleshooting teams that are formed to work on an identified improvement and then are disbanded after the process improvement is implemented and successful. The second type of team, a self-directed work team, is formed to provide for self-governance of a production area. These self-directed teams have members from only the specific process area, and their responsibility includes the full authority to run the process area or production cell. A full description of self-directed work teams is provided in a later section.

TQM Principles for Implementation

TQM includes six principles that must be considered in an implementation: customer focus, process focus, prevention focus, workforce mobilization focus, fact-based decision-making focus, and continuous-feedback focus. These implementation principles are described next.

Customer Focus.   Most employees know that customers are important to the survival of the enterprise; however, few know the full definition of the term *customer* from the TQM perspective. In TQM, the term *customer* refers to the external customers who place the orders and receive the finished goods. In addition, everyone in an organization has internal customers. For example, the internal customers for the heat-treating department are all the other departments that send work for heat treating as part of the production process. Every person in the organization has internal customers who must be served just as well as the external ones who receive the finished products. To implement TQM successfully, every employee must know and serve all the needs of their external and internal customers.

Process Focus.   In the past, the focus of the enterprise was on the results and not on the process used to get the results. In a TQM implementation, the focus is on the process; the results take care of themselves. Improvement in the process is triggered by results that do not meet or exceed the internal and external customers' expectations. This improvement is continuous in the well-oiled TQM system.

Prevention Focus.   Too often, correction of quality problems of internal and external customers occurs after the defective items are identified by inspection. In the worst case, the inspection is a final inspection, and many defective parts are identified as scrap or requiring rework. TQM focuses on prevention of problems, not on inspection of final results; as a result, more defects are prevented before they happen, or problems are identified early in the production cycle before many parts are affected.

Workforce Mobilization Focus.   "Checking your brain at the time clock" was a common practice in organizations that did not recognize and use the expertise present in the workforce. The contribution of every employee toward meeting the enterprise goals changes significantly under TQM. Therefore, the TQM implementation must include a change in the way management views the capability of the workforce and a mobilization of the total workforce for the solution of common enterprise problems.

Fact-Based Decision-Making Focus.   Finger-pointing over problems in processes and products has no place in a TQM implementation. Responsibility for fixing the problem falls on everyone who is associated with the defective process or product. The TQM implementation uses a continuous-improvement process and multifunctional teams to analyze the problems carefully, document the facts, propose and implement solutions to the problems, and verify that the solutions are permanent.

Continuous-Feedback Focus.   The term *feedback* is synonymous with good communication. The success of a TQM implementation is ensured only when open, honest communication occurs from the top of the enterprise to the bottom. Equal access to good information and open communications across the enterprise are necessary for TQM.

An organization moving toward a goal of TQM must work to have all six of these principles in place for the TQM process to be successful and lasting.

TQM Implementation Process

TQM implementation includes a five-step process and an implementation schedule. The five steps are (1) preparation, (2) planning, (3) assessment, (4) implementation, and (5) evangelization. Each of the five steps is described briefly next.

1. The first step prepares the organization for the TQM implementation by starting with the CEO and other top management. The key point at this time is *getting top management commitment* to the principles and changes in the organization required by TQM. Top management must develop a vision statement, set corporate goals, draft a policy to support the strategic plan, and commit the resources to implement TQM successfully.

2. The *corporate council*, the initial planning group, lays the foundation for the TQM process. The vision statements, goals, and strategic plan from step 1 are used to develop the implementation plan and start to commit resources to the TQM process.

3. In the assessment step, the strengths and weaknesses of the organization are studied and documented. The process includes the exchange of information through the use of surveys, evaluations, questionnaires, and interviews from individuals across the enterprise and at all operational levels. This self-assessment provides both individual and group perceptions of the organization and operations.

4. The implementation stage starts execution of the TQM process. All the effort from the previous three stages starts to pay off with a focused training plan for managers and the workforce. The improvement process starts in this step with the formation of *process action teams* (*PATs*) to evaluate processes and implement new processes and changes as necessary.

5. Evangelization is the process of taking the TQM process that was implemented successfully in the initial area to other departments and elements in the enterprise. In the first four steps, much process and organization information and knowledge was acquired. In addition, the PATs recommended and made process improvements on the initial TQM project area. This success record is used as the launching pad for the introduction of TQM into other areas within the enterprise.

The TQM process is vital to the success of a CIM implementation because CIM requires a quality product and quality in the organization's operation. TQM supports the development of both product and organizational quality. In addition, TQM and CIM work well together because both require a change in the management style, the use of teamwork, improved communications, continuous improvement, and more self-management at all levels in the organization.

The quality tools used to measure the current status of the enterprise are critically important to the successful TQM implementation. Their importance becomes apparent when the PATs start to gather data for improvement of a process. The tools used in this process are described in the next section.

13–3 QUALITY TOOLS AND PROCESSES

The goal of every manufacturer is to produce items correctly the first time. The traditional quality assurance and quality control techniques worked to improve product quality but only after the mistakes were made. This traditional approach to quality has a major impact on productivity and product cost. For example, research performed by the American Society of Quality Control indicates that 15 to 30 cents of every sales dollar of products manufactured are lost as a result of the traditional approach of fixing poor production. In the service industries, the cost is closer to 35 cents for every dollar of service delivered. World-class enterprises that work to correct quality problems before the product is manufactured spend less than 5 cents for the same dollar of sales. One of the most frequently used tools in measuring and improving production is *statistical process control.*

Statistical Process Control

The name *statistical process control* (*SPC*) implies that *statistical* analysis of manufactured parts will be used to *control* and *improve* the manufacturing *process*. SPC is a group of statistical methods used to *measure, analyze,* and *regulate* a production process to reduce *defects.* A *defect* is defined as any variation of a required characteristic of the product or part that deviates sufficiently from the nominal value to prevent the product or part from fulfilling the physical and functional requirements of the customer. Product or part variations are grouped into two categories: variable and attribute characteristics. Variable characteristics such as length or weight are measured in physical units. Attribute characteristics such as surface finish are either good or bad. Product and part variations of these two characteristics are monitored and controlled through SPC.

The basis for SPC is the natural tendency for most manufacturing processes to produce products that conform to a *normal distribution.* Although space does not permit an in-depth development of this concept, an overview of the general process is necessary for an understanding of the SPC process. Variable characteristics are specified by the designer as a nominal value (target) and a tolerance about the target (variability). For example, a drawing of a shaft would show the nominal length of the shaft and the upper and lower variation (tolerance) that is acceptable. The manufacturing process attempts to produce the shaft at the nominal value of the length characteristic. No process is perfect, however, and some variation about the nominal length will occur. Some shafts will be longer than the nominal value, and some will be shorter. When the length of each shaft produced by the manufacturing process is plotted on a bar chart, called a *histogram*, the variation usually approximates a normal distribution. Study the normal distribution illustrated in Figure 13–2. The center of the curve represents all the shafts that are at the *mean* (μ), or average, value for all shafts produced. The number of parts larger than the mean are plotted to the right of the mean, and the parts smaller than the mean are plotted on the left. Note that the vertical axis represents the number of shafts at the different lengths, and the horizontal axis represents the mean length and positive and negative variations from the mean.

The variability from the mean, represented by a statistical term called *standard deviation* (σ), is also illustrated in Figure 13–2. The standard deviation is a measure of

Figure 13–2 Normal
Distribution Curve.

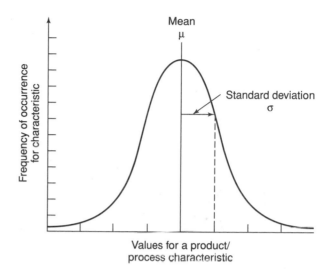

dispersion, or variability, and indicates how far away shaft values lie from the mean value of all shafts. The statistical formula for standard deviation relates the mean value for all shafts, the number of shafts, and the variation from the mean for specific shafts. Therefore, the standard deviation formula determines the number of shafts that lie at some variation from the mean.

The critical property of the normal distribution is that the percentage of parts that vary from the mean is predictable. The curve in Figure 13–3 illustrates this concept. Note that 68.26 percent of the parts are within 1 standard deviation of the mean and that only 0.26 percent fall outside ±3 standard deviations. Therefore, knowing that a process output conforms to a normal distribution, you can predict the number of parts that will fall outside set variation limits. These concepts are used to produce *control charts* that track production variation and lead to process improvement.

Control Charts.   Control charts, derived from statistical analysis of the process, are important SPC tools because they indicate whether a process is statistically predictable. The four basic functions provided by control charts are (1) defining the parameters of the process, (2) predicting when a process is changing as a result of a special cause outside the normal process variability, (3) showing when the state of statistical control for the process has been reached, and (4) monitoring a statistically stable process and alerting operators when a change takes place that could affect cost or quality. All control charts have a centerline and *upper* and *lower control limits* (*UCL* and *LCL*) determined from statistical data. These control limits are usually spaced 3 standard deviations above and below the centerline.

Four control charts are commonly used to track variable and attribute characteristics. The *X-bar* and *R charts*, the most frequently used control charts, track and identify causes of variation in variable data from parts and products. The

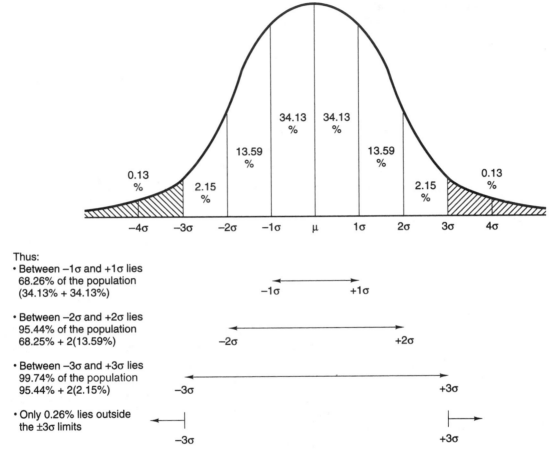

Figure 13–3   Variation from the Mean on a Normal Curve.

charts, usually plotted as a pair, have the X-bar chart above the R chart, as illustrated in Figure 13–4. The X-bar chart plots the averages of a sample group of parts measured across time intervals. The result indicates the center value that the process follows. The R chart plots the range for the samples measured and indicates the variability that exists in the process. For a stable process, the mean, or average, and the standard deviation are used to establish estimates for the center and width of the process. The results are then compared to the desired process center (the nominal value required for the design) and the desired process width (the tolerance established in the design process).

The two control charts that are used for attribute data are the p-chart and the c-chart. The *p-chart*, also called the *percentage chart*, indicates the percentage of parts in the measured group, often called a *subgroup*, that are defective due to a problem associated with an attribute characteristic. A sample of the p-chart is

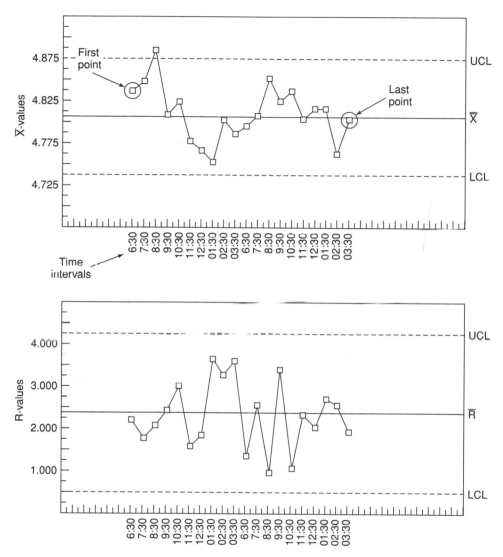

Figure 13–4   X-bar and R Charts (Samples Taken Every 60 Minutes).

provided in Figure 13–5. The *c-chart* is similar to the p-chart because it focuses on defects associated with attribute data. The p-chart data count the number of defective parts and do not differentiate between parts with one defect and those with multiple defects. In contrast, the c-chart graphs the total number of defects present in the subgroup measured, so multiple defects in a part would be included in the data. As a result, the c-chart process is effective only when the sample size of the measured parts is the same in each inspection. An example of a c-chart is provided in Figure 13–6.

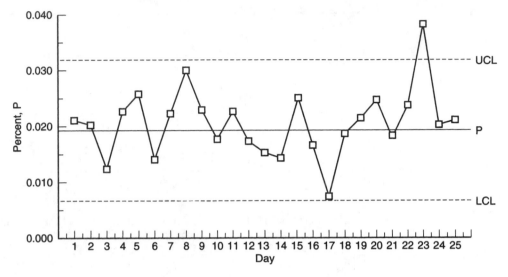

Figure 13–5 P-chart Example (Samples Taken at Daily Intervals).

Solving SPC Problems. The charts are part of the four basic steps used to solve SPC problems. These steps are as follows:

1. *Problem identification.* Problems in quality are detected by the operator, in downstream operations, or at inspection. Techniques used in problem identification include *Pareto analysis, control charts*, and *process capability studies.*

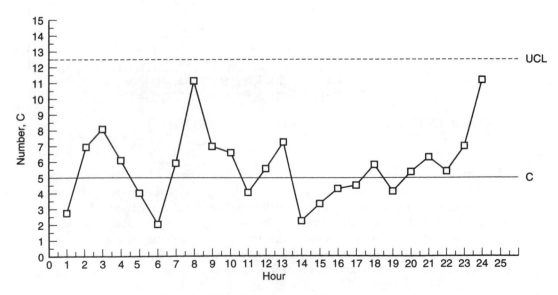

Figure 13–6 C-chart Example (Samples Taken at Hourly Intervals).

2. *Identification of cause(s).* The cause(s) of the problem usually results from a process element, such as people, tools, machines, material, or process methods. Investigation of the problems often includes *cause-and-effect analysis* using an *Ishikawa diagram.* If industrial experiments are needed to help differentiate among interrelated influencing factors, *Taguchi techniques* for experiment design are used.

3. *Problem correction.* The range of possible corrective actions, from a simple machine adjustment to redesign of a complete process, is broad. In addition to the corrective action, a verification that the fix has corrected the problem must be performed.

4. *Maintaining control.* After the problems are corrected, process control is achieved by using either statistical methods or physical systems. The statistical technique uses control charts, and the physical system uses instructions embedded in hardware and software. Constant improvement in the system is possible only if the corrections to the system are maintained.

Other Quality Tools

Several additional techniques are used in concert with SPC to solve production quality problems. Two frequently used tools are Pareto analysis and Ishikawa diagrams. *Pareto analysis* uses the frequency of occurrence of the different problems to set the priority for quality control analysis. Pareto diagrams, named after Vilfredo Pareto, an Italian economist, display the causes of a problem in the relative order of importance. An example of a Pareto diagram is shown in Figure 13–7.

Figure 13–7 Pareto Diagram.

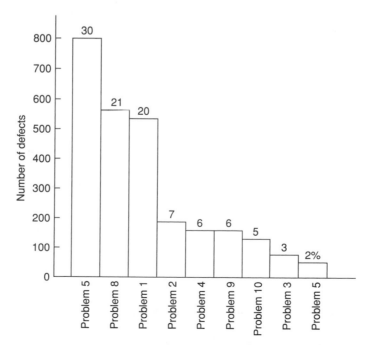

Note that the vertical axis has the number of occurrences of defects, and the horizontal axis has the groups of different problems identified. A Pareto diagram gives a clear picture of the relative importance of each grouping. For example, in Figure 13–7, problems 5, 8, and 1 are present 71 percent of the time when the quality problem occurs. The diagram suggests that correcting these three problems is the place to start. Therefore, the Pareto diagram becomes a prioritization for the efficient and effective use of resources in the solution of a quality problem. From this process, the *80-20 rule* evolved: generally, 80 percent of the results come from 20 percent of the causes. The Pareto diagram helps to focus on the area where the greatest opportunities for improvement exist.

The *cause-and-effect* diagram, also called the *fishbone diagram*, or *Ishikawa diagram*, was developed by Karou Ishikawa to identify the cause of problems in Japanese manufacturing. The cause-and-effect diagram allows a manufacturer to illustrate on a single diagram the relationship between the causes (machine adjustment) and the effect (poor quality). There are three basic types of cause-and-effect diagrams used by industry: *dispersion analysis* (Figure 13–8), *cause enumeration* (Figure 13–9), and *production process classification* (Figure 13–10). Study the diagrams in these figures and note similarities and differences. The first two diagrams have the same basic format, with four branches labeled *workers, materials, inspection*, and *tools* and the effect identified in the box on the right. The majority of the causes fit into one of these four categories.

Dispersion analysis (Figure 13–8) is the most frequently used diagram and allows problem solvers to focus on why variations occur. When a control chart

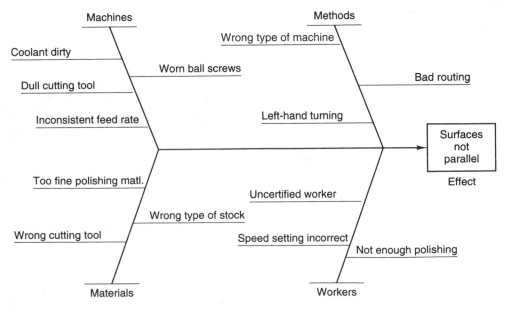

Figure 13–8   Dispersion Analysis Chart.

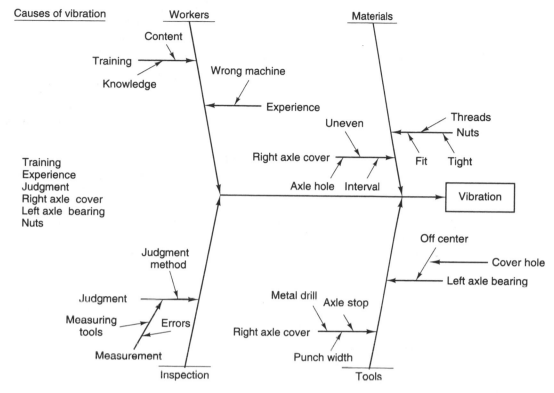

Figure 13–9 Cause Enumeration Chart.

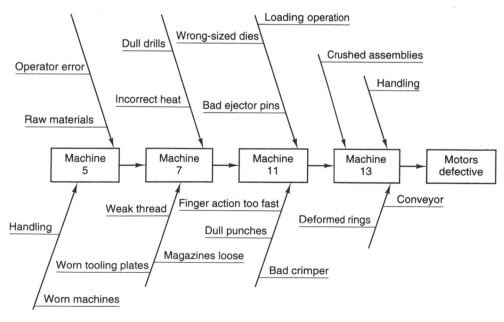

Figure 13–10 Production Process Classification.

indicates that a process is out of statistical control, a dispersion analysis diagram is used to determine the cause.

The cause enumeration type of fishbone diagram (Figure 13–9) is used to list all of the causes of variation that can affect a process. The cause enumeration diagram helps to identify the sets of common causes that inhibit a process from meeting specifications.

The final type of diagram, production process classification, is used to study an entire process. Figure 13–10 shows all machines used in a sequential process and the possible causes at each machine that could result in a defective product.

The graphic diagrams shown in Figures 13–7 through 13–10 are among the most effective problem-solving tools used in manufacturing. A team of problem solvers from different functional areas frequently creates these diagrams. The graphic representations help people visualize the problem and the factors that contribute to it. Working with people involved with a problem to create the analysis graphics is an effective way to encourage everyone's sincere participation.

Quality of Design Versus Quality of Conformance

The importance of the quality of the design has been emphasized throughout the product and process section of this book. A process developed in Japan offers a quantitative approach to analyzing the quality aspects of the design. The technique, named after its inventor, Taguchi, addresses the quality problems that are built into the product and process during their design.

Figure 13–11 shows the TQM and manufacturing cycle from Figure 13–1, with two areas highlighted. The *market research* activity is where customer needs and

Figure 13–11   Quality of Design and Quality of Conformance in the TQM Cycle.

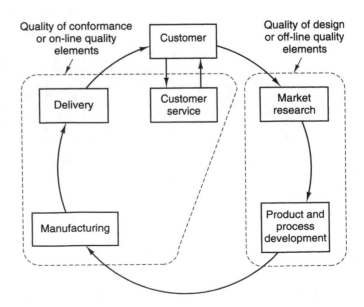

expectations are defined, and the *product and process development* activity is where standards and specifications for the products and processes are developed. Taguchi calls these two areas the *off-line quality system*, and Juran calls them the *quality of design* area. The quality of design area is critical because 80 percent of the quality problems that surface during production of the product are built into the product through bad design and poor process selection and development.

The remaining half of the product and TQM cycle in Figure 13–11 is called the *on-line quality system* by Taguchi. In this area, *quality of conformance* is practiced by the production and quality control departments. The primary quality tools used to check conformance to specifications include testing, inspection, supplier certification, and SPC. However, only 20 percent of the quality problems encountered during production can be eliminated because the balance of the problems are in the design.

The impact of design on quality is well documented, and the following technique illustrates how a well-developed product and process design produces products with near-zero defects.

13–4 DEFECT-FREE DESIGN PHILOSOPHY

In 1798, the U.S. government needed a large supply of muskets, and production at the federal arsenal could not meet the demand. Muskets could not be purchased from Europe because war with the French was imminent. Eli Whitney offered to produce 10,000 muskets for the U.S. government and to meet a twenty-eight-month delivery requirement. It took ten and a half years to complete the project because the musket parts produced in separate batches had a high degree of variability. As a result, musket parts selected at random for assembly would not fit together; therefore, each musket had to have the parts specially selected and assembled individually. After several years of work to reduce the variability of musket parts, Whitney's guns could be built with parts selected randomly from a pile, and parts from different guns were interchangeable. These same issues dominate the manufacture of products today, and the need for a process that produces near-zero defects during production is necessary. Such a process is called the *six-sigma design process*.

Process Capability Index—C_p

Review the normal distribution curve shown in Figure 13–2 and the data associated with the curve shown in Figure 13–3. The products from most manufacturing processes have a normal distribution, which means that 99.74 percent of all production falls within ± 3 standard deviations ($\pm 3\sigma$) of the average center value. The relationship between the normal curve and the product and process design determines how the process variation will affect the final product assembly. The effect of the normal variation is reduced if the range of produced part values, ± 3 sigma, falls comfortably within the tolerance specified for the part. Figure 13–12 illustrates this concept. Six times the standard deviation of the process is a measure of the capability

Figure 13–12   Process
Variation and Design Tolerance.

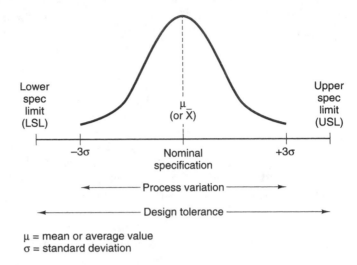

μ = mean or average value
σ = standard deviation

of that process. Defects occur when this capability is wider than the design tolerance or *specification width* marked by the lower specification limit (LSL) and upper specification limit (USL). Anything that causes the process capability to exceed either specification limit (LSL or USL) causes defects in the product.

The *process capability index* (C_p), a measure of the inherent capability of the process to produce parts that meet the design requirement, is found by dividing the total specification width by the process capability. The larger the capability index, the more likely that the parts will satisfy the production requirement. One method of ensuring defect-free parts would be to make the specification width as wide as possible and the process capability narrow. The process capability index is calculated as follows:

$$C_p = \frac{\text{total tolerance}}{\text{six times the standard deviation}}$$

Defects-per-Unit Benchmark

A benchmark used to judge a superior manufacturer, one that is world class, is the number of defects per unit in complex products. The superior manufacturer has a defects per unit that approaches zero, while the more typical manufacturer has a defect level of 5 or higher. The automotive industry can provide a good example of this concept. Many of the Japanese car manufacturers have defects per car in the range of 1.5 to 2, while their American counterparts have levels of 8 or higher. If you consider the production of hundreds of cars per day, with each car having thousands of parts, the concept of less than 2 bad parts per car appears impossible to reach. The key to achieving this level of quality lies in the capability index for the critical design tolerances on each of the parts used in the complex assembly. The superior company enforces a design margin of C_p equal to or greater than 2,

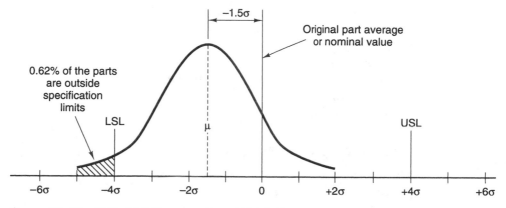

Figure 13–13   Shift of 1.5 Sigma in a Normal Distribution.

while the typical company uses a value of C_p equal to 1.33 or less. The following examples illustrate this important difference.

A company with a C_p equal to 1.33 may feel confident that its process is capable. This capability index appears to be satisfactory for quality production; however, it is not uncommon for the average value, or center point, of a normal distribution to shift ±1.5 sigma. The effect of such a shift is illustrated in Figure 13–13. The area outside the specification limit represents 0.62 percent of the production. Although 0.62 percent is small, it means that for every 1 million parts produced, 6,200 will be defective in the specification measured. This number of defective parts is not acceptable by today's quality standards.

Changing the specification width to ±5 sigma (C_p = 1.66) reduces the defects to 200 parts per million. This level of defects is too high for a company with a world-class quality goal; as a result, a specification width of ±6 sigma is used to ensure a near-zero-defect operation. The normal distribution with a ±6 sigma specification width is illustrated in Figure 13–14. Note that a shift of ±1.5 sigma

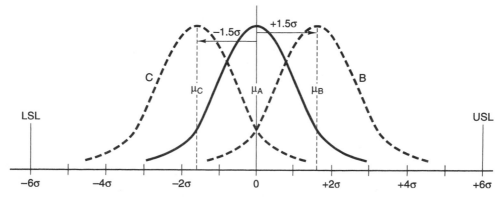

Figure 13–14   Six-Sigma Specification Limits.

in the critical specification has a minimal effect on the product quality, and the number of defects is just 3.4 per million parts produced. There are many more issues associated with achieving this level of quality production; however, for this overview, the key point is that specification limits must be a minimum of two times the process variation. This limit permits the process to shift ±1.5 sigma, a common variation for processes that follow the normal distribution, with little effect on product quality.

Seven Steps to Six-Sigma Capability

Many world-class manufacturers set a performance standard of zero defects in critical product and part parameters. To achieve that goal, many use the following seven-step process:

1. Identify the product or part characteristics critical to satisfying the physical and functional requirements of the customer and any standards agencies.
2. Identify the specific critical dimensional and functional characteristics of the products associated with the critical characteristics identified in step 1.
3. Identify and/or develop the process sequence necessary to control the critical dimensional and functional characteristics.
4. Determine the nominal design value and the maximum dimensional and functional tolerances that permit assembly and operation of the product at a level that guarantees customer satisfaction.
5. Determine the process capability for the processes selected or developed for the critical dimensional and functional characteristics in step 3.
6. Check the capability index, C_p, for each of the critical dimensional and functional characteristics.
7. If C_p is not equal to or greater than 2, change the design of the product and/or the process to achieve a C_p equal to or greater than 2. Another alternative is to institute process control measures that narrow the process capability sufficiently to achieve the necessary C_p for high-quality manufacturing.

The process just described permits a part to be run using standard production processes with the guarantee that 999,997 parts out of every 1 million produced can be used in the assembly process. When this process is applied across all critical dimensions on all parts in the product, defects associated with part compliance and poorly manufactured parts are less than three bad products out of a million produced.

Given the current capability of the production process, the 6-sigma program attempts to make the specification width on critical dimensions and functions twice the process capability. Achieving this two-to-one ratio ensures a near-zero-defects production operation. Design of component parts and processes is a critical part of achieving the capability index of 2 or greater. When economically possible, new process machines capable of holding tighter tolerances on critical dimensions are purchased. Frequently, however, the burden falls on the product design group to produce a design with wider specification widths. In that case, the principles associated with design for manufacture and assembly (DFMA) are used to produce parts with specifications consistent with the 6-sigma program needs. In addition, many of the design tools associated with the new computer-aided design (CAD) software help to create designs that support this production philosophy.

13–5 THE CHANGING WORKFORCE

Management of the manufacturing workforce in a highly automated facility is changing as a result of internal and external factors. Two key internal issues that affect management of the workforce emerge as a result of automation: (1) the interaction of the operator with the machine changed significantly when manual machines were replaced with computer-controlled processes, and (2) the highly integrated nature of the automated production systems and the enterprise requires a new set of skills.

Many external factors are linked to changes in workforce management. The two most significant factors are the *expanding global marketplace* and the *evolution of the workforce.* The issues surrounding these internal and external factors are addressed in the following sections.

Management Change: Internal Factors

The internal factors that forced a change in the management of workers resulted from the introduction of manufacturing automation. In general, automation created an artificial distance between the craftsperson or worker and the finished part or product. For example, when the processes were manual, worker action directly affected the size, shape, look, and feel of the finished product. Workers handled the product and were directly responsible for the quality of the product and the results achieved. Their manual skills as craftspeople were reflected in every finished part.

As the production process was automated, many of the skills of the craftsperson were replaced with a different set of skills as a machine operator. Responsibility for the dimensional integrity of the part shifted from the craftsperson to the automated machine, and the operator's responsibility shifted to control of the automated process. In highly automated work cells, the operator may not even touch the raw material or the finished product. This change in the production process changed the role played by both the machine operators and the first-line supervisors for the area.

In the manual operation, the first-line supervisor frequently was the master craftsperson who brought employees up the productivity curve by teaching and demonstrating techniques on the various machines used in production. The supervisor's information base was a mixture of learned skill and experience. The introduction of automated machines forced the supervisor into the new role of resource provider and coach. With the automated machine in control of the process, the supervisor becomes a manager of the resources required by machine operators. This task includes, for example, acting as an interface to the other departments responsible for machine programming and resolving a more complex set of production problems. The supervisor's production skills must be augmented with skills in communications, negotiation, and technical problem solving, and the ability to resolve human resource and personnel issues, teach, and coach in a team environment. The supervisor and all the production employees are required to have a broader view of the enterprise and the ability to use the wide range of information available in the computer-integrated enterprise database to improve the process constantly and increase productivity.

Management Change: External Factors

The two most significant external factors are the *expanding global marketplace* and the *evolution of the workforce*. While the first challenges the enterprise's ability to survive, the second offers a mechanism to meet market demands that result from worldwide competition. Together, these two external factors demand a change in management at all levels in the enterprise.

The traditional vertical management structure, used by almost every manufacturing organization for the last 75 years, is under pressure from top to bottom. Top management is under pressure to create an organization with few management layers so that response is quicker and overhead cost is reduced. As a result, middle-management positions are eliminated and the associated tasks are absorbed by those managers retained, assumed by automation, or just abolished. Decision making is less centralized; as a result, the authority and responsibility for production decisions are pushed closer to the manufacturing floor. A changing management environment from the top and an evolving workforce at the production level puts first-line supervisors between forces of change that have never been experienced in the history of manufacturing. Global competition is forcing a different management style in U.S. companies, and the evolution that has occurred in the workforce over the last 100 years makes a new style possible. The evolution in the U.S. workforce opens the door for *self-directed work teams* and a major change in the way that manufacturing management functions.

13–6 SELF-DIRECTED WORK TEAMS

The concept behind *self-directed work teams* has been evolving for the last fifty years. The dramatic changes in the global market and the increase in the use of technology in manufacturing since the 1970s has accelerated the interest in team-based management. Global market pressures require manufacturers to respond rapidly to customer needs, frequent new product introductions, increased productivity at all levels in the organization, lower production cost, and higher quality. Achieving these *order-winning criteria* with the workforce management policies of the 1950s and 1960s is not possible. As a result, manufacturers interested in achieving world-class status are ready to consider a change in management style and to embrace the concept of manufacturing teams. The following sections cover a short history of self-directed work teams and a description of their operation and implementation requirements.

Development of Self-Directed Work Teams

Management style in the United States has been moving away from autocratic management toward worker empowerment for many years. A brief overview of this transition puts into perspective the shift to self-directed work teams.

The chart provided in Figure 13–15 describes the development of the management style used since the eighteenth century. Study the chart until you are familiar with the data. Manufacturing started in the United States as *cottage industries*, where families produced products for distribution within the town or village. The radius of impact was usually less than thirty miles, and the management style was autocratic or paternalistic. The head of the household, usually the eldest male, was both the dominant leader and typically the master craftsman for the product produced. For example, if the industry was ironwork, the father or grandfather was usually the master blacksmith. It was not uncommon for everyone in the family to have some job or responsibility in the business; however, all control and power resided with the master craftsman.

In the mid-nineteenth century, several events caused a change in manufacturing: (1) the population of the nation grew, and town markets became regional markets spanning hundreds of miles; (2) the implementation of the power spinning shaft, powered by water wheels, provided the energy to run all the machines in the factory; and (3) the development of the steam engine permitted wider distribution of finished products and mobile power so that factories could be located anywhere. This stage in the transition was marked by a move away from single-family operations to company affiliations. However, the management style did not change. The autocratic leadership moved from the family cottage industry to a position as plant manager or factory owner. The central source of authority was still present with just a larger family of unrelated workers.

The current industrial mode, industrial automation, started with electrification of the country and continues today with the rapid development of high technology. The markets are worldwide, with distances marked in thousands of miles. National companies are now multinational, with raw material sources, labor, and

Work system mode	Economic impact area	Management style
Cottage industry	Town or village	Autocratic and paternalistic
Production plant	Region	Autocratic and paternalistic
Automated manufacturing	World	Worker managed

Figure 13–15 Development of Management Styles.

distribution coming from across the globe. The management style necessary for survival in this climate is *worker managed.*

The transition in the management style during the last fifty years started with the industrial buildup for World War II. The first stage in the empowerment process was the move to *delegation.* Delegation is not empowerment; rather, it is the granting of permission to carry out a task. The next step was the establishment of *quality circle* groups to perform group problem analysis. In the 1970s the transition to empowerment started with quality circle groups given greater authority to work on their own and the initiation of cross-functional corrective, action teams. However, neither of these initial efforts experienced the empowerment of the self-directed work teams used today.

Establishing Self-Directed Work Teams

There are many misconceptions about what constitutes a self-directed work team. A brief discussion of the major misconceptions helps to bring the concept into sharper focus. The most common is that the process is *not about power.* The entire issue is the authority to act and the power to carry the actions to completion. The major variable that affects the implementation of self-directed teams is the importance of power to management. If power is important, the transition to self-directed teams is difficult. Supervisors and workers have a new role when self-directed teams are implemented, and neither group risks a loss of jobs. Job security is a problem only for supervisors who cannot coach without power and for workers who want management to think for them.

Other misconceptions about self-directed teams are that they are easy, take a limited amount of time, are management driven, and are not imperative. Companies are not implementing self-directed teams because management wants to give up power. Teams are installed because management recognizes that the workforce has changed and that to be competitive and survive, the enterprise must have every employee working on the solution to manufacturing problems. In world-class companies, the important issue is how many people are thinking about the enterprise problems. Self-directed work teams are a significant change in management style that provides a real transfer of power from supervision to the work team.

The Transition to Work Teams

One of the first questions deals with the *readiness* of the workforce. No amount of training can change a workforce not ready to accept the responsibility placed on a self-directed team. Stated in the simplest terms, the readiness test identifies workers who understand the operation of the process and want ownership of it. Workers were asked, "Could you do your job better if something were changed?" Those ready for self-directed work teams know specifically what needs to be changed to improve their performance. In contrast, workers who don't know would not operate well in the team environment. The evolution toward self-directed teams is driven by a workforce that is generally more intelligent because of media and information access. In addition, workers are process competent and know how the production system functions. Finally, they have a higher level of motivation and sense of ownership about their jobs. If these conditions are not present, the group is not ready for the self-directed work team concept.

The transition to self-directed work teams is smooth when four rules are followed: (1) management must understand the issues and transfer of power required for functional teams, (2) a plan must be developed for the specific plant location on the basis of local culture and existing readiness, (3) functional teams must be trained in skills required for self-management, and (4) the transition through the three stages of empowerment must be spread across five to eight years. Functional self-directed work teams have been implemented in many companies who have followed these four rules.

Description of a Self-Directed Work Team

The common characteristics for the generic work team include the following:

- Frequently, a self-directed work team is formed from a functional work group currently in a process area. The term *functional* means that all members of the team come from a single process area. Cross-functional work groups composed of employees from several different areas cannot operate in the self-directed mode.
- A self-directed work team meets as a functional group to manage the process area and resolve problems that reduce the effectiveness and profitability of the process.
- Team members are trained to work in self-managed mode and are managed by natural leadership from within the group.
- The team improves the production process in their area by using the following six-step process: (1) identification of a process problem, (2) development of a process improvement plan, (3) implementation of the process improvement, (4) tracking of plan effectiveness through data collection, (5) evaluation of process data to determine performance, and (6) participation in both personnel and financial rewards as a team.

A successful team is identified by the existence of interdependence between the functional team members and exercising *true power* through the ability to take corrective action.

Installing Self-Directed Work Teams

The installation process passes through the following three stages:

1. *Process focus.* In this stage, functional work teams are formed, training is carried out in critical interpersonal skills (communication, problem solving, consensus building, and conflict resolution), technical and process training

is provided, relationships within the group are established, and natural leadership elements are developed. However, the primary focus of the team is on gaining ownership of the process, and this stage takes from eighteen months to two years to complete.

2. *Self-directed work teams with supervision.* In this stage, the work teams begin to perform the duties for the functional area but with the help of supervision. The supervisors start to assume the role of *coach* but retain the right to final approval for work team decisions. The help provided by the supervisor includes budget development and presentation, peer assessment techniques, and vendor interfacing and communications. The time required to work through all of the elements in this stage is usually two to three years.

3. *Full self-managed work teams.* In the final stage, the supervisor is totally removed from the group and acts only as a coach or an adviser. The team assumes full responsibility for operation of the process area. The team is empowered to perform all of the tasks previously performed by the supervisor, which include setting schedule priorities, planning working hours and duties, implementing disciplinary action for team members, selecting and hiring new team members, and terminating current team members. It often takes one to three years to perfect the self-management process.

After the implementation is complete, management becomes the "what" team to decide what and how much to make; the self-directed work team becomes the "how" team that decides how best to produce it. The critical job for management is the development of clear objectives and goals with an openness and patience to resolve the disagreements that occur along the implementation process. For the teams, the critical work includes establishing a climate of support and trust, developing problem-solving skills and leadership, and regularly reviewing tasks and participants' progress.

The implementation process normally moves through the following five steps:

1. Executive training
2. Planning and establishment of the work team method
3. Work team training on a team-by-team basis
4. Implementation of work team control over production processes as the employees are trained and implemented into teams
5. Work team beginning to perform the following tasks: participate in preshift meetings, participate in formal scheduled team meetings, examine the production process, plan corrective action for process problems, and take responsibility for operation of the process

Self-directed work teams have two types of authority: sovereign and negotiated. In the production area, a team has the power to change anything viewed as an improvement in the manufacturing process. However, any change that affects functions outside the process area must be negotiated with the external team before any changes are made. Problems that cross into other areas fall into three general categories: *operational* problems, *broad-based* problems, and *enterprise-wide* problems. In practice, operational problems that affect another team are solved by sending two team members to a team meeting of the other work team. For example, the speed of loading of parts into a machine is slowed by the way the parts are oriented on the pallets at an upstream work center. An improved pallet orientation is developed and presented to the work team by two members at the upstream work center.

When the problem falls into the broad-based category, a joint work team meeting is planned to solve the problem. For example, the plating process has an increase in parts rejected due to a nonuniform finish. The team isolates the problem to handling and processing in another work center. Because the source of the problem is not defined clearly and may affect other parts produced with the same routings, a joint team meeting between the plating department work team and the process center work team is held to define and propose a solution to the problem.

Major problems affect a range of enterprise departments and require a cross-functional meeting with representatives from several work teams. For example, a work center performing parts assembly routinely has a problem assembling mating parts. The cross-functional solution team would have representatives from assembly, material processing, design, and production engineering.

Companies that implement self-directed work teams have an advantage over manufacturers who choose to stay with the traditional supervisor-worker relationship. Consider the following two scenarios. In organizations where the workers don't have to think after they arrive for work, only a handful of employees are working on the solution to enterprise problems. However, the company that successfully adopts the work team concept has everyone working on the solution to enterprise problems. With time, the company with self-directed work teams will win over the "mindless" competition.

13–7 SUMMARY

The lessons learned by manufacturing in the 1970s is that technology alone cannot improve business performance, and a new emphasis on quality and development of the workforce was required. The move to quality has two major elements: a new philosophy on how the business must function and the use of quality tools to achieve the desired results. This broader view of quality is included in a process called *total quality management (TQM)*. TQM has two components: principles and tools. The principles allow management changes that lower the barriers to a successful TQM implementation. The tools permit quantitative and qualitative measurement of the system to determine how well the process is meeting organizational goals.

The definition of TQM has three dominant themes: (1) participative management for everyone in the enterprise, (2) implementation of a successful continuous-improvement process, and (3) efficient use of multifunctional teams. The six important principles that must be considered in a TQM implementation include (1) a focus on customer needs and satisfaction; (2) a focus on the process used to produce a product as opposed to a result or product focus; (3) a focus on the prevention of problems in areas such as quality, production, machine operation, and engineering design changes; (4) a focus on using the brain power of every employee in the organization; (5) a focus on basing decisions on facts about manufacturing and design without finger-pointing and blame when problems occur; and (6) a focus on developing an enterprise where communication channels are always open so that product data and process information flow freely among all levels and all employees.

Implementing TQM is a five-step process: preparation, planning, assessment, implementation, and evangelization. The first step sets the goal for the organization, and in the second step the corporate council builds plans around these enterprise goals and guidelines. The third step identifies the strengths and weaknesses of the enterprise by using a broad-based assessment process. TQM is implemented in the fourth step by using process action teams to look at the weaknesses in the processes and to implement changes to make improvements. In the fifth and last step, the successful TQM processes are moved to other parts in the enterprise for implementation.

The quality tools used to support the quantitative and qualitative needs in TQM include statistical process control (SPC), Pareto analysis, Ishikawa diagrams, and Taguchi techniques. The SPC process uses four different types of control charts (X-bar, R, p, and c) to analyze the performance of a process and suggest changes to reduce the variability present in the process.

The goal of producing with near-zero defects is reached through the adoption of a 6-sigma design and production process. The basis for 6-sigma design and production is a focused effort on making the maximum limits on critical dimensional tolerances of parts twice the width of the process variability. If machines produce parts with the standard variation found in a normal distribution, most of the parts will fall between ±3 standard deviations of the average value. If the allowable specification width for critical part dimensions is ±6 standard deviations, even with normal process variation fewer than 4 defective parts are produced in every batch of 1 million. This process leads to near-zero-defect production.

Management of the automated workforce is changing as a result of two key internal factors: the relationship between the operator and the machine, and the highly integrated nature of automated production systems. External factors affecting workforce management are the expanding global marketplace and evolution of the workforce. The most significant event resulting from evolution of the workforce is the move to self-directed work teams. The move to these teams is a result of two factors: first, for the organization to remain competitive, global competition requires problem solving from everyone in the enterprise; and second, the workforce has evolved to a state where employees are ready to take the responsibility to control the production processes assigned to them by management. After self-directed work teams are implemented, management becomes the "what" team to decide what and how much to make; the teams deal with the "how" issues, where they focus on how to produce the product most efficiently.

BIBLIOGRAPHY

ARNSDORF, D. *Technology Application Guide: Quality and Inspection.* Ann Arbor, MI: Industrial Technology Institute, 1989.

BYHAM, W. C. *Zapp! The Lightning of Empowerment.* Pittsburgh, PA: Development Dimensions International Press, 1989.

COX, J. F., and J. H. BLACKSTONE, eds. *APICS Dictionary.* 10th ed. Alexandria, VA: American Production and Inventory Control Society (APICS)—The Educational Society for Resource Management, 2002.

CROSBY, P. B. *Quality Is Free.* New York: McGraw-Hill, 1979.

DEMING, W. E. *Out of the Crisis.* Cambridge, MA: MIT Center for Advanced Educational Services, 1986.

EVANS, J. R., and W. M. LINDSAY. *The Management and Control of Quality.* 5th ed. Minneapolis–St. Paul, MN: West Publishing Co., 2002.

JABLONSKI, J. R. *Implementing Total Quality Management.* Albuquerque, NM: Technical Management Consortium, 1991.

JURAN, J. M., and F. M. GRYNA. *Quality Planning and Analysis.* 3rd ed. New York: McGraw-Hill, 1993.

LOCHNER, R. H. *Designing for Quality.* Milwaukee, WI: Statpower Associates, 1991.

MOTOROLA. *Design for Manufacturability.* Tucson, AZ: Motorola, Inc., 1988.

SOBCZAK, T. V. *A Glossary of Terms for Computer-Integrated Manufacturing.* Dearborn, MI: CASA of SME, 1984.

QUESTIONS

1. Describe the two different views of quality in the area of manufacturing.
2. Describe the two components present in TQM.
3. Define *total quality management.*
4. Describe the three dominant themes present in the definition for TQM.
5. Name the six principles that are critical for a successful TQM implementation and describe each briefly.
6. Describe the TQM implementation process.
7. Define *SPC.*
8. Describe the normal distribution and the significance of 3 standard deviations.
9. What are control charts, and how are they used to control a process?
10. Compare and contrast the four types of control charts with respect to their process control function.
11. Describe how Pareto charts are used to solve process problems.
12. Describe how Ishikawa diagrams are used to solve process problems.
13. Name and describe three types of cause-and-effect diagrams.

14. Describe the solution process for problems identified with SPC.
15. Compare and contrast quality of design versus quality of conformance.
16. Describe the 6-sigma design process.
17. Why is a specification width of 4 standard deviations inadequate for high-quality production?
18. What is the relationship between DFMA and near-zero-defects production?
19. Why is defects per unit a good method for measuring the quality of a product?
20. Briefly describe the seven steps required for near-zero-defects manufacturing.
21. What are the key internal and external issues affecting the management of the workforce?
22. Describe the evolution of self-directed work teams.
23. Describe the four rules used in the transition to self-directed work teams.
24. What are the common characteristics that describe a self-directed work team?
25. Describe the four stages in the implementation of self-directed work teams.
26. Describe the authority vested in self-directed work teams.

PROJECTS

1. Using the list of companies developed in project 1 in chapter 1, create a matrix that illustrates which company is using TQM, SPC, other quality tools, a 6-sigma design process, and self-directed work teams.
2. Select one company from the list in project 1 that uses TQM, and describe how the process was implemented.
3. Select one company from the list in project 1 that uses SPC, and describe how it implemented the process, what control charts are used, and how it solves problems identified through SPC.
4. Select one company from the list in project 1 that uses self-directed work teams, and describe how it implemented the process, the current stage of implementation, and how it determines when a team is ready for self-governance.
5. Use library and Internet resources to find current information about the teaching of the quality management experts such as Deming, Juran, and Crosby. Identify companies that have begun quality improvement initiatives and look for the influence of these recognized quality leaders.

INDEX

ABC part classification, 247
ABCD checklist, 32–34, 308–318
Aggregate planning, 224–226
Analysis
  CAE, 143, 152
  design, 82
APICS, 221
APT programming, 193
Assemble to order (ATO), 50, 241
Assessment, enterprise, 28
Automated material handling, 461–465
  asynchronous transfer, 464
  continuous, 462
  intermittent transfer, 463
  transfer systems, 461
Automatic guided vehicles, 465–476
  applications, 484
  definition of, 465
  justification, 475, 476
  systems, control, 479–475
  types of, 476–470
Automatic storage and retrieval, 476–480
  carousel systems, 479
  components, 477
  definition of, 476
  types of, 478
Automatic tracking, 525–529
  bar codes, 525
  components, 527
  rf tags, 529
Automation
  functional, 72
  time line, 112

Backward shop scheduling, 252, 256
Baka-yoke, 380
Bar codes, 525
Best of breed systems, 337
BOM (bills of material), 250, 279, 318
Boothroyd Dewhurst, Inc., 150
Boundary representation, CAD, 118
Brainstorming, 80–81
Burden rate, 16

CAD, 107–136
  2-D wire frame, 111
  3-D wire frame, 113
  CAD/CAM links, 191
  CAM, 190
  concept design, 129
  cost of paper-based designs, 110–111
  definition of, 108
  design model, 108
  drafting, 129
  future trends, 126
  hardware, 123–126
  introduction to, 107
  manufacturing applications, 129
  NURBS, 140
  product data management, 134

repetitive design, 129
  selecting hardware and software, 129–134
  software, 111
  software market, 121, 124
  solid models, 116
  surface models, 115
  system, basic, 122, 124
  today, 122
  yesterday, 121
CAD interface, 191
CAE (Computer-aided engineering), 143–218
  assembly analysis, 159
  circuit analysis, 159
  definition of, 144
  design evaluation, 163–177
  design verification, 162
  DFMA, 144–152
  finite element analysis, 152–157
  group technology, 177–184
  mass properties analysis, 157–159
  new design model, 82
  other types of analysis, 159
CALS programming, 196
CAM (Computer-aided manufacturing), 189–198
  CAD, 190
  CAD/CAM links, 191
  definition of, 190
  numerical control, 189
  software, 192–198
Capacity planning
  defined, 285
  rough cut, 229
Capacity scheduling, 233, 240
CAPP (Computer-aided process planning), 184–189
  advantages, 189
  expert systems, 188
  generative and variant, 187–188
  manual process, 184
  METCAPP software, 188
  process plan, 185
  types of, 187
Case studies
  AGV applications at GM, 484
  New United Motors, part I, 376–378
  New United Motors, part II, 378–381
Cell control, 493–500
  hardware, 494
  software, 494–499
Cell control software
  application enablers, 497
  in-house developed, 497
  OSI solutions, 499
Chain code, 182
Challenges
  external, 6–10
  internal, 10–11

Chase production strategy, 236
CIM
  assessment, 28
  axiom, 73
  benefits, 39–40
  CIM as a competitive weapon, 66–68
  data requirements, 27
  definition of, 24, 34
  Enterprise Wheel, 25, 222
  implementation process, 28–35
  importance of, 68
  is not, 27
  learning concepts, 26–35
  obstacles to implementation, 29
  philosophy, 28
  process segments, 27, 28
  simplification, 30
Closed-loop MRP, 289
CNC (Computer numerical control), 521–525
  G codes, turning, 535
  history of, 422
  programming, 522–525
COGS, 21
COMPACT II programming, 193
Competition from Japan, 353
Computer-aided. See CNC
Conceptualization
  concept design, 77, 129
  defined, 75–76
Concurrent engineering
  definition of, 86
  five phases, 89–91
  operational model, 90–92
  in traditional design, 87
Constructive solid geometry, CAD, 117
Continuous
  improvement, 31
  type manufacturing, 47
Control charts, 549–553
Coopers and Lybrand data, 16
Corporate planning process, 13–14
Cost estimating, 101, 202
Cost of goods sold, 21
Cost-added operations, 30
CRM (Customer relationship management), definition, 325
Crosby, Philip, 540–543
Cross-functional teams, 14
CRP (Capacity requirements planning)
  automation of, 287
  definition of, 285
  interface, manufacturing system, 288
  process diagram, 286
Customer lead time, 49
Cycle counting, inventory, 247–248

571

Dassault Systèmes, 345–348
Database
  central data resource repository
    (CDRR), 342
  shared single image, 73
Defects per unit, 558
Delivery time, 46, 50
Deming, W. Edwards, 538
Deming's fourteen points for
    management, 539–540
Design
  analysis, 82
  case study, 105–106
  concept definition of, 77–81, 129
  cradle-to-grave, 92
  data integration and management,
    341–348
  documentation, 84
  engineering, 53
  evaluation, 83
  five-step process, 75–76
  form, fit, and function, 53, 77, 84
  information flow, 75
  model, 75–81
  parametric analysis, 119
  process, 75–86
  product 75–86
  quality conformance, 556–557
  repetitive, defined, 77–81, 129
  synthesis, 81
Device controllers, 499
DFMA
  analysis for manufacturing and
      assembly, 101
  assembly method scoring chart, 148–149
  computer support, 150
  definition of, 144
  design guidelines, 147, 215
  for assembly, 144
  for manufacturing, 144
  history of, 145
  justification for, 145
  manual process, 146
  new design model, 82
Disaggregate plan, 226–228
Dispatching
  defined, 253–254
  four functions, 254
  mechanism, 253
  priority control techniques, 253
  priority rule systems, 265–270
Distance standard, 22
Documentation, design, 84
Drafting, CAD, 129
Drum-buffer-rope system, 228, 364–366

Early manufacturing involvement in
    design, 72
E-commerce, 9, 339–340
Electronic Data Interchange (EDI), 209
Engineer to order (ETO), 49
ENOVIA, 347
Enterprise
  assessment of, 27–28
  data types, 209
  functional areas, 258
  goal, 28
  implementation of, 32
  LAN definition, 204
  LAN model, 204

organizational model, 54, 55
simplification of, 30
ERP
  asset optimization software, 333–334
  control issues, 328–329
  data collection, 328–329
  database issues, 335–338
  database model, 336
  definition of, 324
  e-commerce, 339–341
  example, ERP WinMan, 318–322
  implementation, 330–331
  information requirements, 331
  integration, information, 330
  relationship to MRP, 324–235
  relationship to MRP II, 326–327
  suppliers, identifying, 332–335
  system architecture, 332–333
  virtual manufacturing, 339
Evaluation, design, 83
Exception messages, 251
External challenges, 6–11, 22–24, 34–35

Factory of the future database model, 336
FEA (finite element analysis)
  application areas, 155, 156
  definition of, 152
  in design model, 83
  heat transfer analysis, 156
Feigenbaum, Armand, 540
Financial management, 56
Finishing operations, 404
Finite shop loading, 256
Fish-bone diagram, 554
Fixed automation, 413–417
  in-line, 414
  rotary, 414
  selection criteria, 414
Flexibility standard, 22
Flexible manufacturing cells, 411–413
  definition of, 411
  types of, 411
Flexible manufacturing systems, 405–411
  advantages of, 411
  cells versus systems, 411
  control architecture, 409, 410
  definition of, 405, 406
  example system, 407
  technology levels, 408
Fluid analysis, FEA, 156
Forecasting
  Focus, 232
  future demand, 230–232
Form, fit, and function. See Design
Formal systems
  describing items and products, 293–295
  models, 225, 229
Forward scheduling, 256–257
Functional automation, 72
Fused deposition modeling, 169–173

Gantt charts, 252
Generative, CAPP, 187
Goals, product versus process, 6
Great manufacturing divide, 10
Group Technology (GT), 177–183
  definition of, 177
  part coding, 180
  production cells, 183
  production flow analysis, 177

Harrington, Joseph, 24, 34
Heat transfer analysis. See FEA
Hierarchical code, 180
Hill, Terry, 11
Horizontal shop loading, 256

IGES file interface, 191
Implementing CIM
  assessment, 28
  implementation, 32
  simplification, 30
Industrial robots. See Robots, industrial
Infinite shop loading, 255
Information technology
  data collection and control, 328–329
  implementation decisions, 330–331
  integration nightmare, 329–330
  typical systems environment, 327
Input-output control, shop, 257
Internal challenges
  defined, 10
  description of, 10–13, 15
  implementation steps, 15
Inventory
  ABC part classification, 247
  accuracy, 248
  anticipation stock, 21
  cycle counting, 247
  cycle stock, 20
  definition of, 245
  input-output control, 257
  in-transit, 20
  movement and organization, 20
  physical count method, 246
  pipeline, 20
  residence time, 21
  safety stock, 20
  turns, 21
  velocity, 20
  work-in-process, 20
Ishikawa diagrams, 554
Ishikawa, Kaoru, 540

Jidoka, 378, 379
JIT (just in time)
  definition of, 353
  implementation, 358–360
  kaizen, 14, 379, 380
  Kanban, 14, 360–363, 371, 378, 379
  major elements, 355–358
  manufacturing, 353–363
  structured product flow, 356
  versus MRP II, 363
Job shop type manufacturing, 45
Juran, Joseph, 540

Kanban, 14, 360–363, 371, 378, 379
Kettering, Charles, 220
Kiazen, 14, 379, 380

Laminate object manufacturing, 173–176
Layouts, plant, 46
Lead time
  customer, 47, 49
  manufacturing, 49
Lean manufacturing systems, 369–374
Lean production
  definition of, 368
  five principles, 368
  Henry Ford's view, 367

seven wastes, 354
Toyota's process, 367
transactions in lean manufacturing
    systems, 373
Lean tools in WinMan, 371–374
Leaping, product development, 14
Life cycle, product, 13–15, 51
Line type manufacturing, 47
Lot sizes, 16, 45

Machine programming (CNC, NC), 98
Maintenance automation, 201
Make to order (MTO), 50, 241
    make to stock (MTS), 50, 241
Management review process, 54–59
Manual production operations, 59–62
Manufacturing
    business factors, 5
    characteristics chart, 48
    classifications, 45–49
    decline stages, 4
    definition of, 3
    economic factors, 5
    goal of, 4
    input-output model, 44
    order-qualifying criteria, 14
    orders, 97
    order-winning criteria, 11–22
    organizational model, 55
    planning and control, 57, 223–228
    political factors, 6
    process goals, 6
    processes, 96–97
    product definition groups, 60
    product goals, 6
    rebirth of, 4
    resource planning. *See* MRP II
    retreat of, 4
    software evolution, 325
    space ratio standard, 19–20
    strategy, definition of, 10
    systems, 45–49
    systems and MPC, 227
    systems versus production strategies, 52
Mass property analysis, 157–159
    DFMA data, 159
    in design model, 83
Material and capacity resources, 248–251
Measurement. *See* performance measures
Measures, world-class, 16
Mechanical event simulation, 156
Motion analysis, FEA, 156
Motion and time measurement, 95
MPC (manufacturing planning and
    control)
    aggregate, goal of, 224
    aggregate planning, 224
    automating of, 258–259
    business planning, 230
    definition of, 222
    demand forecasting, 230
    disaggregate planning, 226–228
    model, information flow, 58, 224, 228
    production planning, 231–236
    responsibilities, 57–58
    strategic plan, 224
    system features, 291–292
MPS (master production schedule)
    automating of, 258–259
    definition of, 240–241

order promising, 244
    and production strategies, 241
    techniques, 242
    time-phased record, 242–245
MRP (material requirements planning)
    automation of, 282
    benefits, 284–285
    bill of materials (BOM), 278
    calculations, 275–277
    computer-assisted MRP, 282–283
    definition of, 273
    indented bill of materials, 279
    input data, 249
    logic rules, 274–275
    manufacturing system interface, 283
    net change, 283
    operational model, 249
    product structure diagram, 249–250, 279
    record, 279
    Wight's bicycle MRP example, 306–307
MRP II (manufacturing resources
    planning)
    benefits, 295–296
    cost justification, 296
    education for, 300
    implementation, 296–301
    in ERP, 326–327
    model for, 229
    operations planning, 227–228
Multijet modeling, 176–177

Natural frequency analysis, FEA, 155
Networks communications, PDM, 529
Networks, control, 488–490
    ControlNet, 488
    DeviceNet, 488
    Ethernet/IP, 489
    Networks, 490–493
Networks, data, 204–210
    data conversions, 209
    definition of, 204
    IP addressing, 207–208
    LAN, 204
    model, 205
    types of, 206
New product development, 53
NURBS, CAD, 140

Oliver Wight Companies checklist,
    32–34, 308–318
Open-loop MRP, 288
Operations management, 222–223
Optiz code, 182
Order
    manufacturing, 97
    qualifying criteria, 14
    winning criteria, 11–22
    world-class, 14–21
Organizational model, manufacturing
    enterprise organization, 54–59
    finance and management, 56
    information flow model, 59
    manufacturing planning and control, 57
    product and process definition, 56
    sales and promotion, 55
    shop floor, 58
    support organization, 58

PAC (production activity control)
    defined, 252

dispatch defined, 254
dispatch techniques, 254
Gantt charts, 252
priority control, 253
priority rule systems, 265–270
shop loading, 254–257
Parametric analysis, 78
Pareto analysis, 553
PDES programming, 196
PDM (product data management), 73,
    341–345
    CAD, 134
    data interfaces, 135, 530
Performance measures
    ABCD check list, 32–34, 308–318
    CIM implementation case study, 32–33
    key parameters, 32
Phantom items in a product structure, 282
Planning for manufacturing, 223–228
Planning process, corporate, 10–11
Plant engineering, 101
Plant layouts, 46
PLM (product lifecycle management),
    341, 345–348
Polycode, 182
Primary manufacturing operations,
    96–97, 387
Priority sequencing rules, 265–270
Process capability index, 557–558
Process goals, 6
Process layout, job shop, 46, 47
Process operations
    categories, four, 387
    finishing, 404
    machine processes, 387–403
    physical property, 403
    primary, 387
    secondary, 392
Process planning
    definition of, 93
    manufacturing process planning, 61
    operation sequence, 95
    operation sheets, 95
    routing sheets, 95, 97
    shop packet, 97
    typical system, 94
Product
    data management, 73
    design, 56, 60, 75–86
    design and production engineering
        model, 74
    design group definition, 60
    design process improved, 89
    development cycle, 52–54
    development process, 88
    flow layout, 46–47
    goals, 6
    life cycle, 13, 51
    request for engineering change, 60
Product structure diagrams, 250, 279, 318
Production
    chase production, 236
    cost analysis, 202
    efficiency, 386
    engineering, 53, 56, 72–86, 93
    level production, 238, 380
    manual production operations, 59–62
    mixed production, 239
    model, 44
    planning, 231–240

Production (*continued*)
  planning model, 231
  planning process, 234–235
  sequence, 61
  simulation, 199
  strategy, 45, 49, 51–52
Production engineering
  CAM, 189–198
  CAPP, 184
  DFMA, 101
  elements of, 93
  machine programming, 98
  modeling, 198
  plans, 93–95
  plant engineering, 101, 201
  process planning, 93–98, 184
  production cost estimating, 101, 202
  simulation, 199
  strategies, 184
  tool and fixture design, 99–100
  work standards, 100–101
Production operations, manual, 59
Production strategy classifications, 49–52
Programmable logic controllers, 500–521
  basic operation, 515
  benefits, 517–518
  definition of, 500
  history, 501
  IEC 61131 standard, 500–501
  PC versus PLC, 502
  programming, 513–517
  relay ladder logic, 502
  system components, 511
  types of, 511–513
Project type manufacturing, 45
Prototyping, 163–177
  rapid, 165–177
  virtual, 163
Pull system, 358

Quality
  captured, 18
  control charts, 549–553
  defect free design, 557–561
  Deming's fourteen points for
    management, 539–540
  normal distribution, 548, 550
  other quality tools, 553–556
  SPC, 548–553
  total, 18
  TQM, 543–547
  warranty, 18

Rapid prototyping
  comparison chart, 178
  in design model, 84
  fused deposition modeling, 169–173
  future trends, 177
  laminate object manufacturing, 173–176
  multijet modeling, 176–177
  selective laser sintering, 166–168
  stereolithography, 165–166
  three-dimensional printing, 168–169
REA (request for engineering action),
    60, 75
Reorder point
  in MRP, 291
  systems, 272
Repetitive design

defined, 77
  three-step process, 78
Repetitive type manufacturing, 46
Robots, industrial, 426–465
  applications, 441
  controller, 427, 438
  definition of, 426
  geometry, 428–435
  production tooling, 435–438
  selection and justification, 442–448
  system, basic, 426
  teach stations, 440
Robots, programming, 448–465
  command modes, 454
  commands, 456–460
  coordinate system, 454
  data types, 455
  online and off-line, 453
  process, 448–453
  task point graph, 451
Rough cut capacity analysis, 229
Routing sheets, 61, 95, 98, 251

Sales and field force automation, 340–341
Sales and promotion, 55
Secondary manufacturing operations, 392
Selective laser sintering, 166–168
Self-directed work teams
  description of, 565
  establishing, 564
  history, 562
  installing, 565–567
  transition to, 564
Setup time
  loading, 256–257
  standard, 16
Seven wastes, 354
Shop floor interfaces in MPC, 58
Shop loading
  description of, 254
  finite, 256
  infinite, 255
  vertical and horizontal, 256
Shop packet, 97
Short-cycle manufacturing, 354
Simplification
  definition of, 30
  process in CIM implementation, 30
Simulations, production
  continuous process, 200
  definition of, 199
  discrete-event, 200
Simultaneous engineering. *See*
    Concurrent engineering
Six-sigma design
  capability index, 557–558
  defects per unit, 558
  seven steps to, 560–561
  zero defects, 559–561
Solid models, CAD, 116–121
SPC (statistical process control)
  control charts, 549–553
  Ishikawa diagrams, 554
  Pareto analysis, 553
Standardized work, 380
Static analysis, FEA, 155
Stereolithography, 165–166
Stockless manufacturing. *See* JIT
Strategy, manufacturing, 15

Suppliers, 8
Supply chain management software, 349
Support organizations, 58
Surface models, CAD, 115–116
Synchronous production
  definition of, 228, 364
  drum-buffer-rope system, 365
  goal of, 364
  versus JIT and Kanban, 366
Synthesis, design, 81

Teams, cross-functional, 14
Technology
  changes, 44
  converging and enabling, 335–341
  fundamentals of an operation, 39
  lost, 5
Terry Hill model, 11–12, 15
Thinking outside the box, 80
Three-dimensional printing, 168–169
Time fence, 235
Time studies, direct, 95
Tools and fixtures, 99–100
TQM (total quality management)
  definition of, 544
  implementation principles, 545–546
  implementation process, 546–547
Transient dynamic analysis, FEA, 155
Transient items in a product structure, 282

Up-time standard, 22

Value-added operations, 30
Variable resource management
  chase production strategy, 236
  level production strategy, 238
  mixed production strategy, 239
  resource capacity considerations, 240
Variant, CAPP, 187
Vertical shop loading, 256
Virtual manufacturing, 339
Visual controls, 381

Waste
  five general categories, 30
  rules for elimination, 30–31
  seven wastes in JIT, 354
  in simplification, 30
Weakly integrated manufacturing
    systems, 59
Web-centric PDM, 344
WinMan, 293–295, 318–322, 369–374
Workforce changes, 561–562
World-class metrics/standards
  Coopers and Lybrand data, 16
  description of, 15–22
  distance, 22
  flexibility, 22
  inventory, 20
  manufacturing-space ratio, 19
  measurement system—case
    study, 65–66
  metrics, 23
  quality, 18
  setup time, 16
  up-time, 22

Zero defects. *See* six-sigma design
Zero-inventory manufacturing, 354